Direct Fuel Injection for Gasoline Engines

PT-80

Edited by

Arun S. Solomon
Richard W. Anderson
Paul M. Najt
Fuquan Zhao

Published by
Society of Automotive Engineers, Inc.
400 Commonwealth Drive
Warrendale, PA 15096-0001
U.S.A.
Phone (724) 776-4841
Fax: (724) 776-5760
www.sae.org

Preface

For the past 100 years, the automobile industry has devoted considerable effort to increasing the efficiency and power output of the internal combustion gasoline engine, and for the past fifty years, direct injection has been considered one of the most promising solutions for achieving these aims.

Over the last twenty-five years the operating configuration of the gasoline engine has shifted from being carbureted to being single-point or port-fuel-injected. This shift was driven by the need for an increased level of air-fuel ratio control to meet vehicle system performance requirements for better fuel economy and reduced emissions. Could it be that the gasoline engine is on the threshold of yet another shift to direct-injection operation, spurred this time by the need for increased fuel economy and reduced carbon dioxide emissions?

Considerable research effort has been devoted to the development of combustion systems for direct-injection gasoline engines. With the introduction of the first commercially successful stratified-charge, direction-injection, gasoline-engine-powered vehicle in 1996, research and development activity in this area has intensified. Despite this considerable research and development effort, however, direct-injection gasoline combustion remains a technology that is still evolving. Also, severe challenges remain in the area of exhaust aftertreatment to meet future emissions standards and cost reduction requirements.

The editors hope that this "Progress in Technology" book on direct injection will be of benefit to researchers and developers in the engine development community by offering a current perspective of this still evolving technology.

Table of Contents

I. Combustion System Design and Development of Direct-Injection Gasoline Engines

II. Fuel Spray Characteristics for Direct-Injection Gasoline Engines

III. Multi-Dimensional Modeling of Direct-Injection Gasoline Engine Phenomena

IV. Diagnostic Methods for Direct-Injection Gasoline Engines

Combustion System Design and Development of Direct-Injection Gasoline Engines

Combustion System Design and Development of Direct-Injection Gasoline Engines

The goal of current research on direct-injection gasoline engines is an engine that offers fuel economy comparable to a diesel at part-load and full-load power superior to that of the conventional port-injected gasoline engine. To achieve this goal, the engine must operate unthrottled, lean, and stratified at light loads, but stoichiometric and homogeneous at high loads. When operating stratified at light loads, the combustion system must reliably generate a near stoichiometric mixture at the spark gap for good combustion stability. At the same time, it must provide sufficient mixing to avoid soot-producing rich zones without overmixing the lean zones to avoid high hydrocarbon emissions. And when operating at high loads, the combustion system must thoroughly mix the fuel and air to maximize power output while avoiding fuel impingement on the bore walls and piston scuffing. To meet these multifaceted goals, engine manufacturers are working to combine advanced control system technology with a new generation of electronic fuel injectors, sophisticated computational tools, and advanced diagnostic techniques. There is, however, no consensus on combustion system design. Indeed, each engine developer appears to be pursuing its own unique solution.

The papers in this section address the many combustion concepts being pursued and their distinct advantages and disadvantages. They address systems with side-mounted and center-mounted fuel injectors; with wide and narrow spark plug to injector spacing; with tumble-dominated and swirl-dominated in-cylinder flow; and with air-assist and single fluid injectors. They also address systems that rely on the piston surface to control the mixture preparation process, and that rely on in-cylinder air motion to control the mixture preparation process. The papers also address the various operational strategies being pursued, the development procedures being followed, and fuel injection and exhaust aftertreatment requirements. Taken together, the papers in this section provide a comprehensive assessment of the current state of direct-injection gasoline engine combustion systems, their fuel economy and power potential, and the emissions and stability concerns surrounding them. Collectively these papers clearly describe the strengths and weaknesses of direct-injection gasoline engines and the development issues that remain in the area of combustion system development.

Combustion Control Technologies for Direct Injection SI Engine

T. Kume, Y. Iwamoto, K. Iida, M. Murakami, K. Akishino, and H. Ando
Mitsubishi Motors Corp.

ABSTRACT

Novel combustion control technologies for the direct injection SI engine have been developed. By adopting upright straight intake ports to generate air tumble, an electromagnetic swirl injector to realize optimized spray dispersion and atomization and a compact piston cavity to maintain charge stratification, it has become possible to achieve super-lean stratified combustion for higher thermal efficiency under partial loads as well as homogeneous combustion to realize higher performance at full loads. At partial loads, fuel is injected into the piston cavity during the later stage of the compression stroke. Any fuel spray impinging on the cavity wall is directed to the spark plug. Tumbling air flow in the cavity also assists the conservation of the rich mixture zone around the spark plug. Stable combustion can be realized under a air fuel ratio exceeding 40. At higher loads, fuel is injected during the early stage of the intake stroke. Since air cooling by the latent heat of vaporization increases volumetric efficiency and reduces the octane number requirement, a high compression ratio of 12 to 1 can be adopted. As a result, engines utilizing these types of control technologies show a 10% increase in improved performance over conventional port injection engines.

INTRODUCTION

The automobile industry has devoted considerable effort to meeting inherent requirements for realizing lower fuel consumption and higher performance while maintaining cleaner exhaust gas and greater driving comfort. For the past fifty years, the direct injection spark ignition engine has been considered to be one of the most promising solutions for achieving this aim . Extensive research and development efforts have been concentrated on establishing the combustion concept of the direct injection spark ignition engine [1~7].

A direct injection SI engine has potential to realize the greater fuel economy compared with a diesel engine at partial loads and to realize better performance than the port injection spark ignition engines at high loads.

In order to realize its fuel economy potential at partial loads, the direct injection SI engine should be operated unthrottled in an extremely lean condition by distinctively stratifying the charge and by preparing a rich air-fuel mixture around the spark plug. For that purpose, many of the direct injection concepts proposed so far have adopted the basic configuration of locating the spark plug gap near the fuel spray cone. It has been confirmed that stable combustion can be realized by such configurations. Ignition fouling caused by the impingement of the liquid fuel on the spark plug, however, prevented such a concept from being developed into mass production. Although higher energy ignition systems have sometimes shown the effects of reduced ignition fouling, problem associated with the deterioration of the durability of the spark plug due to the higher ignition energy have not been solved.

In order to achieve its higher performance potential at high loads, on the other hand, the direct injection SI engine should be operated under stoichiometric or slightly rich conditions. When the charge is stratified, soot is generated in the rich zone. Sufficient excess air should be provided

---

* Numbers in parentheses designate references at end of paper.

around the combustion zone containing soot, in order to burn-up the soot which is generated. Therefore, when the average mixture strength is stoichiometric or slightly rich, that is, when the equivalence ratio is larger than unity, the mixture should be homogeneous so as to suppress the formation of soot.

Systematic research has been carried out in the laboratories of Mitsubishi Motors Corporation regarding adequate air-fuel mixing to establish the combustion concept for the direct injection SI engine. At partial loads, the goal was to prepare a rich gaseous mixture around the spark plug, and at high loads, the goal was to prepare a homogeneous mixture.

It was found that these goals could be realized by a novel mixture preparation concept which employs a sophisticated electromagnetic swirl injector, a novel in-cylinder flow referred to as reverse tumble, and a spherical compact piston cavity having an optimized configuration. Mitsubishi Motors Corporation plans to introduce a direct injection SI engine adopting this mixture control concept into the Japanese market in 1996.

In this paper, the basic concept, measures for realizing the concept and the effectiveness of the concept with regard to fuel economy and the engine performance will be described based on the results of single cylinder experiments. This work was followed by extensive and systematic experiments adopting the multi-cylinder engines and vehicles equipped with these direct injection engines. The results of these experiments will be described in a subsequent paper [8].

BASIC STRATEGIES OF MIXTURE PREPARATION

Basic strategies for preparing the air-fuel mixture are illustrated in Figure 1. In the partial load operating conditions, a late injection strategy is adopted, that is, fuel is injected during the later stage of the compression stroke so as to realize distinctively stratified mixing. Although the engine can be operated at an air-fuel ratio exceeding 100 and complete unthrottled operation is possible, the engine is throttled slightly in this zone and the air-fuel ratio is controlled to range from 30 to 40 in order to introduce a large quantity of EGR and to prepare the vacuum for the brake system. Since dependence of the effects of fuel economy improvement effects on the air to fuel ratio is not significant when the air fuel ratio exceeds 30, sufficient fuel economy improvement can be realized by selected control of the air fuel ratio.

An early injection strategy is adopted for the higher

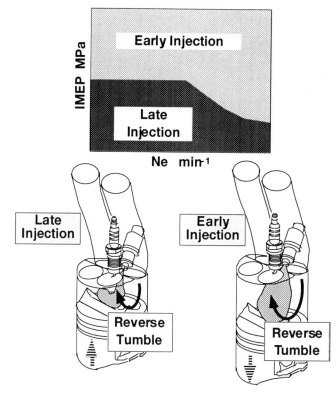

Condition	Injection Timing	Mixing	A/F Ratio
Partial Load	Later Stage of Compression Stroke	Stratified	25 - 40
Higher Load	Early Stage of Intake Stroke	Homo-geneous	Stoich./Rich or 20-25

Fig. 1 Basic concept of mixture preparation

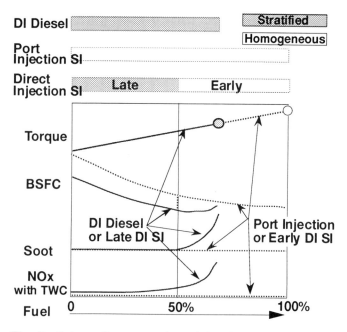

Fig. 2 Schematic expression of the characteristics of early and late direct injection SI

load operating conditions, that is, fuel is injected during the early stage of the intake stroke so as to realize a homogeneous mixture. In most of this zone, the engine is operated under stoichiometric and it is operated at a slightly rich condition at full load. In the lowest load conditions in this zone, the engine is operated at homogeneous lean conditions with a air-fuel ratio of from 20 to 25 for further improvement of fuel economy.

Strategies and the basic characteristics of this direct injection engine are schematically compared with those of diesel and port injection engines in Figure 2. Schematically speaking, late injection zone is extended up to 50% of load, because enough excess air can exist in the cylinder for soot to be burned-up in a short time and not exhausted from the cylinder. When the load exceeds 50%, the equivalence ratio is greater than 0.5 even in case of unthrottled operation. In such a condition, soot is emitted as in the case of the diesel engine. Consequently, the mixture preparation strategy is switched to early injection at this point. Because early injection acts to prepare the premixed homogeneous mixture, combustion characteristics similar to those of the port injection can be realized, that is, higher engine performance, no soot, and sufficient NOx reduction by the conventional three-way catalytic converter. In case of the diesel engine, the amount of fuel that can be introduced into the cylinder is restricted so as to maintain the soot emissions to allowable levels, resulting in lower performance. The homogeneous mixing strategy based on early injection can not be adopted to a diesel engine operated using diesel oil with a significantly lower octane number.

The objectives of the direct injection SI engine can be summarized as follows:
(1) fuel economy equivalent to diesel engines;
(2) performance equivalent to port injection SI engines; and
(3) exhaust gas as clean as that of port injection SI engines.
These objectives are realized by utilizing the inherent characteristics of the direct injection SI engine powered by gasoline which enables switching between early and late injection.

MEASURES TO REALIZE THE CONCEPT

Fundamental technologies developed to realize the mixing control concepts described above are illustrated in Figure 3 and Table 1. The following new technologies are employed:
(1) upright straight intake ports which generate an intense

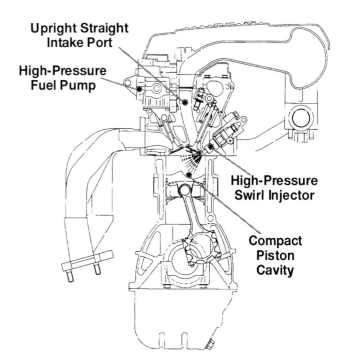

Upright Straight Intake Port

High-Pressure Fuel Pump

High-Pressure Swirl Injector

Compact Piston Cavity

Fig. 3   Major technologies adopted to realize the concept of mixing control

tumble with a direction of rotation which is opposite to that of the tumble generated by conventional horizontal intake ports;
(2) a spherical compact piston cavity which controls the behavior of fuel spray and vaporized gaseous mixture;
(3) a simple high pressure fuel feed pump with low driving loss; and
(4) an electromagnetic swirl injector which realizes the requirements for the atomization, controlled penetration and controlled dispersion of the fuel spray.

One of the objectives of these technologies is to promote vaporization of the fuel and the entrainment of the air into the vaporized fuel before it is carried to the spark plug. Therefore, a conventional ignition system with a moderate ignition energy of 50 mJ having a degree of reliability that have been suitably tested in the field could be adopted. In order to realize the sufficient reduction of NOx emitted in the stratified lean condition, a large quantity of EGR should be introduced. Quick and accurate control of EGR is realized with an EGR valve actuated by a stepping motor. In this engine, the early and late injection operation modes are switched frequently. In order to realize a smooth and comfortable feeling while driving, the torque generated before and after the switching should be kept constant. This is achieved by the quick and accurate control of the air with a

Table 1 Concepts, goals of direct injection SI engine and the measures to realize them

	Concept	Goal	Measures
Flow	❏ Reverse tumble	❏ Fuel vapor transport to plug ❏ Higher port flow rate	❏ Vertical intake port ❏ Enhanced gas-dynamics effect by straight port
Injection	❏ Lower inj. pressure	❏ Lower fuel compres. loss	❏ Swirl injector
	❏ Engine-driven pump	❏ Lower driving loss	❏ Start by feed pump
Mixing — Higher Load	❏ Homogeneous (early injection) ❏ No wall-wetting ❏ Charge-air cooling	❏ Soot reduction in stoich. & rich conditions ❏ Knocking suppression ❏ Higher volumetric efficiency	❏ Suppressed spray tip penetration ❏ Widely dispersed fuel spray ❏ Enhanced reverse tumble
Mixing — Partial Load	❏ Stratified (late injection) ❏ Enhanced evaporation	❏ Stable leanburn ❏ Lower-soot-emission at high load	❏ Compact chamber ❏ Spray motion by reverse tumble ❏ Injection timing control
Ignition	❏ Conventional ignition system with confirmed reliability		
Emission	❏ High EGR	❏ NOx reduction in stratified charge	❏ Electronically actuated EGR valve
Intake	❏ High-speed & accurate air control	❏ Smooth operation ❏ Wider EGR zone	❏ Electronically actuated air control valve

control valve actuated by a solenoid and a stepping motor.

IN-CYLINDER FLOW CONTROL

The authors have proposed the concept of the pre-mixed leanburn engine known as Barrel-Stratification [9,10]. In that concept tumble with the direction flowing the piston surface from the exhaust to the intake side was adopted. In order to intensify the tumble in this direction, intake port flow passing through the upper area of the intake port and directed

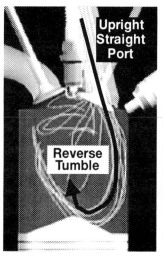
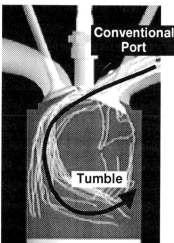

(a) Direct Injection **(b) Conventional Port Injection**

Fig. 4   Generation of intense reverse tumble by upright straight intake port

to the cylinder liner at the exhaust side was enhanced.

In the case of the direct injection engine, the authors considered that the fuel injectors should not be located at the exhaust side with its high temperature and that it would be preferable to locate the spark plug at the center of the cylinder in order to realize improved combustion with lower heat loss. As a result, the location chosen for the injector was on the intake side. Since the fuel spray is injected on the intake side, the piston cavity should also be located on the intake side. This cavity attenuates the tumble in the conventional direction. Therefore, the authors adopted a tumble having a reverse direction. Reverse tumble was also effective in moving the fuel spray toward the spark plug after impingement on the piston cavity.

To realize the reverse tumble, an upright straight intake port was selected. Tumble ratio, that is, the rotating speed of tumble over the engine speed was increased up to 1.8 by the careful optimization of the intake port configuration. Because the upright straight intake port had the inherent characteristics of the higher flow coefficient, the engine performance at the highest engine speed range was improved. This configuration was also effective in preparing the room in the cylinder head to locate the injector. If the conventional horizontal intake port configuration were selected, it would be difficult to direct the fuel spray towards the piston cavity because the angle between the injector and the piston surface would be determined available under the restriction of the room under the horizontal intake port.

PISTON CAVITY

A spherical compact piston cavity illustrated in Figure 5 was located on the piston surface as a combustion chamber. The geometry of the piston cavity is so designed that the fuel spray impinging on it moves toward the spark plug. By the squish flow from the exhaust to the intake side, combustion in the piston cavity is enhanced. During the later stage of combustion, reverse squish flow propagates the flame to the exhaust side. The wall of the spherical bowl at the exhaust side changes the downward liner-side air flow to an upward flow after impinging on the piston. Therefore this design is appropriate for enhancing reverse tumble. Furthermore, the shape of the bowl is suitable for preserving the rotational momentum of the reverse tumble until the end of the compression stroke.

FUEL INJECTION SYSTEM

HIGH PRESSURE FUEL PUMP - In order to simplify the high pressure common rail system, a piston pump without any metering function was selected. The injection pressure was regulated by spilling the high pressure fuel, which increased the fuel compression loss. To reduce the fuel compression work, a relatively low fuel injection pressure of 5MPa was selected. When compared with the fuel pump used for diesel fuel, more elaborate care should be taken with regard to lubrication and cooling, because the gasoline has a lower viscosity and higher volatility than diesel fuel. The reduction in the injection pressure was also effective in establishing the reliability of the pump system. This pump is installed on to the cylinder head and driven directly by one of the camshafts.

BASIC REQUIREMENTS FOR FUEL SPRAY - The behavior of the fuel spray and the requirements for fuel spray characteristics are illustrated in Figure 6. In case of the early injection, widely spread fuel spray is required so as to achieve homogeneous mixing. The impingement of the fuel spray on the piston surface and the cylinder liner should be suppressed, because liquid fuel captured by the wall results in the emission of soot. In order to meet this requirement, a widely dispersed fuel spray should be prepared. It should be injected at

1. **Control of spray impingement**
2. **Control of flame propagation**
3. **Reverse tumble enhancement**
4. **Reverse tumble preservation**

Fig. 5 Spherical compact piston cavity and its role

such a timing that the spray chases the piston so as to avoid any impingement on the piston surface.

A compact spray is required in case of the late injection, because it should be inside the compact piston cavity. At the same time, it should be atomized, because it should be vaporized in a short time so as to promote the entrainment of air into the vaporized fuel before reaching the spark plug.

ELECTROMAGNETIC SWIRL INJECTOR - In order to realize accurate timing and quantity control, the electromagnetic injector shown in Figure 7 was selected. It is approximately the same size as conventional injectors for port injection. In case of the direct injection, the injection should

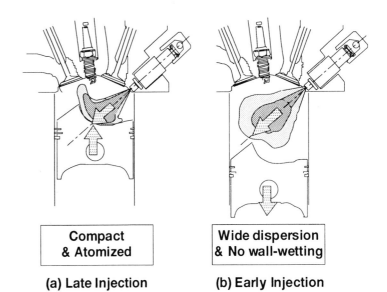

Compact & Atomized	Wide dispersion & No wall-wetting
(a) Late Injection	**(b) Early Injection**

Fig. 6 Behavior of the fuel spray an the requirements for fuel spray characteristics

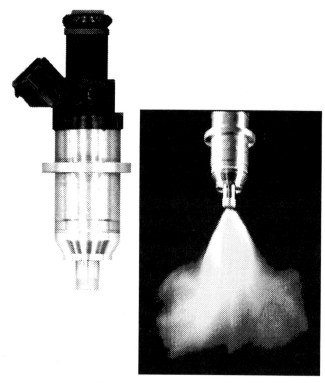

Fig. 7 Electromagnetic swirl injector

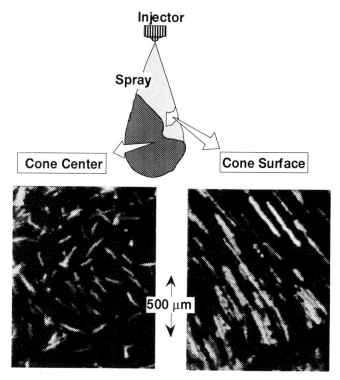

Fig. 8 Fuel spray trajectory measured by the laser light sheet technology

be completed within 180 degCA, that is, during the intake stroke at the maximum engine speed. Therefore, the dynamic range of the injector should be four times as wide as that of the port injection injector which may complete the injection in 720 degCA. This is realized by the high voltage injector driving circuit.

A swirl injector was selected in order to inject a widely dispersed fuel spray and to promote atomization of fuel using relatively low injection pressure. A rotational momentum is given to the fuel droplets by a swirler tip located at the upstream of the injection hole. The geometry of the swirler tip was optimized through extensive study of the spray characteristics, by means of a phase Doppler system to measure droplet size and the droplet velocity. In addition, laser light sheet equipment was used in order to investigate the inner structure of the spray and the trajectory of the fuel droplets. A high speed video system with microscopic adaptor was used to record the structure and trajectory of the droplets.

Figure 8 shows the microscopic droplet trajectory of the optimized injector. At the periphery of the spray, the angles between the vertical and the horizontal velocity components are approximately 45 deg. In other words, the velocity components of the rotating and the penetrating directions are approximately the same. The spray has a basically hollow cone-like structure. An intense air flow is generated at the

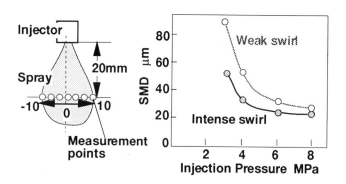

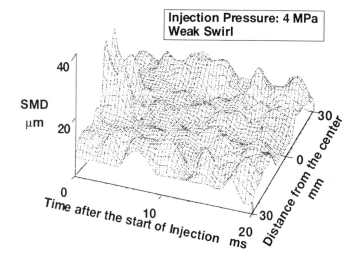

Fig.9 Droplet size and distribution

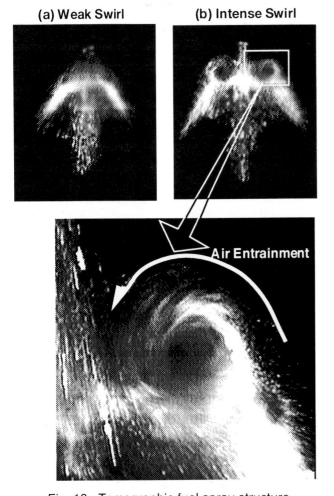

Fig. 10 Tomographic fuel spray structure

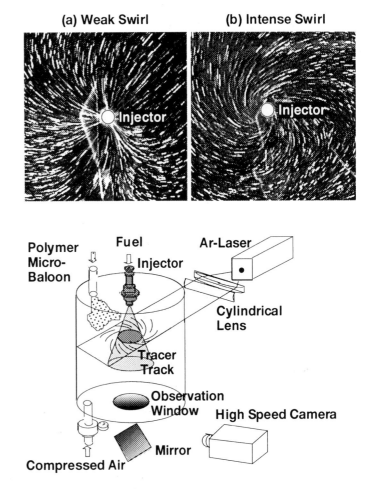

Fig. 11 Ambient air motion induced by movement of the spray

center of the hollow cone due to the movement of the fuel spray. Intense turbulence is also generated due to the interaction of the movements of the air and fuel flows. This turbulence enhances the entrainment of air into the fuel spray.

Figure 9 shows the dependence of the Sauter Mean Diameter of the droplets on injection pressure and swirl intensity. Measurement was performed at a position 20 mm downstream of the injection hole. Data values plotted in the upper field indicate the average value over the entire location and time. It can be concluded from these data that if the swirl can be intensified, an injection pressure of 5MPa is sufficient for atomization. In the lower field of the figure, distribution of the droplet diameters under the condition of low injection pressure and weak swirl are shown. It can be seen that the increase in the average diameter is due to the large droplets observed at the center of the fuel spray during the earliest stage of injection, which can be improved by increasing injection pressure and swirl intensity.

Enhancement of the swirl also promotes air entrain-

ment. Figure 10 shows the tomographic structure of the fuel spray. It can be seen that large scale vortices generated at the upper part of the spray promote the entrainment of air during the later stage of the injection and after the end of injection.

A rotating air motion is induced around the spray by the swirling motion of the spray as shown in Figure 11. In this experiment, fuel was injected into the chamber which was filled with tracer particles of polymer micro-balloon having the specific gravity of 0.05 g/cm^3, and the air motion was tracked by the trajectories of the tracer particles. Figure 12 shows the cone angle, as well as the vertical and the horizontal components of the droplet velocity vectors. The vertical vector, that is, the penetrating velocity component decreases with distance. The horizontal vector, that is, the rotating component, however, remains more or less constant. The decrease in penetrating velocity can be explained by the resistance caused by the drag force by the ambient air. Resistance by the ambient air with regard to the rotating velocity is not significant, because the ambient air rotates with the fuel drop-

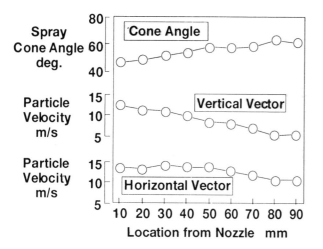

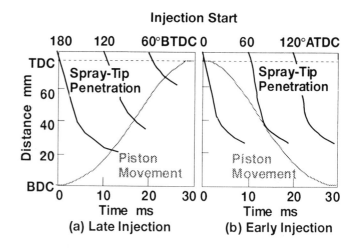

Fig. 12 Cone angle, vertical and horizontal components of the velocity vectors (injection pressure: 5 MPa, intense swirl)

Fig. 14 Spray tip penetration and the piston trajectory (1000 min⁻¹)

lets. Suppression of penetration while maintaining the momentum by the rotating spray motion is the inherent characteristics preferable to meet the requirements of the fuel spray for the direct injection SI engine.

In the case of early injection, fuel is injected into the cylinder during the intake stroke. Consequently, the ambient pressure is approximately the same or slightly less than the atmospheric pressure. In the case of later injection, fuel is

injected during the later stage of the compression stroke into air which have been compressed up to 0.3 to 1 MPa. The drag force acting on the fuel droplets increases as the density of the surrounding air increases. Figure 13 shows the influence of the ambient pressure on spray shape. In the case of low ambient pressure, the fuel spray has a widely spread hollow cone structure. In the case of higher ambient pressure, higher drag force changes the spray shape into a compact and solid cone

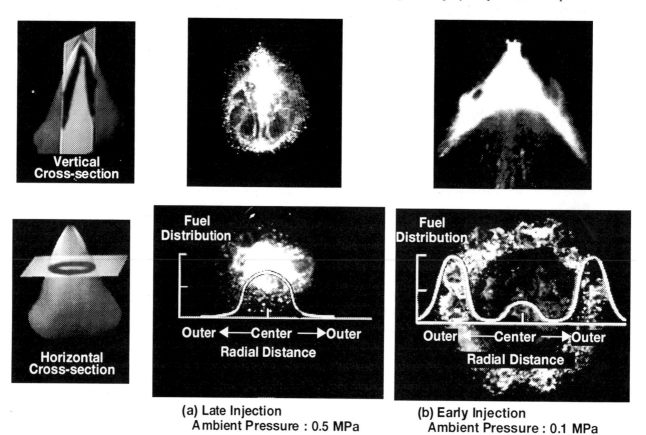

Fig. 13 Dependence of fuel distribution on ambient pressure (injection pressure: 5 MPa, intense swirl)

having suppressed penetration. As indicated in Figure 6, the aim of the spray optimization was wide dispersion for the early injection and compact shape for late injection. This goal can be realized using the inherent characteristics of the interaction between the fuel spray and the ambient air. This characteristic is enhanced, however, by the careful design of the injector so that the kinetic energy of the fuel droplet is shared not only to the penetrating but also to the rotating velocity component. Figure 14 shows the extent of spray tip penetration and the piston surface trajectory at 1000 min^{-1}. In the case of late injection, spray tip penetration is attenuated with a retarding of the injection timing, that is, with an increase in the density of the ambient air. It is necessary to put a sufficient interval between the end of the injection and the spark ignition in order to promote the evaporation and diffusion of the fuel spray for the suppression of overly-rich air-fuel mixtures around the spark plug which may lead to ignition fouling. Consequently, the injection timing should be advanced with an increase in the engine speed. As a result, the point of the impingement of the fuel spray on the piston cavity varies with the engine speed. The injector angle and the cavity geometry have been so designed that the behavior of the fuel spray after impingement is not sensitive to the impinging timing or location.

In the case of early injection, an injection timing was selected such that the fuel spray does not impinge on the piston surface to avoid the surface wetting. When this requirement is satisfied, an earlier injection timing is preferable so as to promote evaporation and mixing of the fuel.

COMBUSTION CHARACTERISTICS

EXPERIMENTS - Systematic experiments were performed using single cylinder engines with the major specifications shown in Table 2, in order to optimize the geometries of the intake ports, the piston cavity and the structure of the fuel spray. A square piston engine with the side length of 81.0mm was adopted so as to observe the phenomena taking place in the cavity.

INTAKE PORT OPTIMIZATION - The aim of the designing of the intake port was to realize a higher reverse tumble intensity and higher intake port flow coefficient at the same time. An upright straight geometry was found to have the inherent characteristics suitable both for the enhancement of reverse tumble and improvement of the flow coefficient. As a result of careful optimization, intense reverse tumble

Table 2   Specifications of the direct injection engine

Bore x Stroke mm	81.0 x 89.0
Displacement cm³	1864
Number of cylinder	IL4
Valve train	Configuration	DOHC
	Number of valve	Intake: 2 ,Exhaust: 2
Compression ratio	12.0 : 1
Combustion chamber	Pentroof (compact piston cavity)
Intake port	Upright straight
Fuel system	Direct injection
Injection pressure MPa	5.0

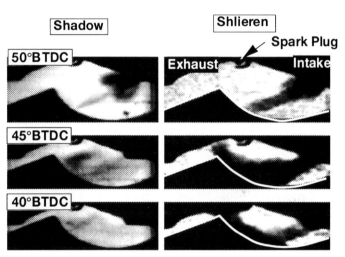

Fig. 15 Liquid and gaseous fuel behavior of the fuel spray in the cylinder
(1000 min^{-1}, firing, Q: 15 mm³/st, end of injection: 50 degBTDC, spar timing: 20 degBTDC)

with a tumble ratio of 1.8 could be realized while maintaining a flow coefficient 10% higher than that of the horizontal intake port of conventional port injection engines.

BEHAVIOR OF FUEL SPRAY - Figure 15 shows the behavior of fuel spray in the cylinder. Both a shadowgraph and schlieren measurements were performed. The sensitivity of shadowgraphy was tuned so that only the image of the liquid phase fuel was captured. In the shlieren photographs, images of both the liquid and the gaseous phase of the fuel were captured. In this experiment, fuel injection was ended at 50 degBTDC, and the fuel evaporation completed within 10 degCA. The vaporized fuel diffused slowly, and moved to

13

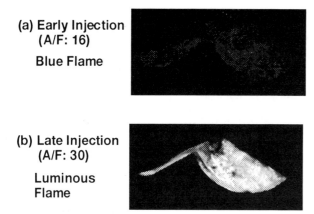

(a) Early Injection (A/F: 16)

Blue Flame

(b) Late Injection (A/F: 30)

Luminous Flame

Fig. 16 Flame radiation (1500 min⁻¹, Q: 15 mm³/st)

the spark plug as a result of being reflected off of the piston surface and by the intense reverse tumble preserved in the cavity. In this situation, upward piston movement, reflection of the spray and reverse tumble flow in the chamber promoted movement of the fuel toward the spark plug.

FLAME RADIATION - Figure 16 shows the spontaneous flame radiations in the cases of early and late injection. Figure 17 shows the result of spectral analysis of the radiation of the flame performed using a newly developed high-speed optical multichannel analyzer [11]. In the case of early injection, luminescence is attributed to OH and CH chemiluminescence as well as CO-O recombination emission. Luminescence in the longer wavelength zone can not be observed. This is a typical characteristic of the premixed lean or stoichiometric flame, showing that in the case of early injection, the goal of mixing control to prepare a homogeneous mixture is satisfied.

In the case of late injection, the major component of the radiation consists of the continuous solid radiation emitted from the soot generated in the cylinder. This is a typical characteristic of distinctively stratified combustion. However, this radiation is attenuated in a short time with the burnup of the soot generated, because sufficient air had been entrained into the fuel before ignition.

IMPROVEMENT IN FUEL ECONOMY - Significant improvement in fuel economy is realized by late direct injection as shown in Figure 18. The index of the improvement in fuel economy is defined as the increase of the IMEP of the engine operated with the same quantity of fuel. The base value consists of the IMEP of the conventional port injection engine having the same specifications but with a compression ratio of 10.5 to 1. When compared with the index based on BSFC of an engine operated at the same load, this

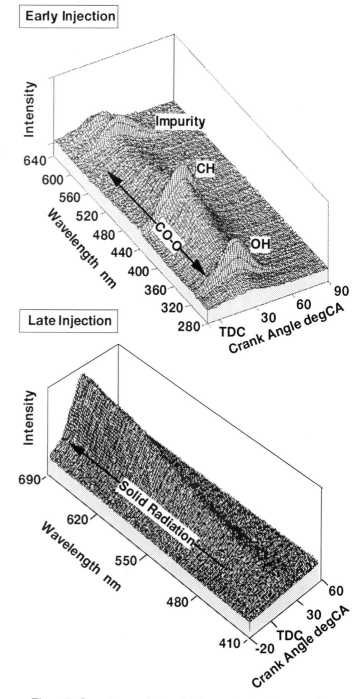

Fig. 17 Spectrum of the visible emission from early direct injection SI engine (condition: see Fig.16)

index underestimates the effect, because the constant mechanical loss is added both to the base value and to the result measured for direct injection engine. At the same time this index overestimates the effect, because the comparison is performed at different loads. These opposite effects are compensated with each other. Therefore the index derived from the IMEP can represent the index based on the BSFC. As shown in the figure, this engine can be operated in a completely unthrottled condition, and in conditions in which the air fuel ratio exceeds 40, fuel economy is improved by 30%. The effect is approximately twice as much as the effect of a premixed leanburn engine using port injection. Thermodynamic analyses were performed to breakdown the factors contributing to the improvement of fuel economy. Reduction of the pumping loss is a principal factor. Reduction of the specific heat ratio caused by an increase in the simple two-atom molecules present, O_2 and N_2, and by the lower thermodynamic temperature, together with a reduction of heat loss also contributes to improved fuel economy. In this direct injection SI engine, octane number requirement is increased by the charge air cooling taking place with early injection, a result being that the compression ratio can be increased to 12.0 to 1. The effect of the higher compression ratio on improved fuel economy is also significant.

Because the charge is distinctively stratified, the mixture strength of the reaction zone is rich, even in the case of an average air fuel ratio exceeding 40. Consequently, the reaction rate is high enough to realize efficient and stable combustion. Figure 20 shows the indicator diagram during idling. In the case of the port injection engine, the combustion rate is low and the combustion stability is penalized primarily because of the large amount of the residual gas present, . In the case of direct injection, the combustion rate is approximately the same as that of the full load condition even during idling. Furthermore, the combustion rate and stability are improved with a reduction in idle speed. Improvement in fuel economy during idling exceeds 50%, due to the lean operation of the engine at lower speed.

NOx REDUCTION - In the case of the premixed leanburn of the port injection engine, an increase in the air fuel ratio results in a decrease in NOx emissions. This is realized by a reduction in the temperature of the reaction zone. In the case of late direct injection, although the thermodynamic temperature is decreased due to the lean operation, the temperature of the reaction zone remains high. However, stratified combustion has the advantage that the combustion characteristics are not deteriorated by the EGR. Consequently, sufficient NOx reduction can be realized by the introduction of a large amount of EGR. Figure 21 shows the effect of EGR on NOx reduction. When compared with the engine-out NOx of the port injection engine operated stoichiometrically, NOx reduction exceeding 90% can be realized while maintaining improved fuel economy. In the case of direct injection diesel

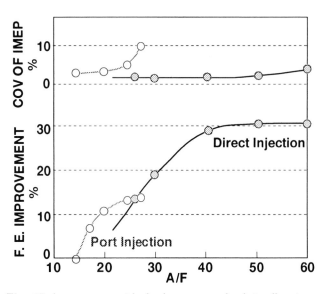

Fig. 18 Improvement in fuel economy by late direct injection (2000 min^{-1}, Q: 15 mm^3/st)

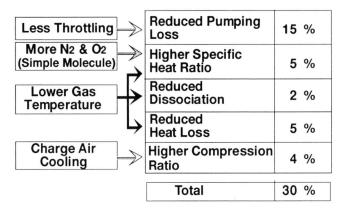

Less Throttling	Reduced Pumping Loss	15 %
More N2 & O2 (Simple Molecule)	Higher Specific Heat Ratio	5 %
Lower Gas Temperature	Reduced Dissociation	2 %
	Reduced Heat Loss	5 %
Charge Air Cooling	Higher Compression Ratio	4 %
Total		30 %

Fig. 19 Factors contributing to improvement in fuel economy
(2000 min^{-1}, Q: 15 mm^3/st, A/F: 40, EGR: 0%)

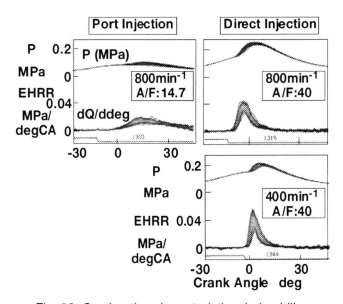

Fig. 20 Combustion characteristics during idling

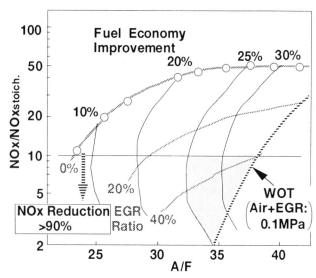

Fig. 21 Reduction of NOx by EGR
(2000 min⁻¹, Q: 15 mm³/st)

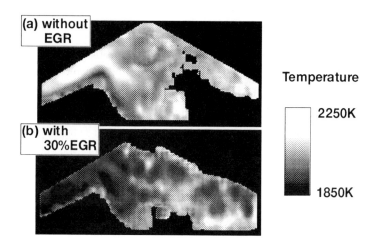

Fig. 22 Influence of EGR on flame temperature
(2000 min⁻¹, Q: 15 mm³/st, A/F: 30)

engine, the sensitivity of NOx reduction to changes in EGR is inferior to that of the premixed gasoline engine. This is caused by the inferior mixing of the EGR gas and fuel, resulting in a higher reaction zone temperature. In the case of direct injection, sensitivity is approximately the same as that of premixed combustion. Figure 22 shows the effect of EGR on the flame temperature measured using the two-color method. It can be seen that the flame temperature is reduced by about 200 K by the introduction of an EGR of 30%. This can be explained as being a result of the mixing of the EGR gas and the vaporized fuel promoted by the spray motion, squish flow and the reverse tumble occurring in the chamber.

FULL LOAD PERFORMANCE

At higher loads, the direct injection SI engine is operated by adopting the early injection strategy. The combustion characteristics are basically the same as those of the premixed port injection engine. However, improved engine performance can be realized through optimized injection . Figure 23 shows the influence of the injection timing on torque, volumetric efficiency and spark advance during trace knock condition. Smoke and the exhaust O_2 are also plotted. In this experiment, the engine was operated at the trace knock condition using stoichiometric fuel at 1500min⁻¹.

In the case of port injection engine, latent heat of evaporation is supplied from the surface of the intake port, intake valve or the cylinder liner. In the case of early direct injection, the fuel spray chases the piston, and the impingement of the liquid fuel on the surface is carefully minimized.

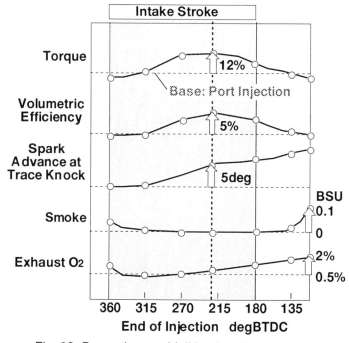

Fig. 23 Dependence of full load performance on injection timing
(1500 min⁻¹, stoichiometric, trace knock)

Therefore, the latent heat is supplied by the intake air. This causes efficient charge air cooling. When the injection timing of the earliest phase of the intake stroke is selected, fuel spray impinges on the piston as shown in Figure 14. In such a case, the volumetric efficiency is the same as that of the port injection engine. When an adequate injection timing is selected, volumetric efficiency is increased by 5%. It can be estimated that the charge air is cooled by about 15K. The gas tempera-

16

Table 3 Summary of the characteristics of the proposed direct injection concept

Characteristics		Key Technologies	Key Components
Fuel Economy	++	❏ Stratified super leanburn using reverse tumble ❏ Higher compression ratio by charge air cooling ❏ Stoich. combustion with high EGR	❏ Electromagnetic swirl Injector ❏ Compact chamber ❏ Vertical intake port
Power	+	❏ Straight port ❏ Charge air cooling by fuel evaporation ❏ Higher compression ratio	❏ Upright straight intake port ❏ Mixing control
Response	++	❏ Torque control by fuel (not influenced by air lag) ❏ Not influenced by fuel transport lag ❏ Transient torque control	❏ Direct injection ❏ Air control valve
Emission	±	❏ Stable combustion under high-EGR condition	❏ Quick & accurate EGR control ❏ Quick & accurate A/F control
Vibration & Sound	±	❏ Compensation of reciprocal inertia by increased gas-force ❏ Torque damping by fuel management	❏ Electromagnetic swirl injector ❏ Air control valve

ture at the end of the compression stroke is estimated to be reduced by about 30 K, and knocking is thus attenuated. As a result of the effect of the increase in volumetric efficiency and the advance of the spark timing, generated torque is increased by 10%. Further retarding of the injection timing can be expected to cause further improvement of the antiknock quality due to a slight stratification of the charge. Such timings are not preferable, because the charge cooling after the intake valve closes does not improve volumetric efficiency and charge stratification acts to penalize air utilization and causes the emission of soot.

By utilizing the effect of the charge air cooling, the compression ratio of the direct injection engine could be increased to up to 12.0 to 1.

DEVELOPMENT OF A PRODUCTION ENGINE

After confirming the potential and the feasibility of this direct injection concept, Mitsubishi Motors Corporation has started development of an engine adopting this concept for production. Development work is currently being performed successfully. The characteristics of this engine are summarized in Table 3. Improved fuel economy of 25% is realized in the Japanese 10-15 Mode. Torque is increased by about 10% in most of its operating range. Response of the vehicle is improved significantly, because engine performance can be controlled by changing the amount of fuel only, and not by controlling the air with transport lag caused by the

volume downstream of the throttle valve. Moreover, it is not affected by the fuel transport lag caused by the formation of the fuel film on the surface of the intake ports. It can meet the Japanese 10 - 15 and the EU Phase 2 emission standards. Although the higher compression ratio and unthrottled operation results in an increase in engine vibration, it can be compensated through careful torque management realized by the improved engine response.

SUMMARY

Novel combustion control technologies for the direct injection SI engine have been developed and briefly described in this paper. By adopting the following technologies, it was possible to achieve super-lean stratified combustion for higher thermal efficiency at partial load and uniform premixed combustion to realize higher performance at full load:

(1) upright straight intake ports to generate reverse tumble,
(2) an electromagnetic swirl injector to realize optimized spray dispersion and atomization, and
(3) a compact piston cavity to maintain stratification.

At partial load, fuel is injected toward the piston cavity during the later stage of the compression stroke. Fuel spray impinging on the cavity wall is directed toward the spark plug. Tumbling air flow in the cavity also assists the fuel to conserve the rich mixture zone thus created around the spark plug. By tuning the interval between the end of injection and ignition of the spark, mixture strength at the spark plug is optimized. As a result, stable combustion can be realized under an air fuel ratio exceeding 40. Significant fuel economy improvement is thus realized.

At higher load, fuel is injected during the early stage of intake stroke. Injection timing to prevent fuel spray impingement onto the piston surface is selected. Air cooling by the latent heat of vaporization taking place during the intake and compression strokes increases volumetric efficiency. It also contributes to a reduction in the octane number requirement, thereby making it possible to adopt a high compression ratio of 12 to 1. As a result of charge air cooling and the higher flow coefficient made possible through the use of upright straight intake ports, engine performance is significantly improved.

REFERENCES

1. Bauler, E.M. et al.: Elimination of Combustion Knock - TEXACO Combustion Process, SAE.Q.T.5-1 (1951)

2. Bishoy, Sinbo: A New Concept of Stratified Charge Combustion, SAE Paper 680041 (1968)

3. Wood, C.D.: Unthrottled Open-Chamber Stratified Charge Engines, SAE Paper 780341 (1971)

4. Schapertons, H., Emmennnthal K. D., Grabe, H. J. and Oppermann, W.: VW's Gasoline Direct Injection (GDI) Research Engine, SAE Paper 910054 (1991)

5. Shiraishi, T., Fujieda, M., Ohsuga, M. and Ohyama, Y.: A Study of the Mixture Preparation Process on Spark Ignited Direct Fuel Injection Engines, Proceedings of 8th IPC, pp.235-240 (1995)

6. Kono, S.: Study of the 'Cardera" Stable Combustion Method in DI Gasoline Engine (ibid.) pp. 223-228

7. Shimotani, K., Oikawa, K., Horada, O. Kagawa, Y.: Characteristics of Gasoline In-Cylinder Injection Engine, Proceedings of the 12th Internal Combustion Engine Symposium, Japan (in Japanese), pp.289-294 (1995)

8. Kiyota, Y., Akishino, K. and Ando, H.: Combustion Control Technologies for Direct Injection SI Engines, FISITA 26 (to be submitted) (1996)

9. Kiyota, Y., Akishino K., and Ando, H.: Concept of Lean Combustion by Barrel-Stratification, SAE Paper 920678 (1992)

10. Kuwahara, K., Watanabe, T., Takemura, J. Omori, S. Kume, T. and Ando, H.: Optimization of In-Cylinder Flow and Mixing for a Center-Spark Four-Valve Engine Employing the Concept of Barrel-Stratification, SAE Paper 940986 (1994)

11. Kuwahara, K., Watanabe, and Ando, H.: Analysis of the flame luminescence by high-speed optical multichannel analyzer, Proceedings of the 13th Internal Engine Symposium of Japan (in Japanese), to be submitted (1996)

Development of Gasoline Direct Injection Engine

Y. Iwamoto, K. Noma, O. Nakayama, T. Yamauchi, and H. Ando
Mitsubishi Motors Corp.

ABSTRACT

The major problems of the various mixture formation concepts for direct injection gasoline engines that have been proposed up to the present were caused by the difficulties of preparing the mixture with adequate strength at spark plug in wide range of engine operating conditions. Novel combustion control technologies proposed by Mitsubishi is one of the solution for these problems. By adopting upright straight intake ports to generate air tumble, an electromagnetic swirl injector to realize optimized spray dispersion and atomization and a compact piston cavity to maintain charge stratification, it has become possible to achieve super-lean stratified combustion for higher thermal efficiency under partial loads as well as homogeneous combustion to realize higher performance at full loads.

GDI™ (Gasoline Direct Injection) engine adopting these technologies is developed. At partial loads, fuel economy improvement exceeding 30 % is realized. At higher loads, since air cooling by the latent heat of vaporization increases volumetric efficiency and reduces the octane number requirement, a high compression ratio of 12 to 1 can be adopted. As a result, 10% increase in performance is realized. NOx emission had been considered as one of the most significant issues of lean burn engines. This problem is solved by using the inherent characteristics of stratified combustion of high EGR tolerance and by the newly developed lean-NOx catalyst.

INTRODUCTION

Throughout the over a century long history of internal combustion engines, gasoline engines with the distinctive feature of higher performance and cleaner exhaust gas and diesel engines with higher thermal efficiency have been used. It goes without saying that an engine realizing the feature of gasoline and diesel engines at the same time is the ideal engine. Therefore, the automobile industry has devoted considerable effort to increasing the thermal effi-

---

* Numbers in parentheses designate references at end of paper.

ciency of gasoline engines to the level of diesel engines. For the past fifty years, the direct injection gasoline engine has been considered to be one of the most promising solutions for achieving this aim. Extensive research and development efforts have been concentrated on establishing the combustion concept of the direct injection gasoline engine.

The history of the research of the gasoline direct injection engines can be divided into three stages.

First Stage (to 1950's) :

Before the invention of sophisticated carburetors, some of the highly boosted air plane engines adopted direct injection systems using the fuel injection technologies for diesel engines. This technology disappeared with the progress of carburetor technologies. In 1954, Benz 300SL adopted direct injection system to solve the problem of the deficiency of performance inherent to carburetor systems [1]. However, this technology was replaced by sophisticated port injection systems in a few years. Direct injection engines during this stage adopted early injection strategy, that is, fuel is injected into the cylinder during the intake stroke to prepare the homogeneous mixture. Consequently, the objective had not been in the fuel economy improvement.

Second Stage (from 1950's to 1980's) :

In order to realize its fuel economy potential, the direct injection gasoline engine should be operated unthrottled in an extremely lean condition by distinctively stratifying the charge and by preparing a rich air-fuel mixture around the spark plug. For that purpose, several stratified charge combustion concepts were proposed in this stage [2-7]. As shown in **Figure 1**, these concepts adopted the narrow spacing configuration, that is, the spark plug gap was located near the fuel spray cone. Although it had been confirmed that stable combustion can be realized by such a configuration, the following problems prevented these concepts from being developed into mass production.

(1) Hydrocarbon emission :

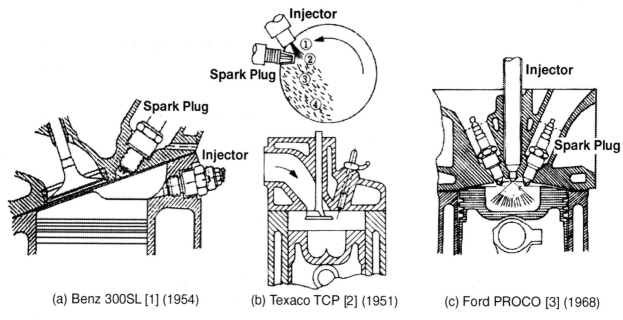

(a) Benz 300SL [1] (1954) (b) Texaco TCP [2] (1951) (c) Ford PROCO [3] (1968)

Fig.1   Example of combustion system for direct injection gasoline engines in the earlier stages

Large amount of hydrocarbon was emitted because it was difficult to complete the combustion.

(2) Limited operation zone :

These concepts adopted swirl for the promotion of fuel-air mixing. The momentum of swirling air is in proportional to engine speed, while the momentum of fuel spray does not depend on engine speed. Therefore, the engine speed range in which the adequate air-fuel mixing is realized was limited.

(3) Spark plug fouling :

In case of narrow spacing, the liquid fuel spray or the over-rich mixture was located at the spark plug, resulting in the formation of soot. Soot was accumulated between plug gap and caused the ignition fouling. Although higher energy ignition systems have sometimes shown the effects of reduced ignition fouling, problem associated with the deterioration of the durability of the spark plug due to the higher ignition energy could not be solved.

(4) Poor performance :

Injection timing variation range realized by the mechanical fuel injection equipment employed in this stage was limited, and the switching from late to early injection strategy was not possible. Consequently, these engines should be operated with the stratified charge even in the highest load range. In order to prevent the soot emission, air-excess ratio should be maintained high, resulting in poor performance.

(5) Dilution of lubricating oil :

It was difficult to prevent the liquid gasoline droplets to impinge on the cylinder liner or on the piston surface. Gasoline on the cylinder liner diluted the lubricating oil on the cylinder liner. Gasoline on the piston surface was captured in the piston crevice and also diluted the lubri-

cating oil.

(6) Soot emission and deposit accumulation :

Over-rich mixture around the spark plug caused the formation and emission of soot. Liquid fuel film layer on the piston surface caused the accumulation of combustion chamber deposit.

By the comprehensive studies performed by many researchers, the difficulties to solve these problems had become a common understanding. Research activities on direct injection gasoline engine had fallen off at the end of this stage.

Third Stage (from 1990's) :

At this stage, fuel economy improvement has become the most important subject for automobile industry, because it is the key factor for the energy saving and the reduction of CO_2, one of the most harmful green house effect gas. In order to meet this requirement, energetic research activities started to establish the direct injection gasoline engine technologies that can be applied to the practical engines in the real world.

It has been a common target of research activities to develop a direct injection gasoline engine realizing greater fuel economy compared with a diesel engine at partial loads and to realize better performance than the conventional MPI (Multi Point Injection) engines at high loads. In order to realize its fuel economy potential, the direct injection gasoline engine should be operated unthrottled in an extremely lean condition by distinctively stratifying the charge. In order to achieve its higher performance potential at high loads, the direct injection gasoline engine should be operated under stoichiometric or slightly rich conditions. When the charge is stratified, soot is generated in the rich zone. Sufficient excess air should be provided around the combustion zone containing soot, in order to burn-up the generated soot. Therefore, when the average mixture strength is stoichiometric or

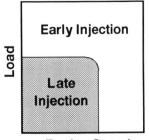

Injection Timing Concept	Early	Late
Benz 300SL	✓	
Texaco TCP		✓
Ford PROCO		✓
Mitsubishi GDI	✓	✓

Fig.2   Operation modes of direct injection engines

slightly rich, that is, when the equivalence ratio is larger than unity, the mixture should be homogeneous so as to suppress the soot formation Accordingly, as shown in **Figure 2**, the direct injection gasoline engines at this stage should realize the following goals ;

(1) stable and distinctive stratification by late injection mode at partial load,
(2) completely homogeneous charge by early injection mode at higher load,
(3) rapid and smooth switching of late and early injection modes.

Comprehensive research has been carried out in the laboratories of Mitsubishi Motors Corporation. It was found that these goals could be realized by a novel mixture preparation concept, Mitsubishi Motors Corporation adopted this concept for its GDI™ (Gasoline Direct Injection) engine, and started the mass production in August of 1996.

In this paper, the basic concept, measures for realizing the concept and the characteristics of Mitsubishi GDI engine will be described.

BASIC STRATEGIES OF MITSUBISHI GDI [8, 9]

The novel mixture formation concept is illustrated in **Figure 3**. In place of narrow spacing layout, it adopts wide spacing layout. Fuel spray is not directed toward the spark plug. It is directed to the piston surface and after impinging on the spherical piston cavity, it is reflected toward the spark plug. By adopting this layout, the interval between the end of injection and the spark ignition long enough for promoting the fuel vaporization and the mixing with the surrounding air can be realized. Consequently, problems of the previous stage caused by the liquid fuel or the over-rich mixture around the spark plug can be solved. The principal factor controlling the mixing is the fuel spray or gaseous mixture reflection on the cavity wall which is subject to the fuel spray momentum. Unlike the methods controlling the mixing by swirl, it is hardly affected by the engine speed. This guarantees the adequate mixing in wide speed range.

Fundamental technologies developed to realize the adequate mixing control by wide spacing layout are illustrated in **Figure 4**. The following new technologies are employed ;
(1) upright straight intake ports generating an intense reverse tumble, that is, a tumble with a rotational direction oppo-

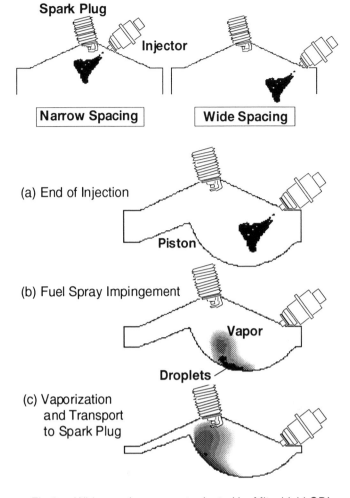

(a) End of Injection

(b) Fuel Spray Impingement

(c) Vaporization and Transport to Spark Plug

Fig.3   Wide spacing concept adopted by Mitsubishi GDI

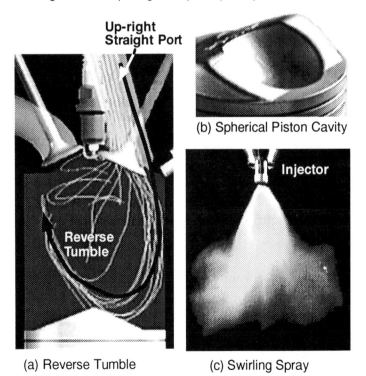

(a) Reverse Tumble   (c) Swirling Spray

Fig.4   Base technologies for Mitsubishi GDI

site to that of the tumble generated by conventional horizontal intake ports,
(2) a spherical compact piston cavity,
(3) an electromagnetic swirl injector.

Mitsubishi Motors Corporation has proposed the concept of the premixed leanburn engine known as Barrel-Stratification [10, 11]. In that concept, tumble with the direction flowing the piston surface from the exhaust to the intake side was adopted. In order to intensify the tumble in this direction, intake port flow passing through the upper area of the intake port and directed to the cylinder liner at the exhaust side was enhanced. In the case of the direct injection engine, the authors considered that the fuel injectors should not be located at the high temperature exhaust side and that it would be preferable to locate the spark plug at the center of the cylinder in order to realize improved combustion with lower heat loss. As a result, the location chosen for the injector was the intake side. Since the fuel spray is injected on the intake side, the piston cavity should also be located on the intake side. This cavity attenuates the tumble in the conventional direction. Therefore, the authors adopted a tumble having a reverse direction. Reverse tumble was also effective in moving the gaseous fuel toward the spark plug after the impingement on the piston cavity. To realize the reverse tumble, upright straight intake ports were selected. Because the upright straight intake port had the inherent characteristics of the higher flow coefficient, the engine performance at the highest engine speed range was improved. This configuration was also effective in preparing the space in the cylinder head to locate the injector. If the conventional horizontal intake port configuration was selected, it would be difficult to direct the fuel spray towards the piston cavity because the angle between the injector and the piston surface would be determined by availability of space under the horizontal intake port.

A spherical combustion chamber was designed so as to capture the fuel spray and the gaseous fuel and to direct the reflected gaseous mixture to the spark plug. Spark plug was located at the periphery of the combustion chamber. The air-fuel mixture and the flame should propagate to the center of spherical chamber where the large amount of air exists. This was realized by the intense squish flow generated by the squish area on the exhaust side. During the later stage of combustion, reverse squish flow propagates the flame to the exhaust side. The wall of the spherical bowl at the exhaust side changes the downward liner side air flow to an upward flow after impinging on the piston. Therefore this design is appropriate for enhancing reverse tumble. Furthermore, the shape of the bowl is suitable for preserving the rotational momentum of the reverse tumble until the end of the compression stroke.

An electromagnetic swirl injector realized the atomization, controlled penetration and controlled dispersion of the fuel spray

SYSTEM CONFIGURATION OF MITSUBISHI GDI

Principal specification of Mitsubishi GDI engine is summarized in **Table 1**. Its cutaway view and front and side

Table 1   Pricipal specification of Mitsubishi GDI

Displacement cm³			1834
Bore x Stroke mm			81.0 x 89.0
Number of Cylinder			in-line 4
ValveTrain			4-valve DOHC
Intake Port			upright straight port
Combustion Chamber			spherical piston cavity
Compression Ratio			12 : 1
Fuel			premium/ regular (RON: 100/92)
Fuel System	Pump		axial piston pump
	Pressure MPa		5.0
	Injector		electromagnetic swirl injector
Air Control			throttle valve with electronic air bypath valve
Ignition System	Energy mJ		60
	Plug		narrow-gap Platinum plug
Exhaust emission Control System			Electric EGR System and lean-NOx catalyst

views are illustrated in **Figures 5 and 6**. System configuration is schematically expressed in **Figure 7**.

A four valve DOHC MPI engine was converted into a GDI engine. Horizontal intake ports was replaced by upright straight intake ports. A spherical cavity was provided on the top of the piston. Compression ratio was increased to 12 to 1 from 10.5 to 1 of the original MPI engine. It can be operated both by premium and regular gasoline with the RON of 100 and 92, respectively. By an adaptive control, air quantity, air-furl ratio as well as spark advance are controlled to bring out

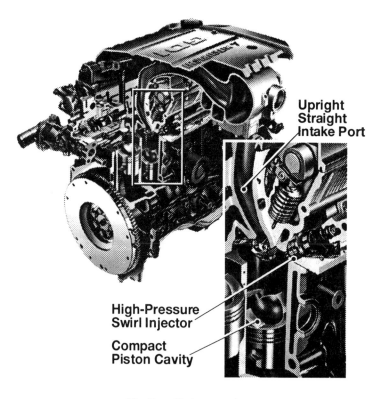

Upright Straight Intake Port

High-Pressure Swirl Injector

Compact Piston Cavity

Fig.5   Cutaway view

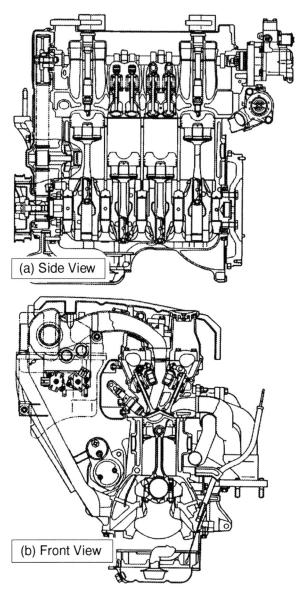

(a) Side View

(b) Front View

Fig.6　Side and front views

the best potential of the supplied fuel. It was equipped with a newly developed high pressure injection pump and electro-magnetic swirl injector. Relatively low injection pressure of 5 MPa was selected. Air bypass valve was used for the smooth switching of the early injection mode stoichiometric operation and the late injection mode lean operation. Ignition energy was slightly increased up to 60 mJ, and a narrow gap Platinum spark plug was employed. For the reduction of NOx, EGR was precisely controlled by an EGR valve actuated by stepper motor. Selective NOx reduction lean-NOx catalyst was newly developed for the further reduction of NOx.

OPTIMIZATION OF INTAKE PORT

For the improvement of the tradeoff relationship between flow coefficient and reverse tumble intensity, upright straight intake port design was optimized. An overhang on the exhaust side wall and a small projection on the liner side wall shown in **Figure 8** was adopted. The overhang attenuates the flow passing through the valve seat area toward the exhaust side. The small projection laminarize the flow upstream of the valve seat area and directed the flow downward. Consequently, flow to the exhaust-side is attenuated and the flow parallel to the intake-side cylinder liner is enhanced. By this intake port, tumble ratio, that is, the rotating speed of tumble over the engine speed can be increased up to 1.8 while maintaining the 10 % higher flow coefficient than that of the horizontal intake port of conventional MPI engines. High flow coefficient is essential for the torque increase at the highest engine speed range, and intense reverse tumble plays a predominant role in the stabilization of the stratified charge combustion.

FUEL INJECTION SYSTEM

HIGH PRESSURE PUMP - A cutaway view of a high pressure fuel pump is shown in **Figure 9**. In order to simplify the system, a piston pump without metering function was se-

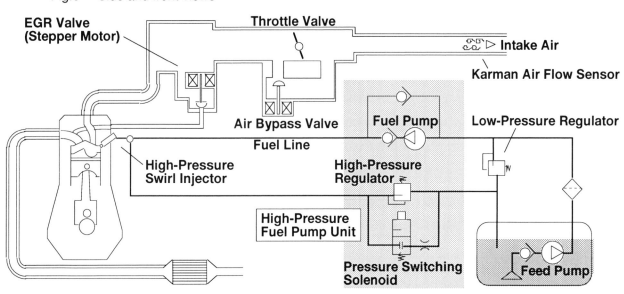

Fig.7　Schematic diagram of system configuration

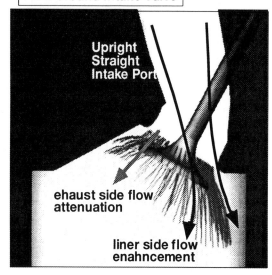

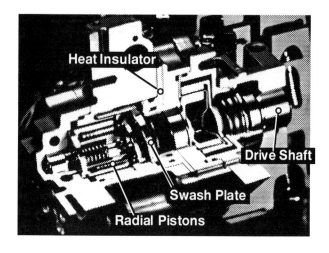

Fig.9　A cutaway view of high pressure fuel pump

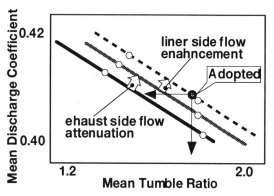

Fig.8　Intake port optimization

lected. The injection pressure was regulated by spilling the high pressure fuel, which increased the fuel compression work. To reduce the fuel compression work, a relatively low fuel injection pressure of 5 MPa was selected. When compared with the fuel pump used for diesel fuel, more elaborate care should be taken with regard to lubrication and cooling, because the gasoline has a lower viscosity and higher volatility than diesel fuel. The reduction in the injection pressure was also effective in establishing the reliability of the pump system. This pump was installed on to the cylinder head and driven directly by one of the camshafts.

The most significant defect of the engine-driven pump is the insufficiency of the fuel during cranking, that is, in the conditions where the engine speed is the lowest. In order to compensate for this shortage, this system uses the in-tank fuel feed pump for the cranking as in the case of the conventional MPI engine. This is realized by a simple bypass valve of the high pressure regulator shown in **Figure 10**. In case of the port injection engine, engine startup process is affected by the fuel transport lag caused by the formation of the fuel film on the wall of the intake ports. On the contrary, in case of the direct injection engine, fuel is injected directly to the cylinder. Consequently, the starting characteristics of this engine

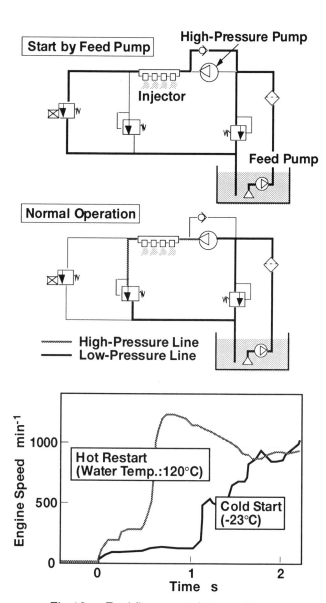

Fig.10　Fuel flow control at cranking

are excellent as shown in Figure 10. This engine can be started within 1.5 seconds under the strict conditions such as after the hot soaking and as the low ambient temperature.

BASIC REQUIREMENTS TO FUEL SPRAY - Basic requirements to fuel spray characteristics are illustrated in **Figure 11**. In case of the early injection, widely spread fuel spray is required so as to achieve homogeneous mixing. The impingement of the fuel spray on the piston surface and the cylinder liner should be suppressed, because liquid fuel captured by the wall results in the emission of soot and the dilution of the lubricating oil. In order to meet this requirement, a widely dispersed fuel spray with low penetration should be prepared. It should be injected at such a timing that the spray chases the piston so as to avoid any impingement on the piston surface.

A compact spray is required in case of the late injection, because it should be inside the compact piston cavity. At the same time, it should be atomized, because it should be vaporized in a short time so as to promote the entrainment of air and EGR gas into the vaporized fuel before arriving at the spark plug.

SWIRL INJECTOR - Basic requirements to fuel spray are summarized as follows ;
(1) controlled dispersion,
(2) suppressed penetration,
(3) atomization.

Basic process of the fuel atomization can be described by the following Weber Number [12].

$$\text{Weber Number} = \frac{\rho \cdot U^2 \cdot d}{\sigma}$$

ρ : density (kg/cm^2) U : velocity (m/s)
d : diemater (m) σ: surface tension (kg/s^2)

This number is a ratio of kinetic energy to surface energy of droplets. The breakup of fuel spray proceeds until Weber Number becomes unity. In other words, fuel atomization process is a process of conversion of kinetic energy into surface energy. Suppose the situation where the droplets have the velocity of 40 km/s and the diameter of 100 micrometers, Weber Number exceeds 100. This means that the fuel atomization proceeds without any external forces. As a result of atomization, it loses the kinetic energy and the velocity is attenuated to zero. As shown in **Figure 12**, however, when the fuel spray has the shape of solid jet, it develops before the start of break up. The break up length, that is, the distance of the droplet travel before the break up is called break up length. The jet flow from the conventional hole nozzle is a turbulent flow, and the break up length exceeds 100 mm when it is injected into the air with atmospheric pressure. Therefore, it impinges on the cylinder liner before it is broken up.

In order to suppress the penetration, swirl nozzles with the configurations shown in **Figure 13** are generally adopted by various stationary combustors. A part of the kinetic energy of fuel flow is converted into surface energy in the nozzle hole, resulting in significantly short break up length.

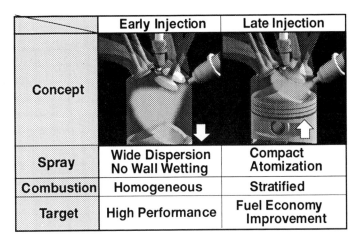

	Early Injection	Late Injection
Concept		
Spray	Wide Dispersion No Wall Wetting	Compact Atomization
Combustion	Homogeneous	Stratified
Target	High Performance	Fuel Economy Improvement

Fig.11   Requirements to fuel spray characteristics

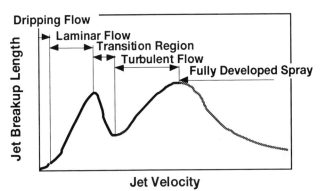

Fig.12   Breakup length of solid spray jets [12]

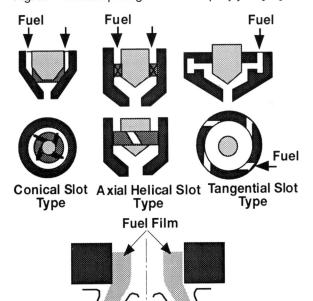

Fig.13   Basic principle of swirl nozzle [12]

In order to meet the basic requirement of direct injection gasoline engine, Mitsubishi Motors Corp. and Mitsubishi Electric Corp. developed an electromagnetic swirl injector. It has approximately the same size as conventional injectors for port injection. A swirler tip is located at the upstream of the injection hole. Three types of swirler tips shown in Figure 13 were tested. When the swirl Reynolds Number that will be described later was tuned to the same value, three types of swirlers gave the similar fuel spray characteristics. Judging from the production feasibility, tangential slot type swirler was selected. It is actuated by a high voltage driving circuit to extend the dynamic metering range.

Fuel swirl intensity was optimized. Swirl intensity can be described by the following swirl Reynolds Number.

$$\text{Swirl Reynolds Number} = \frac{U \cdot r}{\mu}$$

U : Velocity in Swirling Grooves (m/s)
r : Swirling Radius (m)
μ : Viscosity of Fuel (m/s^2)

Dependence of spray cone angle and spray tip penetration on swirl intensity is shown in **Figure 14**. With the increase of swirl intensity, spray tip penetration decreases. When the swirl Reynolds Number is larger than 3×10^4, however, it is not affected by the swirl intensity. 50 mm is the shortest penetration realized by this injector, under the ambient pressure of 0.5 MPa with normal temperature, that is, under the condi-

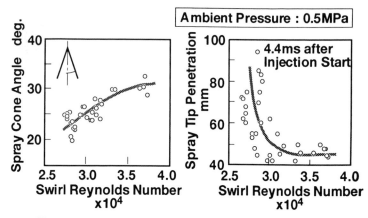

Fig.14 Dependence of spray cone angle and spray tip penetration on swirl Reynolds Number

tion simulating the in-cylinder condition of late injection. In the conditions where the swirl Reynolds Number is larger than 3×10^4, spray cone angle shows slight dependence on the swirl intensity. Precise tuning of the cone angle was performed in this swirl intensity range.

Figure 15 shows the droplet diameter distribution measured by phase Doppler method. Excellent atomization with the SMD less than 15 micrometers is realized at the periphery of hollow cone spray. Large droplets with the SMD exceeding 30 micrometers exist at the center of fuel spray. As shown in the results of modeling in the same figure, these large droplets are captured by the piston cavity, and do not

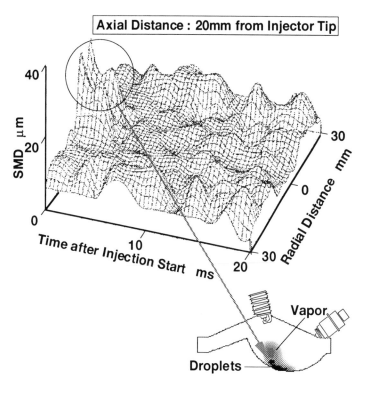

Fig.15   Droplet diameter distribution

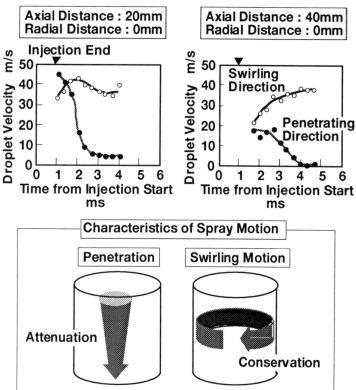

Fig.16   Cone angle, vertical and horizontal componetnts of velocity vectors

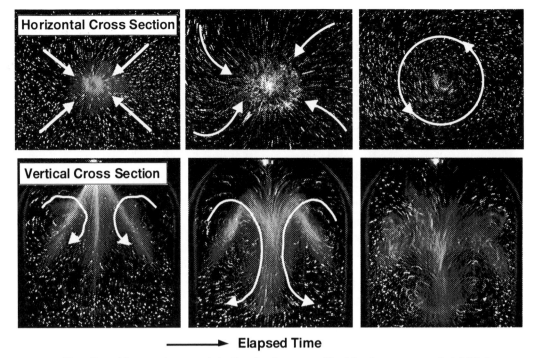

Fig.17　Air entrainment into the fuel spray　(Ambient pressure : 0.1 MPa)

travel to the spark plug. As had been shown in Figure 3, droplets on the piston are evaporated in a short time on the high temperature piston surface.

Figure 16 shows the vertical and the horizontal components of the droplet velocity vectors. The vertical vector, that is, the penetrating velocity component decreases with distance. The horizontal vector, that is, the rotating component, however, remains more or less constant. The decrease in penetrating velocity can be explained by the resistance caused by the drag force by the ambient air. Resistance by the ambient air with regard to the rotating velocity is not significant, because the ambient air rotates with the fuel droplets. Suppression of penetration while maintaining the momentum by the rotating spray motion is the inherent characteristics of the swirl injector. It is preferable to meet the basic requirements to the fuel spray for the direct injection gasoline engine.

Figure 17 shows the interaction between the fuel spray and the ambient air. A rotating air motion is induced around the spray by the swirling motion of the spray. In this experiment, fuel was injected into the chamber that was filled with tracer particles of polymer micro-balloon, and the air motion was tracked by the trajectories of the tracer particles. The spray has a basically hollow cone structure. An intense air flow is generated at the center of the hollow cone due to the movement of the fuel spray. Intense turbulence is also generated due to the interaction of the movements of the air and fuel flows. This turbulence enhances the entrainment of surrounding air and EGR gas into the fuel spray. It can be seen that large scale vortices generated at the upper part of the spray promote the entrainment of air during the later stage of the injection and after the end of injection. These vortices also promote the air and EGR gas entrainment into the fuel

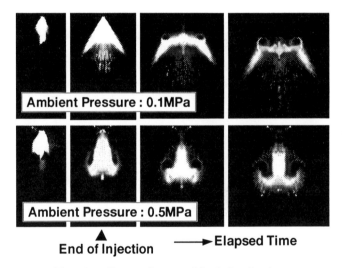

Fig.18　Dependence of fuel distribution on ambient pressure

spray. Mixing of fuel and air was effective for the suppression of the soot formation, and the fuel and EGR gas mixing were effective for the reduction of the local flame temperature, resulting in the lower NOx emission.

In the case of early injection, fuel is injected into the cylinder during the intake stroke. Consequently, the ambient pressure is approximately the same or slightly less than the atmospheric pressure. In the case of late injection, fuel is injected during the later stage of the compression stroke into air that have been compressed up to 0.3 to 1 MPa. The drag force acting on the fuel droplets increases as the density of the surrounding air increases. **Figure 18** shows the influence of the ambient pressure on spray shape. In the case of low ambient pressure, fuel spray has a widely spread hollow cone struc-

27

ture. In the case of higher ambient pressure, higher drag force changes the spray into a compact and solid cone with suppressed penetration. As have been indicated in Figure 11 the target of the spray optimization was the wide dispersion for early injection mode and compact shape for late injection mode. These targets can be realized using the inherent characteristics of the interaction between the fuel spray and the ambient air. These characteristics are enhanced by the careful design of the injector so that the kinetic energy of the fuel droplet is shared not only to the penetrating but also to the rotating velocity component.

Careful durability tests were repeated to get rid of the anxiety that the injection characteristics would be deteriorated by the accumulated deposit. Two kinds of deposit could be predicted. One is the deposit generated in the normal engine operation from the soot or the lubrication oil. The other is the deposit generated during the hot soak period from the olefin or aromatic ingredient of the gasoline. **Figure 19** shows the picture of the injection hole before and after the 400 hours of full load and full speed durability test. Although small amount of deposit accumulation can be observed on the injector surface, injection hole is kept clean.

The so-called injector deposit accumulated in hot soak period could not be observed. **Figure 20** shows the surface temperature of the injector during the continuous high speed full load operation and the hot soaking. Injector temperature does not show the significant difference from that of the conventional MPI engines. This will be the possible reason explaining that the trouble caused by the injector deposit has not been experienced.

Judging from these results, there seems to be no need to worry about the deposit accumulation during the engine operation, because soot, lubricating oil or the deposit is always washed by a high pressure gasoline jet.

IGNITION SYSTEM

One of the objectives of the mixture preparation concept by wide spacing of the injector and the spark plug is to promote the vaporization of fuel and the air entrainment into the vaporized fuel before it is carried to the spark plug. Consequently, the liquid fuel impingement and the delivery of the over-rich mixture to the spark plug that may cause the building up of soot layer on the spark plug do not take place. Therefore, a conventional ignition system with a moderate ignition energy could be adopted. As a conclusion of the deterioration acceleration tests, it was concluded that there seems to be no problem of spark plug fouling peculiar to the gasoline direct injection engine.

AIR CONTROL

Air quantity is controlled by a conventional throttle valve connected to acceleration pedal and a bypass solenoid valve parallel to the throttle valve that is actuated with pulse width modulation. The control system uses the acceleration pedal position as a demand on torque. This engine uses two different combustion modes, early and late injection. Air-fuel ratio in late injection mode exceeds 30, and stoichiometric or

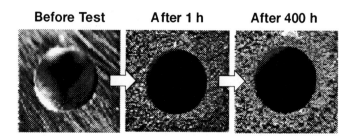

Fig.19 Picture of injection hole after the durebility tests

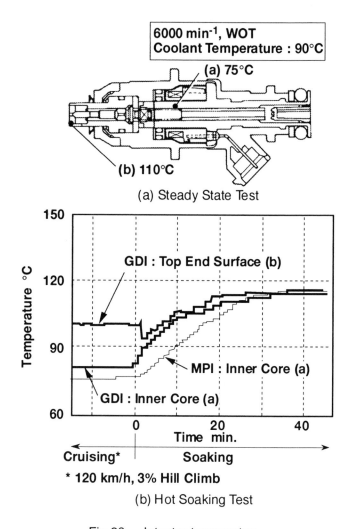

(a) Steady State Test

(b) Hot Soaking Test

Fig.20 Intector temperatue

slightly rich in early injection mode. Roughly speaking, generated torque is determined by the fuel quantity. Therefore, air quantity required in the early injection mode should be at least double of that in the late injection mode to generate the same torque. Air quantity is controlled by the air bypass valve.

Figure 21 schematically illustrates the example of the air and air-fuel ratio control during the transient condition of the switching from the early injection stoichiometric operation to the late injection lean operation. Because a short delay in air quantity control caused by the plenum chamber volume is inevitable, switching should be performed under the condi-

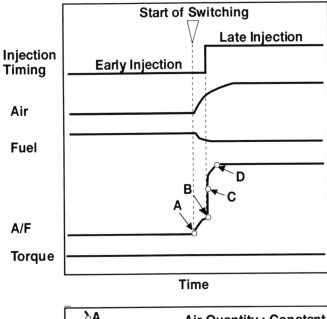

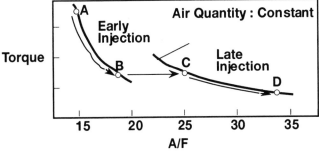

Fig.21 Air and fuel management during the switching from early to late injection mode

tion in which the generated torque of both modes under the same quantity of air is precisely the same to avoid the shock caused by the torque difference before and after the switching. In this example, the early injection with the air-fuel ratio of 18 and the late injection with the air-fuel ration of 25 give the same torque under the constant air condition. Air fuel ratio is gradually increased from stoichiometric to 18 by the increase of air while maintaining the fuel quantity. Then the combustion mode was switched instantaneously to late injection. In the first cycle of the late injection mode, fuel quantity is tuned so as to realize the air-fuel ratio of 25. It is increased gradually to 35, the air-fuel ratio of the cruising condition by decreasing fuel and increasing air. In such a manner, engine is managed to minimize the torque difference at any transient conditions. Rapid and accurate air control by the air bypass valve or electric throttle is one of the essential technologies to actualize the gasoline direct injection system.

LEAN-NOx CATALYST

NOx reduction has been the major subject to be solved in any leanburn engines. In Mitsubishi GDI engine, it is realized essentially by EGR taking the advantage of its high EGR tolerance. Although sufficient NOx reduction can be realized by the introduction of a large amount of EGR it reduces the lean operation range because the introduction of EGR reduces

the air to be supplied into the cylinder. A lean-NOx catalyst newly developed by Mitsubishi Motors Corp. and Nippon Shokubai Corp. was applied for the further NOx reduction maintaining the lean operation range.

Two kinds of lean-NOx catalyst have been proposed, the NOx adsorber catalyst and the selective reduction catalyst. NOx adsorber catalysts [13-15] trap NOx during the lean operation and release adsorbed NOx to reduce it during the short rich excursion period. Selective reduction catalysts [16-18] make use of the direct reduction reaction of NOx by hydrocarbon. The lean NOx catalyst selected is the selective reduction catalyst. The advantage of NOx adsorber catalyst is its high reduction efficiency. However, it is subject to sulfur poisoning. When it is operated supplying the high sulfur gasoline in the European market, it loses the activity in a short mileage. On the contrary, although its reduction efficiency is not remarkable, the selective reduction catalyst maintains its efficiency even after the mileage accumulation durability tests with high sulfur European gasoline. Anyhow, the lean-NOx catalyst is a developing technology and further improvement is expected.

CALIBRATION OF MITSUBISHI GDI ENGINE

INJECTION AND IGNITION TIMING - **Figure 22** shows the conditions in which the stable combustion is realized by late injection on the injection and ignition timing map. The behavior of fuel spray and gaseous mixture is controlled by the impingement on the piston cavity and the reverse tumble. Fuel impingement and reverse tumble are effectively utilized only when the piston locates at the appropriate position at the impingement timing. It is necessary to put a proper interval between the end of the injection and the spark ignition in order to adequately promote the evaporation and diffusion of the fuel spray and the mixing with the surrounding air. When this interval is too short, over-rich mixture is prepared around the spark plug. When it is too long, fuel diffusion makes the mixture around the spark plug too lean. In such a way, appropriate injection and ignition timings are determined. When the engine speed is 1500 min^{-1}, the stable combustion area on the map is the widest. Available interval ranges from 10 to 60 degrees crank angle. With the decrease of the engine speed, the kinetic energy of the reverse tumble decreases. In such a situation, fuel spray impingement should play a more predominant role in the mixture preparation because the assistance of the reverse tumble is reduced. As a result, stable combustion area is cramped. When the engine speed is increased, time required for the fuel spray to penetrate to piston surface is increased when it is expressed by crank angle. Thereby, the crank angle to start the injection should be advanced to keep the impingement location constant. The stable combustion area on the map is also cramped in such a situation because the available mixture formation duration is reduced when it is expressed by the unit of time. When the engine speed exceeds 3500 min-1, available area diminishes. If the injection and ignition timings are selected in this stable combustion area, rapid and stable combustion is realized as shown in **Figure 23**.

29

In the case of early injection, an injection timing was selected such that the fuel spray does not impinge on the piston surface to avoid the surface wetting. When this requirement is satisfied, an earlier injection timing is preferable so as to promote the evaporation and mixing of fuel.

COMBUSTION MODE - Combustion mode calibration is shown in **Figure 24**. In the partial load operating conditions, a late injection strategy is adopted. Although the engine can be operated at an air-fuel ratio exceeding 100 and the complete unthrottled operation is possible, the engine is throttled slightly in this zone and the air-fuel ratio is controlled to range from 30 to 40 in order to introduce a large quantity of EGR. Since the dependence of the fuel economy

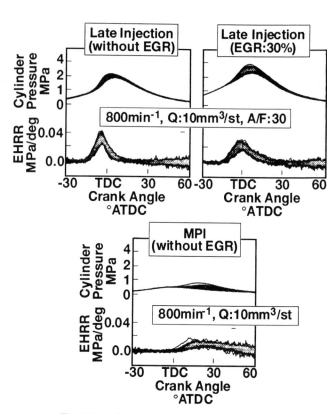

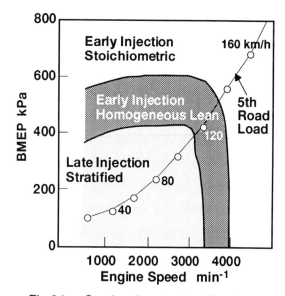

Fig.23   Combustion characteristics

improvement effects on the ai-fuel ratio is not significant when the air fuel ratio exceeds 30, the penalty of slight throttling on the fuel economy is little. As shown in **Figure 25**, manifold absolute pressure in the condition frequently used in the actual operation is higher than 75 kPa and significant pumping work reduction is realized throughout the actual driving conditions.

An early injection strategy is adopted for the higher load operating conditions. In most of this zone, the engine is operated under stoichiometric and it is operated at a slightly rich condition at full load. In the lowest load conditions in

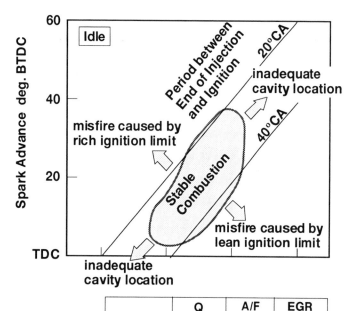

	Q (mm³/st)	A/F	EGR (%)
Idle	6	30	30
1500 min⁻¹	10	30	50
3000 min⁻¹	17	30	20

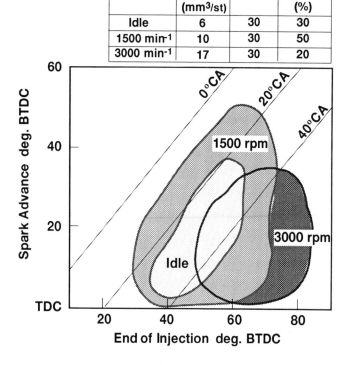

Fig.22   Injection and ignition timing for stable combustion

Fig.24   Combustion mode calibration

30

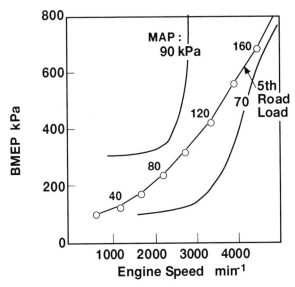

Fig.25   Manifold absolute pressure

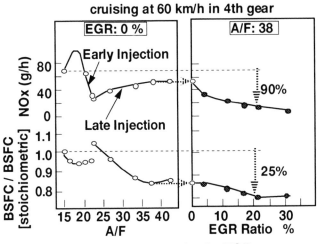

Fig.26   NOx reduction by EGR

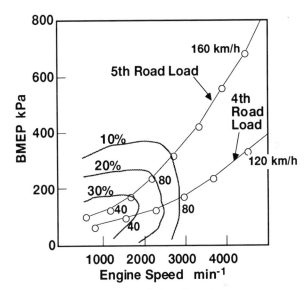

Fig.27   EGR calibration

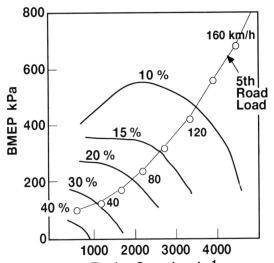

Fig.28   Fuel economy improvement (base : MPI engine with the same displacement and the compression ratio of 10.5 to 1)

this zone, the engine is operated at homogeneous lean conditions with the air-fuel ratio of from 20 to 25 for further improvement of fuel economy.

EGR - In the case of premixed leanburn MPI engines, an increase in the air-fuel ratio results in the decrease in NOx emission. This is realized by a reduction in the temperature of the reaction zone. In the case of late direct injection, although the thermodynamic temperature is decreased due to the lean operation, the temperature of the reaction zone remains high. However, stratified combustion has the advantage that the combustion characteristics are not deteriorated by the EGR as had been shown in Figure 23. Consequently, sufficient NOx reduction can be realized by the introduction of a large amount of EGR. **Figure 26** shows the effect of EGR on NOx reduction. When compared with the engine-out NOx of the MPI engine operated under the stoichiometric condition, NOx reduction exceeding 90 % can be realized while maintaining improved fuel economy. In the case of direct injection diesel engine, the sensitivity of NOx reduction to changes in EGR is inferior to that of the premixed gasoline engine. This is caused by the inferior mixing of the EGR gas and fuel, resulting in a higher reaction zone temperature. In the case of direct injection, sensitivity is approximately the same as that of premixed combustion [19]. This can be explained as a result of the mixing of the EGR gas and the vaporized fuel promoted by the spray motion, squish flow and the reverse tumble. EGR calibration is shown in **Figure 27**. 10 to 40 % of EGR is introduced in the late injection zone. For the quick and accurate control of large amount of EGR, electric EGR valve actuated by a stepping motor is adopted.

CHARACTERISTICS OF GASOLINE DIRECT INJECTION ENGINE

FUEL ECONOMY - Significant improvement in fuel economy is realized by late direct injection as shown in **Figure 28**. The base of comparison is the conventional MPI engine having the same specifications but with a compression

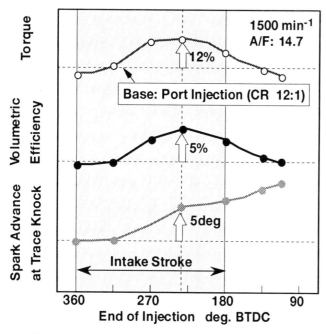

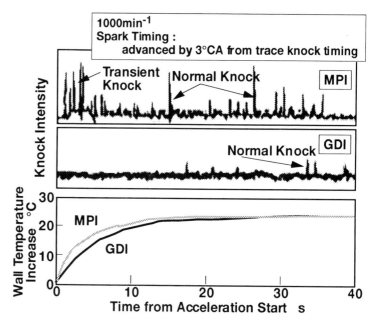

Fig.29 Dependence of full load performance on injection timing

Fig.30 Knock intensity during the acceleration

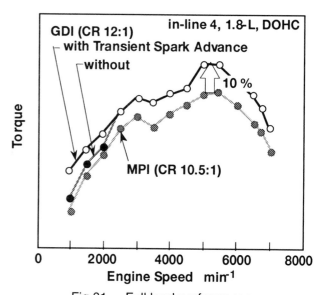

Fig.31 Full load performance

ratio of 10.5 to 1. In the lowest speed and load condition, fuel economy improvement exceeding 40 % is realized. In the road load cruising condition of 100 km/h, fuel economy is improved by 15 %. The 10-15 mode fuel economy of Galant 96 model employing the GDI engine prevails that of formar Galant by more than 35 %.

FULL LOAD PERFORMANCE - At higher loads, the direct injection SI engine is operated by adopting the early injection strategy. The combustion characteristics are basically the same as those of the premixed port injection engine. However, improved engine performance can be realized through the optimized injection. **Figure 29** shows the influence of the injection timing on torque, volumetric efficiency and spark advance during trace knock condition. Smoke and the exhaust O2 are also plotted. In this experiment, the engine was operated at the trace knock condition using stoichiometric fuel at $1500min^{-1}$.

In the case of MPI engines, latent heat of evaporation is supplied from the surface of the intake port, intake valve or the cylinder liner. In the case of early direct injection, the fuel spray chases the piston, and the impingement of the liquid fuel on the surface is carefully minimized. Therefore, the latent heat is supplied by the intake air. This causes the efficient charge air cooling. When the injection timing of the earliest phase of the intake stroke is selected, fuel spray impinges on the piston. In such a case, the volumetric efficiency is the same as that of the port injection engine. When an adequate injection timing is selected, volumetric efficiency is increased by 5 %. It can be estimated that the charge air is cooled by about 15 K. The gas temperature at the end of the compression stroke is estimated to be reduced by about 30 K, and knocking is thus attenuated. Further retarding of the injection timing can be expected to cause further improvement

of the antiknock quality due to a slight stratification of the charge. Such timings are not preferable, because the charge cooling after the intake valve closes does not improve the volumetric efficiency and charge stratification penalizes the air utilization and causes the soot emission.

In the case of MPI engines, significant transient knock takes place during the several cycles at the start of vehicle acceleration. Transient knock is caused by the selective transport of the low boiling point gasoline component with lower octane number [20]. As shown in **Figure 30**, the direct injection engine is not affected by such a transient knock, because all of the gasoline components are transported into the cylinder. A knock suppression period caused by the delay of the combustion chamber surface heating follows the transient knock period. Consequently, in case of the GDI engine that is

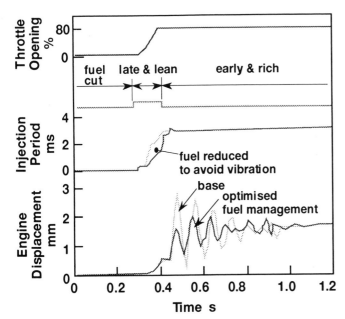

Fig.32 Suppression of engine vibration during acceleration by engine management

Table 2 Summary of durability tests

Engine System	engine endurance vehicle endurance road test (no specific definition)
Injector	WOT heat cycle knock operation continuous late injection continuous idle soak repetition panel coking
Fuel Pump	continuous operation dry operation alcoholic gas (M15) vibrational test anti-contamination actual engine test
Spark Plug	WOT heat cycle no-load racing continuous late injection continuous idle
Catalyst	accelerated deterioration (test bench/vehicle)

number of engines : 40
number of vehicles for endurance test : 20
number of vehicles for road test : 120

Table 3 Specification of gasoline used in the durability tests

		Aroma (%)	Oreffin (%)	MTBE (%)	Sulfur (ppm)
Test Fuel		45	25	9.5	440
Unleaded Premium	Japan	37	15	6	<100
	EU	45	10	9.5	440
	USA	43	10	0	260
Unleaded Regular	Japan	27.5	17.5	0	<300
	EU	37.5	10	-	275
	USA	32.5	12.5	0	350

not affected by the transient knock, ignition timing can be advanced for 10 seconds or more at the start of acceleration [20]. In general, this period is sufficient to complete the vehicle acceleration.

Figure31 shows the full load performance. In almost all of the engine speed range, torque higher than the conventional MPI engine by 10 % is realized. In the lower engine speed range, the primary factors are the increase in volumetric efficiency and the transient spark timing advance. In the higher speed range, the primary factors are the increased volumetric efficiency, higher compression ratio and the improved intake port flow coefficient.

VIBRATIONS - Direct injection gasoline engines have a substantial characteristic of higher vibration, because of the higher compression pressure as a result of higher compression ratio and the unthrottled operation even in the no load operation. However, this defect was compensated completely by the optimization of the engine mounts. These mounts are apt to induce the engine vibration during the transient condition after the rapid change of engine torque. Direct injection engines have the advantage of superior governability. Its torque can be controlled by changing the amount of fuel only, not by controlling the air with transport lag caused by the volume downstream of the throttle valve. Moreover, it is not affected by the fuel transport lag caused by the formation of the fuel film on the surface of the intake ports. Engine vibration during the transient condition is suppressed by the torque management capability inherent to the direct injection engine. **Figure 32** shows an example of the management. It is shown that the engine vibration during the acceleration could be reduced significantly by reducing the generated torque of several cycles after the end of accelerator pedal operation without penalizing the vehicle response.

VERIFICATION OF DURABILITY

In order to confirm the life long reliability of this completely new combustion system, comprehensive durability verification tests listed in **Table 2** were performed. Because it was suspected that the gasoline quality significantly affects the durability of fuel injection system, combustion chamber and the lean-NOx catalyst, a specially synthesized fuel whose composition is listed in **Table 3** was prepared. The representative gasoline ingredients contained in the worst gasoline in Japanese, European and United States market are listed in the table. Test gasoline was composed so that the maximum values of aromatic, olefin, MTBE and Sulfur in the three markets are contained in it.

Throughout the durability tests, problems peculiar to the gasoline direct injection engine had not been experienced.

The presumable explanation will be that in the highest speed and highest load condition, in which many thermal problems may take place, it is operated by the early injection mode. Combustion characteristic of early injection mode is approximately the same as that of the conventional MPI engine whose reliability has been confirmed completely by the life long field experiences in the worldwide markets.

CONCLUSION

As a conclusion of this research and development program, the authors would like to summarize the issues on gasoline direct injection engine that had prevented this technology to be applied to the practical engines in the real world, and the solutions adopted by the Mitsubishi GDI engine.

Major issues to be solved were as follows :

(1) Under a stratified charge condition, it was difficult to complete the combustion. Thereby, considerable amount of hydrocarbon was emitted.

(2) Stable combustion could be realized under very limited condition.

(3) Spark plugs were found malfunctioning.

(4) Under a homogeneous combustion, soot was emitted.

(5) There was no benefit of power increase under a homogeneous combustion.

(6) No fuel injection equipment was offered that enables the rapid switch over of two combustion modes, homogeneous combustion by early injection and stratified combustion by late injection.

(7) Lubrication oil on the cylinder liner was diluted by the injected fuel.

These issues are solved by Mitsubishi GDI engine as follows :

(1) Under a stratified charge condition, it was difficult to complete the combustion. Thereby, considerable amount of hydrocarbons was emitted.

This is caused by the failure of preparing the moderate strength mixture around the spark plug. As a conventional approach, narrow spacing layout locating the fuel spray near the spark plug had been adopted. When such a configuration was adopted, liquid fuel or the over-rich mixture was supplied to the spark plug because of the lack of the time for the vaporization and the mixing. Because the mixture zone was separated by the inflammable over-lean mixtures between the rich mixture layers around the droplets, propagating flame quenched. Although the autoignition occurs independently around each diesel fuel droplet, gasoline droplets isolated from the flame front is not ignited and emitted from the engine.

To solve this problem, a basic target was imposed to locate the sufficiently evaporated and dispersed gaseous bulk of mixture around the spark plug. This target was realized by the wide spacing layout, that is, by locating the injector apart from the spark plug. Fuel was injected towards the piston, and transported to the spark plug after the vaporization and the mixing with the surrounding air is completed.

Upright straight intake port was employed to generate the reverse tumble that transports the gaseous mixture on the piston to the spark plug. In addition, the shape of the piston cavity was carefully designed so that the gaseous fuel impinged on the piston surface could easily turn towards the spark plug. Furthermore, during the late injection mode, fuel is injected in the opposite direction against the upward motion of the piston. At this moment, in-cylinder air already has formed an upward motion that helps to carry the mixture to the spark plug. Thus, all of them, that is, the reverse tumble, fuel droplet reflection on the piston surface and the piston movement cooperate with one another to make the gaseous mixture to turn upward and to form a continuous bulk of the gaseous mixture that realizes stable flame propagation.

(2) Stable combustion could be realized under very limited condition.

As a conventional approach, swirl was used to promote the air-fuel mixing. The momentum of swirling air is in proportional to engine speed, while the momentum of fuel spray does not depend on engine speed. Therefore, the engine speed range in which the adequate air-fuel mixing was realized was limited. In case of the authors' concept, the principal factor controlling the mixing is the fuel spray or gaseous mixture reflection on the cavity wall which is not affected by the engine speed. This guarantees the adequate mixing in wide speed range.

In many cases, fuel spray was directed towards the piston surface. Taking into account that the piston keeps moving, there must be limited range of injection timing for the spray to fit in a cavity on the piston. The injection timing, at the same time, should meet a requirement to allow enough time to promote the vaporization and the mixing. Injection duration on crank angle basis becomes longer at high speed and high load operation. This means that it is difficult to comply with the above mentioned requirement. However, if the spray direction coincident with the cylinder axis could be adopted, the spray would be captured in the cavity regardless of the injection timing. Looking from this standpoint, it is desirable to set the injector as upright position as possible. This is realized by the upright straight intake port layout that leaves larger space for the injector beneath the intake port.

(3) Spark plugs were found malfunctioning.

Spark plug fouling had been one of the most serious problems that prevented the direct injection engine from a practical application. This problem is solved also by the wide spacing layout of an injector and a spark plug to prevent the liquid fuel or the over-rich mixture from locating around the spark plug.

(4) Under a homogeneous combustion, soot was emitted.

It had been pointed out that soot emission was observed even in the case of early injection. By the authors' analysis, it was found that the source of soot is the liquid fuel layer on the piston. During the injector development work, the suppression of the spray tip penetration was considered as priority. It is realized by lowering the injection pressure and enhancing the fuel swirl intensity. By minimizing the fuel spray impingement on the piston surface, soot emission can be suppressed.

(5) There was no benefit of power increase under a homoge-

neous combustion.

To increase the power output, it is essential to take advantage of intake air cooling effect. From this point of view, it was expected to be beneficial to promote the fuel atomization and scatter the fuel droplets all over the cylinder. This is realized by the newly developed swirl injector.

As a conventional approach, air swirl had been adopted for the promotion of mixing and combustion. Generally speaking, the flow coefficient of the intake port that is suitable for the swirl enhancement is low. On the contrary, the upright straight intake port adopted by the authors have a feature of superior flow coefficient. It contributes a great deal in increasing the power output in the highest speed region.

(6) No fuel injection equipment was offered that enables the rapid switch over of two combustion modes, homogeneous combustion by early injection and stratified combustion by late injection.

In the conventional approach, mechanical injectors with the similar design to diesel nozzle had been used. It was quite difficult to meet the requirement of the instantaneous switching of the intake stroke injection and the compression stroke injection.

An electromagnetic injector for MPI engines was modified for the direct injection application. Although the original injector was designed to be used applying a few bars of fuel pressure, a small modification enabled it to be used as a high pressure fuel injector. Comprehensive works were performed to give the swirling momentum to the fuel spray. The conclusion of the development work, however, is very simple. By a simple swirler tip provided with tangential slots and a groove, well-atomized fuel spray with controlled dispersion and penetration can be realized.

(7) Lubrication oil on the cylinder liner was diluted by the injected fuel.

The lubrication oil dilution by injected gasoline had been considered as a serious problem. However, as mentioned before, spray tip penetration of developed injector is suppressed. Consequently, lubrication problem had not been experienced in the development of Mitsubishi GDI engine.

By these solutions, Mitsubishi GDI engine realizes the following characteristics ;
(1) fuel economy equivalent to diesel engines,
(2) performance superior to port injection SI engines, and
(3) exhaust gas as clean as that of port injection SI engines.

REFERENCES

[1] Scherenberg, H. : Ruckbliik uber 25 Jahre Benzin-Einspritzung in Deutchland, MTZ, No.16, Vol.9 (1955)

[2] Bauler, E.M. : Elimination of Combustion Knock - TEXACO Combustion Process, SAE.Q.T.5-1 (1951)

[3] Bishoy, Sinbo: A New Concept of Stratified Charge Combustion, SAE Paper 680041 (1968)

[4] Wood, C.D. : Unthrottled Open-Chamber Stratified Charge Engines, SAE Paper 780341 (1971)

[5] Schapertons, H., Emmennnthal K. D., Grabe, H. J. and Oppermann, W. : VW's Gasoline Direct Injection (GDI) Research Engine, SAE Paper 910054 (1991)

[6] Shiraishi, T., Fujieda, M., Ohsuga, M. and Ohyama, Y. : A Study of the Mixture Preparation Process on Spark Ignited Direct Fuel Injection Engines, Proceedings of 8th IPC, pp.235-240 (1995)

[7] Kono, S. : Study of the 'Cardera" Stable Combustion Method in DI Gasoline Engine (ibid.) pp. 223-228

[8] Kume T., Iwamoto, Y., Iida, K., Murakami, N. , Akishino, K. and Ando, H. : Combustion Control Technologies for Direct Injection SI Engines, SAE Paper 960600 (1996)

[9] Kiyota, Y., Akishino, K. and Ando, H. : Combustion Control Technologies for Direct Injection SI Engines, FISITA 96 (1996)

[10] Kiyota, Y., Akishino K. and Ando, H. : Concept of Lean Combustion by Barrel-Stratification, SAE Paper 920678 (1992)

[11] Kuwahara, K., Watanabe, T., Takemura, J. Omori, S. Kume, T. and Ando, H. : Optimization of In-Cylinder Flow and Mixing for a Center-Spark Four-Valve Engine Employing the Concept of Barrel-Stratification, SAE Paper 940986 (1994)

[12] Lefebvre, A. H. : "Atomization and Sprays", Hemisphere Publishing (1989)

[13] Miyoshi, N., Matsumoto, S., Katoh, K. Tanaka, T., Harada, J. and Takahara, N. : Development of New Concept Three Way Catalyst for Automotive Lean-Burn Engines, SAE Paper 950809 (1995)

[14] Brogan, M. S., Brisley, R. J., Moore, J. S. and Clark, A. D. : Evaluation of NOx Adsorber Catalysts Systems to Reduce Emissions of Lean Running Gasoline Engines, SAE Paper 962045

[15] Strehlau, W., Kreuzer, T., Leyrer, J., Hori., M., Lox, E. S. and Hoffmann, M. : New Developments in Lean NOx Catalysis for Gasoline Fueled Passenger Cars in Europe, SAE Paper 962047 (1996)

[16] Iwamoto, M. Yahiro, H. Mine, Y. and Kagawa, S. : Excessively Cu Ion Exchanged ZSM-5 Zeorites as Highly Active Catalyst for Direct Decomposition of NO, Chem. Lett., p.213 (1989)

[17] Truex, T. J., Searles, R. A. and Sun, D. C. : Catalysts for Nitrogen Oxides Control under Lean Burn Conditions, Platinum Metals Rev. No.36, p.2 (1992)

[18] Schelef, M. : Selective Catalytic Reduction of NOx with N-free Reductants, Chem. Rev. 95, P.209 (1995)

[19] Kuwahara, K., Watanabe, T. and Ando, H. : Analysis of the Flame Luminescence by High-Speed Optical Multichannel Analyzer, Proceedings of the 13th Internal Engine Symposium of Japan (in Japanese), pp.145-150 (1996)

[20] Ando, H., Takemura, J. and Kojina, E. : A knock anticipating Strategy Basing on the Real-Time Combustion Mode Analysis, SAE Paper 890882 (1989)

Mixture Preparation for Direct-Injection SI Engines

M. Ohsuga, T. Shiraishi, T. Nogi, Y. Nakayama and Y. Sukegawa

Hitachi Ltd.

ABSTRACT

Meeting future exhaust emission and fuel consumption standards for passenger cars will require refinements in how the combustion process is carried out in spark ignition engines. A direct injection system decrease fuel consumption under road load cruising conditions, and stratified charge of the fuel mixture is particularly effective for ultra lean combustion. To achieve stable combustion at an ultra lean air fuel ratio, fuel spray speed, angle, droplet diameter and in-cylinder air behavior must be optimized.

In this paper, the engine system which has an injector and spark plug located in the center of combustion chamber, and has a flat type piston was investigated. We adopted a swirl type injector (fuel pressure 5MPa) which gave a droplet diameter of 20 μm. Swirl air motion was applied in the cylinder to concentrate the fuel mixture at its cylinder center. Fuel spray speed was decreased to under 30m/s to keep the mixture at the top of the cylinder. We achieved a lean limit air fuel ratio of about 40 ohsing the flat type piston.

INTRODUCTION

Many methods to get a stratified charge mixture have been developed for direct-injection spark ignition (SI) engines. A side injection type was proposed which has an injector located between two intake valves. [1-4] This type of system has a cavity at the top of the piston. Fuel is injected into the cavity to form the stratified charge late in the compression stroke. A center injection type was also proposed which has an injector located at the center and top of the cylinder. [3-12] This second type also has a cavity at top of the piston. Ignition plug is also located at the center and top portion of the cylinder.

We have been developing an injection system for the stratified charge direct-injection engine. Cavity and flat type pistons were used for the engine tests. These also have an injector which is installed at the center and top of the cylinder. This report describes our study with the flat type piston. Performance of this type is affected by fuel spray characteristics such as speed, diameter and spray angle. We optimize these characteristics to get a stratified charge.

SYSTEM CONFIGURATION

Fig.1 shows a system configuration for the direct-injecton SI engine. An injector and a spark plug are installed at the center of the engine head. Fuel is injected toward the piston from the top of the cylinder. Fuel is pressurized by low and high pressure pumps. Fuel pressure of the former is 0.3MPa, the same as the fuel pump for the intake port injection system. Fuel pressure is increased to about 5MPa by the high pressure pump. The electrode of the spark plug is located near the injector tip. But injected fuel spray does not hit the electrode directly to avoid misfiring. A

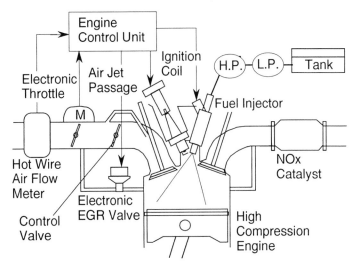

Fig.1 System configuration
(H.P. / L.P. : High Pressure / Low Pressure Fuel Pump)

hot wire air flow meter and electrically controlled throttle valve are included in the intake manifold. The electronic throttle is used for the engine torque control , even if the air fuel ratio changes from rich to lean , in accordance with the accelerator pedal angle. And air jet passage is included in the intake manifold which generates the swirl air flow in the cylinder. This passage can easily produce swirl flow independently of the intake port shape. The electronic EGR valve and NOx catalyst are equipped to reduce NOx emission during operation at a lean air fuel ratio. Compression ratio of the engine is increased to 12 from 9 of base the engine.

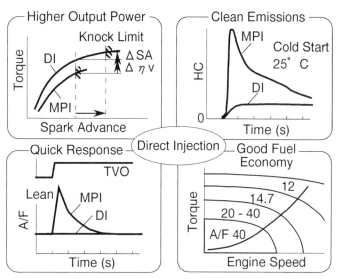

Fig.2 Features of direct injection engines

Fig.2 summarizes futures of direct-injection engines. They get a higher output power because knock limit can be expanded by retarding ignition and increasing volumetric efficiency. Injected fuel during the intake stroke loses heat from intake air. As a result, a larger amount of air is inducted by the cylinder. The direct-injection system can decrease HC emissions during warm up condition after starting the engine. A fuel film is not formed on the surface of the intake port in comparison with the MPI (port injection) system of port injection. As a smaller fuel film is formed on the surface of the piston and cylinder, HC emissions can be decreased. Quick response is another feature of the direct-injection system. Fluctuation of the air fuel ratio after throttle valve opening is much smaller than in the MPI system because of the smaller fuel film. Higher torque performance is achieved by the flat characteristic of the air fuel ratio during acceleration. Pumping loss is decreased by setting the air fuel ratio at lean; with an A/F of 40, fuel consumption can be reduced by about 30% compared with A/F of 14.7. Set point of air fuel ratio control is changed from 40 to 12 to meet torque demand.

CONCEPT OF MIXTURE PREPARATION

Table 1 lists features of side and center injection type systems. For the former, the fuel injector is mounted on the intake valve side, and the spark plug is located at the top of the pent-roof. The piston has a cavity. Fuel is injected into this cavity late in the compression stroke to get the stratified charge mixture. Fuel is vaporized at the surface of the piston, and vaporized fuel is transported to the spark plug and ignited. In this case, vaporized fuel is introduced by in-cylinder air flow and spray momentum. The injector is easily installed and ignition is good because of the vaporized fuel movement. But there are possibilities for a fuel film to form on the piston surface and soot and HC to be emitted.

For the center injection type, the injector and spark plug are located at the center of the cylinder. Fuel is injected downward from the top of the cylinder. Fuel is vaporized during the compression stroke. This type of engine is potentially independent of piston shape, but combustibility is affected by injector performance. We selected the center injection type and optimized its fuel spray characteristics and in-cylinder air flow pattern using the flat type piston. We wanted to be able apply our developments to commercially available engines without their modification.

Fig.3 shows the concept for stratified charge using the flat type piston. This commercial engine uses gasoline fuel and has two intake and two exhaust valves and a displacement of 1.8L . The fuel injector and spark plug are installed at the center of the engine head which is the location of the spark plug in the base engine. The intake system includes an air jet passage. This device generates a swirl flow in the cylinder which is made by a directional air flow through the intake manifold. This directional air flow is

Table 1 Features of side, center injection types

	Side Injection Type	Center Injection Type
Layout	Spark Plug / Injector / Vapor / Spray / Piston cavity	Injector / Spray / Spark Plug / Piston cavity
Principle	□ Vaporization at piston cavity	□ Vaporization during compression stroke
	□ Vapor fuel introduced by air flow and spray momentum	
Features	□ Easier installation of injector □ Good ignition ■ Wall wetting	□ Independent of piston shape ■ Affected by injector performance

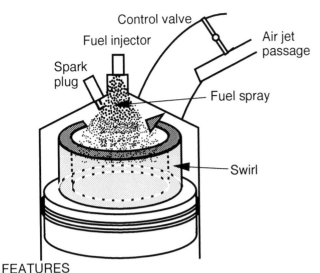

FEATURES
Low spray speed: No inpingiment against piston surface
Swirl: Keeps the fuel spray at the cylinder center
In-cylinder air flow: Swirl due to the air jet passage

Fig.3 Concept for stratified charge

made by the air jet passage which is a bypass for the main intake passage when the control valve is closed. The control valve is located downstream from the throttle valve as shown in Fig.1. Stable swirl flow is generated in the cylinder. Fuel is injected into the center of the swirl flow and concentrated there, assuming that swirl flow is a solid vortex. But the velocity of the fuel spray must be small enough that the mixture is concentrated at the upper side of the cylinder near the spark plug and spray does not impinge against the piston surface. If a tumble component is included in the in-cylinder air flow, fuel spray is transported toward the wall side of the cylinder and not concentrated at the center. So swirl with less tumble is better for getting stratified charge. Swirl flow also spreads the combustion flame widely in the cylinder and increases flame the speed at a lean air fuel ratio. Intake air velocity is changed according to the open angle of the control valve and maintained at the desired value for all driving conditions. As a result of this control by the control valve, the lean combustion region is widely expanded toward both high and low engine speeds.

OPTIMIZATION OF IN-CYLINDER AIR FLOW

Measurement of in-cylinder air flow

Fig.4(a) shows the layout of the air jet passage. Air direction is determined by the air jet passage, so air flow has directivity without regard to the intake port shape of the engine head. When driving conditions are in the lean region, the control valve is closed in order to get intake air flow through the air jet passage. As shown in Fig.3, the aim of the air jet passage is to get the swirl air flow with less tumble. The air

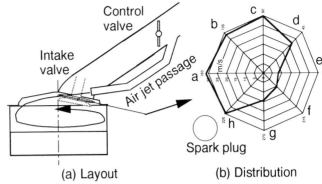

(a) Layout   (b) Distribution

Fig.4 (a) Layout of intake system and (b) distribution of air velocity around the intake valve

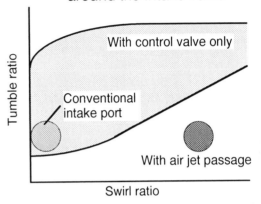

Fig.5 Tumble and swirl ratios as a function of air jet passage arrangement

jet passage is directed towards the outer side of an intake valve. The engine under study has two intake valves, so there are two air jet passages in all. Both two passages are directed towards the same position which is an opening of one intake valve and outer side from the valve stem, reverse side of spark plug against the valve stem. Both two passages aim at only one intake valve side to get the swirl flow independent of intake port shape. Fig.4(b) shows a distribution of air velocity around the intake valve obtained using a hot wire air flow meter with a diameter of 5μ m. Measuring points are the valve opening and eight points around the intake valve. Spark plug position is as shown. Side c is the outer side of the intake valve. Air velocities of points a, b, c and h are higher than at other positions. As a result of this measurement directional air is made by air jet passages. Intake air flows into the cylinder at a direction along the wall side, so swirl flow is generated effectively.

Fig.5 shows tumble and swirl ratios as a function of air jet passage arrangement. The ratios are measured at steady air flow with a constant quantity of air. A conventional(base) intake port without air jet passage and control valve generates tumble flow without swirl because of the straight intake port. Air flow generated with only a control valve which has an opening has more tumble than swirl. So more swirl with less tumble can not be obtained by the conventional intake port and control valve only. When the air jet passage is applied to the intake port, the swirl component is generated with less tumble. The air jet passage with the control valve changes the air direction of the main air flow in the base engine.

Simulation of vaporized fuel in the cylinder

Vaporized fuel distribution at ignition timing which is concentrated around the spark plug is analyzed at a lean air fuel ratio. The simulation takes into account the following phenomena. 1) Droplet movement influenced by the air flow. 2) Heat transfer between air and droplet. 3) Evaporation and droplet diameter reduction. 4) Evaporation of fuel on wall. 5) Vapor transportation. The simulation uses the discrete droplet model with multi-group classification of droplet diameters and Euler's time integration method. For air, conservation laws for compressible and inviscid flow are used and a numerical analysis by the modified FLIC (Fluid in Cell) method is developed and used.

Fig.6 shows simulation results of vaporized fuel distribution when tumble flow is applied in the cylinder using an air jet passage which has a different direction from that in Fig.4. Conditions are as follows. Engine speed, 1400rpm at partial load; average air fuel ratio, 14.0; fuel injection timing, 240° BTDC. At 100° BTDC, a rich vapor zone appears at the center and lower side of the cylinder. Vaporized fuel is transported by the intake air through the intake valve. The direction of this intake air is downward from the intake valve, so vaporized fuel moves to the bottom of the cylinder. As air is rotated vertically in the cylinder by the directional flow from the air jet passage, tumble flow is formed which has a 3 or 4 tumble ratio. Vaporized fuel is transported counter clockwise by the tumble flow as shown in Fig.6(b). But vapor is spread widely and not concentrated in the center of the cylinder. At 30° BTDC, vapor density becomes lean around the spark plug near the top of the pent-roof.

Fig.7 shows the vapor distribution when swirl flow is generated by the air jet passage. Fig.7(a) shows side and top views at 100° BTDC in which fuel is injected at 120° BTDC. As swirl is applied to the cylinder, vapor is concentrated at the center of the cylinder. As this swirl flow has less tumble, vaporized fuel is kept in the center without spread widely. If tumble, it is included, the vapor moves to the wall side and it is not concentrated in the center. This concentration of vapor is maintained until the spark timing at 30° BTDC (Fig.7(b)). Vapor density around the spark plug becomes rich and it is necessary to ignite the fuel mixture in the lean driving region.

Fig.8 shows air fuel ratio around the spark plug within a 5mm square with the plug gap at its center for

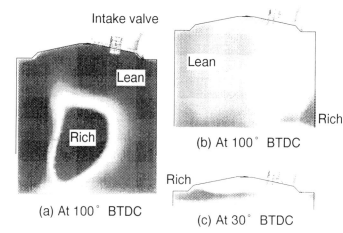

Fig.6 Vaporized fuel distributions in tumble flow

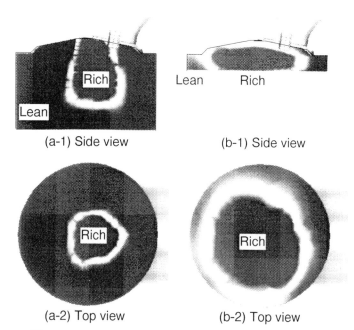

(a-1) Side view   (b-1) Side view

(a-2) Top view   (b-2) Top view

Fig.7 Vaporized fuel distributions in case of swirl flow: (a-1, a-2) At 100° BTDC; (b-1, b-2) At 30° BTDC

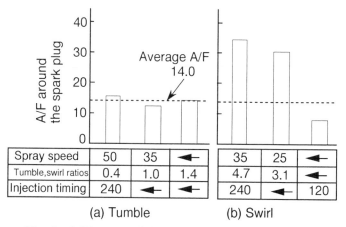

	(a) Tumble			(b) Swirl		
Spray speed	50	35	←	35	25	←
Tumble, swirl ratios	0.4	1.0	1.4	4.7	3.1	←
Injection timing	240	←	←	240	←	120

(a) Tumble   (b) Swirl

Fig.8 A/F around the spark plug for tumble flow and swirl flow

tumble flow and swirl flow. For the former, the best conditions for getting a richer air fuel ratio around the spark plug are: spray speed, 35 m/s; tumble ratio, 1.0; injection timing, 240° BTDC. As spray speed becomes lower, vapor moves towards the cylinder wall due to the tumble flow and it takes more time to transport it to the spark plug position. In all three cases, air fuel ratio around the spark plug is about 14.0 which means the this shows mixture is not stratified charged. For swirl flow, the best conditions for getting a richer air fuel ratio around the spark plug are: spray speed, 25 m/s; swirl ratio, 3.1; injection timing, 120° BTDC. In this case, good ignition is assured even in the lean region.

OPTIMIZATION OF SPRAY CHARACTERISTICS

Optimization of fuel spray speed

Fig.9 shows the relationship between movement of the piston surface and spray tip when engine speeds are 1400 and 4000rpm. The fuel spray must not impinge on the piston surface because impingement forms a fuel film on the piston surface which spreads over it. For a 53m/s spray speed, spray impinges on the piston, but for a spray speed of 22m/s it does not hit the piston and floats through the cylinder. This floating spray stays at the center of swirl flow. If the injection timing is earlier than 240° BTDC, even spray with a speed of 22m/s hits the piston surface. On the other hand, injection timing is later than BDC (Bottom Dead Center), vaporization is not completed because there is not enough time.

Fig.10 shows distributions of fuel spray in the cylinder during different engine operating conditions. The piston and cylinder are made of quartz glass to allow observation. Sheet laser with 238nm wave length is shone into the cylinder to activate the fuel. Fluorescent light occurs from activated fuel is

Top of piston
(a-1) Before BDC

Top of piston
(a-2) After BDC

(b-1) Before BDC

(b-2) After BDC

Fig.10 Distributions of fuel spray in the piston taken by LIF method:
(a-1, a-2) Large spray speed;
(b-1, b-2) Small spray speed

detected by a high speed CCD camera equipped with an optical filter. Fluorescent light intensity is related to the concentration of liquid fuel. Fig.10 (a) shows liquid fuel distribution when spray speed is 53m/s. Spray hits the piston surface directly and after impingement moves to the cylinder wall along the piston surface. After BDC, the mixture stays in the lower part of the cylinder and does not concentrate around the spark plug. Air fuel ratio at the upper side of the cylinder is leaner than that of the lower side. In this case, good ignition is not assured. Fuel spray distribution for a spray speed of 22 m/s is shown in Fig.10(b). Spray speed decreases in the cylinder and droplets float in the air. Flotation of this spray is not disturbed by the swirl flow. If tumble flow is included in the air flow, the fuel spray having a lower speed moves toward the wall side, and as a result, spray concentration is not realized. Concentrated spray in the center moves upwards according to the piston movement, and the air fuel ratio around the spark plug become richer. Due to the combination of a lower spray speed and swirl flow with less tumble, stratified charge can be achieved.

Optimization of spray angle and pattern

Fig.11 shows lean limit air fuel ratio as a function of spray speed, angle and pattern. Spray pattern A is a solid cone and its spray angle at ambient pressure is approximately 45deg. Spray pattern B is a hollow cone type which has 50 and 80deg spray angles. Spray speed with same spray angle and pattern is changed by selecting swirl intensity of the fuel and

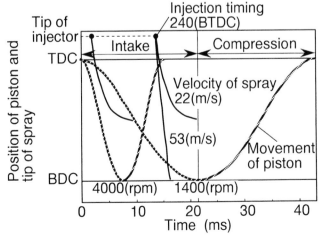

Fig. 9 Relationship between positions of piston and tip of spray

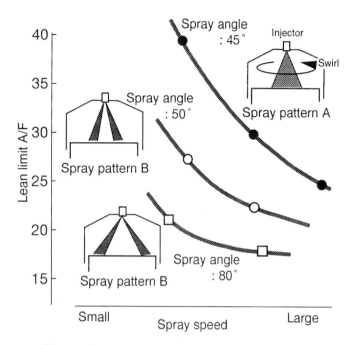

Fig.11 Lean limit A/F as a function of spray speed, angle and pattern

nozzle hole size. As intensity of fuel swirl increases, penetration of spray decreases and the spray angle becomes larger. But changing the diameter and length of the nozzle hole limits the spray angle. A swirl type injector generally forms a hollow cone type spray pattern, but selecting the proper number of grooves and structure of the swirler chip, a hollow cone shape is realized. Pressure of the fuel which is supplied to the injector is changed from 3 to 7MPa in order to get many experimental results. At the same spray speed, for spray pattern B with 80° angle, the maximum lean limit air fuel ratio is about 20. As the spray angle is made smaller, the lean limit air fuel ratio gets larger. Furthermore spray pattern A can reach a lean limit air fuel ratio of 40. With the same pattern of spray, as spray speed increases, a lean limit air fuel ratio decreases. Spray pattern B distributes the mixture to the outer area of the cylinder by its own inertia force. As the spark plug is located at

the center of the cylinder, if there is less fuel mixture around the spark plug at the center, ignition is unsuccessful and misfiring occurs. In the case of pattern A, the mixture exits at the center of the cylinder and is carried upwards according to piston movement if the spray speed is small. Vaporization of fuel spray progresses late in the compression stroke, and reliability of ignition is achieved. Solid cone spray is more effective to get the stratified charge for center injection and ignition engines.

Fig.12 shows smoke emission as a function of spray angle and deceleration rate of spray speed G. G is defined as follows.

G(%) = (V1-V2) / V1
   where V1: Average spray speed at L=10-40mm
          V2: Spray speed at L=40mm

A large G means that spray speed decreases during transportation. G depends on the initial spray speed and fuel droplet diameter if ambient pressure is the same. As shown in the test results, with the same spray angle, smoke emission decreases as deceleration rate G increases. There are two main reasons for smoke emission appearance.

(1) Air fuel ratio of the mixture is smaller than 9.5
(2) Temperature of burned gas is between 1700 and 2100° K

It is important for the fuel spray to be dispersed into the air and not to form a liquid fuel film on the wall which makes a richer air fuel ratio than 9.5. Large spray angle and large G give the desired fuel spray. Spray which has G=80% and a large spray angle does not emit smoke.

ENGINE PERFORMANCE

Response in air fuel ratio

Fig.13 shows response in air fuel ratio for transient operating conditions with an engine speed 1400rpm at partial load and cooling water temperature of 40 ℃. Before and after transient, the engine was running in the steady state and air fuel ratio was set at 14.0. Throttle valve opening was rapidly changed to a large opening value and then returned to the initial opening before the test. Fig.13(a) shows test results of the MPI which injects fuel intake port. The indicated mean pressure Pi, fluctuation of air fuel ratio and throttle valve opening are shown as a function of time. When the throttle valve opening becomes large, the air fuel ratio becomes leaner temporarily and after this approaches 14.0 gradually. This behavior is caused by delay transport of fuel which is injected on the wall surface of the intake port. Injected fuel is attached to the wall surface and makes a fuel film which moves along the wall surface slowly to the intake valve and is not inducted by the cylinder directly. This delay causes a lag fluctuation in the air fuel ratio and slows the change of Pi. At an early stage after the transient,

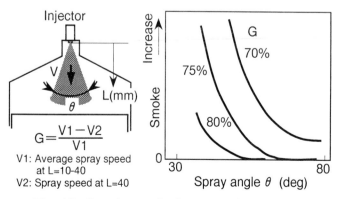

Fig.12 Smoke emission as a function of spray angle and G

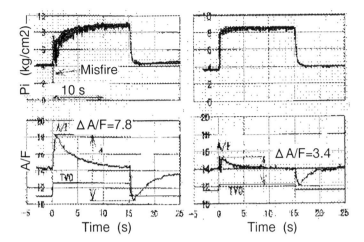

Fig.13 A/F response at cold condition

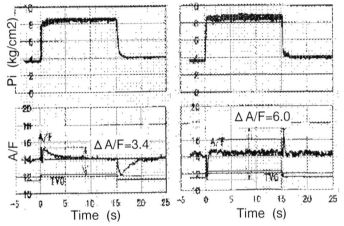

Fig.14 A/F response after warm up condition

a misfire occurs because of hot enough fuel in comparison with the large amount of air. Pi becomes larger gradually according to inflow of fuel film on the wall surface, and the acceleration lag gets worse. When throttle valve opening returns to the previous state, air fuel ratio becomes richer because excess fuel film flows into the cylinder in spite of the decrease in air quantity. Fluctuation of the air fuel ratio during opening and return of the throttle is 7.8. With the direct injection engine, this fluctuation becomes smaller and drops to 3.4. Fluctuation of Pi is also smaller than that of the MPI, and good performance in the cold transient state is assured. Misfires do not occur. The direct-injection method does not form the fuel film on the wall of the intake manifold or on the piston and cylinder surface, when spray angle , speed and pattern are properly selected.

Fig.14 shows air fuel response after warm up with a cooling water temperature of 80 ℃. For the MPI, fluctuation in the air fuel ratio is about 3.4, which is smaller than that at 40 ℃. The amount of fuel film on the wall of the intake port is decreased owing to good vaporization at the higher temperature. But a lean spike remains on acceleration and a rich spike also remains on deceleration. On the other hand, with

direct-injection, small rich spike appears on acceleration and the lean spike appears on deceleration. This behavior is opposite that with the MPI. At acceleration , delay of air flow inducted by the cylinder which is caused by the inertia of air in the intake manifold occurs. Fuel is injected immediately after acceleration according to the signal of the air flow meter. But the quantity of air which flows into the cylinder does not increase immediately. This difference causes a rich spike at acceleration. The acceleration feeling is not poor. Quick response according to acceleration pedal operation is achieved. At deceleration, the lean spike appears. But as a misfire does not occur, the driving feeling is not worsened. These high response are one of the merits of direct injection engines.

<u>Lean burn performance</u>

Fig.15 shows lean limit air fuel ratio as a function of spray speed. This test was conducted with a single cylinder engine at engine speed, 1400rpm; load,

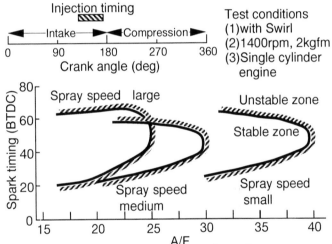

Fig.15 Lean limit A/F as a function of spray speed

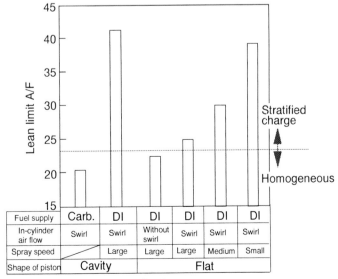

Fig.16 Lean limit test summary

2kgfm; and cooling water and oil temperatures, 80 ℃. Swirl air flow is made by the air jet passage. Injection timing is set to the best position of 240° BTDC. Spark timing is changed from 0 to 80° BTDC. When spray speed is large, injected fuel attaches to the piston surface and spreads widely throughout the cylinder. As the mixture is distributed lower side of the cylinder and becomes nearly homogeneous, the lean limit air fuel ratio is about 25. This is like lean limit performance of the port injection type lean burn engine. As spray speed decreases, lean limit air fuel ratio becomes larger and reaches about 40. This means that stratified charge is achieved. In this case, fuel is injected at the intake stroke, but the mixture can be distributed around the spark plug and good ignition is assured by selecting fuel spray speed, angle, pattern and droplet diameter.

Fig.16 summaries lean limit test results. All tests were conducted with center injection and ignition type of direct-injection engines. Test results of the cavity type piston are listed for reference. When a carburetor is used as a fuel supply device with swirl air flow, lean limit air fuel ratio is about 20. In this case, the mixture is nearly homogeneous. When the fuel supply device is changed to direct injection, lean limit air fuel ratio reaches 40 in spite of the large spray speed. In this test, injection timing and ignition timing were set late in the compression stroke. Spray was injected and kept in the piston cavity, and vaporized at the cavity surface. The cavity type piston is suitable for stratified charge. When the flat type piston, swirl air flow and spray speed with solid cone are more effective to get stratified charge. If the spray speed is large, the lean limit air fuel ratio without swirl is about 22 and with swirl, it reaches about 25. Furthermore, spray speed is decreased to about 22 m/s, and lean limit air fuel ratio reaches about 40.

CONCLUSIONS

A stratified charge concept for direct-injection spark ignition engines was presented. It was developed from the viewpoints of in-cylinder air flow and fuel spray characteristics. Test results using the concept are as follows.
(1) The new method for stratified charge uses a flat type piston and offers a swirl in-cylinder air flow, low spray speed and solid cone spray. It can be applied to center injection and ignition engines.
(2) Intake manifold of this system includes an air jet passages which bypasses the main air passage and generates a swirl flow in the cylinder with less tumble flow.
(3) Fuel is injected at the center of the swirl flow during the intake stroke and is kept at this position. As the piston moves upward in the compression stroke, the mixture also moves upward and is concentrated around the spark plug, so stratified charge is achieved.
(4) Fluctuation of air fuel ratio in the transient operating condition is much smaller than for the intake port injection engine, especially under cold conditions.
(5) This method can expand the lean limit air fuel ratio to about 40. Optimized conditions are: fuel spray speed, 22 m/s; injection timing, 240° BTDC; and ignition timing, 50° BTDC.

ACKNOWLEDGMENTS

We wish to express our gratitude to personnel at Hitachi and its suppliers who cooperated in this study.

REFERENCES

(1) T. Kume, et al.: Combustion Control Technologies for Direct Injection SI Engine: SAE Paper No.960600
(2) S. Kono et al.: Development of the Stratified Charge and Stable Combustion Method in DI Gasoline Engines: SAE Paper No.950688
(3) Y. Iriya, et al.: Engine Performance and The Effects of Fuel Spray Characteristics on Direct Injection S.I. Engines: '96 Autumn Conference of the Society of Automotive Engineers of Japan No.9638031(1996-10)
(4) S. Matsushita, et al.: Mixture Formation Process and Combustion Process of Direct Injection S.I. Engine: '96 Autumn Conference of the Society of Automotive Engineers of Japan No.9638022(1996-10)
(5) T.Shiraishi, et al.: A Study of the Mixture Preparation Process on Spark Ignited, Direct Injection Engine: SAE IPC-8 Paper No.9530409
(6) M.Fujieda, et al.: Influence of the Spray Pattern on Combustion Characteristics in a Direct Injection Engine: '96 Spring Conference of the Society of Automotive Engineers of Japan No.9631911(1996-5)
(7) R.W. Anderson, et al.: Understanding the Thermodynamics of Direct Injection Spark Ignition (DISI) Combustion Systems: An Analytical and Experimental Investigation: SAE Paper No.962018
(8) K. Shimotani, et al.: Characteristics of Exhaust Emission on Gasoline In-Cylinder Direct Injection Engine: 13th Symposium of Internal Combustion Engines (Japan) No.20 (1996-6)
(9) G. Karl, et al: Thermodynamic Analysis of a Stratified Direct Injected Gasoline Engine: VDI 17th International Wiener Motorensymposium No.267 (1996-4)
(10) H.Stutzenberger, et al.: Gasoline Direct Injection for S.I. Engines- Development Status and Outlook: VDI 17th international Wiener Motorensympsium No.2 67 (1996-4)
(11) T.D. Fansler, et al.: Fuel Distributions in a Firing Direct-Injection Spark-Ignition Engine Using Laser-Induced Fluorescence Imaging: SAE Paper No.950110
(12) G.K. Fraidl, et al.: Gasoline Direct Injection: Actual Trends and Future Strategies for Injection and Combustion System: SAE Paper No.960465

Simultaneous Attainment of Low Fuel Consumption, High Output Power and Low Exhaust Emissions in Direct Injection SI Engines

Yasuo Takagi, Teruyuki Itoh, Shigeo Muranaka, Akihiro Iiyama, Yasunori Iwakiri, Tomonori Urushihara and Ken Naitoh

Nissan Motor Co., Ltd.

ABSTRACT

This paper describes simultaneous attainment in improving fuel consumption, output power and reducing HC emissions with a direct injection S.I. engine newly developed in Nissan.

Straight intake port is adopted to increase discharge coefficient under WOT operation and horizontal swirl flow is generated by a swirl control valve to provide stable stratified charge combustion under part load conditions. As a result, fuel consumption is reduced by more than 20 % and power output is improved by approximately 10 %. Moreover, unburned HC is reduced by equivalently 30 % in engine cold start condition. An application of diagnostic and numerical simulation tools to investigate and optimize various factors are also introduced.

INTRODUCTION

Research and development work on direct injection stratified charge S.I. engines dates back as far as the 1960s and led to the development of Texaco TCP engine[1], Ford PROCO engine[2], GM DISC engine[3] and other models. Some of these engines reached the market in early 1980s, but the number of units sold and customers base were limited. Since that time, the developmental goal for direct injection SI engines has been to achieve combustion of an ultra-lean mixture through a stratified charge combustion in order to reduce fuel consumption significantly. One reason why earlier efforts tended to end in limited commercialization success was that the excellent characteristics intrinsic to this type of engine could not be fully elicited due to the immaturity of the peripheral technologies for fuel supplying components and controlling the fuel supply system.

Subsequent to the 80s, the development of these engines fell out of the limelight but proceeded without fanfare until Mitsubishi and Toyota started to market them in

These engines featured combustion technologies that employed stratified charge mixtures to achieve stabilized combustion in a super lean air fuel ratio that were basically the same as previous technologies, but the peripheral technologies drew notice because of the remarkable progress and innovations that had been accomplished.

Subsequently, development work on these engines proceeded without much attention until Mitsubishi and Toyota started to market them in 1996. Their engines are characterized by the use of a stratified charge combustion process that achieves stable combustion of an ultra-lean mixture. While the combustion technologies are basically the same as those used previously, one notable feature of these engines is the remarkable progress and innovation that has achieved in peripheral technologies.

First, the fuel injection system has been improved. Most of the previous direct injection gasoline engines had a jerk pump system, similar to that used for diesel engines, which raised the fuel pressure intermittently. However, in order to use this system with gasoline, which has lower lubricity than diesel fuel, the maximum fuel injection pressure that could be applied was 3 to 5 MPa an the most. As a result, spray atomization was limited. Moreover, there were doubts as to whether the high reliability that would be required under real world conditions could be achieved in a mass-produced engine.

By contrast, the fuel injection systems of the direct injection S.I. engines that have been commercialized in recent years differ greatly from the jerk pump arrangement. Systems have been adopted in which the injection timing and fuel injection quantity are controlled by solenoid valves in the same way as in the electronically controlled fuel injection systems used to date. Established gasoline engine technologies have been continued and high pressure injection up to a maximum of 13 MPa has been achieved.

Similarly, remarkable innovations have also been

achieved in control technologies, especially those for controlling fuel injection. The use of high pressure solenoid-driven injectors in combination with electronic control has made it possible to inject the fuel at precisely the right time for obtaining the engine maximum performance. For example, at part load operation for which a reduction in fuel consumption is required, the injection timing can be set in the latter half of the compression stroke in order to achieve stratified charge combustion. With this injection timing alone, however, it would be very difficult technically to achieve good combustion characteristics without soot formation under high load operation where greater power is required. This is one reason why the previous generation of direct injection stratified charge S.I. engines that employed a jerk pump fuel injection system did not reach the stage of commercial mass production.

In systems with electronically controlled solenoid valves, on the other hand, the fuel can be injected during the intake stroke under high load conditions as is done in conventional gasoline engines, thereby eliminating the problem of soot formation. In addition, the use of other sophisticated technologies for precise control of injection timing and the quantity of fuel injected have also made it possible to reduce exhaust emissions.

This combination of recent new technologies with combustion concept that had been previously developed has opened up avenues for the development of a new generation of direct injection stratified charge S.I. engines. Against this background, this paper will describe potential for improving engine performance and the combustion characteristics of this new generation of direct injection

stratified charge S.I. engines, taking as an example the research and development activities being carried out at Nissan Motor Company.

TEST ENGINE, SPRAY, DIAGNOSTIC TOOL AND SIMULATION MODEL USED IN THE STUDY

Table 1 Major Specifications of Prototype Nissan Direct Injection Gasoline Engine

Engine Type	4-Stroke,4-Cyl,4-Valve DOHC
Bore,Stroke	86mm,86mm
Engine Displacement	1,998 cm3
Compression Ratio	10.5:1
Combustion Chamber	Pentroof

TEST ENGINE AND SPRAY - The engine used in this work was a prototype direct injection stratified charge S.I. engine having the major specifications as given in Table 1. This 2.0 liter 4-stroke water cooled engine with four valves per cylinder is under development at Nissan. The combustion chamber configuration and layout of peripheral components are shown in Figure 1. As shown in the figure, the fuel injector is installed below the two intake ports at an angle of 36 deg. from the horizontal. A round bowl is provide in the piston crown eccentrically to the intake valves. The intake ports have the same straight geometry as that used in conventional 4-valve S.I. engines. A swirl control valve is installed at the inlet of the intake ports and is used to generate swirl aiming swirl ratio of between 3 to 3.5 during the stratified charge operating conditions. The spark plug is installed in the center of the combustion chamber in the same manner as in con

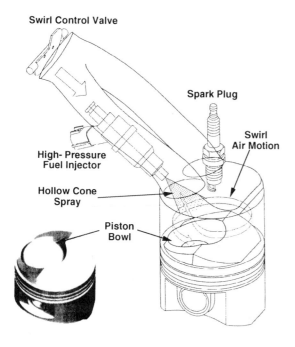

Fig. 1 Combustion Chamber Configuration and Layout of Peripheral Components of Prototype Nissan Direct Injection Gasoline Engine

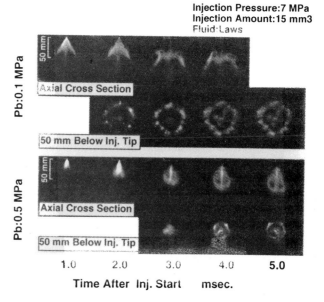

Fig. 2 Spray Geometry used in Prototype Nissan Direct Injection Gasoline Engine and Effect of Back Pressure

Table 2 Specification of Fuel Injection System Used in Prototype Nissan Direct Injection Gasoline Engine

Injector Type	Solenoid Driven Common Rail Type
Line Pressure	10 MPa(Max.)
Spray Geometry	Hollow Cone
Cone Angle	70 deg./100 kPa
SMD(D32)	20 microns/100 kPa

ventional 4-valve S.I. engines. The fuel injector is driven by a solenoid and the injection pressure is up to a maximum level of 10 MPa. A hollow cone spray having cross sectional geometry shown in Fig.2 together with their effect of back pressure, which cone angle is designed to be approximately 70 deg. An outline of the fuel supply system is given in Table 2. In this study, selective use was made of the 4-cylinder direct injection gasoline engine and a single cylinder version of the engine. The same engine with the swept volume reduced to 1.8 liters by changing both bore and stroke was used at times.

DIAGNOSTIC EQUIPMENT - In addition to significant advances in fuel supply systems components and electronic control, the successful development of this new generation of direct injection stratified charge S.I. engines has also been supported by the progress achieved in techniques for measuring and diagnosing combustion characteristics, air motion and spray. These techniques have played an important role in enabling the new combustion technologies to be developed with high accuracy. The techniques used in this research are described below.

Fig. 3 shows a schematic diagram of an optical and data acquisition system of the laser-induced fluorescence (LIF) system used for visualizing behavior of fuel in the combustion chamber in this research. A Kr-F excimer laser operating at a wavelength of 248 nm was used as the light source for irradiating the fuel in a combustion cham-

ber. The fluorescence of wavelength between 280 nm and 400 nm produced by the fuel was passed through a band pass filter and then the entire image was captured by a CCD camera fitted with an image intensifier. Fluorescence can be obtained easily from liquid phase gasoline. With a multi component fuel like gasoline, however, there is a problem that the fluorescence cannot be observed when fuel is gaseous phase especially under the high temperature and high pressure conditions as is in the compression stroke. This is due to increase in probability of absorption of emitted fluorescence from one component with another component of gasoline constitutions and optical quenching caused by the presence of oxygen. In this study, this problem was overcome by using iso-octane as the fuel and by adding 0.2% of dimethyl aniline(DMA) to the fuel as the fluorescent tracer. This made it possible to visualize of the gaseous phase fuel even under high temperature and pressure conditions and thereby confirmed the applicability of LIF to direct injection gasoline engines. A detailed explanation is given in the references[4]. A sapphire cylinder was used to facilitate visualization inside the cylinder under firing operation. A quartz window was provided in he pentroof section of the combustion chamber to facilitate visualization inside the chamber.

Laser Doppler velocimetry(LDV) was used to measure in-cylinder air motion. In addition to improving the measuring system by adopting a frequency shifter with a rotating gratings, a sampling rate above 10 kHz was attained to make it possible to measure cycle-resolved turbulence characteristics in the cylinder[5]. Measurements made at multiple points in the cylinder were expressed as mean flow velocity vectors in order to identify the flow fields in the combustion chamber of the direct injection gasoline engine.

The use of these types of diagnostic tools for visualizations facilitated efficient optimization of the spray pattern, piston crown and bowl geometry. In addition, tomography based on the use of a laser light sheets was employed to measure the spray pattern and cone angle.

CFD BASED NUMERICAL SIMULATION - In addition to the laser based measuring and diagnostic technologies, simulation performed with a computational fluid dynamics (CFD) code represent another important and fundamental techniques supporting R&D work on direct injection gasoline engines. Thanks to the advances achieved in computer performance and computational techniques in recent years, it now possible to perform calculations of the combustion process that takes into account in-cylinder air motion, the spray formation process, mixture formation process, interaction between the fuel spray and the wall and the evaporation process on the wall, among other factors. These advances have also been utilized in developing our direct injection gasoline engines.

Although KIVA[6] is widely used as a code for per-

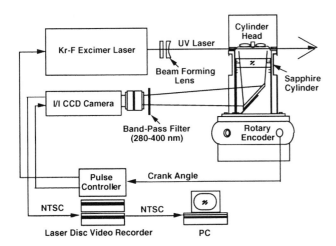

Fig. 3 Schematic Diagram of Optical Setup and Data Acquisition of the LIF System

Fig. 4 Streamlines Obtained with a CFD Based Simulation Code in the Intake Port and Cylinder During the Intake Stroke at 160 deg. ATDC

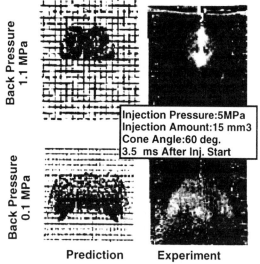

Back Pressure 1.1 MPa

Back Pressure 0.1 MPa

Injection Pressure:5MPa
Injection Amount:15 mm3
Cone Angle:60 deg.
3.5 ms After Inj. Start

Prediction   **Experiment**

Fig. 5 Effect of Back Pressure on Spray Geometry used in DI Gasoline Spray Obtained with Simulation

forming these simulations, the authors have developed an original code based on the finite differential method (FDM) in order to achieve greater computational flexibility than KIVA. This code is being used in the development work of this new combustion engine.

In the hollow cone spray used in direct injection S.I. engines, a liquid film is formed immediately after injection and subsequently breaks up into liquid droplets. Since this dispersion process is affected by the injection pressure, the back pressure, the injection rate and other parameters, it is necessary to have a sub-model that takes their influence into
consideration. One feature of the simulation model used in this study is that the synthesized spheroid particle (SSP) method was used to model the break-up process, which is indispensable to the formation of the hollow cone spray, and this submodel was added to the droplet discrete model (DDM) traditionally utilized to model the spray. An original oval parabola trajectory (OPT) model was also developed and used to calculate the quantity of fuel adhering to the wall, taking into account the interaction between the fuel and the wall at the time of impingement. These innova-

tions made it possible to apply the simulation model to the hollow cone spray used in the prototype direct injection S.I. engine. Detailed explanations of SSP and OPT models are given in the references.[7,8]

As an example of the computational capability of the simulation model, the instantaneous streamlines, not time averaged, in the intake port and cylinder during the intake stroke of the 4 valve engine calculated with the method are shown in Fig. 4. As another example, the calculated results for the effect of back pressure on the external shape of hollow cone spray pattern are shown in Fig. 5 in comparison with the experimental data. Characteristics and importance of the results are shown in the references[7,9].

MEASUREMENT OF SPATIAL FUEL DISTRIBUTION IN THE SPRAY BY RING PATTERNATER - A ring patternater shown in Fig. 6 was used to measure the spatial mass distribution of the fuel in spray. The patternater consists of 37 collectors, with 10 divisions in the radial

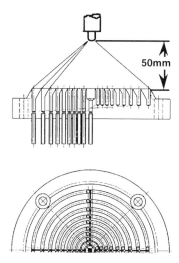

50mm

Fig. 6 Geometry of the 37-ring Fuel-Mass Patternater

direction and 4 divisions in the circumferential direction. Intermittently injected fuel was collected in each collector in the plane 50 mm below injector nozzle tip and weighed. A certain pattern of distribution is required in the radial direction and a more homogeneous distribution is required in the circumferential direction.

POTENTIAL OF DIRECT INJECTION ENGINES FOR PERFORMANCE IMPROVEMENTS

REDUCED FUEL CONSUMPTION WITH STRATIFIED CHARGE COMBUSTION - One of the key objectives set for the direct injection S.I. engines is to achieve stratified charge combustion. A major feature with conventional homogeneous charge S.I. engines is that engine output is controlled by throttling the intake air quantity, which results in poor fuel economy under part-load operation due to the throttling loss. This throttling loss is one fundamental reasons why the fuel economy of S.I.

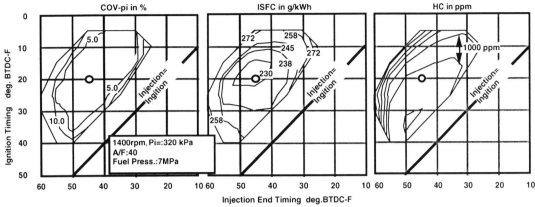

Fig. 7 BETOS Map Showing Stable Combustion Region, ISFC, HC and Points of Best Fuel Consumption

engine is inferior to that of their diesel counterparts.

Stratified charge combustion is a process in which engine output is controlled only by regulating the amount of fuel supplied without throttling the intake air quantity. An essential condition for achieving stratified charge combustion under part-load operation when a small quantity of fuel is injected is to have a combustible mixture in the vicinity of the spark plug at the time of ignition. To accomplish that, it is necessary to optimize numerous parameters, including the relative positions of the spark plug and the fuel injector, spray properties such as the droplet diameter and spray orientation, the air motion that transports the injected fuel to the spark plug and the piston bowl geometry, which can interfere with diffusion of the mixture. As a result of optimizing these various parameters, a region of stable combustion defined as coefficient of variation of pi (COV-pi) of less than 5% was obtained

in relation to combinations of injection end timing and ignition timing, as shown in Fig.7. It is clear that the best fuel consumption point indicated by the open circles and emission characteristics can be achieved in this region. This region is named as the Best Torque or Stability (BETOS) map and the best fuel consumption point as BETOS-IT/AD, which means the optimum injection timing and spark advance for BETOS map. The area of the map and the performance at the BETOS-IT/AD point are indicators of the performance of a stratified charge combustion engine.

Fig.8 compares the combustion performance of the prototype direct injection gasoline engine relative to the A/F ratio at BETOS-IT/AD with that of a conventional MPI lean burn engine at MBT. It is clear from the figure that the lean limit of the A/F ratio of the MPI lean burn engine was around 25:1. In contrast to this, the achievement of stratified charge combustion in the direct injection gasoline engine allowed stable combustion at A/F ratios leaner than 40:1. This means that fuel consumption can be reduced by approximately 20% compared with that of the MPI engine at the stoichiometric A/F ratio. Although an analytical study to clarified the reasons for this reduction is not carried out so far, the primary factor is the reduction of throttling losses. Consequently, a larger reduction will be obtained under low load operation and a smaller reduction under high load operation. Since the A/F ratio that influences combustion is closer to the stoichiometric A/F ratio than the supplied A/F ratio, the NOx emissions level of the direct injection S.I. engine would be considerably higher than that of the MPI engine. However, it is clear from the figure that the application of exhaust gas recirculation (EGR) can easily reduce NOx emission to the level achieved with homogeneous charge lean burn MPI engine without degrading fuel economy or combustion stability. Although not indicated in the figure, HC emissions would be also higher than those from existing MPI engines, but these can kept to levels that satisfy regulatory standards through the use of improved catalyst. By using newly developed de-NOx catalysts, both NOx and HC

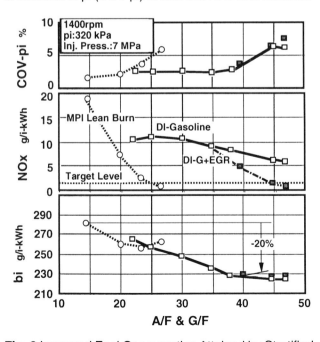

Fig. 8 Improved Fuel Consumption Attained by Stratified Charge Combustion (G/F is the Ratio of Air and EGR Gas to the Fuel)

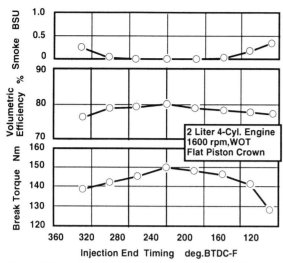

Fig. 9 Effect of Injection Timing on Output Torque and Volumetric Efficiency in Direct Injection Gasoline Engines

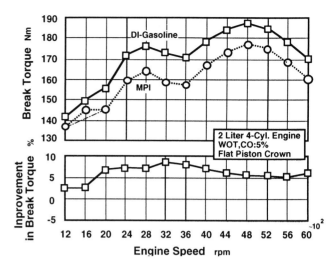

Fig. 11 Improvement in Torque Output Torque in Relation to Engine Speed

emissions can be maintained at levels that will be in compliance with the stricter regulations set to be enforced in Japan in the future.

IMPROVED OUTPUT UNDER WOT OPERATION
- Under WOT operation, the fuel is injected during the intake stroke in direct injection gasoline engines in the same manner as in MPI engines in order to accomplish homogeneous mixture. When finely atomized fuel is injected directly into a combustion chamber, most of the latent heat of evaporation is obtained from the intake air, which has the effect of cooling the air for improved charging efficiency. As shown in Fig.9, the maximum improvement in charging efficiency is obtained at around 120 deg. after the onset of induction, which is the point where the relative velocities of the intake air and fuel spray are thought to be the highest. In port injection MPI engines, the fuel is injected while the intake valves are closed and heat from the cylinder wall evaporates the fuel. Clearly, similar improvement in charging efficiency can not be expected. In addition to the improved charging efficiency obtainable in direct injection gasoline engine, the lower temperature of the intake air also has a noticeable effect on suppressing knock. As shown in Fig.10, ignition timing can be advanced by approximately 2 deg.CA, which together with improved charging efficiency, results in approximately a 6.5 % increase in power output under these operating conditions. Such improved output performance can be achieved under all operating condition, as seen in Fig.11, with a maximum improvement being approximately 9 %.

The results shown in Figs. 9 to 11, however, were obtained with a flat piston-crown configuration. When a bowl is provided in the piston crown to facilitate stratified charge combustion, it became clear that the improvement effect reduced to approximately 50 %, as shown in Fig. 11. This reason is estimated to be that mixture formation deteriorated because some of richer mixture was trapped in the bowl and intake air motion was killed in compression process by the bowl cavity. Effective means for preventing this reduction have to be developed in future research.

REDUCTIONS OF COLD START HC EMISSIONS
- Another excellent performance characteristics of a direct injection gasoline engine is a low level of cold start HC emissions. In MPI engines, injection of the fuel into the intake port results in wall wetting at cold start, making it necessary to enrich the fuel supply in order to start the engine in a short cranking interval. Because of this, excess fuel is evacuated immediately after engine start, resulting in a sharp increase in the HC emissions. By contrast, in a direct injection gasoline engine there is no need for enrichment, since finely atomized fuel can be supplied directly to the combustion chamber in suitably metered quantity. As is obviously observed in Fig.12, more finely atomized fuel is distributed more widely in the cylinder in a direct injection than in MPI engine under cold operating conditions. This makes it possible to provide the optimum

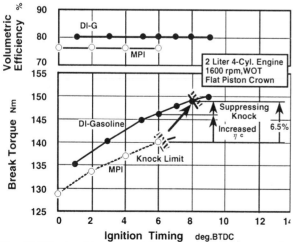

Fig. 10 Effects of Increased Charging Efficiency and Suppression of Knocking on Improvements in Torque Output

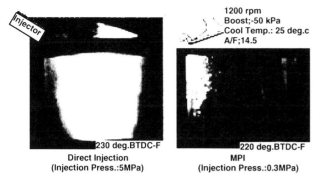

Fig. 12 Comparison of Fuel Behavior in MPI and Direct Injection Gasoline Engines under Cold Engine Conditions as Visualized by LIF

mixture to the combustion chamber even under cold start conditions compared with the situation for MPI engines. As a result, it is possible to achieve stable combustion from the first ignition cycle, which works to prevent any increase in cold start HC emissions, as shown in Fig. 13. In an engine system that meets the LEV regulations, above mentioned effect corresponds to approximately 30% reduction in HC emissions. However, the results in Fig. 13 were obtained in tests conducted with an externally driven high-pressure fuel pump. It is clear from these results that this type of pump to create high injection pressure from the onset of cranking is necessary in order to achieve this reduction in cold start HC emissions.

COMBUSTION OPTIMIZATION AND MIXTURE FORMATION CHARACTERISTICS

As mentioned above, numerous parameters must be optimized in order to achieve stratified charge combustion in a direct injection S.I. engine. Furthermore, these parameters must be optimized over a wide range in order to satisfy the requirements for engine performance under homogeneous charge combustion as well.

REQUIRED INTAKE AIR MOTION - Two different

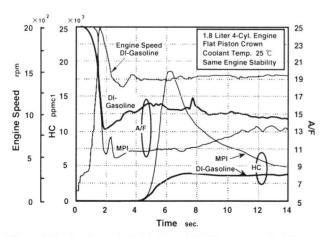

Fig. 13 Reduction of Cold Start HC Emissions in Direct Injection Gasoline Engines

requirements must be satisfied with regard to the intake air motion in a direct injection gasoline engine. The first is to obtain a level of charging efficiency under WOT operation that is no lower than that achieved by conventional MPI engines. The second requirement is the provision of suitable air motion during the compression stroke to accomplish stratified charge combustion under part load operating, which is not necessarily identical to the flow fields required for WOT conditions. In order to satisfy these two requirements, a straight intake port configuration like that of conventional MPI engines has been adopted to generate strong forward tumble for securing high charging efficiency under WOT operation. Since it was found that swirling air motion without a tumble component was necessary for stratified charge combustion, a swirl control valve with a proven record of performance in lean burn MPI engine system was adopted to generate horizontal swirl.[10]

Fig.14 compares the behavior of the fuel injected in the latter half of the compression stroke under stratified charge operation as visualized by LIF when air motion with forward tumble and air motion with horizontal swirl were provided. The visualized results indicate that the injected spray was transported directly to the spark plug by air motion before evaporated fuel in bowl reach to the spark plug when forward tumble was present at the time of fuel

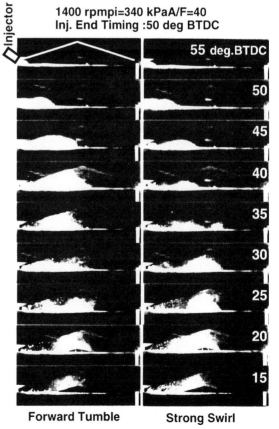

Fig. 14 Changes in Evaporated Fuel Behavior in the Combustion Chamber with Swirl and Forward Tumble Air Motion as Visualized by LIF

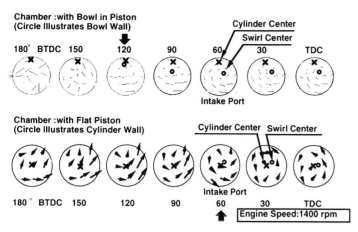

Fig. 15 Comparison of Flow Field in the Combustion Chamber with and without Piston Bowl as Measured by LDV

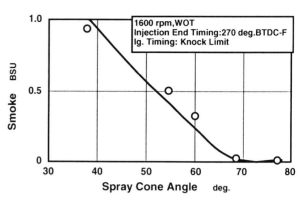

Fig. 16 Effect of Spray Cone Angle on Soot Formation Under WOT Operation

injection. Since the spray still contained a large quantity of droplets owing to insufficient evaporation, which results in unstable combustion with large cycle to cycle fluctuation. By contrast, when swirl air motion was generated direct routing of the spray to the spark plug was suppressed and almost all the supplied fuel was transported to the spark plug by way of the piston bowl, where fuel was fully evaporated and resulted in reducing cycle to cycle fluctuation of the combustion. From this comparison, it can be concluded that swirl is effective in suppressing direct routing of the spray to the spark plug.

A serious of pictures in the upper of Fig.15 shows the changes over time in the horizontal flow field in the combustion chamber of a direct injection S.I. engine as measured by LDV. Measurements was made in the horizontal plane 5 mm above the bottom of the bowl when the piston was at top dead center(TDC). Since the lowest part of the bowl depth is 10 mm, this plane is inside of the bowl when piston is at TDC. As is indicated in the figure, the swirl center was definitely evident from around 120 deg. BTDC. This indicates that the use of a swirl control valve is clearly effective in promoting the formation of horizontal swirl flow fields in the compression stroke from around 120 deg.BTDC in the direct injection engine. As seen in the lower part of the figure, when a bowl was not provided in the piston crown, the flow fields were not transformed to horizontal swirl until around 60 deg.BTDC even if the same swirl was provided in the air motion. This result confirms that the piston bowl functions to transform flow fields into horizontal swirl at an earlier crank angle.

As a result of these studies, it was found that specification of swirl control valve and piston bowl in the direct injection gasoline engines were selected so that a horizontal swirl must be stabilized before the fuel injection.

SPRAY REQUIREMENTS IN DIRECT INJECTION GASOLINE ENGINE - The spray used in direct injection gasoline engines must have properties that simultaneously satisfy two different combustion processes of homogeneous charge and stratified charge combustion. When the

fuel is injected during the intake stroke for the homogeneous charge combustion under high load operation, it is necessary to have a relatively wider spray cone angle in order to form a more homogeneous mixture. As shown in Fig. 16, smoke was generated in the prototype direct injection engine when the spray cone angle was less than approximately 65 deg. A relatively wide spray cone angle was therefore adopted under atmospheric pressure in order to meet the requirement for suppression of smoke formation.

For stratified charge combustion, on the other hand, a relatively smaller spray cone angle is required in order to transport the spray to the spark plug via the piston bowl without the injected fuel diffusing excessively. A requirement in the cone angle clearly differs from that needed for homogeneous charge combustion.

The initial center spray has been effectively utilized to meet these two conflicting cone angle requirements. Fig. 17 compares the changes in the spray pattern under different back pressures for two types of spray having dif-

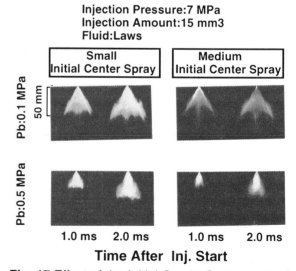

Fig. 17 Effect of the Initial Center Spray on the Spray Pattern Under Different Back Pressures(Small Initial Center Spray: Outward Geometry, Medium Initial Center Spray: Cross Sectional Geometry)

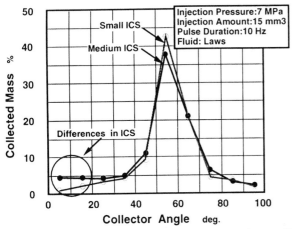

Fig. 18 Fuel Mass Distribution Measured by 37 Ring Patternater in Sprays Having Different Initial Center Spray

ferent amount of initial center spray, small and medium, which are identified on the basis of the amount of fuel present to an angle within 20 deg. from the center measured with 37 ring patternnater as is shown in Fig.18. The spray having medium initial center spray is the same one shown in Fig.2. The results shown in the Fig.17 indicate that the spray cone angle is smaller under the higher back pressure than it was at atmospheric pressure, which is a characteristics of the hollow cone spray. This tendency became more pronounced when the quantity of the initial center spray was increased, resulting in an even smaller cone angle. The reason for this is that the pressure difference between the cone interior and exterior is proportional to the product of the density and squares of the velocity inside the cone. Therefore, a higher back pressure with a higher air density increases this pressure difference, causing the spray to be squeezed inwardly. This effect is augmented with a increased quantity of initial center spray because the air flow velocity inside the spray cone increases due to the initial center spray. It is seen in Fig.19 that increasing the quantity of the initial center spray extended the region of stable combustion on a BETOS IT/

AD map. This effect is more dominant under high engine speed conditions.

For these reasons, spray used in this research has been designed with a relatively large cone angle and with an appropriate quantity of initial center spray.

FUEL INJECTOR MOUNTING ANGLE - Fig. 20 presents simulation results showing the effect of the injector mounting angle on the quantity of fuel adhering to the cylinder head, piston crown and cylinder wall for fuel injection during the intake stroke. In the direct injection gasoline engine used in this work, the results indicate that the amount of fuel adhering to the cylinder wall, which is estimated to be a cause of cylinder wall scuffing, was reduced when the fuel injector mounting angle was greater than 36 deg. On the other hand, the quantity of fuel adhering to the piston crown ,which is a cause of smoke formation, increased. Therefore, a mounting angle of 36 deg. was selected to offer a good balance between these two phenomenon. This example illustrates how the simulation code was used to confirm the effect of different factors in much less time than what would have been required to conduct corresponding experiments.

REQUIREMENTS FOR THE PISTON BOWL GEOMETRY - This prototype engines have a round bowl that is provided in the piston crown on the fuel injector side. In addition to promoting evaporation of the spray for good mixture formation, as mentioned earlier, the bowl also functions to control the swirl motion of the intake air.

CHARACTERISTICS OF MIXTURE FORMATION PROCESS - The simulation model was used to visualize the mixture formation process for fuel injection in the latter half of the compression stroke under the conditions required for stratified charge combustion. The results shown in Fig. 21 indicate that presence of horizontal swirl in the cylinder in the compression stroke generates a much stronger vertical swirling

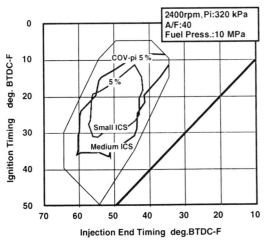

Fig. 19 Effect of the Initial Center Spray on Stable Combustion as Shown on a BETOS Map

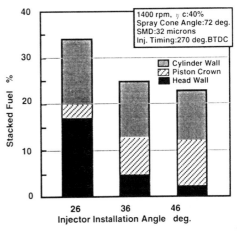

Fig. 20 Effect of the Fuel Injector Mounting Angle on Residual Fuel Adhering to the Cylinder Head, Piston Crown and Cylinder Walls Obtained with Simulation

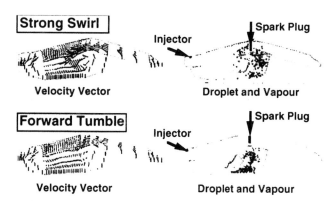

Strong Swirl

Velocity Vector | Injector | Spark Plug | Droplet and Vapour

Forward Tumble

Velocity Vector | Injector | Spark Plug | Droplet and Vapour

Fig. 21 Visualized Flow Field and Distribution of Evaporated Fuel in the Combustion Chamber in a DI Gasoline Engine Obtained by Nissan Simulation Code Based on CFD

motion inside the bowl which rises upward to transport the evaporated fuel in the bowl to the spark plug than presence of forward tumble. This vertical flow is understood to be a resultant vector of swirling air motion and squish one generated by piston crown geometry and correlates with the results shown in Fig.14, where it was seen that the evaporated fuel is transported toward the spark plug with the assistance of this vertical swirling air motion. This characteristics of the mixture formation process is a distinguishing feature of direct injection gasoline engine used in this study.

CONCLUSION

The Nissan prototype direct injection S.I. engine adopting the same geometry surrounding combustion chamber as that of conventional MPI engines and using a swirl control valve to control in-cylinder air motion in order to meet the requirements for stratified charge combustion provided performances noted below.

Under part load operation, stable lean burn based on a stratified charge mixture as lean as 40:1 is accomplished to reduce fuel consumption by 20 % compared with that of conventional homogeneous charge engines. The slightly higher NOx and HC emissions can easily be reduced to an acceptable levels through the application of EGR and by using improved catalysts.

Even under WOT conditions, it was confirmed that the engine has the potential to improve power output by as much as 9 %. However, the improvement in power output under WOT operation is reduced by approximately 50% when a bowls is provided in the piston crown to accomplish stratified charge combustion.

It was also shown that it has the potential to reduce cold start HC emissions substantially. A fuel pump capable of producing high injection pressure from the onset of cranking must be used to achieve reduction in cold start HC emissions. These are issues that will have to be addressed in future research.

Finally, the use of laser based diagnostic tools such as LIF and LDV and numerical simulations made it possible to identify the requirements for stratified charge combustion and the associated roles played by the in-cylinder air motion, fuel spray, piston bowl geometry and other factors.

ACKNOWLEDGMENTS

Authors wish to acknowledge to all the personnel in Nissan Research Center who collaborate with in the research and development of our prototype direct injection S.I. engine, especially to Mr. K.Ebina, Mr. Y. Amenomori, Mr. T.Ishikawa and Mr. A. Teraji who contributed by achieving skillful visualization experiment and numerical computations.

REFERENCES

1. Barber E. M., Reynolds B., Tierney W. T., "Elimination of Combustion Knock-TEXACO Combustion Process", Vol.5, No.1,Jan., SAE Quarterly Trans.,1951
2. Scussel A. J., Simko A. O., Wade W. R., "The Ford PROCO Engine Update", SAE Paper 780699, 1978
3. Lancaster D. R., "Diagnostic Investigation of Hydrocabon Emissions From a Direct Injection Stratified Charge Engine with Early Injection", I Mech E C379/80, 1980
4. Itoh, T., Kakuho A.Hishinuma H., Urushihara T., Takagi Y.,Horie K., Asano M., Ogata E., Yamashita T.,"Development of New Compound Fluid and Fluorescent Tracer Combination for Use with Laser Induced Fluorescence", SAE Trans. 952465,1995
5. Urushihara T., Murayama T., Takagi Y., Lee K., "Turbulence and Cycle-by-Cycle Variation of Mean Velocity Generated by Swirl and Tumble Flow and Their Effects on Combustion", SAE Trans. 950813, 1995
6. Amsden A. A., Ramshaw J. D., O'Rourke P., J., "KIVA: A Computer Program for Two-and Three-Dimensional Fluid Flows with Chemical Reactions and Fuel Sprays", Los Alamos National Laboratory LA-10245-MS, 1985
7. Naitoh K., Takagi Y., "Synthesized Spherical Particle(SSP) Method for Calculating Spray Phenomena in Direct-Injection SI Engines", SAE. Trans. 962017, 1996
8. Naitoh K., Takagi Y., Kokita H., Kuwahara K., "Numerical Prediction of Fuel Secondary Atomization Behavior in SI Engine Based on the Oval-Parabola Trajectories(OPT) Model", SAE Trans. 940526, 1994
9. Naitoh K., Fujii H., Urushihara T., Takagi Y., Kuwahara K., "Numerical Simulation of the Detailed Flow in Engine Port and Cylinders", SAE Trans. 900256, 1990
10. Urushihara T., Nakada T., Kakuhou A., Takagi Y., "Effects of Swirl/Tumble Motion on In-Cylinder Mixture Formation in a Lean-Burn Engine", SAE Trans. 961994,1996

Combustion and Emissions Characteristics of Orbital's Combustion Process Applied to Multi-Cylinder Automotive Direct Injected 4-Stroke Engines

Rodney Houston and Geoffrey Cathcart
Orbital Engine Company

ABSTRACT

Orbital have been developing their stratified combustion process (Orbital Combustion Process OCP) for direct injection gasoline engines over the last 15 years, with successful production releases of the system in both the marine and automotive 2-stroke applications in 1996. This paper discusses how the same basic qualities of the air-assist fuel system and combustion process have been applied to automotive 4-stroke engines. The inherent qualities of the air-assist fuel system in combination with careful design of the combustion chamber has enabled high charge stratification with late injection timings and very stable combustion over a wide range of operating conditions. Experimental test data from a 4-cylinder, 16 valve 4-stroke development engine demonstrates the ability of this low pressure system to operate at very lean air/fuel ratios, with part load fuel economy improvements of up to 34 % at an operating condition equivalent to a vehicle speed of 40 km/hr.

Results presented from steady state simulation of the New European Drive Cycle also demonstrate an overall base engine fuel economy improvement of over 20%, whilst achieving NOx emissions reductions of over 85% and HC emissions comparable to the baseline port injected (MPI) engine.

As a result of the combined NOx, fuel economy and HC control of the Orbital Combustion process, the authors have proposed an alternative exhaust aftertreatment strategy to the still unproven and expensive lean NOx catalyst options. In particular, steady state results show how the low raw emissions of the air-assisted DI system developed could enable a conventional 3-way catalyst system to be utilised to meet Stage 3 European emissions levels with a minimal impact on the DI fuel economy advantage.

INTRODUCTION

The current resurgence of interest in direct injected (DI) 4-stroke gasoline engines has been primarily due to the promise of an improvement in fuel economy with the ability to run under very lean conditions. In particular, an optimized stratified charge, lean combustion engine has reduced levels of pumping work losses, and improved thermodynamic efficiency by virtue of higher levels of diluents in the combustion charge. Early researchers in this field (1-3)* have highlighted the possible benefits of such engines with respect to fuel economy gains, however difficulties with precise control of the stratified mixture formation ultimately lead to poor soot emissions, HC emissions and limited operation zones.

However, further environmental pressures on the global goal for reduced CO_2 and ultimately improved fuel economy has produced a number of new emerging direct injection technologies for automotive application. In general, two types of direct injection technologies have been developed to a level where they are in series production today:

- In early 1996, Mercury Marine went into series production with the Orbital Combustion Process applied to its flagship V6 200 horsepower outboard marine engines (4). In addition, in 1996 Orbital released a limited [100] number of DI two-stroke powered vehicles for the Australian market, which were fully validated and certified to Australian automotive production design rules.

- The application of high pressure, single fluid systems that achieve fuel atomisation mainly by virtue of high fuel system pressures (50 to 120 bar) were released in 1996 for automotive applications in Japan by both Mitsubishi and Toyota (5-9).

The development of the combustion process has been discussed in many previous publications (10-13), with pioneering work in the development of an air assist fuel system for a stratified charge combustion system. These previous publications have until now concentrated mainly on 2-stroke applications for automotive, motorcycle and marine. As discussed in these previous publications a combustion and fuel system was developed with the

---

Numbers in parentheses designate references at end of paper

following important characteristics:

- Finely atomized, low penetration spray
- Highly stratified operation with late injection timings
- Air injected with the fuel charge enabling very high EGR tolerance
- Stratification control insensitive to the in-cylinder flow conditions
- Homogeneous charge operation at high loads and speeds
- Smooth transition between stratified and homogeneous operation
- Injector deposit control strategies

As a result of the above characteristics, the transfer of the basic process and hardware to the initial 4-stroke application was simplified. In addition, the system has now been successfully applied to a number of different multi-cylinder 4-stroke engine configurations, with only minimal changes to the cylinder head as shown in Figure 1. The series production development engine utilised for the results presented here has retained the standard intake and exhaust valve geometry with no changes to the intake or exhaust manifold.

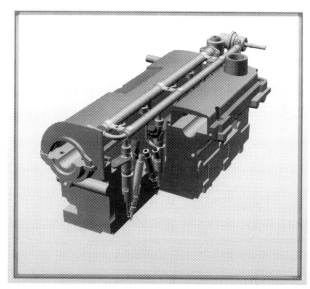

Figure 1: Air-assisted DI4S DOHC 16 Valve installation.

Interestingly both Iwamoto et al (7) and Tomoda et al (8) discuss the major qualities of optimal DI fuel spray characteristics in a similar manner. This can be summarised in simple terms to a requirement for small particle size, controlled penetration and controlled fuel charge dispersion. As discussed in previous publications (10-12) and elaborated on further in this paper, it is believed that the above desirable characteristics are inherent in the basic design of a low pressure air assist fuel system. In particular the very small particle size of the air assist fuel spray enables very late injection timings (as late as 25-30° BTDC), therefore minimising the time between end of injection and ignition, as shown in Figure 2. These late timings enable very good charge containment and very lean stratified charge operation

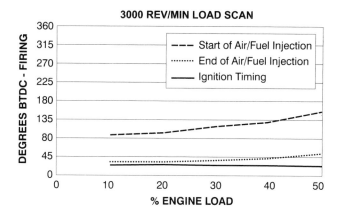

Figure 2: Direct injection event timing in stratified operating zone.

Despite the most recent developments, the essential challenge for all new generation DI systems is the delivery of the fuel economy benefits under the constraints of the current and future emissions standards in both Europe and USA. This will require further development in the simultaneous control of HC and NOx emissions under lean stratified conditions. This paper will describe how the Orbital Combustion Process has addressed this challenge as applied to automotive 4-stroke engines.

SYSTEM DESCRIPTION

Figure 1 shows a solid model view of the development engine cylinder head utilized for the results discussed here, and Table 1 shows the engine's basic specifications.

Base Engine	Ford Zetec 1.8L DOHC
Displacement (cm3)	1796
Bore * Stroke (mm)	80.6 * 88.0
ValveTrain	4-valve DOHC
Compression Ratio	10.0 : 1 (MPI engine)
	10.4 : 1 (DI engine)

Table 1: Development engine specifications.

The engine features the following major items:

- Air-assist fuel system mounted centrally in the camshaft valley, utilising a conventional multi-point fuel injector mounted axially to the top of the direct injector.
- Spark plug with extended electrode mounted close to the direct injector with a specific geometric relationship between the plug and injector.
- Piston with customised bowl geometry to enhance fuel charge containment and stratification.

The Ford Zetec engine used for development purposes of the air-assisted direct injection combustion system maintained the standard intake and exhaust geometry, with no additional swirl or tumble motion introduced, which corresponds to a low tumble and zero swirl in-cylinder flow condition. The compression ratio for the DI engine was 0.4 of a ratio higher than the MPI baseline engine, although this has not been fully optimized at this stage. The valve timings for the DI engine were modified slightly to increase the standard overlap by 25 to an overlap duration of 40°. This type of modification increases the levels of trapped residuals (or internal EGR). This is a very efficient way to introduce the extra diluent as opposed to increasing the level of external EGR as the trapped residuals are at a significantly higher temperature. For four-stroke applications, the greater the temperature of EGR, the greater the volume for a given mass flow (or EGR dilution ratio) which results in a greater reduction in the intake manifold pressure and thus pumping work. Good combustion stability at idle has been obtained on this DI converted engine with valve overlaps greater than 80°. Without employing variable valve timing control, however, 40° was found to give the best compromise between part load fuel economy and full load performance. The manifold port injection system is unable to operate with acceptable combustion stability when operated using the 40° valve overlap period.

The application of lean, stratified charge direct injection to any internal combustion engine requires recognition of the complex interactions between the fuel system, combustion system and the control system. These three critical parts make up the complete Orbital Combustion Process (OCP), with each area being discussed in detail below.

FUEL SYSTEM - At the heart of the fuel system is a solenoid actuated outwardly opening direct injector, which delivers the finely atomised air/fuel charge directly into the combustion chamber, as shown schematically in Figure 3.

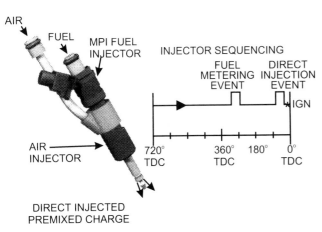

Figure 3: Air-assisted direct injection system (axial type).

The direct injector utilizes the same basic design as the series production 2-stroke injector with small changes to the solenoid and leg diameter for reduced package size. The fuel is metered and delivered to the top of the air injector by a conventional multi-point fuel injector which has a supply pressure regulated to 7.2 bar. The orientation of the fuel injector to the air injector can be easily modified depending on the packaging constraints of the application. Indeed, significant experience has also been with the fuel injector mounted perpendicular to the direct injector, as shown in Figure 4.

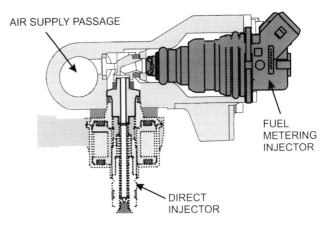

Figure 4: Lateral type air assisted fuel system.

The independent delivery of the fuel to the holding chamber inside the direct injector enables separate control of the fuel metering (typically 1.5 to 10 millisecond duration pulse widths) from the direct injection event. As such the fuel metering is separated from the possible influence of deposit build-up on the direct injector. In addition, the time available for fuel metering is separated from the direct injection event and hence the dynamic range of the fuel injector is similar to that of a conventional manifold injection system.

The direct injector delivers the combined air and fuel mixture at the regulated supply air pressure of 6.5 bar, with injection durations in the range of 3 to 6 milliseconds. At these low pressures with a nozzle flow area of approximately 3.5 mm² for the direct injector, the mass of injected air is only a small percentage of the total engine airflow. At a typical light load condition for example, the injected air is approximately 6 mg/cylinder/cycle which constitutes only 3.5% of the total engine airflow at a delivered air/fuel ratio of 21:1. The overall injected air/ injected fuel ratio characteristic over the load range is illustrated in Figure 5.

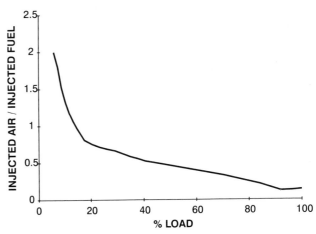

Figure 5: Typical injected air-to-fuel ratio.

As shown, at low load the injected air/injected fuel ratio is in the range of 1 to 2:1, with a gradually decreasing characteristic down to 0.2:1 at full load. Further fine-tuning of the quantity of injected air can also be made independent of the metered fuel by adjusting the direct injector opening duration or adjusting the supply air pressure.

Figure 6 illustrates the typical prototype installation of the air-assisted DI system on the development engine. The compressed air required for the injection process is supplied for this prototype development purpose by a small (40 cm^3) conventional single cylinder-reciprocating compressor, driven by the front accessory drive belt. Many alternative low-pressure compressed air supply systems are also possible for the air-assisted system, depending on the application and packaging requirements. Please note that in all cases the results presented in this paper include the full parasitic loss of the reciprocating compressor.

ENGINE MANAGEMENT SYSTEM - A schematic of the typical air-assisted DI engine management system is shown in Figure 7. In this case, the intake airflow is controlled by an electronic drive by-wire system, with a high flow electronic EGR valve controlling the EGR flow rate to the intake plenum. The 4-stroke development programs utilise an in-house development ECU, employing a Motorola embedded 32-bit controller based on the Motorola 683xxx family of micro-controllers. This hardware provides a very flexible platform

Figure 6: Prototype air-assisted direct injected 4-stroke multi-cylinder test cell installation.

for the development of new algorithms and calibrations. In general, the ECU requirements for the system are well matched to existing automotive specifications, with injector driver requirements for the complete fuel system being similar to current series production multi-point fuel systems.

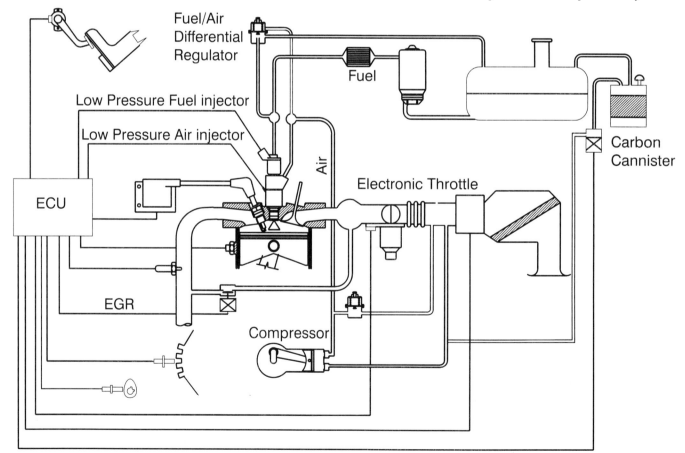

Figure 7: OCP direct injected 4-stroke system schematic.

58

In control system terms, the driver demand or accelerator pedal position and engine speed are utilised as the primary inputs to the control algorithms. A detailed description of the operation of the control system is described by Worth et al (4).

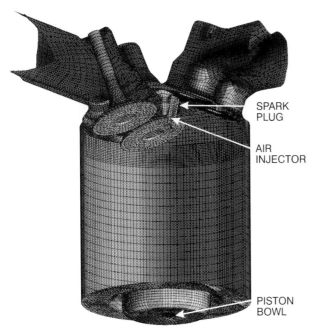

Figure 8: Direct injector and spark plug orientation in DOHC, 4 valve per cylinder engine.

COMBUSTION SYSTEM   The configuration of the combustion chamber design utilised for the development work discussed here is illustrated in Figure 8. This basic geometry, with the spark plug electrode in the edge of the spray, is similar to Orbital's series production 2-stroke engine development and validation. The exact location of the plug is controlled by both practical limitations of the cylinder head geometry and the location of the ignitable stratified charge in the cylinder. This latter location is controlled primarily by the complex interaction of the direct injector nozzle geometry and the reflective/containment characteristics of the piston bowl and combustion chamber. An important tool in understanding these critical interactions is the application of computational fluid dynamics (CFD). As discussed previously by Houston et al (11), the STAR CD code in combination with the moving piston model meshing techniques developed by Orbital can provide a very good insight into the in-cylinder flow conditions and the impact of these conditions on the spray characteristics.

Figure 9 illustrates a typical result from the application of CFD modeling a 4-valve DI 4-stroke engine. In this instance, the boundary conditions at the inlet and exhaust ports correspond to those measured on the firing engine running at 1500 rev/min and 2 Bar BMEP. The full intake and compression strokes are simulated to enable the intake airflow characteristics of the engine to be included. Experience has shown that although the model employs a simple single-phase injection model, the qualitative insight into the influence of both combustion geometry and in-cylinder flow is very valuable in developing the combustion system. In this particular case the ignition timing is approximately 30° BTDC, and even allowing for ignition delay the initial reaction zone is controlled primarily by the exiting spray from the injector rather than the reflected charge. This capability to reliably ignite the fuel cloud exiting directly from the charge injector is important in enabling late injection timings to be utilised and in reducing injection timing sensitivity to engine speed.

By comparing Figure 9a and 9b it can be seen how subtle changes to the piston bowl geometry can effect the fuel cloud containment and the stratification gradient. By increasing the containment of the injected charge, as shown in Figure 9b, the proportion of fuel that migrates to the extremities of the combustion chamber (resulting in very dilute mixtures in these regions) is reduced. At very high dilution rates, the flame may be extinguished, leaving the fuel in the extremities unburned. This is a major source of unburned hydrocarbons for very lean operation, stratified charge engines. Experience has shown that by increasing the containment of the injected charge, via changes typical to the designs shown in Figure 9, raw HC emissions can be reduced by up to 30% over the European drive cycle with additional benefits to both fuel consumption and combustion stability.

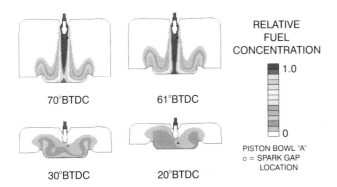

Figure 9a: CFD modelling of air-assisted direct injection; 1500rpm, 2 Bar BMEP.

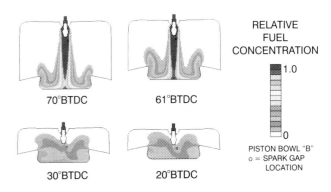

Figure 9b: CFD modelling of air-assisted direct injection; 1500rpm, 2 Bar BMEP.

FUEL SPRAY CHARACTERISTICS

As discussed by previous publications (7 - 9) the basic requirements for an optimal gasoline direct injection spray are as follows:

- Small Particle size (<25 microns), which is important for reduced evaporation time and hence charge preparation time.
- Control of spray penetration to prevent wall wetting and loss of containment at light load, without compromising high load mixing and air utilisation.
- Control of spray dispersion, requiring low dispersion at light load for optimal stratification and containment, complimented by high dispersion at full load conditions.

These characteristics are satisfied well by the characteristics of the low pressure air-assisted direct injection system utilised here. As shown in Figures 10 and 11, the spray is highly atomised with a Sauter Mean Diameter (SMD) less than 8 microns in the main spray. These very small particle sizes enable very fast evaporation of the fuel, minimizing the need for further preparation time inside the cylinder. It is also important to be able to achieve these particle sizes in the real engine environment where the downstream pressure at late injection timings is increasing, hence decreasing the differential pressure across the nozzle of the direct injector.

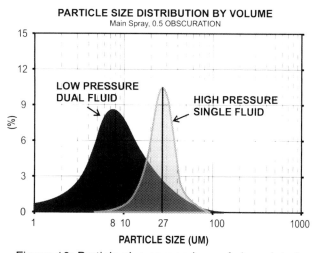

Figure 10: Particle size comparison of air-assisted and single fluid systems.

As illustrated in Figure 12, the air-assisted injection system is able to maintain very small particle size with only minimal pressure differential. This capability is primarily due to the mechanism of utilising air to shear through the metered fuel rather than high pressure alone, therefore enabling the system to maintain good charge preparation even at very late injection timings (or low differential pressures).

Central to realising the fuel economy of DI combustion, is the ability to run a highly stratified, lean combustion process which in turn is highly dependent on the control of the penetration and dispersion of the fuel cloud in the combustion chamber. In order to minimise wall wetting and maximise

stratification at partial loads, a "soft" spray characteristic i.e. a low penetrating and low dispersion spray is desirable. This is an inherent quality of the low-pressure air-assist injection system, as illustrated in the photographs of the spray in Figure 11.

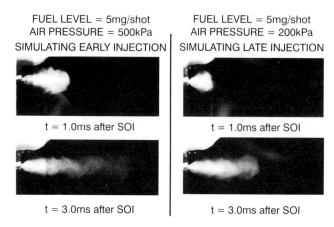

Figure 11: Fuel spray characteristics of DI4S air-assisted injector.

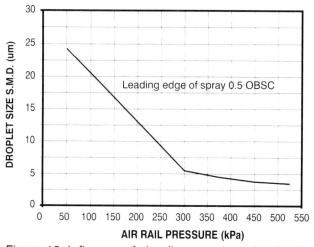

Figure 12: Influence of air rail pressure on droplet size.

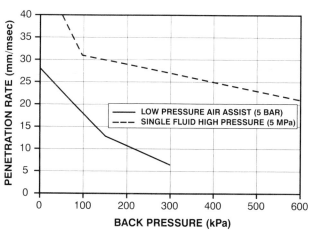

Figure 13: Fuel spray penetration of air-assist and high pressure single fluid direct injectors.

In addition, due to the low pressure differential across the direct injector the penetration is easily controlled by small adjustments to the start of injection timing. Comparison of Figures 11 & 13 also highlight how the "soft" spray characteristic at late injection timings is transformed into a highly penetrating spray at early injection timings (atmospheric downstream pressure). This latter spray characteristic is particularly important for full load operation where good air utilisation and homogeneous combustion is essential for maximum performance capability.

FUEL ECONOMY AND EMISSIONS CHARACTERISTICS

The emissions and fuel consumption characteristics of a light load operating point for the DI Zetec engine are shown in Figures 14 and 15.

2000 RPM, 2 BAR BMEP, 1.8L DOHC ZETEC

$\sigma_{imep} \leq 0.2$ Bar

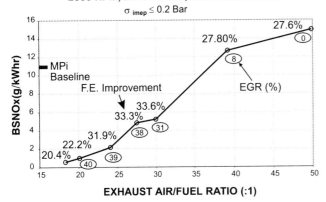

Figure 14: Fuel economy and NOx emissions of air-assisted DI 4-stroke engine.

2000 RPM, 2 BAR BMEP, 1.8L DOHC ZETEC

$\sigma_{imep} \leq 0.2$ Bar

Figure 15: HC emissions of air-assisted DI 4-stroke engine.

These figures illustrate well the typical characteristics of the air-assisted DI 4-stroke system, in particular the influence of air/fuel ratio and EGR rates on the fuel consumption and emissions. The fuel economy improvement shown is relative to the port injected homogeneous charge stoichiometric base engine. Note that all of the data from the DI engine includes the parasitic loss of the compressed air supply.

The ultimate fuel economy potential (ignoring emissions) of single fluid and air-assisted DI 4 stroke engines is most likely very similar, especially at light loads which are dominated by the potential reduction in pumping work. However, it is the real world simultaneous achievement of low engine-out emissions of NOx and hydrocarbons without detriment to the fuel economy capability, which is important to the introduction of DI technologies. As shown in Figures 14 and 15, the DI engine can be operated at WOT without external EGR (corresponding to an exhaust air/fuel ratio of 50:1 at this operating point), down to an air/fuel ratio of 18:1 with hydrocarbon emission levels comparable to the baseline MPI engine, while reducing NOx emissions by up to 95%. In addition, in the region of best fuel economy improvement of 32-34% the hydrocarbons were equal to the MPI baseline engine while running up to 40% by mass of EGR. As also noted the operating stability of the DI engine at all air/fuel ratios corresponds to standard deviations of IMEP less than 0.2 Bar.

As discussed previously, the air-assist system has undergone significant development on 2-stroke engines, which has dictated that the system displays a high EGR diluent tolerance. The injection of air with the fuel results in both fuel and air stratification within the cylinder. It is this stratification which enables a charge, with high overall dilution rates of EGR, to be successfully and repeatably ignited. This has enabled the development engine to be tested with valve overlap periods in excess of 80 degrees, while still maintaining acceptable combustion stability, even at idle. It is this tolerance of large proportions of EGR, which facilitates the reduction in NOx emissions generated during the combustion event, as illustrated in Figure 14 and Table 2.

NEDC 10-POINT SIMULATION, 1360kg ITWC, 2 LITRE, 4 CYL ZETEC ENGINE

	FE [km/l]	FE Saving %	HC * [g/km]	CO * [g/km]	NOx * [g/km]
MPI VEHICLE CALIBRATION, NO EGR	11.70	0.0%	0.85	5.46	1.64
DI SINGLE INJECTION, WITH EGR	13.64	16.6%	1.10	6.68	0.25
DI DUAL INJECTION, WITH EGR	13.84	18.3%	1.17	5.39	0.25
DI DUAL INJECTION, HIGH FLOW EGR SYSTEM	14.05	20.1%	1.30	3.88	0.26

* Emissions data measured before catalyst ie feedgas emissions.

Table 2: Air-assisted DI 4-stroke fuel consumption and emissions capability.

The small preparation time of the injected fuel required and control of the air/fuel mixture injected through the charge injector enables precise control of the air/fuel ratio distribution near the spark plug. This helps to ensure a closer to ideal air-fuel gradient exists around the plug to promote low NOx production, i.e. a progression from rich to lean air/fuel ratio.

As the air/fuel ratio is reduced, the specific heat of the EGR becomes greater leading to further reduction in combustion temperatures, as well as less re-circulated excess oxygen in the trapped charge. These mechanisms, as well as the development of a richer mixture near and around the plug, lead to reductions in NOx emissions as the overall air/fuel ratio is reduced, as shown in Figure 14. This reduction in NOx emissions with reduced air/fuel ratio does not necessarily come at the expense of fuel economy improvement. In particular the large amount of re-circulated exhaust gas through the inlet manifold offsets the increase in pumping work normally associated with reducing air/fuel ratio. For the data presented in Figure 14, the manifold vacuum does not increase until the A/F ratio is less than 24.0:1. As the EGR in the inlet manifold is hotter than freshly induced air, although the manifold vacuum is not increased, the total gas-to-fuel dilution ratio is reduced as the air/fuel ratio is reduced. This leads to a benefit in burn rate and gross indicated efficiency, which can ultimately lead to an improvement in the fuel economy as the air/fuel ratio is reduced. These characteristics of the combustion system enable very low NOx emissions to be produced while providing significant fuel economy improvements. At the particular speed/load shown in Figures 14 & 15, an improvement of some 32% in fuel economy is possible in conjunction with an engine-out NOx emissions level of 2.0 g/kWh.

Table 2 illustrates the overall fuel economy and emissions capability of the development engine, as tested over a ten-point steady state simulation of the new European drive cycle (NEDC). The test points, as illustrated in Figure 16, were suitably weighted to produce a time-averaged simulation of a 2.0 Liter engine fitted to a typical European C class vehicle in an inertia test weight class (ITWC) of 1360 kg.

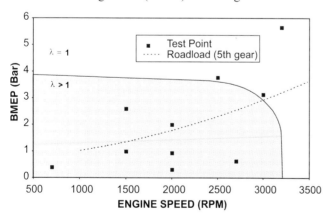

Figure 16: NEDC test points, roadload curve and calibration mode for catalyst strategy.

As shown in Table 2, with an engine-out NOx constraint of 0.25 g/km (85% reduction from MPI) the OCP system has demonstrated up to a 20.1% improvement in fuel economy.

Also of importance is the very good control of HC emissions over the complete drive cycle, especially in comparison to the non-EGR baseline MPI calibration. The engine-out HC control is especially important in the lean operation zone where the inherently lower exhaust gas temperatures can significantly reduce catalyst efficiency.

The key characteristic of the air-assisted DI system is the very fine particle size of the air/fuel spray combined with the containment of a readily combustible charge. The very small particle size minimizes the need for further preparation inside the cylinder. This is especially true at light loads, as shown in Table 3, where very late end of injection timing (25-30° BTDC) is utilized, while still achieving efficient and stable combustion characteristics. In particular the 50% mass fraction burn for the DI data is well positioned at 5-10° ATDC, with the end of injection at 30° BTDC and EGR rates as high as 28% by mass.

	ENGINE CONFIGURATION	
	MPI MBT IGN, WITH EGR	DI NOx TARGET CALIBRATION
SA (deg BTDC)	33	29
10% MFB (deg ATDC)	-3.7	-3.6
50% MFB (deg ATDC)	9.6	7.5
90% MFB (deg ATDC)	27.9	23.1
BURN DUR. (deg)	31.6	26.7
EGR%	9	28

Table 3: MFB locations at 2000rpm, 2 Bar BMEP.

Further enhancements of the control of the air/fuel ratio around the plug have been observed by utilising a dual injection strategy in the medium to high load range. This involves the double opening of the direct injector in a staged sequence during one complete cycle of the engine (see Figure 17).

The two injection events are typically staged such that the first injection occurs early during the intake stroke of the engine while the second occurs late during the compression stroke. The first event allows sufficient time for full mixing of the injected charge with the combustion chamber contents, forming a lean (or highly dilute) homogenous mixture. The second event promotes an ignitable charge around the spark plug, providing increased combustion stability. At moderate fueling levels, it is important to have sufficient fuel mixing to ensure adequate in-cylinder air utilization. On the other hand, a fully premixed charge at high dilution ratios leads to poor combustion stability. The dual injection technique therefore enables operation at higher overall dilution ratios at these moderate fueling levels. This results in a greater level of EGR to be introduced which also results in the ability to run to a higher a/f ratio for the same level of NOx emissions. This increased EGR flow and airflow increases the indicated efficiency of the cycle when compared to single injection operation. When applied to the 3 highest load points in the 10pt drive-cycle simulation, as illustrated in Table 2, this strategy provides a further 3.5 % improvement in fuel economy over the complete test cycle for the same overall raw NOx emissions. As such an overall fuel economy of up to 20.1 % has been realised for this vehicle/engine combination, with cycle averaged raw NOx emissions of 0.25 g/km, a reduction of 85% compared with the MPI base engine. It should also be noted that these figures have been achieved without optimisation of the compression ratio or the intake geometry of the standard engine. In addition, the low raw emissions levels of the air assist system enable several different options for exhaust gas aftertreatment to be considered in order to meet future proposed emissions standards.

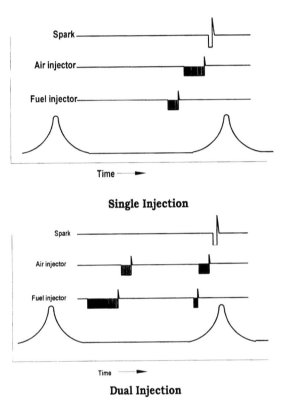

Single Injection

Dual Injection

Figure 17: Air assisted DI4 dual injection strategy.

FULL LOAD PERFORMANCE CHARACTERISTICS

Figure 18 shows the full load performance and volumetric efficiency of the air-assisted DI engine compared with the MPI baseline engine. Both engines were optimized for best torque with air/fuel ratio constraints of greater than or equal to 12.5:1 and ignition timing at MBT or 2 degrees from the onset of knock, whichever was most retarded. The DI and MPI results were obtained with the same cam timing and a similar compression ratio. As shown, the air-assisted DI system displays increased full load performance and increased volumetric efficiency across the speed range. This increase in volumetric efficiency can be mainly attributed to the charge cooling effect of the direct injected fuel. A minor influence will also be from the air injected through the direct injector, which at full load is typically only 8.0 mg of air per cylinder

per cycle, or less than 2.0% of the intake air flow. Further, at typical full load injection timings and low to moderate engine speeds, the injected air can actually displace fresh charge entering through the inlet valves, as the cylinder is not super-charged while the valves are open and there is sufficient time for pressure equalization. Please note that the increase in full load performance displayed by the DI system takes into account the parasitic loss of the air compressor.

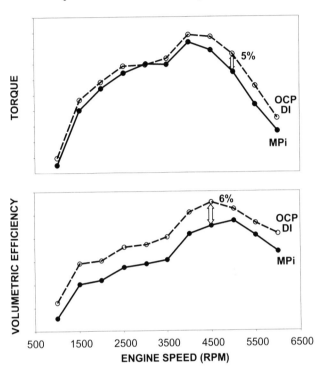

Figure 18: Full load performance and volumetric efficiency.

Figure 19 shows the typical direct charge injector timings for the air assist system at full load for the development engine. As shown, the injection timings corresponding to maximum performance change significantly with engine speed. At low engine speeds, the calibration represents a compromise between air charge cooling/mixture preparation time (corresponding to early injection timings), and reduced fuel resident time for increased knock resistance (retarded injection timings). As the engine speed increases, and the knock tendency reduces and mixing time reduces, the injection event becomes more advanced.

The air-assisted DI system also enables an increase in the ignition timing before the onset of knock as shown in Figure 20. This is again due to the charge cooling effect and the reduced resident time of the injected charge. This increase in knock resistance could enable further increases in compression ratio to be adopted without loss in low speed torque, and potential benefits in part-load fuel economy due to the increased cycle efficiency. This testing is part of the future development work for the DI Zetec engine, however experience from other 4-stroke applications have confirmed the ability to increase compression ratio significantly.

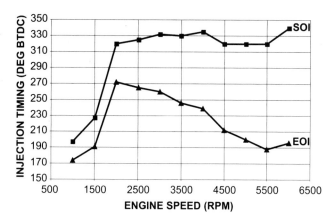

Figure 19: Full load air-assisted direct injection timing (single injection).

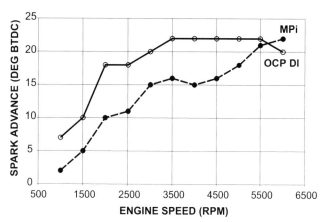

Figure 20: Comparison of full load ignition timings.

FUTURE STRATEGIES

COMBUSTION SYSTEM CONFIGURATION As discussed in detail in this paper the spray characteristics of the air-assisted fuel system enable the spark plug to be mounted directly in the spray without fouling. This capability, in combination with good charge preparation, enables stratified, lean combustion to be achieved without reliance on specific in-cylinder flow conditions.

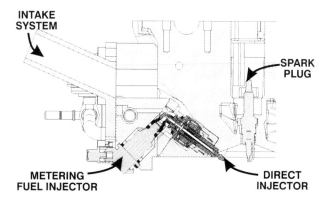

Figure 21: Prototype side injection installation.

As shown from the CFD analysis in Figure 9, the central injection system relies heavily on the combination of the readily ignitable fuel spray and the reflective characteristics of the piston bowl in achieving a stratified lean combustion process. This has the advantage of greater flexibility of the application of the system to existing engine designs without the reliance on complex re-designs of the intake system or port de-activation systems. More recent experience has shown that the Orbital combustion process can also be designed to work effectively with the direct injector and spark plug spaced apart, that is with the direct injector mounted on the intake side of the cylinder head. A prototype design of this type of system, currently under assessment on a modified Zetec engine is shown in Figure 21.

HC/NOx CONTROL As discussed previously the air-assisted DI 4-stroke system enables very low engine-out NOx emissions to be achieved without sacrificing HC emissions control. However, unlike the OCP 2-stroke engines (11), the 4-stroke requires some exhaust aftertreatment to achieve European Stage 3 NOx levels in conjunction with a significant improvement in fuel economy. Up to this point there have been two proposed solutions for existing high pressure DI vehicle releases involving either a Lean NOx catalyst or a NOx storage/release catalyst system. As described by Ando (14), the preferred option in terms of peak conversion efficiency and operating temperature window is the NOx storage/release system. However, Mitsubishi have adopted an Iridium lean NOx catalyst system for the European release of their DI gasoline vehicle due to the poor durability of the NOx storage catalysts when exposed to the typical European fuel sulphur levels (100 ppm). The lean NOx system by contrast is quite temperature sensitive and requires a specific HC/NOx ratio feedgas, which ultimately dictates the installation must be underbody and cannot utilise a close-coupled catalyst. These latter requirements have necessitated the development of a catalyst light-off routine, which utilizes the 40-second portion prior to bag sampling of the current European Stage 2 emissions test. However, the underbody installation is unlikely to be suitable for future Stage 3 and 4 European legislation which employs a new European test cycle, i.e. sampling from crank and only a 10 second idle before drive away. As such, the current lean NOx solutions for future European emissions legislation are still not fully developed and may require significant changes to current European sulphur levels (<50 ppm). Due to the changing legislation, Orbital has been investigating an alternative strategy which proposes to take full advantage of the air-assisted DI engine tolerance to exhaust gas diluent in the light load area to control the engine-out NOx, i.e. up to 90% reductions as illustrated in Figure 14. Further, at the higher load region the engine could be calibrated at Stoichiometry enabling a conventional 3-way reducing catalyst to be utilised. The basics of this strategy are illustrated in Figure 16, with the top three of the ten steady state test points being operated in the Stoichiometric region. Due to the ability of the air-assisted DI engine to handle high

3-WAY CATALYST STRATEGY APPLIED TO TOP THREE LOAD POINTS OF NEDC 10-POINT SIMULATION
2 litre, 4 cyl ZETEC engine at 1360kg ITWC.

OCP - Dual Injection Stratified Lean Calibration

Speed (rpm)	2500	3000	3200
Load (Nm)	60	50	90
BSFC (g/kW.hr)	298.5	327.7	308.5
BSNOx (g/kW.hr)	2.009	1.824	2.515
NOx % contribution	27.1	14.8	15.6 (57.5 Total)

Drive cycle Totals

FE (km/l)	13.93
HC (g/km)	1.343 Raw
NOx (g/km)	0.26 Raw

OCP DI - Stoich. calibration at high loads.

Speed rpm)	2500	3000	3200
Load (Nm)	60	50	90
BSFC (g/kW.hr)	315.4	339.3	279.3
BSNOx (g/kW.hr)	3.48	6.96	6.29
NOx % contribution	25.5	30.3	21.2 (77 Total)
postcat BSNOx	0.174	0.348	0.314

FE (km/l)	13.85
HC (g/km)	1.296 Raw
NOx (g/km)	0.47 Raw
NOx (g/km)	0.14 Tailpipe.

Note 3 way NOx reduction = 90% at Stoichiometric A/F Ratio.

Table 4: Comparison of stoichiometric calibration/3 way catalyst strategy and low NOx calibration techniques.

levels of EGR across the speed/load range, the sacrifice in fuel consumption is only marginal for this type of calibration. As shown in Table 4, calculations based on a 90% NOx catalyst efficiency applied to the three highest load test points only, illustrate the capability to meet the overall European stage 3 NOx targets with similar capability in fuel consumption. Therefore, by taking advantage of the very low raw NOx emissions and with HC emissions comparable to the baseline MPI engine, the air-assist system has the potential to meet future European and US emissions legislation without relying on undeveloped, potentially expensive and complex lean NOx catalyst technologies.

CONCLUSION

In conclusion, this paper has shown the following:

1. Utilisation of the Orbital air-assisted direct injection enables very lean, stratified combustion to be achieved on conventional 4-stroke engine designs without modification or enhancement of the intake airflow characteristics.
2. Good charge preparation and late injection timings of the system enable fuel economy improvements of up to 20.1% to be achieved over the European drive cycle with simultaneous control of both HC and NOx emissions.
3. The very high tolerance of the air-assisted DI system to EGR and the control of the air/fuel ratio around the plug enables very low overall raw NOx emissions to be achieved across the speed/load range of the European drive cycle. This is typified at the 2000 rev/min, 2 bar bmep test point, with NOx reductions of up to 95% and fuel consumption improvements of up to 34%.
4. HC emissions over the European test cycle from the DI development engine have been maintained close to the MPI baseline engine, even with the addition of up to 40% EGR in the light load operating region.
5. An alternative calibration strategy utilising the low engine-out NOx and HC emissions of the air-assisted DI combustion process in combination with a 3-way conventional catalyst system has demonstrated the potential to meet European stage 3 emissions under steady state simulations.
6. The air-assisted DI system has demonstrated up to a 5% improvement in full load performance across a wide engine speed range, without fully optimizing the compression ratio or intake system of the baseline engine.

ACKNOWLEDGMENTS

The authors would like to thank all of the dedicated personnel at Orbital Engine Company and Corporation, who through their tireless efforts and endeavor have all contributed in part to this paper.

REFERENCES

1. Alperstein, M., Schafer, G., Villforth, F., " Texaco's Stratified Charge Engine Multifuel, Efficient, Clean and Practical," SAE 740563.
2. Scussel , A.,Simko, A., Wade, W., "The Ford PROCO Engine Update," SAE 780699.
3. Schäpertöns, H., Emmenthal, K.-D.,Grabe, H.-J., Oppermann, W., " VW's Gasoline Direct Injection (GDI) Research Engine," SAE 910054
4. Worth, D.R., Coplin, N., Stannard, J.M., McNiff, M., " Design Considerations for the Application of Air Assisted Direct In-Cylinder Injection Systems," SAE Small Engine Technology Conference, Japan 1997.
5. Kume, T., Iwamoto, Y., Iida, K., Murakami, M., Akishino, K., Ando, H., " Combustion Control Technologies for Direct Injection SI Engine," SAE 960600.
6. Kiyota, Y., Akshino K., Ando, H., " Combustion Control Technologies for Direct Injection SI Engines," FISITA 96.
7. Iwamoto, Y., Noma, K., Nakayama, O., Yamauchi T., Ando, H., " Development of Gasoline Direct Injection Engine," SAE 970541.
8. Tomoda, T., Sasaki, S., Sawada, D., Saito, A., Sami, H., "Development of Direct Injection Gasoline Engine- Study of Stratified Mixture Formation," SAE 970539.
9. Harada, J., Tomita, T., Mizuno, H., Mashiki, Z., Ito, Y. " Development of Direct Injection Gasoline Engine," SAE 970540.
10. Schlunke, K. " The Orbital Combustion Process Engine," 10th Vienna Motorsymposium 1989, VDI No 122, pp 63-68.
11. Houston, R.A.R, Archer, M.D., Moore, M., Newmann, R., " Development of a Durable Emissions Control System for an Automotive Two-Stroke Engine," SAE 960361.
12. Smith, D.A., Ahern, S.R., " The Orbital Ultra Low Emissions and Fuel Economy Engine", 14th Vienna Motorsymposium, 1993, VDI No182, pp 203-209.
13. Leighton, S.R., Ahern, S.R., " The Orbital Small Engine Fuel System (SEFIS) for Direct Injected Two-Stroke Cycle Engines ", 5th Graz Two-Wheeler Symposium, T-U Graz, 1993.
14. Ando, H., "Mitsubishi GDI Engine- Strategies to meet the European requirements,", AVL conference GRAZ, 4-5TH September 1997.

Gasoline DI Engines: The Complete System Approach by Interaction of Advanced Development Tools

**M. Wirth, W. F. Piock, G. K. Fraidl, P. Schoeggl
and E. Winklhofer**
AVL List GmbH

ABSTRACT

Gasoline direct injection is one of the main issues of actual worldwide SI engine development activities. It requires a comprehensive system approach from the basic considerations on optimum combustion system configuration up to vehicle performance and driveability.
The general characteristics of currently favored combustion system configurations are discussed in this paper regarding both engine operation and design aspects. The engine performance, especially power output and emission potential of AVL's DGI engine concept is presented including the interaction of advanced tools like optical diagnostics and 3D-CFD simulation in the combustion system development process. The application of methods like tomographic combustion analysis for investigations in the multicylinder engine within further stages of development is demonstrated.
The system layout and operational strategies for fuel economy in conjunction with exhaust gas aftertreatment requirements are discussed.
For the optimization of the complex transient functions as e. g. switching between stratified and homogeneous operation and catalyst purging, a new evaluation tool providing objective driveability assessment is applied.

INTRODUCTION

Within worldwide automotive industry, the technology of gasoline direct injection is clearly rated as a decisive step towards the future of SI engine technology. This technology will have to meet significant reductions in fleet fuel consumption and, simultaneously, comply with lowest emission limits. After a first phase of development work on the subject that was dominated by design considerations and conversion of existing engine designs, it is now obvious that the potential of this technology can only be unveiled by a complete system approach including considerations from the first design steps until vehicle performance. Regarded that way, direct injection forms a networked system with multiple interactions as shown in Fig.1.

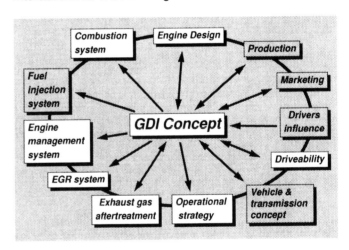

Fig. 1 : GDI Concept interactions

As it is already indicated by highlighting in the figure, this paper will focus on these elements of the complete network which are actually most important for the development process as an engineering service while some of the remaining elements are discussed in [1;2;3;11;12].

COMBUSTION SYSTEM AND ENGINE DESIGN

The first stage of a GDI system layout is always the combustion system consideration in conjunction with the general engine design.
For the GDI combustion system, a broad variety of completely different approaches has been discussed in the past [7;8] that were predominantly distinguished by

different component arrangement (**close spacing** vs. **wide spacing** [8] of the injector with respect to the spark plug position) and different charge motion concepts (swirl, tumble, reverse tumble, squish).

Meanwhile, a clear focus of interest on wide spacing concepts can be stated due to the following reasons:

- The valve size of a comparable MPFI engine can usually be maintained in order to fulfill the demand for improved power output of the GDI engine.
- The durability requirements for injection system and spark plug are reduced since spray impingement on the spark plug can be avoided.
- Mixture transport to the spark plug by the combustion chamber geometry supported by the charge motion leads to increased mixture preparation time scales resulting in a stratified combustion with prevailing premixed regime and minimized particulate emission.
- Wide stability range of stratified combustion enables reduced precision requirements for injection and ignition timing.

Regarding injection system technology, the concept of swirl assisted high pressure liquid fuel injection [11] represents the actual mainstream while gas-assisted systems remain an interesting alternative for some applications like future lowest emission concepts. The work presented in this paper is therefore focused on high pressure liquid injection technology.

The decision on a suitable charge motion concept is dependent on the design and production constraints of a specific engine. The impact of a major parameter - the valve angle - is demonstrated in Fig. 2.

An important base of the combustion system is an efficient charge motion generation during the intake stroke and its preservation during compression for late turbulence generation and accelerating the end of combustion [5;7;8]. Especially for tumble intake flow, this requires a sufficient remaining height of the combustion chamber at TDC which leads to unfavorable deep piston bowl geometries, if medium or even small valve angles are considered as they can be seen as an actual trend in high performance multi-valve SI engine design (compact cylinder heads).

In contrast, the preservation of an intake-generated swirl motion is more efficient due to reduced momentum dissipation and transformation into secondary flows. The height of the combustion chamber can be reduced, leading to favorable piston crown designs even for small valve angles, which can be integrated in a suitable high performance piston design.

Another important aspect for GDI combustion system integration into the cylinder head design is the injector mounting. For the wide spacing concept, an injector location on the intake side is clearly favorable for thermal and packaging reasons. Since the spray inclination in the combustion chamber is one of the most decisive parameters for the functionality of the combustion system, it has to be achieved with the best compromise between injector mounting and intake port design as shown in Fig. 3.

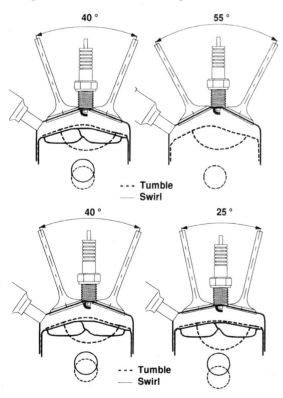

Fig. 2 : Impact of valve angle on combustion chamber and piston design

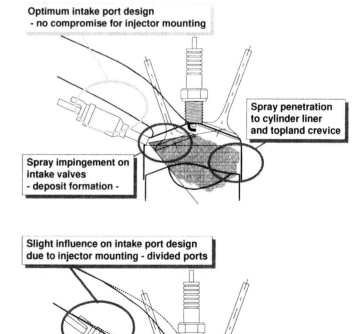

Fig. 3 : Aspects of injector and spray inclination

For spray inclination with respect to the cylinder axis, a medium inclination in the range of 30° to 50° yields the best conditions for homogeneous operation with a minimum of possible cylinder wall wetting (especially important for small bore sizes), a limited spray impingement on the intake valves and optimized homogenization. This injection direction can result in the requirement for either divided intake ports or injectors with inclined spray direction which can then be mounted beneath the intake ports. Regarding combustion system performance, a limited inclination of the spray axis to the cylinder axis is favorable in order to increase the mixture preparation time scale and to facilitate the interaction between spray (constant penetration for all engine speeds) with the upward moving piston in stratified operation, resulting in an extended engine speed range for stratified operation. The adaptation of the GDI injector to the combustion chamber in the cylinder head in its radial position between the intake valves (in case of a multivalve engine) can be further optimized, if the combustion chamber provides a recessed position of at least the intake valves with respect to the cylinder head face, enabling almost flush mounting of the injector nozzle in the combustion chamber surface for best access of the charge motion as well as further reduced intake valve impingement of the spray during early injection at homogeneous operation (Fig. 4)

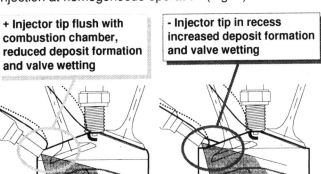

| + Injector tip flush with combustion chamber, reduced deposit formation and valve wetting | - Injector tip in recess increased deposit formation and valve wetting |

Fig. 4 : Aspects of combustion chamber height in the cylinder head

Integrating the aforementioned conditions for combustion system layout and cylinder head design, AVL favors the swirl assisted combustion system for multivalve engines up to about 40° valve angle as well as for 2- and 3-valve engines with their typically small valve angles. A tumble-based approach seems most suitable for some applications with wide valve angle clearly above 45°.

AVL's swirl-based DGI™ combustion system is based on the concept of wall guided mixture transport to the central spark plug with minimized spray impingement by maximum advance of injection timing. The central mixture stratification is supported by an asymmetric piston bowl design and the swirl flow. The bowl asymmetry with a radial flow entry is used for direction of the swirl flow into the bowl area and to achieve sufficient compression volume while maintaining an overall shallow piston crown design. The combustion system is shown in principle in Fig. 5.

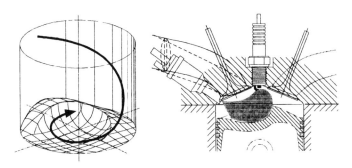

Fig. 5 : Swirl-based DGI™ combustion system

With this configuration, a well centered mixture stratification can be achieved providing an exceptional cyclic combustion stability and extended ignition delay times (time delay between start of injection and ignition timing) combined with moderate swirl demand.

The system has been developed as an engine demonstrator based on a state of the art 4V 2.0l 4-cylinder production engine and is actually mounted in-vehicle for current development activities (Fig. 6)

Fig. 6 : AVL DGI™ engine in vehicle application

The swirl-assisted DGI combustion system in 4V configuration requires an intake port layout with separated ports within the cylinder head up to a flange about 100 mm upstream in the intake manifold comprising a tangential port for swirl generation and a neutral port for maximum volumetric efficiency under full load conditions. The neutral port is equipped with a deactivation slider mechanism for easy variability of the charge motion between the two positions of optimized swirl level for stratified and homogeneous lean part load operation as well as maximum volumetric efficiency at full load. Careful development of the port geometry enabled a port quality of the complete intake port system that reaches at least the level of the MPFI base engine port (Fig. 7) despite its high potential for swirl generation with port deactivation.

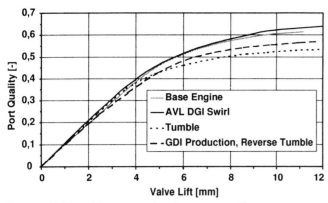

Fig. 7 : DGI swirl concept intake port quality

This intake port quality together with an optimized intake manifold system and the typical thermodynamic advantage of direct injection vs. manifold injection (improved charge cooling, extended knock limit) enables a clear improvement of the engine's full load performance as shown in Fig. 8. The comparison has been performed here with unchanged compression ratio.

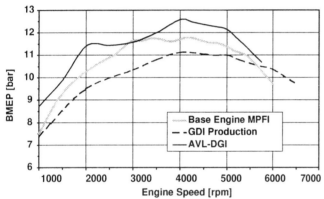

Fig. 8 : DGI full load performance vs. base engine at the same compression ratio and a GDI production engine

The dominant performance parameter of the GDI system is its fuel economy potential in stratified part load operation. The swirl based DGI combustion system yields an outstanding fuel economy potential within the entire stratified part load range as it is shown in Fig. 9. Even compared at same compression ratio, the stratified operation shows a significant BSFC improvement with respect to the homogeneous stoichiometric operation. In comparison to a MPFI engine, even larger improvements due to the higher compression ratio possible with the DGI engine, is obtained.

At lowest part load the actual fuel economy potential of a multicylinder engine mapped for emission compliance is limited due to the requirement to maintain a minimum exhaust gas temperature of about 250°C for exhaust gas aftertreatment by partial throttling and engine friction aspects.

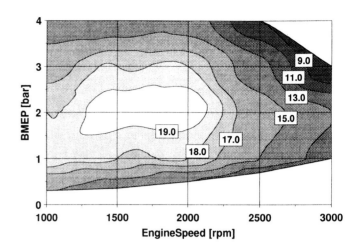

Fig. 9 : Fuel economy potential of the DGI 4-cylinder engine - relative improvement (in %) of stratified versus homogeneous stoichiometric operation at same compression ratio

The emission potential of the combustion system is demonstrated in the best way in conjunction with its EGR acceptance, since high EGR is one key component for low engine-out NOx capability as a base for compliance with future steps in emission legislation (EURO III → EURO IV ; LEV → ULEV) using reasonable exhaust gas aftertreatment concepts. In Fig. 10 a - b, EGR sweeps from the 4-cylinder engine are presented comprising the major emission components as well as fuel consumption and combustion stability (mean value of all four cylinders).

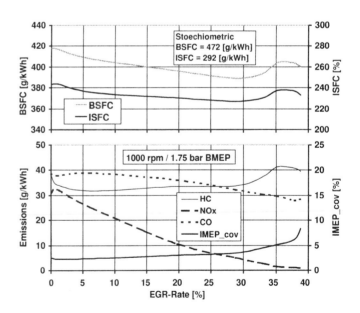

Fig.10a : EGR acceptance and emission potential of the DGI 4-cylinder engine at 1000 rpm / 1.75 bar bmep

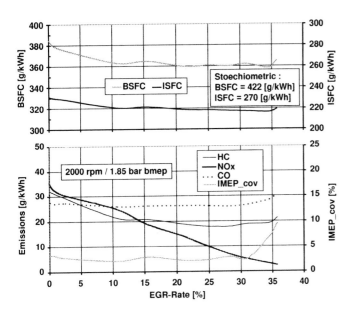

Fig.10b : EGR acceptance and emission potential of the DGI 4-cylinder engine at 2000 rpm / 1.85 bar bmep

In all stratified operating conditions, a high EGR acceptance can be denoted, combined with further reductions of fuel consumption and HC emission with increasing EGR rate up to levels which enable NOx engine-out emissions as low as about 3 g/kWh (brake specific) in the most relevant part load operation range. It is evident that based on this low NOx level, only slight increases of EGR acceptance can result in important further reduction of the NOx emission.

An important characteristic of a stratified combustion system regarding its calibration and control by the EMS is the sensitivity of combustion performance to variations of injection and spark timing. A broad stability range ensures increased flexibility in calibration and limited requirements to the timing precision of the control system. The DGI swirl assisted combustion system provides a broad stability both with respect to injection and ignition timing. Over a range of at least 12 deg. CA, stable combustion without misfires and with exceptional low imep variations is achieved as it is shown in Fig. 11.

The favorable influence of EGR not only on engine out NOx emissions but also on fuel consumption and HC emissions is based on the thermodynamic combustion characteristics of unthrottled stratified combustion. The heat release of an entirely unthrottled stratified combustion is usually situated at very early timing within the engine cycle [7] due to combustion stability reasons and to avoid a delayed end of combustion whereas the end of combustion can favorably be improved by the in-cylinder charge motion concept. An increase of the EGR causing an overall lower rate of heat release shifts the heat release to a later timing in the engine cycle, leading to improved thermal efficiency of the cycle and thus reduced fuel consumption. Improved mixture preparation due to increased charge temperatures and smaller exhaust mass flow contributes to lower HC emissions.

Fig. 12 shows characteristic integral heat releases for the swirl assisted DGI combustion system with and without EGR.

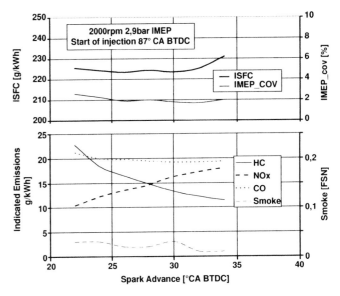

Fig.11a : DGI sensitivity of stratified combustion to spark timing variations (without EGR)

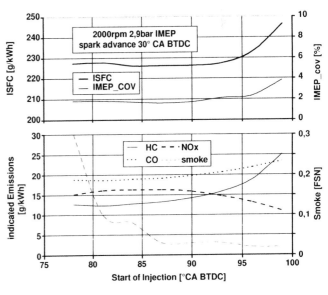

Fig.11b : DGI sensitivity of stratified combustion to injection timing variations (without EGR)

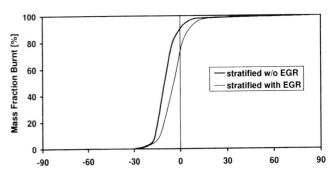

Fig.12 : Integral rate of heat release for DGI stratified part load operation

71

The final effect of stratified operation and EGR on effective fuel economy improvement has to be regarded in conjunction with engine friction aspects. While the stratified unthrottled operation clearly reduces the pumping losses as its major advantage, the overall pressure level in the cylinder is increased in comparison to homogeneous throttled operation. This pressure increase in conjunction with the shift of heat release to earlier timing in the engine cycle causes an increase in engine friction comparable to the friction differences between part load and full load operation which can be in the range of 0.1 to 0.15 bar fmep. Fig. 13 shows the fuel economy improvement for stratified part load compared to homogeneous lean and stoichiometric operation in indicated and brake specific values revealing the slight reduction of the fuel economy improvement in brake specific values vs. indicated values due to the fmep increase.

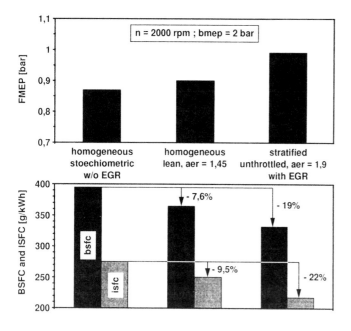

Fig. 13: Impact of different engine friction on fuel economy improvement of stratified operation

These results are obtained with remachined prototype pistons without further friction optimization. The careful development of the piston (weight distribution, piston profile) therefore will be of importance in order to reduce the differences in fuel economy improvement between indicated and brake specific values.

DEVELOPMENT PROCESS

The development process for an advanced GDI system like AVL's DGI engine and vehicle is based on a complete set of advanced development tools including basic measurement techniques for subsystems, simulations and optical diagnostics in research engines as well as in the real multicylinder engine. All these tools are integrated into a comprehensive main process as it is shown in Fig. 14. Optical diagnostics and simulation

techniques are extensively used in the first phases of the development process for the assessment of injection systems to be applied as the most important subsystem. The adaptation of the combustion system to different engine configurations as well as the conception of new approaches is heavily relying on simulation (1D cycle and 3D-CFD) as well as the detailed analysis of prototype geometries in the transparent single cylinder research engine.

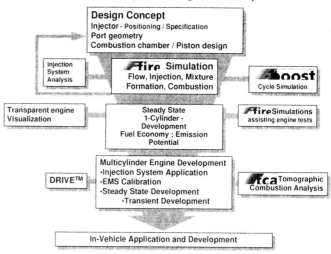

Fig. 14 : GDI System development process

Since actually a wide variety of injection systems is under development and its characteristics are strongly influencing the combustion system performance, spray analysis under realistic engine conditions is constantly being performed in order to assess new systems or major steps in injection system technology. For this purpose, a compression machine is applied comprising a large displacement 2-stroke port-scavenged single cylinder engine with the cylinder head replaced by a special spray measurement chamber (Fig. 15). The motored engine efficiently provides engine conditions regarding backpressure and temperature within the polytropic compression.

Fig. 15: GDI spray research facility for fast injector analysis under engine conditions

Measurements in this chamber deliver information about the spray shape and its penetration behavior from laser sheet imaging and more detailed spray data regarding droplet size, number density, and velocity from PDA measurements. In Fig. 16, an example of such a spray analysis is presented. These data are subsequently use on one hand for combustion chamber layout, where especially the spray shape and penetration is of interest. On the other hand, the more detailed PDA data (Fig. 17) are an essential input for a CFD spray model.

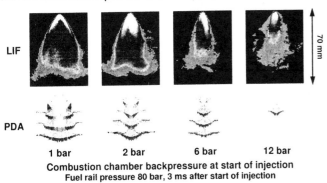

Combustion chamber backpressure at start of injection
Fuel rail pressure 80 bar, 3 ms after start of injection

Fig. 16: GDI spray visualization and PDA velocity field

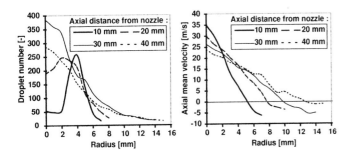

Fig. 17: Spray parameters from PDA measurements at 6 bar combustion chamber backpressure

A spray model based on these experimental data is introduced into the simulation of the combustion system. The instationary 3D-CFD simulation of the complete engine geometry is actually used primarily for the following purposes :

- Assessment of the detailed instationary in-cylinder flow field, providing detailed insight into the behavior of combustion chamber geometries designed to guide flow and fuel spray / mixture.
- Determination of the spatially resolved turbulence distribution and its evolution during the engine cycle in order to further optimize it for improved mixture formation and combustion acceleration.
- Evaluation of the stratified mixture plume evolution including effects of spray-wall interaction and the convective transport of the mixture by the flow field.
- Determination of the time history of relevant parameters like fuel-air ratio, mean velocity, turbulent velocity fluctuation etc. at reference points like the spark location for rapid comparison of different system variants.

- Analysis of the flow field in critical areas like spray impingement surfaces.

Fig. 18 shows the evolution of the mixture distribution during the stratified injection process. The sequence displays the distinct core jet in the initial phase of injection with transition to the hollow cone of the main injection phase. Superimposed are the distributions of fuel air ratio and droplets revealing the predominantly wall guided mixture transport to the spark plug as a combination of spray momentum deviation and convective transport of the fuel vapor.

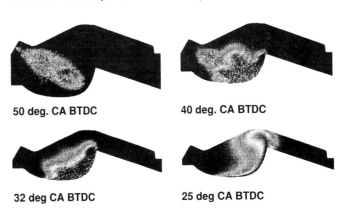

50 deg. CA BTDC   40 deg. CA BTDC

32 deg CA BTDC   25 deg CA BTDC

Fig. 18: 3D CFD-simulation of stratified mixture formation

In the transparent single cylinder research engine, the DGI combustion chamber geometry is represented with a maximum of optical access to the upper cylinder liner, the pentroof in the cylinder head and partially through the piston shape. Fuel spray visualization by fuel PLIF (Planar Laser Induced Fluorescence) yields information about spray penetration and the mixture plume evolution. A special focus is pointed on cyclic fluctuations which can not be fully covered by simulation and on further insight into the injection process like injector dynamic performance (possible secondary injections and large droplet formation at end of injection) in conjunction with the real combustion chamber geometry. In that way, simulation and Laser sheet imaging can be favorably combined with the simulation as the optimum tool for determination of the mean flow field structure and mixture formation process. The laser sheet imaging delivers on one hand the verification of simulations and additional information about the effect of fluctuations. For this purpose, a statistical evaluation of the (qualitative) single cycle images is applied, providing the probability distribution of fuel vapor and liquid portions to be found within the stratified mixture plume. Fig. 19 shows a characteristic sequence of injection and mixture formation in the DGI swirl assisted combustion system in stratified operation revealing its high cyclic stability with a broad central area of 100% fuel vapor probability as well as limited surrounding fluctuation regions indicated by low fuel vapor probability at ignition timing of about 30 deg. BTDC.

Probability distributions for the low density (fuel mist and vapor) and the high density (liquid portion) are superimposed here with priority for the high density distribution using the same color scale due to the limitations of b/w representation. The steep white/black transition indicates the outer limit of the high density distribution which is always superimposed to the low density distribution.

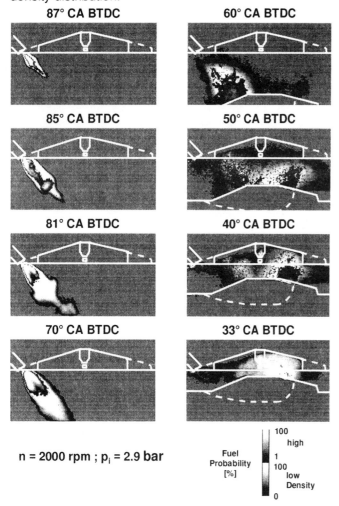

n = 2000 rpm ; p_i = 2.9 bar

Fig. 19: Laser Induced Fluorescence visualization of the injection process (central vertical Laser sheet) during stratified operation

A similar investigation as it was used for the assessment of stratified mixture formation was also applied to the homogeneous mode. Here the interest is focused on the spray evolution, its distribution within the combustion chamber volume and possible critical interactions with the opening intake valves and the combustion chamber walls. Fig. 20 displays a PLIF sequence of early injection during the intake stroke.

The sequence reveals an almost undisturbed evolution of the fuel spray with negligible impingement on the intake valves due to the limited spray inclination with respect to the cylinder axis. Furthermore, no critical spray penetration to the exhaust side of the cylinder liner can be observed.

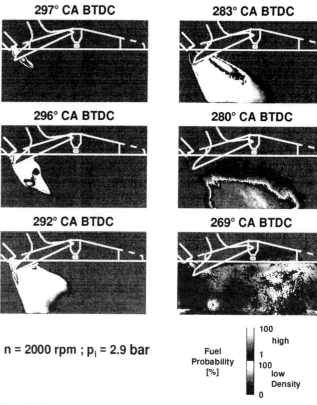

n = 2000 rpm ; p_i = 2.9 bar

Fig. 20: Laser Induced Fluorescence visualization of the injection process during homogeneous operation

A second information about the DGI combustion system is derived in the transparent engine by flame photography where statistical evaluation is applied as well. The result as presented in Fig. 21 yields the probability distribution of flame radiation which is comparable to the mean progress variable in simulation. The evaluation is twofold for low intensity blue flame radiation (premixed regime) and high intensity black body radiation from locally over-rich premixed combustion or diffusion flame regimes. The high intensity distribution is again superimposed to the premixed flame distribution. The probability distribution of the DGI system stratified combustion reveals almost entirely premixed combustion with very low probability of randomly distributed diffusion flame occurrence (black areas at later crank angle) as a result of very complete mixture formation within the stratified mixture plume.

Transparent engine investigations are usually carried out under boundary conditions which are slightly different from the real engine situation for the following reasons :
- single component fuel (iso-octane)
- reduced combustion chamber wall temperatures
- different surface characteristics
- limited engine speed range

Therefore optical investigations on the DGI combustion system were completed by tomographic combustion analysis in the real multicylinder engine using the TCA (Tomographic Combustion Analysis) optically accessible cylinder head gasket [9;10].

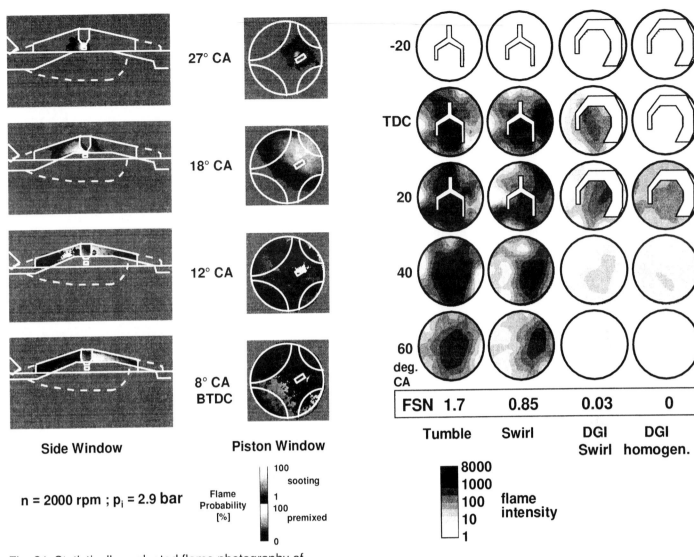

27° CA

18° CA

12° CA

8° CA
BTDC

Side Window

Piston Window

n = 2000 rpm ; p$_i$ = 2.9 bar

Flame
Probability
[%]

100
sooting
1
100
premixed
0

Fig. 21: Statistically evaluated flame photography of stratified part load combustion

This system was used for investigation of stratified part load combustion as well as full load operation regarding engine knock origins - a critical issue if complex GDI piston geometries are used. Stratified part load investigation in the multicylinder engine reveals critical areas of highly radiating sooting combustion from over-rich zones as well as from incomplete wall film evaporation. These are the most critical issues where differences can occur between the transparent engine due to the above mentioned limitations and real engine conditions. In addition, flame propagation is visualized cycle-resolved within the entire bore cross section. In Fig. 22, four TCA sequences are compared :

- A GDI combustion system with tumble charge motion
- The same combustion system with swirl charge motion.
- The swirl based DGI combustion system in stratified operation.
- The swirl based DGI combustion system in homogeneous operation.

Also presented are the different heat release characteristics in Fig. 22b.

-20			
TDC			
20			
40			
60 deg. CA			

FSN	1.7	0.85	0.03	0
	Tumble	**Swirl**	**DGI Swirl**	**DGI homogen.**

8000
1000
100 flame
10   intensity
1

Fig. 22a : Combustion system optimization for stratified part load operation using **T**omographic **C**ombustion **A**nalysis in multicylinder engine

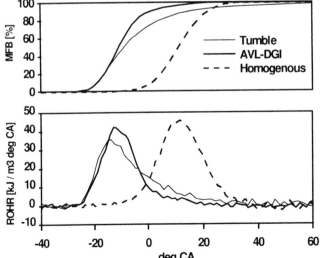

Fig. 22b: Heat release evaluation of combustion processes shown in Fig. 22a, part load, no EGR

The wall-guided tumble combustion system in stratified operation (1st column in Fig. 22a) reveals a high degree of black body radiation from wall film combustion in the spray impingement area which leads to high soot emission (FSN 1.7) and also to a delayed end of combustion. This piston was modified in a first development step for acceptance of swirl charge motion by asymmetric bowl design. The tomographic analysis of this configuration (2nd column in Fig. 22a) in stratified part load reveals a reduced area of highly radiating wall film combustion but also a significant deviation of the flame kernel to on side of the bowl with part of the flame reaching the periphery of the combustion chamber.

The piston geometry for swirl assisted operation was therefore redesigned with a more central bowl position surrounded by the asymmetric flow entry area and increased squish areas especially on the exhaust side as shown in Fig. 5. Stratified operation of the optimized swirl assisted DGI combustion system (3rd column in Fig. 22a) reveals a light intensity evolution which differs only slightly from homogeneous operation (4th column in Fig. 22a) - an observation which is in good agreement with the transparent engine investigations. Homogeneous part load combustion is characterized by a fast flame propagation with smooth and even distribution of a low intensity flame radiation originating from the premixed flame with its characteristic blue CH band radiation. Tomographic combustion analysis was used also for investigation of another critical issue of the combustion system - the engine knock characteristics in view of complex piston crown geometries. Here the statistical evaluation of knock origins (evaluated origins of spherical detonation waves) yields the knock origin probability distribution as it is shown for 4 different piston geometries in Fig. 23.

This distribution is an important input for further development of the piston geometry regarding reduced knock sensitivity which can cause sometimes opposing demands. In Fig. 23, GDI pistons 1 and 2 show a distinct accumulation of knock origins in the vicinity of piston geometries approaching the cylinder liner while piston 3 features a knock origin distribution that is almost as randomly distributed as the flat top piston.

EGR SYSTEM

EGR induction is a significant subject in GDI system layout especially in view of the overall high EGR rates of up to 50% in lower part load and stratified operation. For precise control of these rates, a linear EGR valve has to be employed and an even distribution of the EGR mass flow on the engine cylinders must be provided. The second main requirement for the EGR system is a fast dynamic response in order to cope with transient changes between high EGR stratified operation and homogeneous stoichiometric or rich operation without EGR. To meet this requirement, EGR line volumes between the EGR valve and the induction location should be minimized. The induction location should be close to the engine intake port to further improve dynamic response and reduced intake manifold heating. A possible solution is a close coupled EGR rail with defined induction cross sections to the intake ports and a central EGR supply and EGR valve. The general setup of the EGR system is shown in Fig. 24.

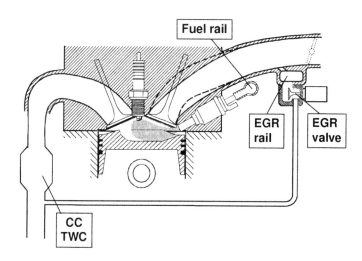

Fig. 24: EGR system layout with close coupled EGR rail for multi point induction

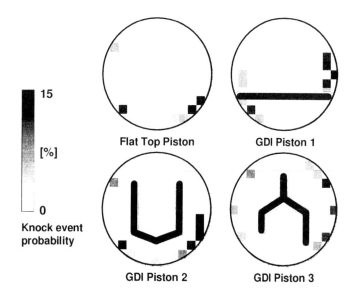

Fig. 23: Development of knock characteristics of different GDI piston geometries

Possible disadvantages due to gas dynamic effects (cross flows) between the cylinders at full load can be eliminated by means of a deactivation mechanism as it is shown in the design view of the system in Fig. 25.

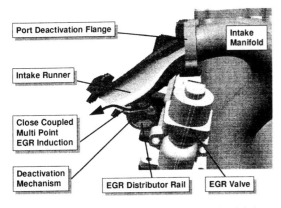

Fig. 25: Multipoint EGR induction system for high dynamic response (AVL *ADECS*)

For further reduction of the effects of gas dynamic cross flows between the cylinders, a divided EGR rail respecting the firing sequence of the engine is used in the ADECS (Advanced EGR Control System).

EXHAUST GAS AFTERTREATMENT

The exhaust gas aftertreatment concept of a GDI is one key component for fulfillment of the next emission legislation steps (EURO III and IV, LEV, ULEV) and here especially the lean NOx reduction is the major challenge. Actually, 2 concurring approaches are followed up : The selective reduction DeNOx catalyst [1] and the NOx storage catalyst or NOx trap [3;13]. A comparison of the conversion efficiencies of these different systems clearly favors the NOx trap. Here the significant problem is the fast aging of the catalyst under high sulfur oxide concentration as it is the case with current European and US fuel qualities and a high temperature aging mechanism under lean conditions (platinum sintering). A sulfur regeneration strategy is possible using high exhaust gas temperatures (>650°C) under rich conditions but there are remaining problems with catalyst durability. In view of the stringent future emission limits, AVL favors the approach of the NOx storage catalyst in conjunction with high EGR strategy due to its higher NOx conversion potential as it is shown in Fig. 26.

The storage catalyst does not require constant presence of HC concentrations and therefore can be combined with the concept of a close coupled three way catalyst for fast light off and HC conversion after cold start. This is essential for the next steps in European emission legislation including reduced warm-up periods and sampling from first engine crank. This approach leads to an aftertreatment system setup as it is shown in Fig. 26.

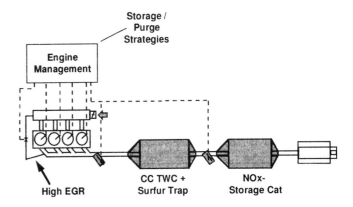

Fig. 26: Catalyst system configuration

An attractive GDI specific light off strategy for such an aftertreatment system can be found in the operational mode of stratified stoichiometric operation. This mode enables a substantial increase of exhaust gas temperatures with simultaneous strong reductions of engine out HC and NOx emissions. These effects are achieved by moderate stratification of the stoichiometric charge resulting in a reduced and retarded rate of heat release as it is demonstrated in Fig. 27.

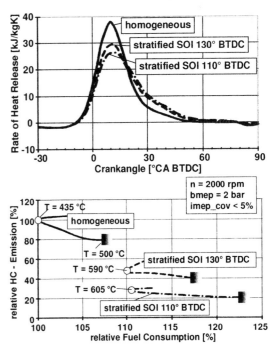

Fig. 27: Stratified stoichiometric operation as a catalyst light off strategy for CC TWC + NOx trap systems

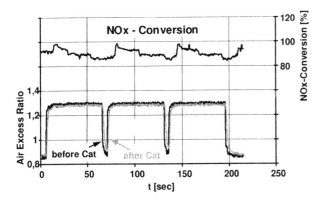

Fig. 26: NOx conversion characteristics of a NOx storage catalyst [15]

OPERATIONAL STRATEGY

The GDI engine with its different modes of operation enables a variety of concepts from homogeneous stoichiometric operation [4] for low emission up to a variable concept with the target of maximum fuel economy improvement [1;2;3;5;6;7;8].

The concept of interest here is the fuel economy-oriented system layout. Since the dominant fuel economy potential of the GDI engine is based on the reduction of pumping losses at part load by stratified operation, this mode should be applied to the highest possible extend in the engine speed and load map. Limitations of stratified operation are imposed by its reduced fuel economy advantage at increasing engine loads as well as by the difficulty to design the stratified combustion system for the entire speed range, a fact that leads to increasing combustion stability problems at high engine speeds (>3500 rpm). An evaluation of the fuel economy potential of the engine as well as its NOx emission level in all 3 operational modes (stratified lean, homogeneous lean and homogeneous stoichiometric) yields clear guidelines for the distribution of modes in the engine map. Fig. 28 shows fuel economy potential and NOx emission level in a load section at 2000 rpm for a specific GDI engine, revealing the following characteristics: stratified operation has the maximum fuel economy potential in the low part load up to about 4 bar bmep.

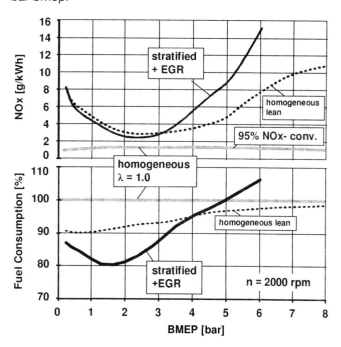

Fig. 28: Fuel economy benefit and NOx emission for different operational modes

An intermediate range between 3 and 5 bar, stratified and homogeneous lean operation provide almost the same fuel economy potential while homogeneous lean operation is clearly superior in the entire upper part load.

The decision between stratified and homogeneous lean operation in the medium part load range therefore has to be guided by the NOx emission characteristics. These are favoring homogeneous lean operation above 3 bar bmep. A maximum fuel economy would therefore be achieved with stratified lean operation up to 4 bar bmep, homogeneous lean operation above 4 bar bmep with its extension to higher engine loads only limited by exhaust gas aftertreatment demands, exhaust gas temperature limitation or dynamic aspects. The situation in Fig. 28 is characteristic for a specific engine and combustion system. Similar investigation at different engine speeds yields the final stationary engine map of the operational mode (Fig. 29), indicated by the air excess ratio.

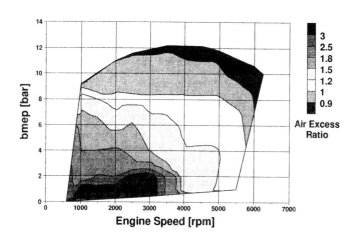

Fig. 29: Air excess ratio map of a GDI engine as a result of operational strategy determination

ENGINE MANAGEMENT

The engine management system for the GDI engine plays the key role to make the best possible use of the available operational modes under any driving condition. A powerful base to meet this target is a torque based engine management system [11;12], which fully decouples the drivers demand by the accelerator from any load control parameter like the throttle position. The drivers demand becomes only one of various inputs to the torque management besides other systems like anti skid systems and vehicle dynamics control systems or speed cut-off. The management of transients can therefore take into account the requirements of the engines operational modes as well as the transitions between these modes which usually need dedicated functions in order to maintain good vehicle driveability. A criterion for the quality of such a system regarding fuel economy is the amount of stratified lean operation that can be achieved in a test cycle. In Fig. 30, the accumulated distribution of released energy over different air excess ratios is presented as it could be measured on a GDI production vehicle.

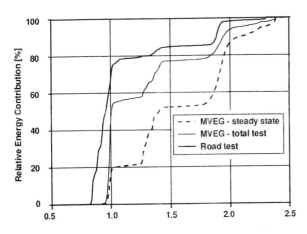

Fig. 30: Influence of transients and driver characteristics on GDI operational mode control

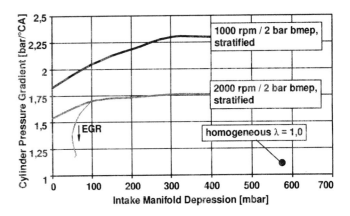

Fig. 31: Cylinder pressure gradient for stratified vs. homogeneous operation

This figure clearly reveals on one hand the contributions of the different operational modes which are detected as distinct steps over the air excess ratio (homogeneous mode at a.e.r. = 1.0, homogeneous lean at a.e.r. = 1.3 - 1.4 and stratified lean at a.e.r. = 1.8 - 2.5).

On the other hand, significant changes of this distribution can be observed from evaluation of only steady state phases with almost 50% stratified lean operation to the complete MVEG test cycle with a remaining 22% of stratified lean operation finally to real road test with only 16% stratified lean operation and a detectable contribution of homogeneous rich operation during accelerations and high speed phases.

Here it is evident that one big challenge for the engine management system is to maintain stratified lean operation to the highest possible extend during all kinds of driving conditions for further reduction of vehicle fuel consumption and drivers influence on fuel consumption.

NVH

In contrary to the MPFI engine, GDI engine acoustics are influenced by the typical non-linearity between engine load and throttle position, especially during instantaneous air flow changes due to switching between operational modes without load step as it is required for instance during DeNOx catalyst purge. Acoustic consequences related to the operational mode of the GDI engine are caused by the strong difference in air mass flow between throttled and stratified part load operation as well as the differences in combustion characteristics that end up in increased pressure gradients for stratified combustion. This can cause an increased structural noise excitation. As shown in Fig. 31, withour EGR pressure gradients can reach more than twice the level of throttled homogeneous operation while high EGR rates reduce the maximum pressure gradient again to moderate levels only slightly above homogeneous values. As such high EGR rates are mandatory from emission requirements, the increase of combustion noise is not a critical issue for GDI engines calibrated for low emissions.

A similar tendency can be evaluated for the intake orifice noise of the GDI engine. Here, completely unthrottled operation can increase the noise by up to 3 dB(A) and only with substantial throttling, the noise level approaches the value of throttled homogeneous operation (Fig. 32).

The impact of these aspects on final engine operation is reduced : Entirely unthrottled operation in stratified mode is more or less unrealistic since there are other arguments like minimum exhaust gas temperature and EGR requiring a minimum throttling even under stratified part load operation. Remaining acoustic effects are treated by careful layout of the intake manifold system by means of gas dynamic simulation.

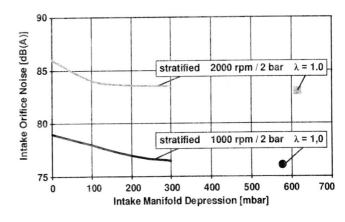

Fig. 32: Intake orifice noise for stratified and homogeneous operation

DRIVEABILITY

A dominant issue in GDI engine and vehicle development is the calibration of engine management functions related to the switching between operational modes during driver-demanded load changes and mode transitions due to DeNOx catalyst purge. With constant throttle positions strong torque variations occur during purging. Drive by wire engine control units change the throttle position while purging and try to keep the torque on a constant level. Especially these mode transitions have to be calibrated with a minimum of driveability influence since they are most relevant during constant driving periods where engine smoothness is recognized to an especially high extend.

For such purposes, AVL applies a newly developed measurement and evaluation system AVL - DRIVE™ [14] which enables the detection, registration and rating of relevant real driving events on line and in real time. Integration and individual averaging of all single events leads to an overall rating of the vehicle that can be carried out load and speed-specific. This leads to a typical driveability map containing a rating between a maximum of 10 (excellent for all drivers) and 5 (very disturbing for all drivers). In Fig. 33. the driveability map for a GDI production vehicle is shown where the critical area of stratified operation can be clearly identified while it has to acknowledged that the overall rating is on a good level within the entire engine map.

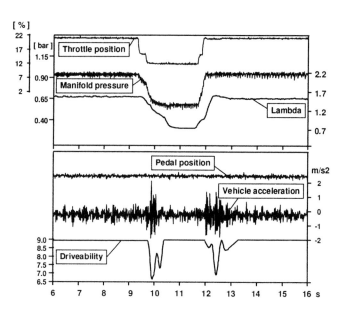

Fig. 34: Driveability measurement of a DeNOx catalyst purge event

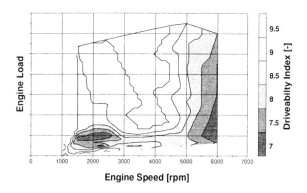

Fig. 33: Driveability rating of a GDI production vehicle in the engine map

The detection and evaluation of driving events is based on the permanent acquisition of vehicle parameters like speed, throttle position, acceleration and the longitudinal shuffle as one of the most important parameters. From these vehicle parameters driveability relevant driving modes are detected. In a second step the driveability is calculated.

In Fig. 34, a typical trace for a DeNOx catalyst purge is shown, demonstrating the transition in air excess ratio and throttle position together with the related longitudinal shuffle of the vehicle caused by the event. During purge process longitudinal shuffle can be detected by the driver. Constant speed driveability is shown as a calculated on line channel, which can be stored equal to other usual measuring values.

CONCLUSION

Gasoline direct injection engine development requires the complete system approach from design considerations to vehicle driveability, incorporating the intense application of advanced development tools. This process has been performed within the development of AVL's DGI™ engine. Evaluations of the variety of different combustion system approaches have led to the concept of a wall guided swirl based combustion system for high pressure liquid fuel injection with minimized spray wall impingement and increased mixture formation time scales. The system has been developed for high fuel economy potential at stratified part load as well as for high EGR acceptance for low engine-out emissions. Sensitivity on ignition and injection timing could be kept extremely low as a base for simplified engine management. Development tools like CFD simulation, PLIF visualization and tomographic combustion analysis were applied for assessment of the combustion system performance.

For exhaust gas aftertreatment, the approach of a NOx storage catalyst combined with close coupled three way catalyst is favored for the next steps of system integration and vehicle calibration. For engine management, a torque based EMS is applied due to its advantages in operational mode utilization. A new driveability measurement system is applied for the investigation of EMS calibration.

REFERENCES

[1] Ando, H.; Noma, K.; Iida, K.; Nakayama, O.; Yamauchi, T.: *Mitsubishi GDI Engine - Strategies to meet the European Requirements*, Congress: Engine and Environment, Graz, 4.-5-9. 1997

[2] Iwamoto, Y.; Noma, K.; Yamauchi, T.; Nakayama, O.: *Development of Gasoline Direct Injection Engine*, SAE 970541

[3] Nohira, H; Ito, S.: *Development of Toyota`s Direct Injection Gasoline Engine*, Congress: Engine and Environment, Graz , 4.-5-9. 1997

[4] Anderson, R.W.; Brehob, D.D., Yang, J.;Vallance; J.K.; Whiteaker, R.M.: *A New Direct Injection Spark Ignition (DISI) Combustion System for Low Emissions*, P0201 XXVI. FISITA Kongress Prag, 17.-21.6. 1996

[5] Grigo, M.; Schwaderlapp, M.; Wolters, P.: *Charge Motion controlled Combustion System for a Direct Injection SI Engine*, 18.Int. Vienna Engine Symposium, 24.-25.4.1997

[6] Jackson, N.S.; Stokes, J.; Lake, T.H; Whitaker, P.A.: *Research and Development of Advanced Direct Injection Gasoline Engines*, 18.Int.Vienna Engine Symposium, 24.-25.4.1997

[7] Karl, G.; Kemmler, R.; Bargende, M.; Abthoff, J.: *Analysis of a DI Gasoline Engine*, SAE 970624

[8] Fraidl, G. K.; Piock, W. F.; Wirth, M.: *Gasoline Direct Injection : Actual Trends and Future Strategies for Injection and Combustion Systems*, SAE 960465

[9] Philipp, H.; Plimon,A.;Fernitz,G.; Hirsch, A.; Fraidl, G.K.; Winklhofer, E.: *A Tomographic Camera System for Combustion Diagnostics in SI Engines*, SAE 950681

[10] Philipp, H.; Fraidl, G.K.; Kapus, P.; Winklhofer, E.*: Flame Visualization in Standard SI Engines - Results of a Tomographic Combustion Analysis*, SAE 970870

[11] Moser, W; Mentgen, D.; Rembold, H.: *Gasoline Direct Injection - a New Challenge for Future Engine Management Systems, Part I and II*, MTZ 58 (1997) 9/10

[12] Stocker, H.: *Gasoline Direct Injection and Engine Management - Challenge and Implementation*, Congress: Engine and Environment, Graz , 4.-5-9. 1997

[13] Strehlau, W.; Leyrer, J.; Lox, E.S.; Kreuzer, T.; Hori, M.; Hoffmann, M.: *New Developments in Lean NOx Catalysis for Gasoline Fuelled Passenger Cars in Europe*, SAE 962047.

[14] List, H.; Schoeggl, P.: *Objective Evaluation of Vehicle Driveability*, SAE 980204

[15] Holy, G.: *Nox-Adsorber Evaluation*, AVL Internal Report 1997

Direct Injection Gasoline Engines – Combustion and Design

José Geiger, Michael Grigo, Oliver Lang and Peter Wolters
FEV Motorentechnik GmbH & Co. KG, Aachen. Germany

Patrick Hupperich
FEV Engine Technology

ABSTRACT

The charge motion controlled combustion concept for SI engines with direct fuel injection exhibits an excellent fuel economy and emission potential in comparison with other DI combustion concepts. It realizes a stable combustion behavior all over the engine map. Because injection and ignition timing has little bearing on emission and ignition safety, the new concept can be easily applied under DI specific operational conditions. The combination of fired engine tests and optical investigations with CFD calculations enables an efficient process optimization under the boundary conditions as imposed by the respective design. The high EGR tolerance enables a large reduction of NO_x emission, which is the expected basic requirement to meet future emission standards.

In addition to favorable part load behavior, the new combustion concept also displays all of the characteristics for a good full load behavior. Its compact central combustion chamber design perfectly meets the requirements of homogeneous operation. Further, the use of roller finger followers enables a favorable cylinder head design. meeting all requirements of the combustion concept and ensuring minimal valve train friction.

INTRODUCTION

Since the first vehicles powered by direct injected gasoline engines were introduced into the Japanese market. the worldwide engine industry has been working to further develop the technology. These activities cover a wide field of different development tasks.

The first task has been the development of fuel injection equipment. The main goals of such programs have been to improve spray quality, metering accuracy and durability of the fuel injectors. Moreover, the dimensions of all necessary system components have to be minimized and their production has to be enabled at low costs.

The engine controller has to be modified in order to fulfill the special requirements of the different operational modes. Besides the claimed fuel economy improvement, the operational strategy must take into account both driveability and comfort, as well as on-board diagnostics feasibility.

The DeNOx aftertreatment technology of lean exhaust gas is an additional field of research and development to be addressed. The NOx adsorption technology has reached a high state of development. Yet, there is no final confirmation, that this solution will fulfill all future emission standards under projected boundary conditions (i.e., fuel qualities, typical traffic situations, long term durability requirements).

Also in terms of engine design the DI technology is creating new tasks. The application of DI technology has resulted in changes to the overall process of designing an engine. For example, the process of internal mixture formation and charge stratification results in low tolerances of the intake port and combustion chamber shape. Similarly, the application of the fuel pump, the fuel rail, and the injectors in accordance with DI system requirements leads to new design solutions.

Within this framework, DI combustion system specialists are pursuing different concepts in terms of in-cylinder charge motion or spray orientation. Currently, at least by all published accounts, there is no single common design trend. However, the most noticeable differences revolve around the process of internal mixture formation and charge stratification. This paper reports on a new charge motion controlled DI combustion system, that has recently been developed at FEV Motorentechnik. The new system is based on a 4-valve design with pentroof combustion chamber, applying tumble in order to support the mixture preparation and its transport towards the central spark plug.

CLASSIFICATION OF DI COMBUSTION SYSTEMS

Figure 1 shows a classification of the different DI combustion systems. It is based on various conceptual models which have been investigated at FEV. The three basic systems are called jet controlled, wall controlled and charge motion controlled, per the different mixture formation processes. Although in practice there often is an overlap between these basic systems, this classification can contribute to a better understanding of the mixture formation and stratification process.

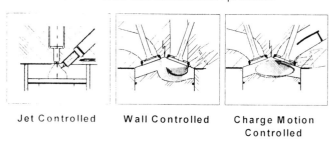

Jet Controlled   Wall Controlled   Charge Motion Controlled

Figure 1: Combustion System Classification

/1/ and /2/ are reporting on concepts that are typical of jet controlled systems. There is a close arrangement of spark plug and fuel injector that results in a direct jet of liquid fuel directed toward the spark plug electrode, with negative impacts on spark plug life. Detailed investigations show very steep stratification profiles of such jet controlled concepts. Consequently, only a small region exists where an ignitable air fuel mixture is furnished near the spark plug /3/. This not only causes problems related to ignition stability, but also extended areas of extremely rich air-fuel mixture which results in debilitating soot formation. Compared to wall controlled systems, the mixture formation process cannot be assisted by means of charge motion as this would create additional problems. Since the spray geometry has to be kept in a small range to be optimized in the interest of ignition stability over a wide speed load regime, stratified operation is restricted to a relatively small area of the map due to limited soot levels in jet controlled systems.

Recently introduced Japanese solutions represent the group of wall controlled systems. From a general point of view, there is no difference between swirl and tumble based systems. The mixture formation process of such wall controlled systems is characterized by fuel injection initially directed toward a piston bowl, and a consecutive interaction with an intensive charge motion that assists the mixture's formation and its transport to the spark plug.

Kume et al. /4/ depict both the formation and oxidation of soot during the internal combustion process. This paper explains in detail the interactive mechanisms of wall wetting and fuel vaporization, assisted by an intensive in-cylinder charge motion. FEV's own research work on wall controlled DI combustion systems confirms this experience and shows that excessive soot formation can be avoided by an extensive optimization of the fuel injection process and combustion chamber design. However, the typical wall wetting, in combination with a large piston bowl surface leads to high levels of THC

emissions. For wall controlled concepts the position of the spark plug is near the edge of the main combustion chamber. In this configuration, the piston bowl is off-center. This boundary condition leads away from the optimal combustion chamber design of SI engines. Consequently, there are disadvantages related to the full load behavior, when the DI engine is running with early injection and a homogeneous mixture.

The new FEV DI concept follows the principle of charge motion controlled mixture formation. Figure 2 shows a schematic cross-section of the combustion chamber of a 4 valve engine. The distance between the fuel injector and the spark plug is significantly larger compared to the jet controlled concept. The direction of the fuel injector points to the center of the combustion chamber and closer to the spark plug, although this does not results in a direct spray toward the electrodes. At stratified operation, the injected fuel spray is deflected by a distinct in-cylinder charge motion. As an additional effect, this charge motion is assisting the mixture of the fuel spray with air, which helps to create an optimized stratification profile. The high tumble intensity is realized by means of a variable tumble device.

Stratified Operation

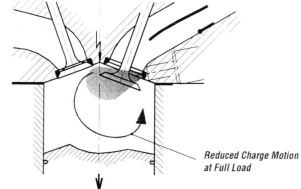

Charge Motion Individually Optimized for Different Operating Conditions

Long Free Spray Length Minimized Wall Impingement

Compact Combustion Chamber in Center

Homogeneous Operation

Reduced Charge Motion at Full Load

Figure 2: Charge Motion Controlled Combustion System

The potential intake port/combustion chamber configuration as shown in Figure 3 features a continuous variable tumble system (CVTS), which is mounted at the intake port entry. A horizontal sheet divides the intake port. This configuration enables intensified charge motion during stratified operation through deactivation of the lower part of the port. The CVTS system had originally been the key issue for an FEV lean burn engine /5/.

The arrangement of the fuel injector below the intake ports is favorable for the assembly and package features. This position also allows maximum flow capacity of the intake port. Consequently, an optimal free spray length can be realized in terms of wall wetting. The related HC emission behavior represents a main advantage compared to wall controlled systems.

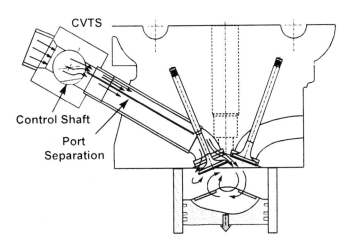

Figure 3: Continuous Variable Tumble System

At homogeneous operation the tumble intensity has to be reduced. The deactivated CVTS then allows a high volumetric efficiency, which contributes to an excellent full load behavior. The compact combustion chamber design, coupled with the central spark plug position, are typical features of the new FEV DISI concept and result in exceptional full load performance.

Compared to alternative DI engine concepts, the new charge motion controlled combustion system shows an excellent fuel economy and emission potential. In Figure 4, the state of development of the different combustion systems is shown in comparison to an optimized "Best of Class" conventional SI engine with multipoint injection. Since all three investigated DI combustion systems were realized on the same single-cylinder test engine, the results can be compared without any restrictions. The figure shows a 10% fuel economy advantage of the new charge motion controlled system over the jet controlled and wall controlled concepts. The following chapter reports on a more detailed analysis of the new combustion system, assisted by means of thermodynamic analysis of the combustion process.

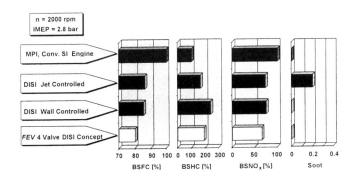

Figure 4: Comparison of Operational Behavior

CHARGE MOTION CONTROLLED COMBUSTION SYSTEM

The following results are based on single-cylinder engine tests, carried out on a FEV Systemmotor (modular basic research test engine). This engine was equipped with a modified series production cylinder head (4-cylinder 16-valve engine). The fuel injector was integrated below the intake port. The intake port was realized by a sleeve design solution with a port configuration adapted in detail to the DISI combustion system requirements.

Figure 5 shows the indicated fuel consumption, the standard deviation of imep and the HC and soot emissions as a function of tumble intensity. In this instance, tumble intensity is controlled by means of the CVTS. The CVTS position "tumble min." represents an intake port flow, where the tumble motion starts at valve lifts exceeding 7 mm, while "tumble max." shows this effect at a mere 2.5 mm valve lift.

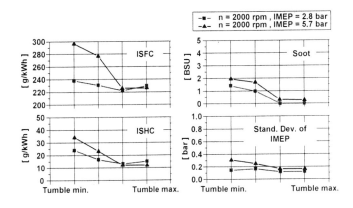

Figure 5: Variation of Tumble Intensity

The results show that the operational behavior is significantly influenced by the tumble intensity. For both investigated part load operation points, the results show a common optimum. Further investigations at different engine speed and load are confirming the same optimum tumble intensity. Therefore, it is not necessary to use a continuous, but rather a simple two-stage variable tumble system that can be realized by means of a simple control device for easy application. A variation of the fuel rail pressure shows that the optimum level of fuel pressure is not largely affected by the engine operation point. Therefore, pressure control as a function of engine torque is not necessary and adaption to engine speed is done mainly due to exact fuel metering purposes.

As a part of combustion system development, the analysis and optimization of the in-cylinder charge motion is supported by transient, three-dimensional calculations. The required meshes are much more complex than those for conventional CIDI and SI engines. Also, for the simulation of transient processes, the dynamics of piston and intake and exhaust valves has to be considered.

Due to inaccuracies in the simulation of injection processes, realistic boundary conditions for the CFD calculations have to defined initially. Based on the results of injection tests in a stationary pressure chamber, essential injector characteristics (spray angles, spray cone, spray dispersion, spray penetration) can be identified. Additional injector characteristics are requested from the supplier, such as frequency rankings of droplet size and droplet velocity, impact of rail pressure on droplet size, as well as opening and closing times of the injector.

As an example, the results of CFD calculations of in-cylinder charge motion and AFR distribution are shown in Figures 6 and 7 /6/. The operation at high tumble intensity (right part of Figures) causes a quicker vaporization of the liquid fuel and the transport of vaporized fuel toward the spark plug, which is located in the top of the combustion chamber.

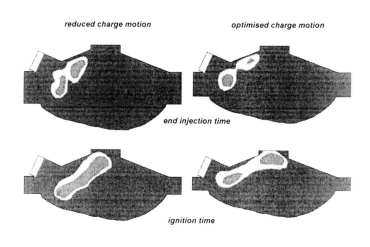

Figure 7: Comparison of AFR Distributions

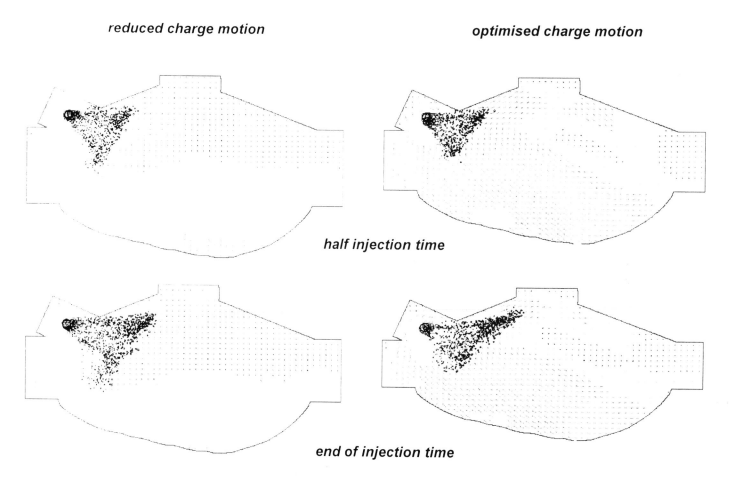

Figure 6: Comparison of Velocity Distributions

n = 2,000 rpm, imep = 2.8 bar, spark advance = 30 deg CA

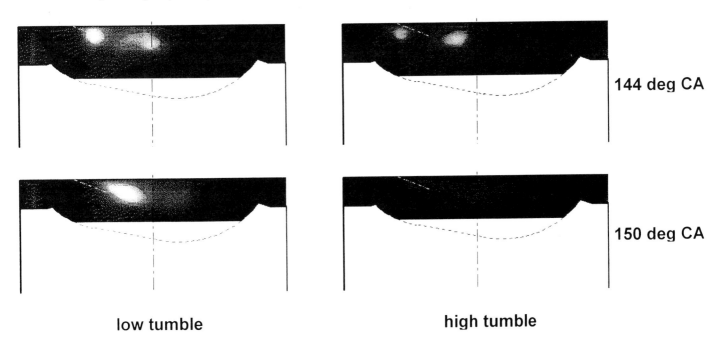

low tumble   high tumble

Figure 8: Tumble Influence on Mixture Preparation

Figure 8 shows the results of a spray visualization by the Planar Laser Induced Fluorescence (PLIF) method. This method uses a fluorescent fuel additive that has only a minor impact on the fuel characteristics themselves. A two-dimensional picture of the mixture distribution with high resolution is obtainable by the application of the PLIF method. A comparison between pictures 6, 7 and 8 reveals a high level of conformity between CFD results and optical measurements.

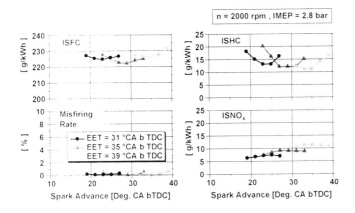

Figure 9: Operational Stability

Figure 9 shows the results of a variation in injection timing. The end of injection is characterized by the end of electrical energizing (EET) of the fuel injector. The misfiring rate is based on 1,000 consecutive cycles. The flat curves are documenting the stable operational behavior of the new engine concept and show that this concept is easy for application under the boundary conditions of vehicle operation.

The fuel economy potential of the charge motion controlled combustion system is shown in Figure 10 within the engine map. Compared to a well optimized conventional MPI engine, there is a benefit of up to 25% in fuel consumption due to throttle free and stratified operation. At a bmep of 2 bar and within a range of engine speed from 1,000 to 3,000 rpm, the bsfc of the FEV DI engine is within a range of 310 to 330 g/kWh (assumption: fmep @ 2,000 rpm = 0.83 bar). This is even better than the fuel consumption level of today's IDI Diesel engines. Moreover, the application of variable charge motion enables a homogeneous lean burn operation at higher part load, offering additional fuel savings up to a bmep level of 7 bar.

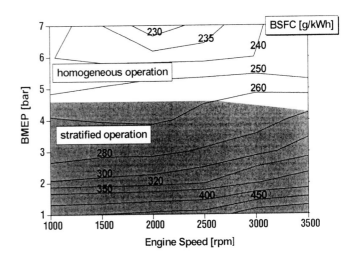

Figure 10: BSFC Map at Part Load Operation

The measured fuel economy advantages of the new combustion system (see Figure 4) were further investigated by means of thermodynamical analysis of the combustion process. Figure 11 shows the average pressure trace of 300 consecutive cycles, the calculated burning function and the burning velocity for all three investigated DI concepts.

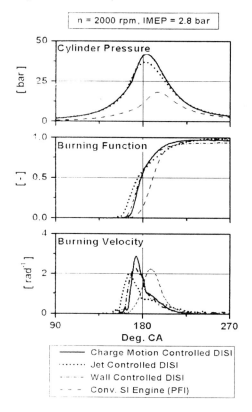

Figure 11: Combustion Analysis

For the jet controlled combustion system, there is an intensive heat release at the very beginning of the combustion cycle. Once half of the fuel is burned (x_B = 50%), however, the combustion is significantly retarded. FEV's experience has been that this behavior is typical for this type of DI combustion system. The wall controlled concept is showing a similarly frozen combustion process which remains uncompleted. This result correlates with the measured higher HC emission level (Figure 4), which is typical for the wall controlled concept.

The burning function of the charge motion controlled combustion system shows the best result. During the second half of the combustion process, heat release is continuous. Compared to the two alternative combustion systems, FEV's new engine concept has the latest position of 50% mass fraction burned. Nevertheless, there would be additional potential if the position of 50% mass fraction burned could be shifted to a later phase of the process, as it is the case with conventional MPI engines.

A more detailed analysis of the process is given in Figures 12 and 13. In addition to cylinder pressure traces, thermal efficiency is depicted as a function of degree crank angle during the complete process cycle.

This value is defined as the heat content of the added fuel per cycle divided by the integrated piston work p*dV. In the lowest part of the figures this value is related to the reference process.

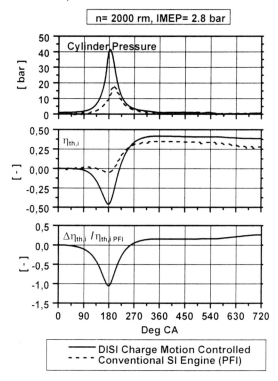

Figure 12: Thermodynamical Process Analysis

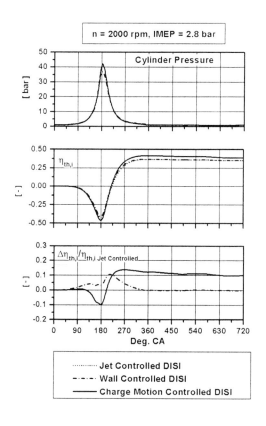

Figure 13: Process Analysis of Different DI Systems

Figure 12 shows the analysis of the new charge motion controlled DI system, while the reference is represented by the conventional MPI engine's process. During the compression stroke, the throttle free DI operation causes increased compression work (negative values of $\eta_{th,i}$). The advantages of the stratified lean burn operation become apparent during the combustion stroke. Upon completion (360 deg. CA), the DI process reaches an efficiency benefit of 16%, due to superior heat release. Minimized pumping losses further pushes this figure to 27% during the charge exchange process.

In Figure 13, all three investigated DI concepts are compared to each other, while the charge motion controlled system utilized as a benchmark. The FEV concept shows a higher rate of compression due to the higher volumetric efficiency of this concept, which in turn is a result of a better intake port design. The main advantage of this combustion system is revealed during the combustion stroke, and is caused by the better heat release function. The result is an efficiency benefit after the combustion phase of approximately 12%.

In order to fulfill actual and future emission standards the NOx engine out emissions must be minimized by means of exhaust gas recirculation. Figure 14 shows the influence of an external EGR at part load (engine speed = 2,000 rpm, imep = 2.8 bar). The FEV concept demonstrates an excellent EGR tolerance. EGR ratios of up to 40% show no significant disadvantage in thermal efficiency and running stability, although a slight throttling has to be applied in order to create differential pressure. This allows a reduction of NOx emissions by more than 70%.

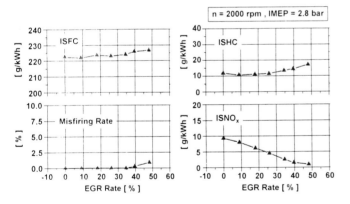

Figure 14: Exhaust Gas Recirculation

DESIGN SOLUTIONS

Integration of the DI technology according to the new charge motion controlled concept mandates some changes of the cylinder head design. There are different design solutions possible. All modification concepts discussed below are convenient with current design and production requirements. However, there are different impacts on cylinder head machining and assembly.

Version A (Figure 15) shows a concept requiring minimal change in the existing processing line. Based on an existing DOHC cylinder head layout with valve tappets, hydraulic lash adjusters and a valve angle of 25°, a modification of the intake port arrangement allows the integration of the fuel injector below. The advantage of only small production modifications being necessary has to be evaluated towards the disadvantages related to the intake port design and the fuel spray distribution within the combustion chamber. Due to the unfavorable angle and position of the fuel injector, a partial wall wetting of the cylinder head bottom cannot be avoided.

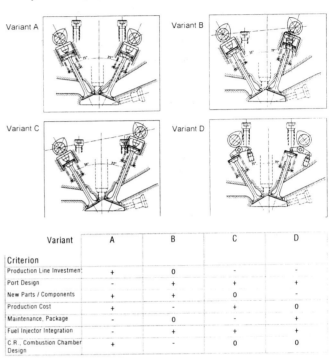

Variant Criterion	A	B	C	D
Production Line Investment	+	0	-	-
Port Design	-	+	+	+
New Parts / Components	+	+	0	-
Production Cost	+	-	+	0
Maintenance, Package	-	0	-	+
Fuel Injector Integration	-	+	+	+
C.R., Combustion Chamber Design	+	-	0	0

Figure 15: Cylinder Head Design (Table 1)

In variant B, the intake valve angle has been reduced to 19°. This allows an improved arrangement of the intake ports. The resulting non-symmetrical combustion chamber enables a flush-mounted integration of the fuel injector without restrictions to the free spray length. As the cross section shows, there are some problems with the pre-assembly of the cylinder head, because the cylinder head bolt axis is covered by the camshaft. In order to avoid this disadvantage, in variant C the intake camshaft is shifted to a different position by means of an optimized valve angle (22°). However, this solution results in the cylinder head gasket plane and the valve cover plane being incapable of achieving a position parallel to one another during the machining process. Moreover, the cylinder head height of variant C is slightly increased, which can create some problems with packaging.

A valve train with roller finger follower (variant D) offers better conditions for the pre-assembly of the cylinder head. In this configuration, the camshafts can be shifted to a position, that allows full access to the cylinder head bolts. Moreover, a parallel orientation of both the cylinder head gasket and the valve cover plane can be achieved.

Table 1 shows a comparison of the different aspects:

- At symmetrical valve angle, there are necessary only small modifications of the production line.
- This solution has disadvantages concerning the optimal intake port geometry and injector position.
- Starting from a basic valve train design with tappets, the roller finger follower variant requires numerous new parts.
- All variants with valve tappets are preferred in terms of production costs.
- The best compromise related to assembly and package features is furnished by the roller finger follower variant.
- Non-symmetrical valve angle arrangements allow optimal fuel injector integration.
- Combining high compression ratio with optimal combustion chamber design at non-symmetrical valve arrangements requires a special piston crown shape.

The material presented shows that the new engine concept can even be realized on a series production engine with only small design modifications. However, the geometric integration of the fuel injector mandates a general design review. With respect to the above mentioned requirements the variant with roller finger follower valve train is the optimal solution, because thermodynamics, design and mechanics can be optimized simultaneously.

CONCLUSIONS

The new charge motion controlled combustion system for gasoline engines with direct fuel injection shows an excellent fuel economy potential compared to other DI concepts. It realizes a stable operation all over the entire speed/load regime. The influence of injection timing and spark advance on emissions and ignition stability is low compared to other investigated concepts. This enables an easy application of the new engine under the specific boundary conditions of lean burn stratified operation.

The high EGR tolerance allows a significant reduction of the engine out NOx emissions, which promises to meet future emission standards.

The new engine concept combines both excellent part load operation behavior and full load performance features. The compact and centrally-positioned combustion chamber enables thermodynamically optimal and knock-free combustion. Therefore, a vehicle application with a modified gear ratio can exhibit further fuel economy benefits without losses in on-road performance.

A comprehensive understanding of the in-cylinder flow and injection process phenomena during intake and compression strokes is essential to optimize the DISI combustion system. Besides optical investigation methods (PLIF), CFD calculations are a useful development tool to accomplish this goal. The combination of fired engine tests and optical investigations with CFD calculations enables an efficient process optimization under the boundary conditions as imposed by the respective design.

The new concept can be realized by means of proven design solutions. A valve train layout with roller finger followers is the preferred solution, taking into account friction losses as well as production aspects.

The new charge motion controlled DI combustion system has been demonstrated on a 4-valve engine layout. It can easily be transferred also to 3-valve engines. FEV has also successfully realized a 2-valve solution that is based on the same mechanisms as the presented engine concepts.

REFERENCES

1. H. Stutzenberger, Ch. Preussner, J. Gerhardt
   Benzin-Direkteinspritzung für Ottomotoren -
   Entwicklungsstand und Ausblick
   17. Internationales Wiener Motorensymposium, 1996
2. G. Karl, J. Abthoff, M. Bargende, R. Kemmler, M. Kühn, G. Bubeck
   Thermodynamische Analyse eines direkteinspritzenden Ottomotors
   17. Internationales Wiener Motorensymposium, 1996
3. J. Geiger, P. Wolters, M. Grigo
   Meßmethoden für die Entwicklung ottomotorischer Brennverfahren mit innerer Gemischbildung
   Tagung im HdT "Motorische Verbrennung, aktuelle Probleme und moderne Lösungsansätze", III. Tagung in Essen, März 1997
4. T. Kume, Y. Iwamoto, K. Iida, M. Murakami, K. Akishino, H. Ando
   Combustion Control Technologies for Direct Injection SI Engine
   SAE Paper 960600, 1996
5. H.-J. Neußer, J. Geiger
   Continuous Variable Tumble - A New Concept for Future Lean Burn Engines
   SAE Paper 960607, 1996
6. O. Lang, W. Willems, M. Grigo
   Theoretical and Experimental Investigations on a Direct-Injected S.I. Engine
   7. Aachener Kolloquium Fahrzeug- und Motorentechnik, 1998

Fuel Spray Characteristics for Direct-Injection Gasoline Engines

Fuel Spray Characteristics for Direct-Injection Gasoline Engines

Fuel spray characteristics are of great importance to direct-injection gasoline combustion systems. Parameters such as the spray cone angle, mean droplet size, spray penetration, and fuel delivery rate are known to be critical, and the optimum matching of these parameters to the air flow field and spark location usually are at the heart of any direct-injection gasoline combustion system development project. This is because of the vital dependence of the subsequent combustion and emissions formation processes on the primary spray characteristics, that is, the ability of the system to prepare and distribute the fuel optimally, both spatially and temporally. A large database has been established correlating diesel fuel sprays to combustion. However, this database cannot be used to predict the characteristics of DI gasoline sprays because of significant differences in fuel properties, injection pressure levels, droplet velocities and size ranges, ambient pressure and temperature levels, and droplet drag regimes.

The required spray characteristics for a DI gasoline engine vary significantly with engine operating conditions. For fuel injection during the induction event, a widely dispersed fuel spray is generally required to achieve increased specific power through improved air utilization. For injection during the compression stroke, a compact spray with reduced penetration is preferred to achieve a highly stratified mixture distribution. At the same time, the spray should be well atomized, because the fuel must vaporize in a short time. Meeting these kinds of diverse requirements over the entire engine-operating map is very difficult in practice. It may thus be seen that the correlation and predictive characterization of the fuel spray from direct-injection gasoline injectors represents a new and important research area. In fact, during the last decade, many R&D efforts have been directed toward this area.

This section of the book provides a selection of SAE technical papers that address fuel spray issues concerning direct-injection spark-ignition engines. The papers are categorized into the following topics: experimental characterization of direct-injection gasoline sprays from both single-fluid high-pressure swirl injectors and pulse-pressurized air-assisted injectors using various diagnostic techniques; modeling of nozzle flow and its subsequent impact on spray formation and development; and CFD analysis of spray structure under various operating conditions. These papers provide state-of-the-art technical data and observations of direct-injection gasoline sprays for researchers and engineers who are engaged in R&D in the area of direct-injection gasoline combustion systems.

970629

Transient Spray Characteristics of a Direct-Injection Spark-Ignited Fuel Injector

S. E. Parrish and P. V. Farrell
University of Wisconsin-Madison

ABSTRACT

This paper describes the transient spray characteristics of a high pressure, single fluid injector, intended for use in a direct-injection spark-ignited (DISI) engine. The injector was a single hole, pintle type injector and was electronically controlled. A variety of measurement diagnostics, including full-field imaging and line-of-sight diffraction based particle sizing were employed for spray characterization. Transient patternator measurements were also performed to obtain temporally resolved average mass flux distributions.

Particle size and obscuration measurements were performed at three locations in the spray and at three injection pressures: 3.45 MPa (500 psi), 4.83 Mpa (700 psi), and 6.21 MPa (900 psi).

Results of the spray imaging experiments indicated that the spray shapes varied with time after the start of injection and contained a leading mass, or slug along the center line of the spray. The penetration of the leading mass was faster than the penetration of the leading edge of the main body. Spray shapes were found to be generally similar over the pressure range examined. As expected, penetration of both the leading mass and the leading edge of the main body increased with injection pressure.

Particle size measurements indicated that the Sauter Mean Diameter (SMD) of the leading mass and leading edge of the main body of the spray were significantly larger than the SMD in other regions of the spray. Peak SMD values were smaller for higher injection pressures, and the change in peak SMD with pressure was more gradual as injection pressure increased.

Transient patternator measurements indicated that transient affects significantly influenced average spatial mass distributions, suggesting that short duration injections are likely to have different spatial distributions from longer duration injections.

INTRODUCTION

There is considerable interest in internal combustion engines that combine some of the best features of compression ignition (CI) and spark ignition (SI) engines. One of these features is direct injection of fuel into the combustion chamber. Advantages of this strategy for SI engines include control of in-cylinder fuel quantity and quality, control of timing of fueling, and benefits due to charge cooling.

Direct injection in SI engines has been studied under a variety of themes, most commonly for stratified charge engines [1-3]. A more recent configuration is a Direct Injection Spark Ignited (DISI) engine. In this type of engine, air is inducted into the engine cylinder, and fuel is injected directly into the cylinder. At the appropriate time, the fuel charge is ignited by a spark discharge. This type of engine offers the possibility of multi-mode operation in which early injection can be used to produce a homogenous charge while late injection can be used to produce a stratified charge. Kume et al. have reported recent work on this type of engine [4]. Operating in a stratified mode, this type of engine has some unique advantages compared to conventional SI and CI engines. A few of these advantages include: higher compression ratios than conventional SI engines, control over the ignition process, good cold starting characteristics, the possibility of multi-fuel capabilities, and the ability to operate with very lean mixtures [1].

Due to the short time available for fuel-air mixing in this type of engine, the characteristics of the fuel spray are likely to play an important role in the DISI engine concept. Recent advancements in injector technology have yielded a variety of new injectors specifically designed for use in DISI engines. This paper describes transient spray characteristics of a high pressure injector which is likely to be typical of those intended for use in DISI engines. While details of the sprays will vary with some injector details, there should be considerable similarity in the spatial and temporal variation of spray characteristics for this injector and other pintle type injectors. The main focus of this work is to characterize the transient characteristics of this type of injector which might be of concern in a DISI engine. The results presented are part of a study involving fuel charge preparation in DISI engines.

EXPERIMENTAL SET UP

The experimental set up consisted of a fuel supply and an injection system, diagnostics for direct imaging of the spray, diagnostics for measuring droplet sizes, and a system for measuring mass flux. A brief description of each of these systems is given in the following sections.

A high pressure, single fluid injector was used for all experiments. The injector was a single hole, pintle type injector and was electronically controlled. Stoddard solvent (a low volatility fluid) was used as the working fluid because of its similarity to gasoline, with reduced safety risk. The injector line pressure was controlled and held constant for any experimental series using a gas charged accumulator. Nitrogen was used as the charging gas, and the gas pressure could be accurately regulated at the selected fuel injection pressure. A series of short, flexible, high pressure hoses connected the accumulator fuel pressure reservoir to the injector.

All measurements reported here were performed by injecting Stoddard fluid into atmospheric air. The injected sprays were collected by a collection receptacle for recycling of the working fluid. The injection pressures used were 3.45, 4.83, and 6.21 MPa (500, 700, and 900 psig).

SPRAY IMAGING

IMAGING SYSTEM - Figure 1 shows a schematic of the imaging system used for acquiring spray images. Images were captured with a noninterlaced CID camera and a frame grabber. A 20 W white light strobe, with a flash duration of approximately 50 msec, was used to illuminate the spray from the side. Control and synchronization of the injector, camera, frame grabber, and strobe was accomplished with a timing card and some logic circuitry.

For each image, a specified time after the start of injection was selected. The timing circuit triggered the strobe at the selected time, and an image was recorded. Each image was recorded from a different injection, although the injector operating conditions were held constant. By selecting an array of times after the start of injection, a set of images indicating how the spray developed, was generated.

SPRAY SHAPES AND PENETRATION - Figure 2 shows spray images at five measurement times after the start of injection, for the three injection pressures. Injection pressures for the images in the top, middle and bottom rows of the figure were 3.45, 4.83, and 6.21 MPa, respectively. The injected mass was the same for all three pressures, 13.1 mg, and is representative of a medium load injection. The rectangles, with three circles within, are visual aids which indicate regions in which droplet sizing and obscuration experiments were

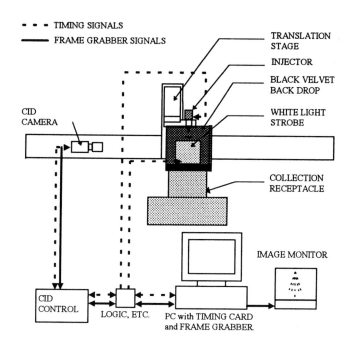

Figure 1. Imaging system schematic.

to be performed. The centers of the three circles are located 38.75 mm (vertically) from the injector tip. These experiments will be discussed in the particle sizing and obscuration section of this paper.

As can be seen from the images, the spray is characterized by a leading mass or slug of liquid which is then followed by the main body of the spray. The leading mass appears to be a coherent slug of liquid at early measurement times; but by 1.5 msec after the start of injection (SOI) it loses its form. A vortex becomes evident at the outer edges of the spray. The vortex development occurs sooner in time and its strength appears to increase as injection pressure increases. The spray becomes "detached" from the injector sooner in time as the injection pressure increases. This is due to the fact that for the injected fuel mass to remain constant, the injection duration has to decrease with injection pressure. The injection duration for each of the injection pressures is noted in figure 2.

Tracking the progress of the leading mass of the spray and the leading "edge" of the main body provides penetration and velocity information. Using images like those shown in figure 2, penetration lengths of the leading slug and the leading edge of the main body were measured. Figure 3 shows the penetration curves for the leading mass (tip) and the leading edge of the main body for the three injection pressures. Each curve terminates between 90 and 100 mm due to the spray feature leaving the field of view. The penetration of the leading mass is significantly faster than the leading edge of the main body. At a given time after the SOI, the spray penetration is greatest for the highest pressure. The smooth lines on the plot are curve fits of the data. Second and third order polynomial regression fitting was used for the tip and leading edge of the main body, respectively. All fits had correlation coefficients, R, greater than 0.999.

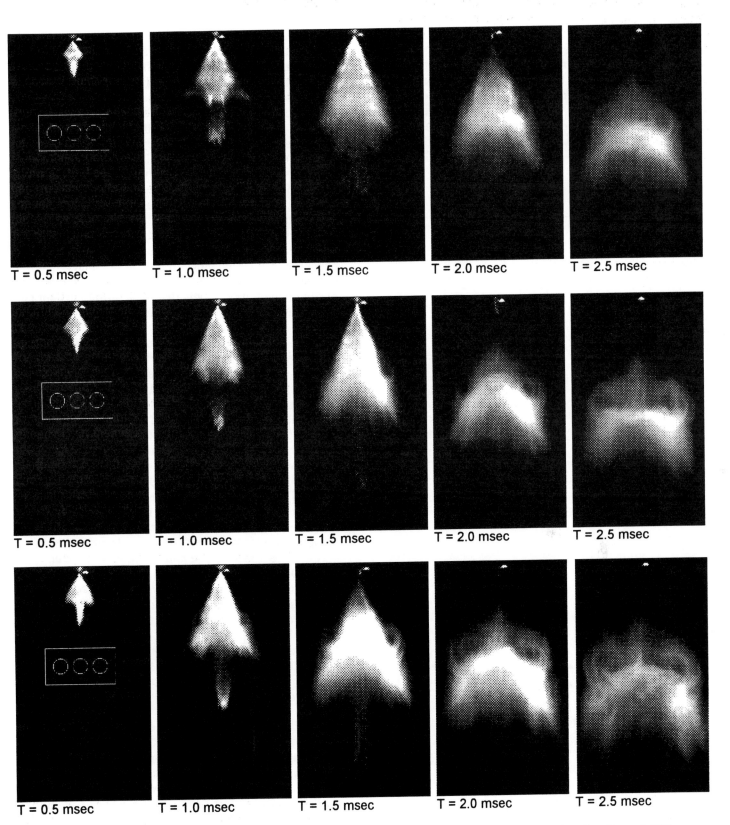

Figure 2. Spray images at various times after the start of injection. Top row - 3.45 Mpa, injection duration - 1.265 msec, Middle row - 4.83 Mpa, Injection duration - 1.060 msec, Bottom row - 6.21 Mpa, Injection duration 0.973 msec. Injected fuel mass: 13.1 mg.

An estimate of the velocity of leading mass and leading edge of the main body is obtained from the slopes, or derivatives of the displacement curves. Figure 4 shows the derivatives of the curve fit penetration curves. The velocity of the leading mass, near the injector tip, ranges from approximately 68 m/sec to 86 m/sec for the pressure range examined. The leading edge of the main body of the spray exhibits velocities of 40 to 53 m/sec for the injection pressures used. The velocities of the leading mass and the leading edge of the main body decrease rapidly with time.

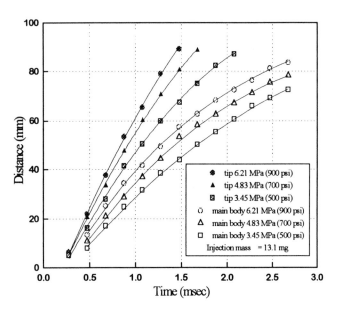

Figure 3. Penetration curves.

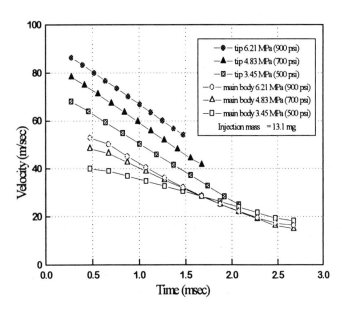

Figure 4. Velocity curves.

DIFFRACTION BASED PARTICLE SIZING AND OBSCURATION

Line-of-sight-particle sizing measurements were performed using a diffraction based system similar to the system developed by Swithenbank et al [5], and later made into a commercial device. The primary differences between the system used here and that of Swithenbank, et al. are the detector and the inversion method. The present system utilizes a CID camera as the detector and employs a Chin-Shifrin analytical inversion [6]. Particle size distribution is the end result of the measurement, which can then be integrated to determine various derived diameters such as the SMD.

In addition to particle size measurements, obscuration measurements were performed. Obscuration is the ratio of unscattered light intensity with and without a spray sample. Measurements of obscuration quantify how much of the laser beam is scattered by the spray sample. For example, an obscuration of 30% means that 30% of the incident light is scattered by particles. Large values of obscuration indicate optically dense sprays.

A 10 mW HeNe laser beam was expanded and collimated to a 7 mm diameter beam, which went through the spray field. A transform lens collected the light scattered by the spray field and imaged it onto the CID detector. The laser was shuttered by an acousto-optic modulator which allowed the "on" time of the scattered laser light to be controlled for timing and duration. Typical duration for a single sample with this system was 20 msec.

The CID camera electronics were controlled by a timing board to ensure that any time the AOM was on, allowing the laser to penetrate the spray field, the CID was integrating the image. Readout and reset of the camera were forced at the end of the laser exposure.

The resulting transform images appearing on the CID were captured with a frame grabber, and a sector of the scattered image was saved and used for subsequent analysis, providing about 250 radial scattered intensity values for one image.

The scattered light data was "inverted" using the Chin-Shifrin inversion to produce the droplet size distribution associated with the scattered light signature. This particle size distribution could be used directly, or integrated to provide weighted size averages, like the SMD.

Obscuration measurements were performed by collecting the unscattered portion of the probing beam. This was accomplished by using a pin hole aperture to block the scattered light and pass the unscattered light.

PARTICLE SIZING VERIFICATION - The particle sizing system was verified using a photomask reticule prior to performing spray measurements. This verification technique has been described in detail by Hirleman et al. [7]. A photomask reticule is a two-dimensional projection of a three-dimensional particle population. The photomask reticules are generally composed of a two-dimensional array of opaque circular chrome discs (termed particle artifacts) deposited on a

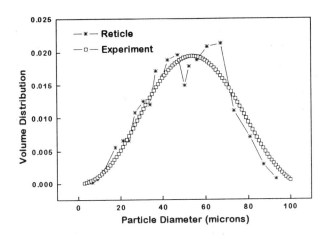

Figure 5. Particle sizing verification.

transparent substrate. A discrete size distribution can be represented by a number of discs of different sizes.

A photomask reticule consisting of a Rosin-Rammler distribution of particle artifacts, having a total of approximately 10,500 artifacts represented by 23 different size discs, was used for verification of the present particle sizing system. A comparison of the experimentally determined volume distribution and the volume distribution given with the reticle standard is shown in figure 5. The experimentally determined distribution agrees fairly well with the reticule distribution. Both distributions have similar shapes and the tails of the distributions are in good agreement.

This method has previously been tested for monodisperse samples and shown to be able to reproduce single particle sizes quite well [8].

TRANSIENT SPRAY MEASUREMENTS - Measurements of particle size and laser beam obscuration were performed at three locations in the spray. Figure 6 illustrates the locations of the measurement regions. All measurements were

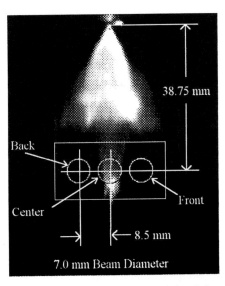

Figure 6. Illustration of the particle sizing and obscuration measurement regions.

performed 38.75 mm downstream of the injector tip with a 7 mm diameter laser beam. One location was on the injector axis, while the other two were off axis, 8.5 mm to either side. The three locations are identified in figure 6 as the back, center, and front locations. These locations were chosen because they are coincident with the window plane of an optical engine which will be used in future experiments.

The droplet sizing and obscuration measurement results are shown in two groups of graphs. The first group, figures 7 and 8, were made for relatively short times after SOI. For these values, data was taken at step intervals of 25 msec. The second set of graphs, figures 9 and 10, show SMD and obscuration results for the same conditions but for a much longer overall range of delay times. For these figures, the step interval was 100 msec for delays less than 2 msec, and 250 msec for delays greater than 2 msec.

For dense sprays, multiple scattering of the probing laser beam is likely. This multiple scattering creates larger scattering angles than a dilute particle field of the same size would produce. The larger angles are interpreted in a diffraction based measurement system as smaller particles. In general, a dense particle or droplet field produces a broader size distribution with smaller mean sizes than a dilute field with a similar distribution. Experiments by Dodge [9] and by Paloposki and Kankkunen [10] show that measured sizes using a commercial diffraction based particle sizer with a ring detector, err significantly for transmissions of less than 20% (obscuration > 80%). For example, Dodge shows that for particles of about 30 mm SMD, for an obscuration of 80%, the measured SMD is about 75% of the value obtained in a dilute field. For particles of about 60 mm SMD, at 80% obscuration, the measured SMD is about 85% of the value for a dilute field.

In these experiments, obscuration values exceeding 80% occurred in nearly every spray, in the densest portion of the spray. As the data for these sprays is discussed, the measured SMD values need to be considered in concert with the obscuration values taken for the same conditions at the same time after SOI.

Figures 7 and 8 show leading edge SMD and obscuration characteristics as a function of time for the three measurement locations. Each data point represents the average of five measurements; the error bars on each curve indicate the absolute value of standard deviation of the five measurements. The three curves on each plot correspond to different injection pressures. The injected mass was the same for all three pressures, 13.1 mg, and is representative of a medium load injection. Note that this is the same operating condition used for the images in figure 2. In order to resolve the features of the leading mass and leading edge of the main body, measurements were performed every 25 μsec.

From the obscuration and SMD curves it is apparent that as injection pressure is increased, sprays arrive at the measurement regions sooner and have smaller SMD associated with them. In the case of the

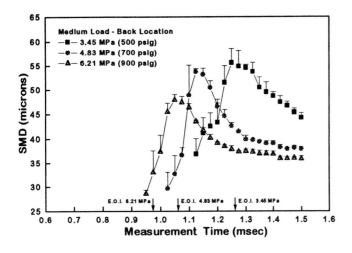

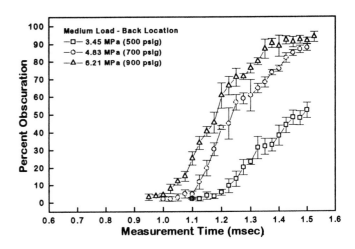

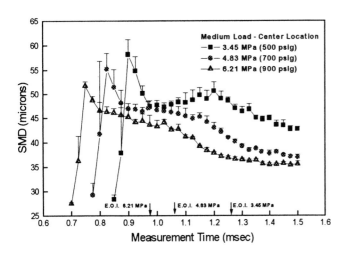

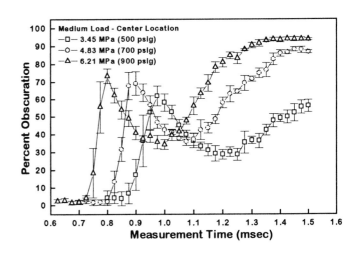

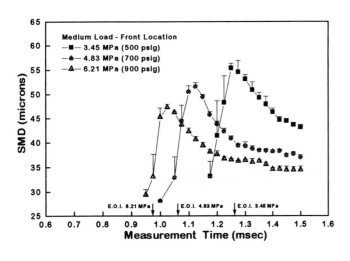

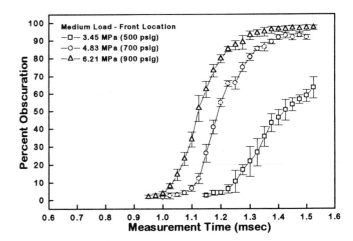

Figure 7. Medium load leading edge SMD characteristics. SMD VS Measurement time. Injected mass: 13.1 mg; Injection durations were 1.265, 1.060, and 0.973 msec for the 3.45, 4.83, and 6.21 MPa injection pressures, respectively. Error bars show plus one standard deviation of five measurements.

Figure 8. Medium load leading edge obscuration characteristics. Obscuration VS Measurement time. Injected mass: 13.1 mg; Injection durations were 1.265, 1.060, and 0.973 msec for the 3.45, 4.83, and 6.21 MPa injection pressures, respectively. Error bars show plus and minus one standard deviation of five measurements.

front and back locations, the SMD and obscuration curves look fairly similar to one another which suggests some degree of symmetry in the spray.

In the case of the center measurement location, the SMD rises very rapidly as the initial mass of fuel enters the 7 mm diameter sample region. A peak SMD value is recorded in this initial slug; the SMD then decreases rapidly. The 3.45 MPa injection pressure case exhibits a second SMD peak as the main body of the spray enters the sample volume. The higher pressure cases do not show a clear second peak of SMD in the main body of the spray.

The off-axis SMD values (back and front) show significantly later and slower rates of increase in SMD. From figure 2 it appears likely that the leading slug of liquid completely misses these measurement volumes, so the SMD values and timings shown are representative of the off-axis arrival of the main fuel spray.

The obscuration curves associated with the center measurement region show the passing of the initial mass. At early measurement times, less than 0.7 msec, the obscuration is very low because the initial mass has not yet reached the measurement region. Later, obscuration values increase rapidly after which they fall, and then increase again. The first increase in obscuration is associated with initial mass entering the measurement region. The dip in each obscuration curve is due to the leading mass leaving the measurement region and the leading edge of the main body entering the measurement region. This can be visualized with the help of the images shown in figure 2. For example, at a measurement time of 1.0 msec the image for the 4.83 MPa injection pressure shows that the majority of the leading mass has passed through the center measurement region and the leading edge of the main body has not yet reached the center measurement region. In agreement with this image, the dip in the obscuration curve for this condition, shown in figure 8, is centered at the 1.0 msec measurement time. The corresponding particle size data, show that the largest SMDs occur in the beginning of the initial mass.

Note that the obscuration for these cases exceeds 80% at about 1.2 msec after SOI for the 6.21 MPa case, at about 1.4 msec after the SOI for 4.83 MPa, and after 1.5 msec for the 3.45 MPa case.

In the case of the 3.45 MPa injection pressure, the SMD curve displays a secondary maximum. The corresponding obscuration curve shows that this secondary maximum is associated with the leading edge of the main body.

The peak in SMD values at the center position occurs before the peak in obscuration, suggesting that the largest droplets occur at the leading edge of the leading slug of liquid.

Figures 9 and 10 show SMD and obscuration characteristics at longer measurement times for the three measurement locations. Measurements were performed every 100 μsec up to 2 msec after the start of injection, and every 250 μsec thereafter.

For the pressure range examined, the SMDs associated with the main body of the spray are between 33 and 43 mm. Particle size reductions due to increased injection pressure are more gradual as injection pressure value is increased. As the measurement time is delayed, the SMD curves for each location seem to converge to one value.

There is a similarity in the data when comparing the back and front measurement regions. The small dip and hump in the SMD curves at measurement times between 1.5 and 3.0 msec for the two highest pressures may be due to the formation and passing of the vortex cloud through the measurement region. The fact that this behavior occurs sooner in time for the higher pressure provides some evidence to support this.

The obscuration curves shown in figure 10 show that the majority of the spray pulse has passed through the measurement region by a measurement time of 5.0 msec. At this measurement time, the trailing cloud of the injection is passing through the measurement region. (see figure 2) It is interesting to note that even though the injection duration is shortest for the highest injection pressure, the obscuration values at the longer measurement times are larger for the highest pressure. This is most likely due to the persistence of the induced vortex shown in the images in figure 2. As noted earlier, the vortex strength appears to increase with injection pressure. Air entrainment, associated with the vortex, transports particles upstream, which maintains high values of obscuration at the sample locations at later times.

The peak obscuration values shown in figure 10 are quite high, exceeding 80% from 1.5 msec after SOI to about 2.5 msec after SOI for the center sample position. For low obscuration values, accurate sizing using diffraction based methods is possible, and the obscuration can be loosely interpreted as an indicator of optical density. At higher values of obscuration, multiple scattering in the sample volume is likely, and the droplet size distribution interpreted from the collected scattered intensity signal is likely to underestimate SMD in the sample volume.

TRANSIENT MASS FLUX MEASUREMENTS

A transient patternator device, developed by Hoffman et al. [11-13], was used to measure average mass flux distributions at various times after the SOI. Transient measurements were made by physically "cutting" the spray with a shuttering device - a rotating disk with an aperture in it. The timing of the disk relative to SOI could be adjusted to enable average mass flux distributions to be measured at various times after the SOI. The distributions were obtained by measuring the average mass flux through 23 collection tubes located along a line which passes through and is perpendicular to the injector axis. The sample tubes were spaced 3.81 mm apart (center to center) with inner diameters of 1.65 mm. Several thousand injections were required to collect a sufficient amount of sample for measurement. Figure 11 shows an illustration of the operating principles of the transient patternator.

Transient patternator experiments were performed with the injector operating at a 4.0 msec

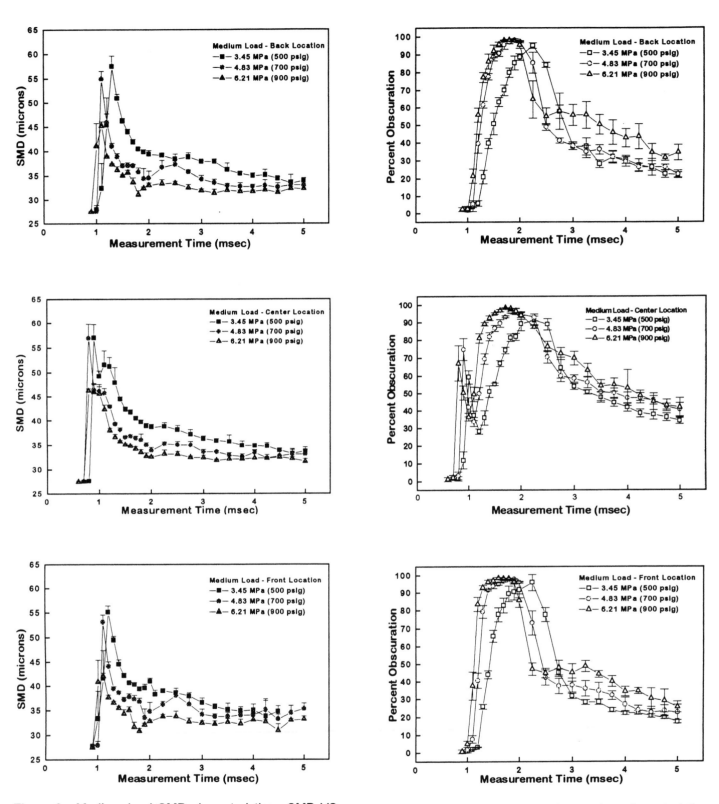

Figure 9. Medium load SMD characteristics. SMD VS Measurement time. Injected mass: 13.1 mg; Injection durations were 1.265, 1.060, and 0.973 msec for the 3.45, 4.83, and 6.21 MPa injection pressures, respectively. Error bars show plus one standard deviation of five measurements.

Figure 10. Medium Load obscuration characteristics. Obscuration VS Measurement time. Injected mass: 13.1 mg; Injection durations were 1.265, 1.060, and 0.973 msec for the 3.45, 4.83, and 6.21 MPa injection pressures, respectively. Error bars show plus and minus one standard deviation of five measurements.

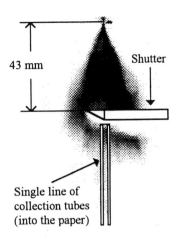

Figure 11. Transient patternator shutter detail for sampling a portion of an injection.

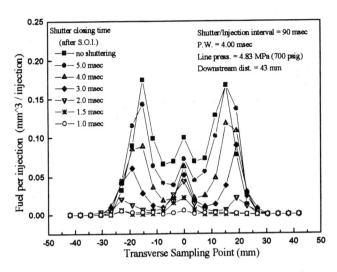

Figure 12. Transient patternator measurements.

duration, injection mass ≈ 55 mg, and an injection pressure of 4.83 MPa. The results of the experiments are shown in figure 12. The shutter closing times noted on the figure correspond to the time, after the SOI, that the collection tubes were covered by the shutter which prohibits further collection of the spray. The curve marked no shuttering was generated by leaving the shutter open throughout the entire injection event. This data, when integrated, matches the total measured injected mass within 5%.

At the 1.0 msec closing time, the spray has just reached the shutter location and very little mass is collected. As the closing time is retarded, more and more mass is collected. Part of the leading mass is captured in the 1.5 msec closing time curve. At a closing time of 2.0 msec, outer regions of the spray begin to be collected. By the 4.0 msec closing time, the collected mass off the injector axis is greater than the on-center collection. The distribution at the closing time of 5.0 msec looks similar to the "no shutter distribution". At this point the majority of the spray pulse has passed the shuttering region.

This data illustrates the transient mass flux behavior of the spray. Short duration, average mass flux distributions are likely to look different from longer duration distributions, due to transient start-up effects. The shorter the injector duration the more influential the transient effect will be. In accordance with figure 12, short duration (low load) injections would deposit most of the fuel mass centrally near the axis of the injector. Longer duration injections (medium load) would deposit fuel centrally near the axis of the injector and at a radial location off axis of the injector. For long duration injections (high load) the largest percentage of fuel mass would be deposited along a radial location off axis of the injector.

SUMMARY

Transient spray characteristics of a high pressure injector typical of those intended for use in a DISI engine, have been examined. A variety of measurement diagnostics, including solid state imaging and line-of-sight diffraction based particle sizing were employed for spray characterization. Transient patternator measurements were also performed to obtain spatially and temporally resolved average mass flux distributions.

Results of the spray imaging experiments indicate that the start-up spray shapes were highly dynamic and contain a leading mass, or slug. The penetration of the leading mass was faster than the leading edge of the main body. As expected, penetration of both the leading mass and the leading edge of the main body was found to increase with injection pressure.

Particle size measurements indicated that the Sauter Mean Diameter (SMD) of the leading mass and leading edge of the main body of the spray were significantly larger than the SMD in other regions of the spray. As injection pressure increased, sprays arrived at the measurement regions sooner and had smaller SMDs associated with them. Peak SMD values were smaller for higher injection pressures, and the change in peak SMD with pressure was more gradual as injection pressure increased. The SMD and obscuration curves for the front and back locations were similar to one another, indicating some degree of symmetry in the spray.

Patternator measurements illustrated that transient start-up effects significantly influence the spatial mass flux distributions. Short duration mass flux distributions are likely to look different (more central mass) than longer duration distributions (more like a hollow-cone spray).

ACKNOWLEDGMENTS

This investigation was funded by Ford Motor Company. Additional support was provided by the US Army Research Office under Grant No. DAAH04-94-G-0328.

REFERENCES

1. Abata, D., "A Review of the Stratified Charge Engine Concept," Automotive Engine Alternatives, International Symposium on Alternative and Advanced Automotive Engines, 1986.

2. Wood. C. D., "Unthrottled Open-Chamber Stratified Charge Engines," SAE 780341 (1978).

3. Haslett, R. A., Monaghan, M. L., and McFadden, J. J., "Stratified Charge Engines," SAE 760755 (1976).

4. Kume, T., Iwamoto, Y., Iida, K., Murakami, M., Akishino, K., and Ando, H, "Combustion Control Technologies for Direct Injection SI Engine,: SAE 960600 (1996).

5. Swithenbank, J., Beer, J. M., Taylor, D. S., Abbot, D., and McCreath, G. C., "A laser diagnostic technique for the measurement of particle and droplet size distribution", in Experimental Diagnostics in Gas Phase Combustion, AIAA vol. 53, *Progress in Astronautics and Aeronautics*, (B. T. Zinn, Ed.), pp. 421-446 (1976).

6. Chin, J. H., Sliepcevich, C. M., and Tribus, M., "Particle size distributions from angular variation of intensity of forward-scattered light at very small angles," *J. Phys. Chem. 59*, 841 (1955).

7. Hirleman, E.D., Felton, P.G., and Kennedy J., "Results Of The ASTM Interlaboratory Study On Calibration Verification Of Laser Diffraction Particle Sizing Instruments Using Photomask Reticules," Proceedings ICLASS, Gaithersburg, MD, 611-619 (1991).

8. Albert, R. and Farrell, P. V., "Droplet sizing using the Shifrin inversion", *Journal of Fluids Engineering 116*, pp. 357-362 (1994).

9. Dodge, L. G., "Change of Calibration of Diffraction-Based Particle Sizers in Dense Sprays", *Optical Engineering 23*, pp. 626-630 (1984).

10. Paloposki, T. and Kankkunen, A., "Multiple Scattering and Size Distribution Effects on Performance of a Laser Diffraction Particle Sizer", Proceedings ICLASS, Gaithersburg, MD, pp. 441-448 (1991).

11. Hoffman, J. A., Eberhardt, E., Martin, J, K., "Comparison between Air-Assisted and Single-Fluid Pressure Atomizers for Direct-Injection SI Engines Via Spatial and Temporal Mass Flux Measurements.", SAE 970630 (1997).

12. Hoffman, J. A., Martin, J, K., and Coates, S. W., "Spatial and Temporal Mass Flux Measurements of a Pulsed Spray - a Review of the Hardware and Methodology," Review of Scientific Instrumentation, submitted, November 1996.

13. Hoffman, J. A., Martin, J, K., and Coates, S. W., "Spray Photographs and Preliminary Spray Mass Flux Distribution Measurement of a Pulsed Pressure Atomizer," ILASS-Americs 96, 9th Annual Conference on Liquid Atomization and Spray Systems, San Francisco, 1996.

A Review of Mixture Preparation and Combustion Control Strategies for Spark-Ignited Direct-Injection Gasoline Engines

Fu-Quan Zhao and Ming-Chia Lai
Wayne State University

David L. Harrington
General Motors Research and Development Center

ABSTRACT

The current extensive revisitation of the application of gasoline direct-injection to automotive, four-stroke, spark-ignition engines has been prompted by the availability of technological capabilities that did not exist in the late 1970s, and that can now be utilized in the engine development process. The availability of new engine hardware that permits an enhanced level of computer control and dynamic optimization has alleviated many of the system limitations that were encountered in the time period from 1976 to 1984, when the capabilities of direct-injection, stratified-charge, spark-ignition engines were thoroughly researched. This paper incorporates a critical review of the current worldwide research and development activities in the gasoline direct-injection field, and provides insight into new areas of technology that are being applied to the development of both production and prototype engines.

The advantages and disadvantages of the emerging technologies that are being utilized to develop gasoline direct-injection combustion systems, such as electronic common-rail injection, variable swirl and tumble control, computational fluid dynamics, and laser diagnostics of fuel sprays and combustion are discussed in detail within the context of the total industry effort to enhance the attainable brake specific fuel consumption and emissions. Significant emphasis is placed upon extracting the key findings of each investigation and discussing the relative potential of the technique for meeting the major requirements of the gasoline direct-injection engine. Areas of general consensus that are evident among the hundreds of researchers who are currently investigating this field are noted, as are apparent conflicts that must be resolved in the near future by obtaining additional data. Also noted and discussed are a number of critical research areas for which it is obvious that insufficient effort and resources are being devoted. One example of such an area is the development of a proven laboratory test for quantifying the deposit formation tendencies of direct-injection gasoline injectors.

Based upon the review of all available technical publications, the trends and directions of the current effort on gasoline direct-injection engine are noted, and the remaining difficult developmental areas are discussed. For planning purposes, the totality of the literature is used to formulate likely engine scenarios regarding the development of specific engine configurations, and to outline technological path options and trends that may be expected in the displacement of the port fuel injection engine by the gasoline direct-injection engine.

INTRODUCTION

The injection of gasoline directly into the cylinder of a four-stroke, spark-ignition engine is a concept that offers a number of unique potential advantages over even the most sophisticated current production engines that inject fuel into the intake port. A comparison of the basic gasoline direct-injection (GDI) and port-fuel-injected (PFI) combustion systems is illustrated schematically in Fig. 1. It may be seen that the major difference between the PFI engine and the GDI engine is the additional mixture preparation stage comprised of a port film of liquid fuel. The thermodynamic potential for reducing the specific fuel consumption, coupled with the advantages of quicker starting, enhanced transient response, and more precise control of the mixture air/fuel ratio, have resulted in a renewed worldwide interest in the GDI engine as an automotive powerplant. The primary developments since the extensive examination of these engines during the 1976-1984 time period include the replacement of pressure-activated poppet injectors by the electronic fuel injector, the incorporation of the common-rail injection system, and the ready availability of extensive new capabilities for computer-controlled optimization of the in-cylinder flow field, cam-phasing, throttle positioning and exhaust-gas-recirculation (EGR).

There are two current strategic objectives in the continued long-range development of the four-stroke, gasoline, spark-ignition engine for automotive applications; the first is associated with fuel economy and the second deals with emissions. The fuel economy objective is the attainment of a step-change improvement in the level of brake specific fuel consumption (BSFC) that is provided by current spark-ignition engines. The emission objective is to significantly reduce the levels of unburned-hydrocarbon (UBHC) and NOx emissions to production-feasible mean levels for ultra-low-emission-vehicle (ULEV) regulatory compliance.

There can be varying degrees of design complexity associated with direct-injection gasoline engines. Perhaps the simplest embodiment of a GDI system is an engine that

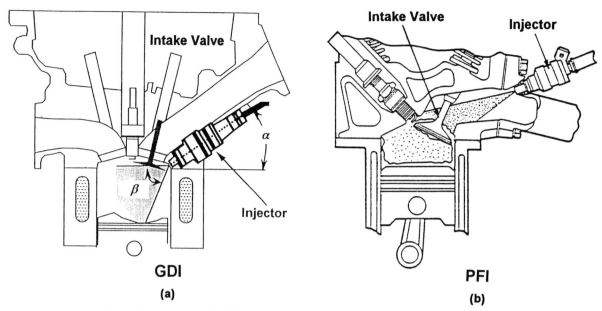

Fig. 1 Comparison of GDI and PFI mixture preparation systems; (a) GDI engine (Zhao *et al.*, 1997a); (b) PFI engine (Ohyama *et al.*, 1992)

operates in the early-injection, homogeneous, stoichiometric mode. This engine must utilize throttling for load control, thus the full fuel-economy potential that is associated with the elimination of throttling will not be realized. Engines that operate exclusively in this mode do, however, experience the thermodynamic advantages of intake charge cooling, as well as the advantages of more rapid starting and fuel cutoff on deceleration. The next level of complexity in GDI engine design is to utilize leaner homogeneous mixtures with reduced throttling for some degree of load control, which permits the realization of additional potentials. Making use of the full potential of the GDI concept requires a design that generates a stable, stratified mixture using an overall-lean air/fuel ratio for operating at part load, and which can transition smoothly to heavy or full-load operation by injecting increased volumes of fuel at earlier times in the cycle. The most complex GDI engine would operate in all three modes at various times, with transitions from one mode to another. This would require not only a complex control system, but an array of fast and reliable sensors and a sophisticated set of algorithms in order to maintain acceptable levels of driveability during mode transitions.

The PFI engine has evolved since 1983, with many incremental improvements having been incorporated on an annual basis. Enhancements such as sequentially-timed injection, computer algorithms for transient metering, four-valves per cylinder, multiple roller camshafts, variable cam phasing, turbocharging, and even supercharging have become features on PFI engines. In spite of these significant developments, the current high-tech PFI engine still requires, and will continue to require, throttling for basic load control, and still has, and will continue to have, an operating film of liquid fuel in the intake port. These two basic PFI operating requirements represent major impediments to achieving significant breakthroughs in PFI fuel economy or emissions. Continuous incremental improvements in the older PFI technology will be made, but it is unlikely that the long-range fuel economy and emission objectives can be simultaneously achieved.

Even though throttling is a well-established and reliable mechanism of load control in the PFI engine, the thermodynamic loss associated with throttling is very substantial. Any system that utilizes this method to adjust load levels will experience the inherent loss, and will have significant efficiency limitations at low levels of engine load. The GDI engine, in theory, does not have these two significant limitations, nor the performance boundaries that are associated with them. The theoretical advantages of the GDI engine over the current PFI engine are summarized as follows :

- improved BSFC (up to 30% improvement)
  ◊ less pumping loss (unthrottled, stratified mode)
  ◊ less heat losses (unthrottled, stratified mode)
  ◊ higher compression ratio (charge cooling with injection during induction)
  ◊ lower octane requirement (charge cooling with injection during induction)
  ◊ increased volumetric efficiency (charge cooling with injection during induction)
  ◊ fuel cutoff during a deceleration
- improved transient response
  ◊ less acceleration-enrichment required
- more rapid starting with less cold enrichment required
  ◊ starting on first or second cranking cycle
- selective emission advantages
  ◊ reduced cold-start UBHC emissions
  ◊ reduced CO_2 emissions
  ◊ more precise air/fuel ratio control.

In spite of the important potential advantages noted above, the practical GDI engine does have certain inherent problems that are similar to those of the direct-injection stratified-charge (DISC) engines. The commercial success of GDI as an automotive production powerplant has been delayed by the following areas of concern :

- difficulty in controlling the stratified-charge combustion over the required operating range
- complexity of control and injection technologies required for stepless load changes

- excessive rate of formation of injector deposits and/or ignition fouling
- excessive light-load UBHC emissions
- excessive high-load NOx emissions
- high local NOx production under part-load, stratified-charge operation
- three-way catalysts cannot be utilized to full advantage
- increased fuel system wear due to the combination of high fuel pressure and low fuel lubricity
- increased rates of cylinder bore wear
- soot formation for high-load operation
- increased electrical power and voltage requirements of the injectors and drivers.

If ULEV emissions regulations and corporate-averaged-fuel-economy (CAFE) requirements can be achieved using PFI engines without the requirement of complex and expensive new hardware, the market penetration rate for GDI engines will be reduced, as the GDI engine will require relatively sophisticated fuel injection hardware and control systems. Two key technical limitations for GDI engines have been excessive UBHC emissions and the fact that three-way catalysts could not be used. Operating the engine under lean conditions reduces the engine-out NOx emissions, but this generally cannot achieve the 90% reduction level that can be attained using a three-way catalyst. Much work is underway worldwide to develop lean-NOx catalysts, but at this time the attainable conversion efficiency is still much less than that of the three-way catalyst. The excessive UBHC emissions at light load represent a significant research problem to be solved.

In spite of very real concerns and difficulties, the GDI engine offers an expanded new horizon for future applications as compared to the well-developed PFI engine. New technologies and computer-control strategies are currently being invoked by a number of automotive companies to re-examine the extent to which the potential benefits of the GDI engine can be realized in a production engine (Kume *et al.*, 1996; Harada *et al.*, 1997; Anderson *et al.*, 1996; Takagi, 1996; Shimotani *et al.*, 1996; Karl *et al.*, 1997; Lake *et al.*, 1996; Buchheim and Quissek, 1996; Douaud, 1996; Fansler *et al.*, 1995; Hauser *et al.*, 1995; Seiffert, 1996). Mitsubishi has very recently introduced a production GDI engine into the Japanese market, as has Toyota, and Mitsubishi plans to introduce a version of their GDI engine into the European market in 1997 (Ando, 1996b).

The information in this paper will provide the reader with a comprehensive review of the mixture dynamics and combustion control strategies that may be utilized in four-stroke, spark-ignition, direct-injection, gasoline engines. The current state of knowledge, as exhibited in more than one hundred recent key publications, many as yet untranslated, is discussed in detail, and the critical research and development needs for the near future are identified.

FUEL SYSTEMS FOR GASOLINE DIRECT-INJECTION

Unthrottled operation with the load controlled by the fuel quantity has been shown to be a very efficient operation mode for the internal combustion engine, as the pumping loss is significantly reduced and the volumetric efficiency is increased. This combustion approach is very successful in the diesel engine, as ignition occurs spontaneously at points within the combustion chamber where the mixture is well prepared for autoignition. The fixed location of the ignition source in the spark-ignition (SI) engine, however, makes it quite difficult to operate in the unthrottled mode for other than full load. This imposes a critical additional requirement on the mixture formation process of this type of engine in that the mixture cloud that results from fuel vaporization and mixing must be controlled both spatially and temporally in order to obtain stable combustion. Preparing the required mixture distribution inside the combustion chamber for a wide range of engine operating conditions is quite difficult, as the fuel/air mixing process is influenced by many time-dependent variables. The development of a successful combustion system depends upon the optimized design of the fuel injection system and the proper matching of the system components to control the in-cylinder flow field and burn rate.

The fuel injection systems of early DISC engines were derived from the basic diesel injection system (Baranescu, 1983; Duggal *et al.*, 1984; Enright *et al.*, 1988; Iida, 1992; Wood, 1978). For example, the Texaco TCCS engine (Alperstein *et al.*, 1974) utilized a diesel-type injector that produced a spray with relatively poor atomization and fuel/air mixing quality, and with high penetration rates relative to sprays from current pressure-swirl atomizers. The Ford PROCO engine (Scussei *et al.*, 1978) used an outwardly opening pintle atomizer with vibration to enhance the fuel atomization. However, the poppet opening pressure was on the order of 2 MPa, which is quite low. Gasoline injection using single-hole or multihole, narrow-angle sprays from diesel injection systems nearly always result in substantial UBHC emissions due to the compactness of the spray and the high penetration velocity. In order to avoid this problem, combustion systems using direct wall impingement and fuel film formation such as the MAN-FM (Urlaub and Chmela, 1974) were developed. For early DISC engine experiments using multi-hole injectors, the singularity of the ignition source proved to be a definite problem, as compared to a diesel combustion system where multiple ignition sites occur simultaneously. Another severe problem for the diesel-based DISC injection system was the lack of variability of injection characteristics between part load and full load. For full-power DISC engine operation, injection of maximum fuel quantities early in the intake stroke resulted in significant wall impingement of the fuel.

In recent years, significant progress has been made in the development of advanced, computer-controlled fuel injection systems, which has had much to do with the expansion of research and development activities related to GDI engines. In this section, the key components that are utilized in the fuel system, and their important roles in the mixture preparation and combustion processes, are described in detail.

General Fuel System Requirements - The fuel injection system in a GDI engine is a key component that must be matched with the specific in-cylinder air flow field to provide the desired mixture cloud over the entire operating range of the engine. For all operating conditions a well-atomized fuel spray must be produced, and, for the efficient combustion of a stratified mixture, a stable and compact spray geometry is necessary (Kume *et al.*, 1996). A GDI fuel system needs to provide for at least two, and possibly three, distinct operating modes. For unthrottled, part-load operation, the injection system should provide the capability for late injection at high rates during the compression stroke into a cylinder pressure of up to 2 MPa, which requires a relatively high fuel injection pressure. The injection pressure is also very important for obtaining both effective spray atomization and the required level of spray penetration. A higher fuel injection pressure is effective in reducing the mean droplet diameter of the spray approximately as the inverse square root of the pressure differential, whereas the use of a lower pressure generally reduces the pump parasitic load and injector noise, and increases the reliability of the fuel pump system. The use of a very high fuel injection pressure, such as 20 MPa, will enhance the atomization but will most likely generate an overpenetrating spray, resulting in fuel wall wetting. The fuel pressures that have been selected for most of the current prototype GDI engines range from 4 to 7 MPa, which are quite low when compared with diesel injection systems of 50 to 140 MPa, but are relatively high in comparison with typical PFI injection pressures of 270 to 450 kPa. A constant fuel-line pressure is utilized in a common-rail system for most of the current GDI applications; however, a strategy using a variable fuel injection pressure does offer an alternative method of obtaining the required flow range while reducing the dynamic-range requirements of the injector itself (Pontoppidan et al., 1996) and for meeting differing fuel spray requirements corresponding to a range of engine loads (Matsushita et al., 1996).

Fuel injection systems for full-feature GDI engines must have the capability of providing both late injection for stratified-charge combustion at part load, as well as early injection for homogeneous-charge combustion at full load. At part load, a well-atomized compact spray or mixture plume is desirable to achieve rapid mixture formation and controlled stratification. At full load, a well-dispersed fuel spray or mixture plume is desirable to ensure a homogeneous charge even for the largest fuel quantities. This is generally achieved by early injection at low cylinder pressure, similar to the mode of open-valve fuel injection in the PFI engine. These fuel system requirements are more comprehensive than those of either the diesel or the PFI injection systems. Based upon recent experimental investigations, it has been concluded that common-rail injection systems with electromagnetically-activated injectors can meet these requirements (Pischinger and Walzer, 1996; Buchheim and Quissek, 1996). In principle, these systems are comparable to those utilized in current PFI and advanced diesel engines, with the GDI application having an intermediate fuel pressure level, high-frequency noise level, and associated cost. A subdivided or split injection during the intake and compression strokes may also be a viable option. In the Toyota D-4 GDI combustion system (Matsushita et al., 1996) a two-stage injection strategy is utilized to improve the transition between part-load and full-load operation.

During the engine cold crank and start, the high pressure fuel pump generally cannot deliver fuel at sufficient pressure due to the short time available and the low engine cranking speed. A bypass valve is used in the Mitsubishi GDI engine to allow the fuel to bypass the high-pressure regulator under such a condition. As a result, the electric feed pump supplies the fuel directly to the fuel rail at a fuel pressure of 335 kPa. After the engine speed increases and the high pressure pump fully primes the system, the bypass valve is closed and the high pressure regulator begins to regulate

the fuel pressure to 5.0 MPa.

Gasoline has a lower lubricity, viscosity, and a higher volatility than diesel fuel, generally resulting in more concerns regarding system friction and wear, greater potential leakage and the need for enhanced cooling in fuel pumps and injectors. It should be noted, however, that hydrodynamic lubrication may be used at high fuel pressures to compensate for the low viscosity (Iwamoto et al., 1997). A very wide range of fuel quality is available in the field, thus it is important that a robust GDI fuel system be developed and proven for production engines.

Fuel Injector Requirements - The fuel injector may be said to be the most critical element in the GDI fuel system, and should have the following attributes. Many of the required characteristics of the GDI injector also correspond to those of the port fuel injector, which are (Zhao et al., 1995):

- accurate fuel metering
- desirable spray pattern for the application
- minimal spray skew
- good spray symmetry over the operating range
- minimal drippage and fuel leakage, particularly for cold operation
- small pulse-to-pulse variation in fuel quantity and spray characteristics
- small sac volume
- good low-end linearity between the dynamic flow and the fuel pulse width
- minimal variation in the above parameters from unit to unit.

In certain critical areas, the requirements of the GDI injector significantly exceed those of the port fuel injector. These requirements are:

- significantly improved level of atomization
- expanded dynamic range
- enhanced resistance to deposit formation
- ability to operate at higher injector body and tip temperatures
- stable operation at elevated fuel and ambient pressures
- avoidance of needle bounce that creates unwanted secondary injections.

The GDI injector should be designed to deliver a precisely metered fuel quantity with a symmetric and highly repeatable spray geometry, and must provide a highly atomized fuel spray having a Sauter mean diameter (SMD) of generally less than 25 μm, and with a droplet diameter corresponding to the 90% volume point (DV90) not exceeding 45 μm. The DV90 statistic is a quantitative measure of the largest droplets in the spray. Smaller values than these are even more beneficial, provided sufficient spray penetration is maintained for good air utilization. The SMD is also denoted by the symbol D32. The fuel pressure required is at least 4 MPa for a single-fluid injector, with 5 to 7 MPa being more desirable if the late-injection, stratified mode is to be invoked. Even if successful atomization could be achieved with fuel pressures less than 4 MPa, significant metering errors could result from the variation of metering pressure differential with cylinder pressure. In general, the smaller the injector sac volume, the fewer large peripheral drops that will be generated when the injector opens. The sac volume within the injector tip is basically a volume of fuel that is not at the fuel-line pressure, therefore it retards the acceleration of the main portion of the injected fuel and degrades both the fuel atomization and the resulting combustion. Needle bounce is to be avoided, as a secondary injection generally results in uncontrolled atomization consisting of larger droplets of lower velocity. It also reduces the fuel metering accuracy and contributes to increases in the UBHC and particulate emissions.

The ability to deliver the required fuel with a short fuel pulse, which corresponds to a higher rate of injection, is much more important for the GDI engine than for the PFI engine, particularly for light-load stratified-charge operation. Therefore, much more significance is attached to the low-pulse-width region of the GDI injector, effectively increasing the importance of the injector dynamic range requirement. The optimal design of the injector to resist coking is also one of the important requirements of the GDI injector, as is discussed in the section on injector deposits. Often over-looked are the voltage and power requirements of the injector solenoids and drivers. A number of prototype GDI injectors have power requirements that would be unaccept-able for a production application. It is also worth noting that it is advantageous to injector packaging to have the body as small as possible. This provides more flexibility in optimizing the injector location and in sizing and locating the ports and valves.

In spite of decades of continuous development on diesel multi-hole injectors, it has been shown that these nozzle-type injectors are not readily adaptable to GDI applications. A multi-hole nozzle used in a GDI engine application generally results in an unstable flame kernel when ignited by a single fixed spark gap. The rich mixture zones are close to the lean mixture zones, thus the flame front does not propagate uniformly through the combustion chamber. With the multi-hole nozzle, the number of nozzle holes is an important factor in determining engine combustion perfor-mance. The hole distribution that is effective in ensuring good spray dispersion and reliable flame propagation between the sprays generally provides the best engine performance. The effect of the cone angle of individual spray plumes on engine performance was studied by Fujieda et al. (1995). A multi-hole GDI injector similar to that used in diesel engines was designed and tested. It was found that reducing the individual nozzle flow area and increasing the number of holes can extend engine lean limit.

Currently, the most widely utilized GDI injector is the needle-type, high-pressure, swirl-spray unit, which delivers a conical spray. This type of injector can be regarded as a multi-hole nozzle with an infinite number of holes, and a uniform distribution of the fuel over the cone circumference may be obtained. As a consequence, wall wetting at full load can be minimized for an appropriate injector position and an optimized spray cone angle (Fraidl et al., 1996). The needle-type, swirl-spray injector is designed to apply a strong rotational momentum to the fuel in the injector nozzle that is in addition to the axial momentum. In a number of nozzle designs, liquid flows through a series of tangential holes or slots into a swirl chamber. The liquid emerges from the discharge orifice as an annular sheet that spreads radially outward to form an initially hollow-cone spray. The initial spray cone angle ranges from a minimum of 25 to almost 180°, depending on the requirements of the application, with a delivered SMD ranging from 15 to 25 µm. In the swirl-type injector, the pressure energy is effectively transformed into rotational momentum, which

enhances atomization, but limits spray penetration. The swirl nozzle generally produces a spray having a narrower distribution of drop sizes (DV90-DV10) than is obtained with the standard hole-type nozzle, with the best atomization occurring at high delivery pressures and wide spray angles. An additional advantage is that the injector designer can customize the spray penetration by altering the swirl ratio with only small changes in the atomization level, thus providing the necessary variability of spray configuration in order to meet different stratification requirements. However, the surface roughness of the orifice wall tends to generate streams or fingers of fuel in the fuel sheet exiting the nozzle, resulting in the formation of a locally rich air/fuel mixture. In order to minimize such mixture inhomogeneity, precise control of the nozzle tip quality is required (Tomoda et al., 1997).

In regards to the relative advantages of inwardly-opening versus outwardly-opening needles, the inwardly-opening needle generally provides better pulse-to-pulse repeatability of the spray cone geometry, especially when a flow guide bushing is present at the needle tip, although the outwardly-opening geometry has enhanced leakage resistance (Schapertons et al., 1991; Pontoppidan et al., 1997).

Engineers at Mitsubishi successfully applied the pressure-swirl injector to meet the requirements of their GDI engine, with the swirl generated by a swirler tip located upstream of the injector hole. The effect of the swirl level on the droplet size and spray structure was studied extensively by Kume et al. (1996) and the results are shown in Fig. 2. It was found that if the swirl can be intensified, a comparatively lower fuel pressure will be sufficient for achieving an acceptable level of atomization. Enhancement of the swirl also promotes air entrainment, and a vortex ring initially generated near the injector tip grows to a large scale toroidal vortex during the last portions of the injection event. Droplet velocity measurements using a phase Doppler anemometry (PDA) showed that the axial velocity component decreases with distance from the injector tip, whereas the swirl component remains fairly constant. The decrease in the axial velocity is caused by drag on the fuel droplets that are moving relative to the ambient air. In contrast, the swirl–component drag is not significant because the ambient air rotates with the fuel droplets. In the case of reduced ambient pressure as would occur with early injection, the fuel spray has a wide hollow-cone structure. In the case of higher ambient pressure, the higher drag force changes the spray into a narrow solid-cone shape having a reduced tip penetration.

Fuel Spray Characteristics - As is the case for diesel combustion, the fuel spray characteristics are of significant importance to GDI combustion systems. Parameters such as the spray cone angle, mean drop size, spray penetration, and fuel delivery rate are known to be critical, and the optimum matching of these parameters to the air flow field and spark location usually constitute the essence of the GDI engine development project. The primary fuel spray characteristics of a port fuel injector generally have much less influence on the subsequent combustion event, mainly due to the integrating effects of the residence time on the closed valve, and due to the secondary atomization that occurs as the induction air flows through the valve opening. For direct injection in both GDI and diesel engines, however, the mixture preparation time is significantly less

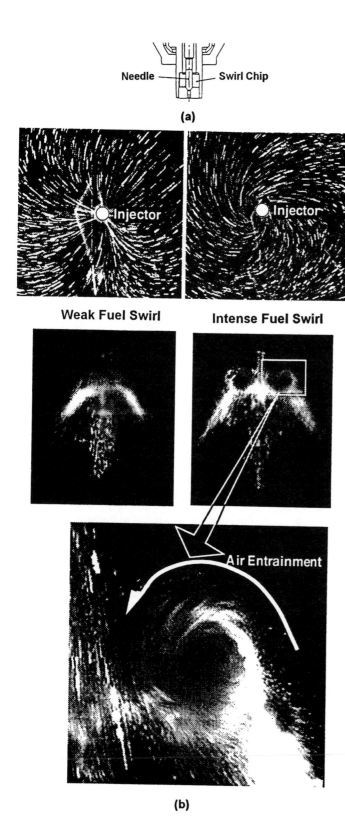

(a)

Weak Fuel Swirl   Intense Fuel Swirl

(b)

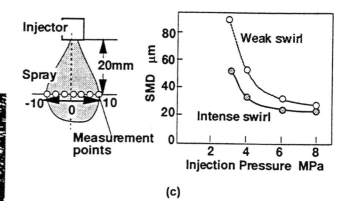

(c)

Fig. 2 Spray structure and atomization characteristics of the Mitsubishi GDI injector (Kume et al., 1996; Iwamoto et al., 1997); (a) nozzle geometry; (b) effect of fuel swirl on spray structure; (c) effect of fuel swirl on spray atomization characteristics

emissions and coefficient-of-variation (COV) of indicated mean effective pressure (IMEP). The diesel engine, of course, generally requires a fuel spray having an SMD that is less than 10 μm.

Even though a relatively complete correlation data base has been established for diesel sprays (Hiroyasu, 1991), the bulk of the correlations unfortunately cannot be applied to predict the characteristics of GDI sprays. This is the result of significant differences in fuel properties, injection pressure levels, droplet velocities and size ranges, ambient pressure and temperature levels, and droplet drag regimes. In many ways, the correlation and predictive characterization of the fuel sprays from GDI injectors represent a new and important research area.

An important operating criterion of a well-designed GDI engine is that the fuel must be vaporized before the spark event occurs in order to limit UBHC emissions to an acceptable level. Moreover, the complete evaporation of the fuel can make the ignition process more repeatable. For a gasoline droplet with a diameter of 80 μm, vaporization under typical compression conditions takes tens of milliseconds, corresponding to more than a hundred crank angle degrees at an engine speed of 1500 rpm. By contrast, the vaporization of a 25 μm droplet requires only several milliseconds, corresponding to tens of crank angle degrees. This is the essence of the degradation of GDI engine combustion characteristics for droplet mean diameters that exceed 25 μm. The rapid vaporization of small droplets helps to make the direct fuel injection concept feasible (Anderson et al., 1996). Therefore, many techniques have been proposed for enhancing the spray atomization of GDI injectors. The most common technique for GDI combustion systems is to use an elevated fuel pressure in combination with a swirl nozzle (Kume et al., 1996; Matsushita et al., 1996). The fuel injection pressure level is generally on the order of 5 MPa, or in some cases up to 12 MPa (Harada et al., 1997), in order to atomize the fuel to the acceptable level of 25 μm SMD or less. The performance of such high-pressure swirl injectors will be discussed in detail in the following separate section. The pulse-pressurized, air-assisted injector has also been applied to direct-injection gasoline engines (Miyamoto et al., 1996) and certainly

than is available for port injection, and there is much more dependence on the spray characteristics to prepare and distribute the fuel to the optimum locations. A port-injected gasoline engine can operate acceptably using a spray of 200 μm SMD, whereas a GDI engine will generally require an order of magnitude finer atomization. Most GDI applications will require a fuel spray having an SMD of less than 25 μm in order to achieve acceptable levels of UBHC

provides a spray with an SMD of less than 20 μm; however, numerous considerations such as the high-pressure, secondary air compressor and the use of two solenoids per cylinder limit its wide application to GDI combustion systems (Fraidl *et al.*, 1996).

The evaporation of droplets injected into the cylinder was evaluated by Dodge (1996) using a spray model, and it was recommended that a mean droplet size of 15 μm or smaller be utilized for GDI combustion systems. A differential fuel pressure of approximately 4.9 MPa is required for a pressure-swirl atomizer to achieve this SMD. It was also noted from the calculations that the additional time available with early injection does not significantly advance the crank angle locations at which complete droplet vaporization is achieved. This is because the high compression temperatures are very influential in vaporizing the droplets, and these temperatures occur near the end of the compression stroke. It was also noted that the atomization level that is utilized in some widely-studied GDI engines may not be sufficient to avoid some excessive UBHC emissions due to reduced fuel evaporation rates and fuel impingement on solid surfaces.

The SMD may not, in fact, be the single best indicator of the spray quality required for the GDI engine, as a small percentage of large droplets will be enough to degrade the engine UBHC emissions, even though the SMD is quite small. The small percentage of 50 μm fuel drops in a spray having a SMD of 25 μm not only have eight times the fuel mass as the mean drop, but are the last to remain as liquid. An injector that delivers a well-atomized spray, but has a wide spread in the drop-size distribution may require an even smaller SMD than quoted above to operate satisfactorily in a GDI engine combustion system. This spread may be quantified by the parameter (DV90-DV10). DV90 may be a parameter that is superior to SMD (D32) in correlating the UBHC performance of different fuel sprays in GDI combustion systems. A comparison of the droplet size distributions between swirl-type and hole-type high-pressure fuel injectors is shown in Fig. 3 (Tomoda *et al.*, 1997). It is clear that even though the difference of the SMD between the sprays from these two injectors is only 4 μm, the hole-type nozzle produces a wider distribution having many larger droplets. It is these larger droplets that are theorized to be responsible for the observed increase in the UBHC

emissions. It should be noted that the use of finer atomization may or may not reduce the UBHC emissions, depending upon the in-cylinder turbulence level due to small pockets of very lean fuel/air mixtures (Zhao *et al.*, 1993, 1994; Iiyama and Muranaka, 1994). A strong turbulence is required to enhance the fuel/air mixing process by eliminating small pockets of very lean mixture.

A significant number of references exist for the application of air-assisted injectors to gasoline direct-injection in two-stroke engines (Das and Dent, 1994; Diwakar *et al.*, 1992; Emerson *et al.*, 1990; Schechter *et al.*, 1991; Ikeda *et al.*, 1992; Kim and Kim, 1994; Lee and Bracco, 1994; Laforgia *et al.*, 1989; MacInnes and Bracco, 1990) and PFI engines (Zhao *et al.*, 1995a; Iwata *et al.*, 1986; Saito *et al.*, 1988; Sugimoto *et al.*, 1991; Zhao *et al.*, 1995b, 1996c); unfortunately, the number dealing with four-stroke GDI engines is much more limited (Meyer *et al.*, 1997), although a large portion of the basic information is applicable to both types of engines. The spray structure of an air-assisted injector that was developed for four-stroke GDI engines was analyzed by Miyamoto *et al.* (1996). The injector geometry that was used is illustrated in Fig. 4(a). As shown in Fig. 4(b), three main regions were identified for the spray structure. These were denoted as the unsteady, steady and stagnant-flow regions, respectively. In the unsteady region, the flow can be characterized by a starting vortex moving downstream. In the steady region, a fixed vortex forms below the poppet valve and air is entrained from outside of the spray cone. Also, small droplets form a solid-cone structure, due to the fact that small droplets are significantly influenced by the gas flow. The large droplets, on the other hand, maintain their trajectories due to their larger inertia, yielding a hollow-cone structure. Thus, the mean droplet size was found to be larger at the spray tip and near the surface of the spray cone, and smaller within the cone. A decrease in the spray cone angle was found to be associated with a slight improvement in the level of atomization.

The injection characteristics obtained using both pressure atomization and air-assist atomization were compared by Hoffman *et al.* (1997) by means of mass flux measurements using a spray patternator. The air-assisted atomizer utilized an outwardly-opening poppet, with the injection cycle starting with the opening of a fuel solenoid and charging the mixing cavity with fuel. Pressurized air then further charges the cavity until the poppet is forced to open at a specified cracking pressure. Due to the counteracting forces of spring tension and pressure, the poppet was observed to oscillate during the entire fuel injection process. This air-assist system operating with a total measured fuel quantity of 69 mm³ delivers 83% of the fuel within a 2.5 ms period even though the poppet oscillated for a total of 7.25 ms. This suggests that fuel delivery occurs during the first three poppet cycles. The air-assist system also exhibits a wide hollow-cone spray structure. By comparison, the pressure atomizer was found to provide a constant fuel delivery rate over the main part of the injection pulse-width, with the flow rate decaying only during the very last portion of the pulse due to the pintle closing. The high-pressure atomizer generates a solid-cone spray at the beginning of fuel injection, which then develops into a hollow-cone structure that eventually collapses. The mass flux data suggest that the interaction of the spray with the entrainment vortex

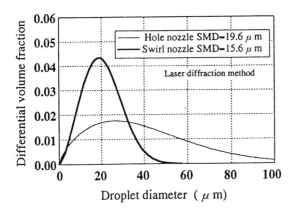

Fig. 3 Comparison of droplet size distribution between hole-type and swirl-type high-pressure injectors for an injection pressure of 20 MPa (Tomoda *et al.*, 1997)

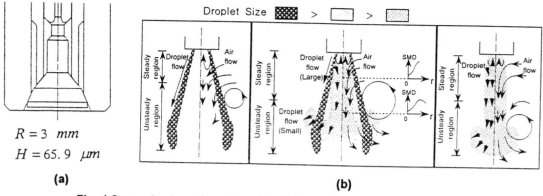

$R = 3 \ mm$

$H = 65.9 \ \mu m$

(a)

(b)

Fig. 4 Spray structure of an air-assisted injector (Miyamoto *et al.*, 1996); (a) schematic of the air-assisted injector; (b) schematic representation of the air-assisted hollow-cone spray structure

contributes to the spray sheet dispersion and the final collapse.

Negative Effects of Sac Volume - The sac volume within the injector tip plays an important role in the transient spray formation process because it contributes to the formation of large droplets at the initiation of fuel injection. This volume is actually the volume of liquid fuel within the injector tip that is not at the fuel line pressure. Therefore, a large sac volume degrades the mixture formation process and increases the UBHC emissions. Parrish and Farrell (1997) used laser diffraction and backlighting techniques to characterize the spray structure of a high-pressure GDI injector that was injecting into an ambient pressure of one atmosphere. A clear leading mass, or slug, which is mainly comprised of fuel that was in the sac volume was observed along the center line of the spray. For the fuel pressure range of 3.45 to 6.21 MPa, the initial axial velocity of the leading mass near the injector tip was in the range of 68 to 86 m/s, which is higher than the velocity of 45 to 58 m/s for the main body of the spray. Both the leading-mass and main-spray velocities decrease rapidly with time. The droplet size measurements in Fig. 5 show that the SMD initially rises rapidly as the leading mass of fuel first enters the measurement location of 38.75 mm from the injector tip along the center line of the spray. A peak SMD value is recorded for this initial slug before the instantaneous mean droplet size decreases rapidly for the main spray body. Patternator measurements of the transient mass flux show that a short injection duration concentrates more fuel mass along the injector axis than is achieved for longer durations.

The initial fuel slug associated with the sac volume of a fuel injector was also observed by Salters *et al.* (1996) inside a firing engine having a four-valve pentroof head and a centrally-mounted injector location. It was found that this slug, consisting of relatively larger droplets with higher velocity, penetrates tens of millimeters prior to the formation of the main spray cone, and impacts directly on the piston crown for early injection. It was also observed that at the fuel injection timing of 80° ATDC on intake, there is no major impingement of the fuel spray upon the piston except for this initial slug of fuel from the sac volume. However, for earlier injection timings, a large portion of the spray directly impacts on the piston crown. During the transition from the initial on-axis slug to a hollow-cone geometry, the spray cone-angle can be as large as 95°, which may result in the

fuel impingement on the edges of the open intake valves. Spray-edge impaction on the open intake valve for early injection is an interaction that should be monitored and avoided.

Performance of High-Pressure Swirl Injectors - The development of the spray from a GDI injector may be divided into discrete stages. The first is the initial atomization process that occurs at or near the injector exit. This is mainly dependent on the injector design factors such as nozzle geometry, opening characteristics, and fuel pressure. The second stage of spray development is the atomization that occurs during the spray penetration process, which is dominated by the interaction of the fuel droplets with the surrounding air flow field. The required spray characteristics change significantly with the GDI engine operating conditions. In the case of fuel injection during the induction event, a widely-dispersed fuel spray is generally required in order to achieve good air utilization for the homogeneous mixture. However, the impingement of the fuel spray on the piston surface and cylinder wall should be minimized. For injection that occurs during the compression stroke, a compact spray is preferred in order to achieve a stratified mixture distribution. At the same time, the spray should be very well atomized since the fuel must vaporize in a very short time (Kume *et al.*, 1996). It may be seen that a suitable control of spray cone angle and

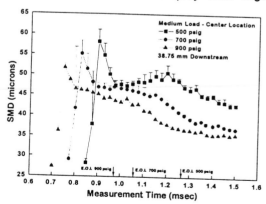

Fig. 5 Time history of the mean droplet size of a high-pressure swirl injector at an injection duration of 1.06 ms and an injection pressure of 4.83 MPa (Parrish and Farrell, 1997)

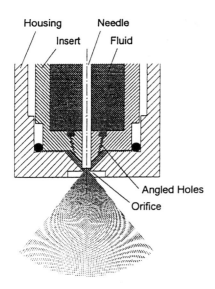

Housing   Needle
Insert   Fluid

Angled Holes

Orifice

Fig. 6 Schematic of the AlliedSignal high-pressure swirl injector (Evers, 1994)

penetration is advantageous, but is difficult to achieve in practice. The spray cone angle is an important parameter that is nominally determined by the injector design; however, in the actual application the spray cone angle also varies with the fuel injection pressure and the in-cylinder pressure. With the pressure-swirl injector, the spray angle generally decreases with an increase in ambient gas density until a minimum angle is reached (Zhao *et al.*, 1996b, 1997c). The ambient gas density also has a strong influence on the minimum atomization level produced by pressure-swirl atomizers. With non-swirl atomizers, however, an increase in ambient gas density generally yields a wider spray cone angle. This results from increased aerodynamic drag on the droplets, which produces a greater deceleration in the axial direction than in the radial direction. Under higher in-cylinder pressure conditions, corresponding to late injection at part load, a more compact droplet plume is required for a higher degree of stratifica-tion. In comparison, the high-pressure swirl injector is more applicable to GDI late-injection applications. Fraidl *et al.* (1996) studied the effect of ambient pressure on the spray cone angle for a range of injectors and injection pressures and concluded that variations in the ambient pressure have a greater impact on the spray cone angle than does the fuel injection pressure.

The spray structure of an AlliedSignal high-pressure swirl injector which utilized a fuel injection pressure of 6.9 MPa was characterized by Evers (1994). The injector design is shown schematically in Fig. 6. Four regions denoted as leading-edge, cone, trailing-edge, and vortex-cloud, were identified for the hollow-cone spray formed by this injector. The leading-edge region was found to show the largest droplet size due to the low fuel velocity at the beginning of the fuel injection pulse. After the pintle is opened fully, the fuel attains a steady velocity and a conical region of small droplets is formed. The trailing-edge region is produced as the pintle closes, whereas the vortex cloud region is formed by the circulating air that carries small droplets from the spray. It was found that the size of the droplets entrained into the vortex cloud is determined by the ambient air

properties. As the ambient air density is increased, the droplet size in the vortex cloud region also increases due to changes in the entrainment characteristics of the ambient air. The droplet size in the vortex cloud is not changed when the fuel injection pressure is increased, as the ambient air properties are not altered.

The detailed spray structure of Siemens high-pressure fuel injectors was studied by Zhao *et al.* (1996a, 1996b) using laser-light sheet photography and PDA techniques. The injector design is illustrated in Fig. 7(a). The spray photographs corresponding to injection into ambient air are shown in Fig. 7(b) for a fuel pressure of 5.5 MPa. Cross-sectional views of the spray structure at an axial distance of 50 mm are shown in Fig. 7(c). The spray photograph for the elevated ambient pressure of 1.48 MPa is shown in Fig. 7(d). A toroidal vortex was observed late in the injection event at reduced injection pressures or short injection durations. The circumferential distribution of the spray, as shown in the cross-sectional view, is fairly irregular. The spray tip penetration and the spray cone angle were found to decrease monotonically with an increase in the ambient pressure.

The characteristics of hollow-cone sprays generated by a high-pressure swirl injector with an exit-angle of 70° were predicted using the generalized-tank-and-tube (GTT) code with spray models, and were validated using PDA measurements by Yamauchi and Wakisaka (1996). The schematic representation of the injector nozzle geometry is illustrated in Fig. 8(a). The predictions were for a swirl angle of 40° and an initial droplet velocity of 60 m/s to simulate a spray injected at 7 MPa into air at one atmosphere. The interaction between the droplets and the gas flow were found to vary with the detailed spray structure. The interaction was found to be more pronounced for smaller droplets. For a monodisperse spray of 40 μm droplets, the droplets do not form a torus. The sprays that have different droplet size distributions have considerably different structures, although the fuel flow rates are nearly the same. It was also found that the spray shapes at the transition between cone growth and torus formation are quite different with and without fuel swirl. The spray cone angle for the case with swirl was found to be significantly larger than that for the non-swirl case. However, the spray-penetration characteristics for the swirl and non-swirl cases were found to be similar. It was reported that both the initial droplet diameter and the initial swirl momentum have more direct impact on the hollow-cone spray structure than the injection mass flow rate. As expected, the mean droplet size at the downstream region of the injector was found to increase with an increase in the ambient gas pressure. The mean droplet size also varies with time, increasing for the short duration just after the end of fuel injection.

As is schematically illustrated in Fig. 8(b), droplets in the mid-size range are found inside the coarse droplet region while droplets smaller than 10 μm do not form a hollow cone and are observed to concentrate towards the injector axis. The instantaneous spray cone angle was found to increase from nearly zero to the steady value in proportion to the needle opening time. Apparent vortex formation occurs only for droplets in the range of 10 to 25 μm, not for other droplet diameter ranges. Moreover, the total number of droplets in the diameter range of 10 to 25 μm was found to be larger than those in the other ranges. It was predicted that the size of droplets forming the toroidal vortex is about

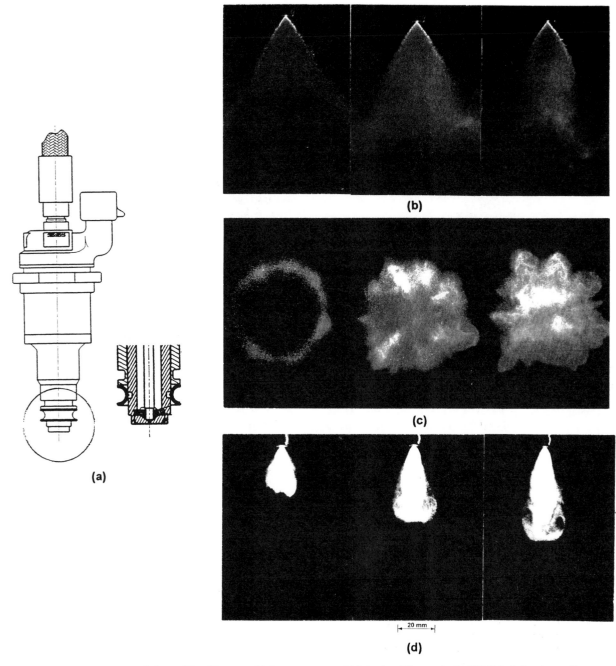

Fig. 7 Spray characteristics of the Siemens high-pressure swirl injector (Zhao *et al.*, 1996b); (a) schematic of injector nozzle design; (b) spray structure at 4 ms after injection start and injection pressures of 2.7, 5.5, 8.3 MPa (ambient pressure : 0.1 MPa; injection duration : 5 ms); (c) cross-sectional view of the spray structure at 4 ms after injection start and an axial distance of 50 mm and injection pressures of 2.7, 5.5, 8.3 MPa (ambient pressure : 0.1 MPa, injection duration : 5 ms); (d) spray structure at 2, 4, 6 ms after injection start and an ambient pressure of 1.48 MPa (injection pressure : 5.5 MPa; injection duration : 5 ms)

20 µm, and the droplets that move directly down the cone have a diameter of the order of 50 µm.

The spray atomization process of a Zexel high-pressure swirl injector was analyzed by Naitoh and Takagi (1996) using the synthesized-spheroid-particle (SSP) method developed at Nissan Motor Co.. The configuration of this nozzle is shown in Fig. 9. It was found that both the initial spray cone angle and the spray penetration decrease with

an increase in the ambient pressure. The mean droplet size in the spray cross section 50 mm from the injector tip also increases with the ambient pressure. The SMD of the cross section at 50 mm from the injector tip is larger than the total spray-averaged SMD during the injection event, indicating that the smallest droplets do not penetrate 50 mm.

In order to quantify certain spray characteristics that are not obtainable by other means, Hoffman *et al.* (1996)

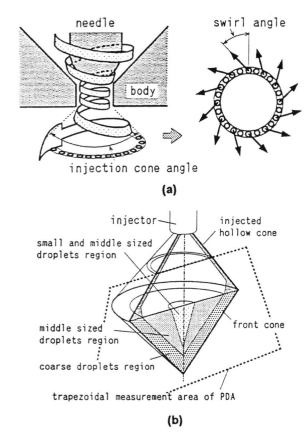

(a)

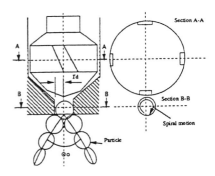

(b)

Fig. 8 Schematic and injector performance for a high-pressure swirl injector (Yamauchi and Wakisaka, 1996); (a) schematic of the nozzle geometry for a high-pressure swirl injector; (b) schematic of the droplet size distribution inside a hollow-cone spray at 0.9 ms after injection start and an injection duration of 0.8 ms

Fig. 9 Configuration of the nozzle of a Zexel high-pressure swirl injector (Naitoh and Takagi, 1996)

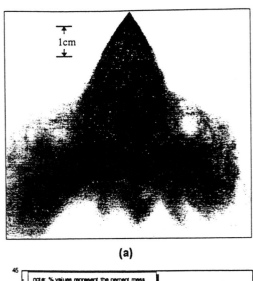

(a)

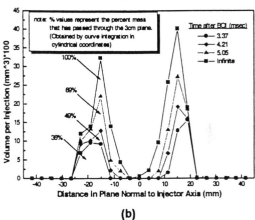

(b)

Fig. 10 Spray structure of a high-pressure swirl injector (Hoffman *et al.*, 1996); (a) spray structure at 4.21 ms after injection start; (b) averaged liquid fuel volume per injection passing through each of the 23 sample points at a plane of 30 mm below the nozzle tip

measured the spray mass flux distributions using a timed-sampling spray patternator. Stoddard solvent at 6.9 MPa was used instead of gasoline, with an injected quantity of 90 mm^3 and an injection duration of 5 ms. Figure 10(a) illustrates the spray structure at 4.21 ms after the start of fuel injection. Figure 10(b) shows the average liquid volume per injection that passed through each of the 23 measurement points equally spaced on a line 30 mm from the nozzle exit. The measurement results are plotted for four different times after the initiation of the injection. It is

evident that the distribution of the liquid mass within the spray is that of a hollow-cone, with over 99% of the liquid mass located in an angular ring with an inner radius of 5 mm and an outer radius exceeding 20 mm. The highest liquid flux was observed at a radius of about 15 mm. The substantial changes in the spray geometry with time can be easily observed from the figure. The thickness of the spray cone walls was found to increase with time, while the outer boundary of the spray, which would typically be used to identify the spray cone angle, does not change at all. The spray structure from the estimates of the average liquid mass flux passing through various points below the injector is similar to that observed by the laser-light-sheet method. The spray core was found to remain intact late into the injection period by the visualization. However, the liquid mass flux attributed to this spray core ends at 20 mm from the tip, after which the spray develops into a hollow cone. At a distance of 30 mm from the injector tip, an indentation ring is observed to form on the wall of the hollow cone. When compared with the spray structure shown in Fig. 10(a), the indentation position corresponds to the area where the toroidal vortex interacts with the hollow cone spray.

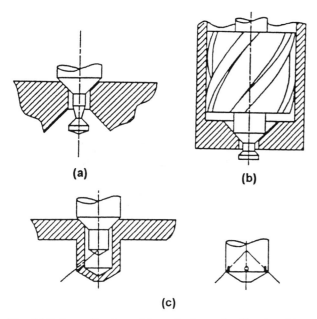

Fig. 11 Schematic view of three test nozzles (Pontoppidan et al., 1997); (a) Type I; (b) Type II; (c) Type III

Pontoppidan et al. (1997) studied the effect of the nozzle-tip design on the resulting spray characteristics. Three different nozzle configurations that were tested are shown in Fig. 11. Type I is a simple pintle-type nozzle having a single cylindrical metering hole, which produces a single jet solid-cone spray. Type II nozzle has a swirl channel upstream of the needle, which is designed to have a 45° main axis intersection between the flow vector and the nozzle axis, and generates a hollow-cone spray. Type III has a single cylindrical metering hole in which the fuel is injected into a sac volume having three non-metering outlet holes. This produces three individual atomized streams, each forming a solid cone. Variation in the circular symmetry of the spray from the Type I nozzle was observed, and it was reported that the spray quality was directly related to the surface quality of the pintle and its attachment to the needle. It was claimed that a GDI combustion system using this type of nozzle is more injector-dependent. In comparison, the Type II nozzle has a reduced spray penetration and improved mixture dispersion characteristics, which are more sensitive to the in-cylinder air flow. The mixture preparation of this type of injector is balanced between injector-dependent and chamber-design-dependent. The Type III nozzle produces a three-jet spray, which has a low spray tip penetration. It was emphasized that such a low-momentum spray requires an optimum in-cylinder air flow to achieve an adequate mixture preparation. As a result, the fuel/air mixing characteristics of this type of nozzle are more dependent upon the particular combustion-chamber-design. Also, the inherent sac volume may generate a spray with a larger droplet size. The Type III nozzle, however, is less sensitive to injector deposits, as the metering function is performed upstream of the outlet holes. This director-plate design has been utilized for more than a decade in PFI injectors to improve the deposit resistance.

The Toyota GDI combustion system uses a fuel line pressure that exceeds that of most of the current prototype GDI engines. A fuel line pressure of 5 to 7 MPa is quite common; however, the Toyota GDI engine uses 12 MPa

(Harada et al., 1997). The high-pressure swirl injector used in this application is illustrated in Fig. 12(a), and the resulting spray and atomization characteristics are shown in Figs. 12(b) and 12(c). It is evident that the spray is not well dispersed from the axis of the injection direction. The study of the effect of fuel injection pressure on the improvement in fuel economy indicates that at slow engine speeds a lower fuel injection pressure with a comparatively longer fuel injection duration is optimum. This can avoid very high rates of fuel injection, thus providing longer spray lengths that promote fuel evaporation and dispersion. At higher engine speeds, a higher fuel injection pressure combined with a shorter fuel injection duration is reported to be effective in avoiding the overdispersion of the fuel, which degrades stratification (Matsushita et al., 1996). It was claimed that for part-load operation at higher engine speeds, the injection duration must be as short, and the fuel pressure as high, as the dynamic range of the injector and the fuel pump design will allow.

In summary, a working design goal for the GDI fuel injection system should be to achieve a fuel spray of 25 microns or less in SMD, with 90% of the injected fuel volume (DV90) in drops smaller than 45 microns, with good spray symmetry, while requiring fuel pressures that are on

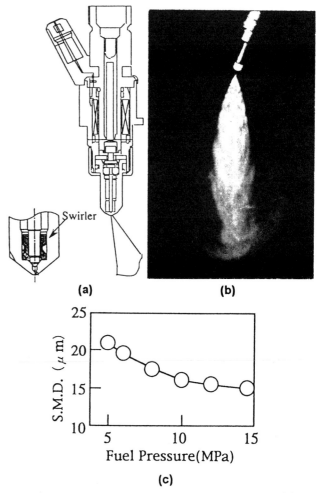

Fig. 12 Injector geometry and spray characteristics of the Toyota GDI injector (Harada et al., 1997); (a) nozzle geometry; (b) spray structure (12 MPa, 3 ms after injection start); (c) atomization characteristics

the order of only 4 MPa. This fuel pressure will generally provide a longer pump life than is obtained using the current 5 to 12 MPa, with less pump noise and a more rapid pressure rise during starting. It may also permit a fuel system without an accumulator to be used. The spray cone angle should ideally become smaller as the required engine air utilization is reduced. This can be achieved mechanically or electrically, or may occur due to the inherent collapse of a swirling cone of droplets that are injected into elevated air densities.

IN-CYLINDER FUEL/AIR MIXING PROCESS

Flow Field Characteristics - The transient in-cylinder flow field that is present during the intake and compression strokes of a GDI engine is another key factor that determines the operational feasibility of the system. The magnitude of the mean components of motion, as well as their resultant variations throughout the cycle, are of an importance that is comparable to that of the fuel injection system. On a microscopic scale, a high level of turbulence is essential for enhancing the fuel/air mixing process; but additionally, a controlled mean or bulk flow is generally required for the stabilization of the stratified mixture plume. For SI engines it is known that the turbulence velocity fluctuations near TDC on compression can attain the same order of magnitude as the mean velocity, and the turbulent diffusive transport and convective transport can be of equal influence in determining the initial state of the combustion process (Wirth et al., 1996; Kono, 1995). The integral scale of the mixture concentration fluctuation inside the combustion chamber can be as large as that of the velocity fluctuation (Zhao et al., 1993). This results in strong concentration fluctuations at a fixed position, such as the spark gap location, which can lead to difficulties in obtaining a stable flame kernel.

There are four key controlling features of the in-cylinder flow field, which are the mean flow components, the stability of the mean flow, the temporal turbulence evolution during the compression stroke, and the mean velocity near the spark gap at the time of ignition. For homogeneous combustion in the SI engine, the combination of high turbulence intensity and low mean velocity at the spark gap is desirable. This is generally achieved for PFI engines, and for GDI engines that operate exclusively in the early-injection mode. Therefore, a flow structure that can transform the mean-flow kinetic energy into turbulence kinetic energy late in the compression stroke is considered desirable. However, the GDI engine using late injection requires a flow field with an elevated mean velocity and a reduced turbulence level, which aids in obtaining a more stable stratification of the mixture. This indicates that the optimum flow field depends upon the injection strategy that is being used. For the GDI combustion system, control of the mixing rate by means of the bulk flow seems to have more potential than the scheduling of turbulence generation. This is not to imply that turbulence is not important to the combustion process. In fact, turbulence is known to be an important factor in entraining the EGR into the local combustion area (Kuwahara et al., 1996).

In general, a rotating flow structure exists within the cylinder and the combustion chamber, and this coherent structure has an instantaneous angle of inclination between the cylinder axis and the principal axis of rotation. The rotational component having an axis that is parallel to the axis of the cylinder is denoted as swirl, and the component having an axis that is perpendicular to the axis of the cylinder is denoted as tumble. The magnitudes of both the swirl and tumble components are very dependent on the particulars of the intake port design, intake valve geometry, bore/stroke ratio, and the shape of the combustion chamber (Hill and Zhang, 1994). Both swirl-dominated (Matsushita et al., 1996) and tumble-dominated (Kume et al., 1996) flow structures are used to achieve stratified combustion in GDI engines. For the tumble case, the fuel plume is deflected from a shaped cavity in the piston, and the vapor and liquid fuel is then transported to the spark plug. For the swirl-dominated flow field, the mixture cloud is concentrated at the periphery of the piston cavity (Daisho et al., 1990).

The swirl component of the in-cylinder motion generally experiences less viscous dissipation than the tumble component, therefore it is preserved longer into the compression stroke and is of greater utility for maintaining mixture stratification. It is usually combined with a squish flow that imparts a radial component to the motion as the piston approaches TDC on compression. A shaped cavity, or a reentrant or cylindrical bowl in the piston, may also be utilized to obtain the required turbulence production late in the compression stroke. The combined effects of squish and swirl lead to intensified swirl and an augmented turbulence intensity during the early portion of the combustion period.

The tumble-motion component of the flow field is transformed into turbulence near TDC by tumble deformation and the associated large velocity gradients, and can only be completed if the combustion chamber geometry is sufficiently flat. Otherwise, an incomplete tumble transformation will occur, which generally results in an elevated mean flow velocity at the spark gap. Also, tumble-dominated flow fields in GDI engines generally yield larger cycle-by-cycle variations in the mean flow than those obtained for swirl-dominated flows. These variations influence both the centroid and the shape of the initial flame kernel following ignition, but do not produce significant changes in the combustion period or flame speed (Takagi, 1996a). Furthermore, the tumble component of the motion tends to decay into large-scale secondary flow structures due to the effect of the curved cylinder wall, which makes maintaining a stable mixture stratification more difficult. With respect to turbulence generation, the presence of a significant tumble component is effective in enhancing the turbulence intensity at the end of the compression stroke, which is essential to compensate for the reduced flame speed of a lean stratified mixture. The tumble motion that is present early in the compression stroke rapidly decays into multiple vortices which have a size on the order of the turbulence length scale. This rapid transformation of kinetic energy into turbulence is not generally observed for swirl-dominated flow fields. The swirl flow continues to rotate, usually with precession, around the vertical cylinder axis for the entire time period from the beginning of the compression stroke to the end of TDC. The cylindrical geometry of the chamber is obviously quite favorable for maintaining a swirling flow with little viscous dissipation. It should be noted, however, that high-swirl-ratio flows can centrifuge the largest droplets from the fuel spray onto the cylinder wall, causing an increase in fuel wall wetting.

Many and varied approaches to GDI combustion systems

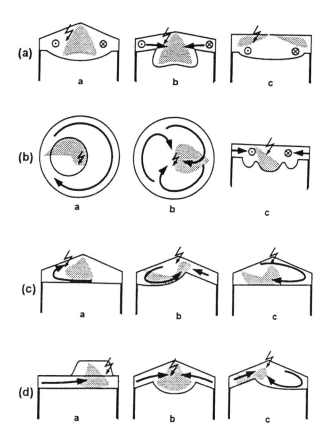

Fig. 13 Mixture preparation strategies for GDI combustion systems (Fraidl et al., 1996); (a) swirl-based systems with centrally-mounted injector; (b) swirl-based systems with centrally-mounted spark plug; (c) tumble-based systems; (d) squish-based systems

have been proposed with different combinations of in-cylinder charge motion (swirl, tumble, squish), combustion chamber shape, piston geometry, and spark plug and injector locations. The majority of GDI combustion systems that have been developed and reported to date, including the classic DISC engine concepts, utilize swirl as the basic in-cylinder air motion. The swirl-dominated flow is combined with either a simple open combustion chamber or a cylindrical or reentrant bowl in the piston. Some key examples of GDI combustion systems, as summarized by Fraidl et al. (1996), are illustrated in Fig. 13(a). All of these systems use a swirling in-cylinder flow to stabilize the mixture stratification. Ignition stability is maintained by positioning the spark gap at the periphery of the fuel spray. This arrangement generally requires special spark plugs with extended electrodes, which has led to some durability problems at higher engine power. Some special designs using a central spark plug and a non-central injector position are shown in Fig. 13(b). The concept of an off-axis piston bowl, fuel injection onto the bowl wall and central ignition in the main cylinder is depicted in Fig. 13(b)-a. An adaptation of the flow-collision concept to the GDI engine is shown in Fig. 13(b)-b. This concept invokes the flow-to-flow collision at the center of the combustion chamber where the ignition occurs. An open chamber that is designed to generate a quasi-divided chamber near TDC is illustrated in Fig. 13(b)-c.

As a result of its inherent rotational acceleration during compression, the tumble motion can be effective in creating high levels of near-wall flow velocities even relatively late in the compression stroke. This can promote the evaporation of a wall film that results from an impinging fuel spray. The transport of fuel vapor to the point of ignition may also be enhanced by this flow structure. In recent years, the tumble-dominated flow field is being intensively applied to GDI combustion systems. Engineers at Mitsubishi (Kume et al., 1996) and Ricardo (Jackson et al., 1994) first proposed the reverse tumble concept in conjunction with a specially-designed piston cavity which creates a stratified charge near the spark gap. The cavity is designed to control the spray impingement and flame propagation by enhancing the reverse tumble flow throughout the compression stroke, with the squish flow from the exhaust to the intake side of the chamber increasing the flame speed inside the piston cavity. Reverse tumble as the dominant in-cylinder air motion may be effective for designs in which the spark plug is centrally located and the injector is positioned below the intake valve. In such designs, the reverse tumble can be effective in moving the vapor and liquid fuel toward the spark gap after spray impingement on the walls of the piston cavity. The reverse tumble is achieved by a straight, vertical intake port having a high flow coefficient, thus enhancing the maximum engine power. This design is also effective in providing additional space in the cylinder head to accommodate the injector. Three additional examples of GDI combustion systems using tumble, as proposed by Fraidl et al. (1996), are shown in Fig. 13(c).

A comparison of the engine performance of a GDI engine that operates using both the swirl and reverse-tumble concepts show that these two types of flow fields provide similar light-load engine performance for the air/fuel ratio range of 35 to 40. However, for the high-load region in the air/fuel ratio range of 20 to 30, problems of combustion stability and smoke emissions are encountered for the swirl-dominated engines. Also, the required control system for this type of engine will generally be more complex in order to accommodate engine load transients (Iiyama, 1996).

Based upon experience gained in developing lean-burn engines, Yamada (1996) proposed a GDI engine concept using an "inclined swirl". This in-cylinder flow contains both swirl and tumble in a design that ostensibly combines the best features of these two flow structures. As reported by Furuno et al. (1990), an inclined swirl at an angle of $45°$ significantly enhances the turbulence intensity and provides a reduction in the COV of IMEP. Some examples that employ squish as the dominant motion for charge stratification are illustrated in Fig. 13(d) (Fraidl et al., 1996). The principle is to use the late squish generation for turbulence production to improve the mixture preparation and enhance the evaporation of fuel on the combustion chamber walls. For these squish concepts the amount of squish area needs to be carefully determined in order to control the initiation of knock by the extended crevice regions for full-load homogeneous operation.

Finally, it is important to note that in GDI engines the spray-induced flow field can exert a significant influence on the in-cylinder flow structure. Han et al. (1997) used the KIVA code to predict the effect of a spray-induced flow field on the flow structure inside the combustion chamber of a GDI engine having a centrally-mounted injector. For the early injection case, it was found that the momentum generated by the injected stream of liquid droplets is partially transferred to the surrounding gases, which increases the kinetic energy of the charge soon after the fuel is injected. This spray-induced flow enhances the in-cylinder fuel/air mixing; however, the increased kinetic energy rapidly decays as the piston moves up during the compression stroke, and the increase in the kinetic energy at TDC on compression over the non-injection case was found to be relatively insignificant. The spray-induced motion affects the large-scale, in-cylinder flow structures. In particular, it increases the mean velocities of the gases in the spray region, and significantly suppresses the intake-generated bulk flow for all of the injection timings considered in the study. For injection later than 150°ATDC on intake, the turbulence intensity as enhanced by fuel injection is substantially higher than that of the non-injection case. About 10% extra turbulence intensity is generated by the typical GDI spray when the initiation of fuel injection is retarded past 150°ATDC on intake. It was concluded that for operation in the early-injection, homogeneous mode, the later the start of the injection, the higher the turbulence intensity at TDC on compression.

Lake et al. (1996) predicted the variation of the turbulence intensity of a GDI engine for the injection timings corresponding to both homogeneous and stratified operation, using the Ricardo VECTIS CFD code. These predictions are shown in Fig. 14. A marked difference in the turbulence level was found between the early- and late-injection strategies. Early injection contributes to the generation of turbulence with an intensity twice that of late fuel injection, which exhibits the typical turbulence history of a typical four-valve PFI engine. The turbulence generated directly by the spray during the intake stroke was found to be sustained well into the compression stroke. The subsequent decay of the mean flow at a higher initial level of turbulence results in a correspondingly greater peak turbulence level near TDC on compression. It was also noticed that the peak of the turbulence intensity for the case of charge stratification with late injection is slightly higher than that of the conventional PFI engine. This is considered to be directly attributable to the turbulence induced by the spray.

Fuel/Air Mixture Preparation - The conditions inside the engine cylinder, such as the temperature, pressure and the air flow field exert a very substantial effect on the spray

atomization and dispersion, the air entrainment in the spray plume, and upon the subsequent fuel/air mixing process. The complex and time-dependent spray/air-flow-interaction process will determine the rate of fuel/air mixing and the degree of mixture stratification. The mixture preparation process is strongly dependent on the spray geometry, the in-cylinder flow structure, and the fuel injection strategy. Han et al. (1996, 1997a, 1997b) examined the effect of the in-cylinder flow field on the spray dispersion process inside the combustion chamber of a GDI engine using a modified version of the KIVA-3 code to simulate early fuel injection. The configuration analyzed was a center-mounted injector that injects fuel vertically into the cylinder during the intake stroke. It was found that the hollow-cone spray structure that was observed for bench tests in a quiescent environment is also obtained in the engine when injection occurs during the intake stroke. However, the intake-generated flow field does influence the trajectory of the injected spray, and the spray is deflected, with the spray-tip axial penetration being increased. Due to the combined effect of deflection and increased penetration, spray impingement on the cylinder liner occurs when the fuel is injected between 90° to 120°ATDC on intake, even though the spray is injected vertically. The intake flow has a greater influence on the spray for injection timings that are earlier than 90°ATDC in the intake stroke. The details of the spray/wall impingement depend not only on the injection timing, and upon the relationship between the spray and piston velocities, but also on the details of the instantaneous flow field. The amount of liquid fuel that impacts the wall may be as high as 18% of the total injected, leading to the formation of relatively rich vapor regions near the piston surface late in the compression stroke. The mixture formation histories vary with the fuel injection timings; however, the distributions of the mixture in the three different equivalence ratio ranges ($\phi > 1.5$, $1.5 > \phi > 0.5$, $\phi < 0.5$) are not significantly different for an injection timing later than 90°ATDC on intake. This was considered to be due to the fact that the remaining liquid in these cases, which is about 3% of the total fuel injected, is located near the piston surface, and the rich mixture in these regions is less affected by the in-cylinder flow field. Although variations in the injection timing result in different levels of charge stratification, the general trend with respect to the locations of rich and lean regions is not modified. It was concluded that the gross features of the charge distribution are determined by the injection and flow-field orientations. For the early-injection cases considered, the mixture is generally leaner in the main-chamber region and richer in the squish region, with the air/fuel ratio ranging from 8 to 24.

Optimizing the spray tip-penetration and cone-angle is one of the most important steps for minimizing the fuel/wall wetting. With substantial fuel impingement on the wall, improved fuel atomization can only partially enhance the mixture preparation. Pool-burning of the wall film will occur, along with the associated negative effects of increased heat loss and UBHC emissions. Dodge (1996a, 1996b) calculated the droplet penetration that is associated with fuel wall impingement in GDI engines by computing the drag coefficients of the droplets. The worst case that was analyzed was for a spray from a pressure-swirl atomizer with a cone angle of 52° which was found to maximize the penetration distance while preserving spray impact on the piston crown and cylinder wall. For the case of early injection of a spray

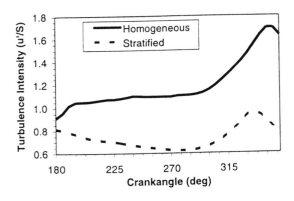

Fig. 14 The effect of the spray-induced flow field on the variation of turbulence intensity for both homogeneous- and stratified-charge operation (Lake et al., 1996)

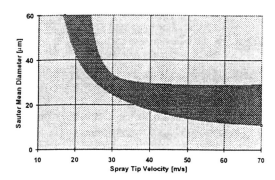

Fig. 15 The effect of spray-tip velocity on the mean droplet size for a wide range of injectors and fuel pressure levels (Fraidl *et al.*, 1996)

having a SMD of 15μm, it was found that most of the droplets decelerate to the very low air velocity prior to reaching the piston crown. Similar results were also found for late injection, in spite of the significantly reduced penetration distance that is available before the spray impacts the piston. This distance was 20 mm for late injection versus 80 mm for early injection. The rapid droplet deceleration is due mainly to the higher air densities that are encountered for the late injection condition, which results in increased droplet drag and enhanced vaporization.

The general relationship between the initial spray-tip velocity and the spray SMD is shown in Fig. 15 (Fraidl *et al.*, 1996) for a wide range of GDI injectors and fuel pressure levels. It is evident that there is a limitation on the spray-atomization benefit that can be obtained by increasing the fuel pressure. Any further increase in the initial spray-tip velocity may aggravate the problem of excessive spray penetration, but will not significantly enhance atomization. Currently most fuel systems for GDI engines utilize fuel pressures in the range of 5 to 7 MPa, although there are some systems that use fuel pressures of 10 to 12 MPa. The curve shows that optimum atomization occurs for initial spray-tip velocities that are in the range of 40 to 50 m/s, which is approximately twice the typical peak flow velocity of the in-cylinder flow field. The momentum of the fuel spray will therefore be a substantial addition to the momentum of the flow field. Since the clearance height is generally quite small in GDI engines, the spray of swirl injectors will likely result in some degree of wall impingement even though the initial spray-tip velocity is significantly less than is obtained with a diesel system.

Lake *et al.* (1996) obtained numerical predictions of the details of fuel/air mixing inside the cylinder of a GDI engine. For early injection, fuel was introduced into the cylinder between 170° and 190°ATDC on intake, whereas the corresponding values for late injection were 20° to 40°BTDC on compression. It was reported that for early injection the initial development of the spray is largely unaffected by the air motion that is generated by the high spray momentum. Spray impingement on the cylinder wall was predicted to occur at 185°ATDC on intake, and a rich mixture remains near the piston crown surface throughout most of the compression stroke. At a crank angle of 25°BTDC on compression, the bulk tumble motion begins to decay rapidly and a fairly homogeneous stoichiometric mixture is produced. It was claimed that over 90% of the injected fuel

is evaporated by 20° BTDC on compression. For the case of late injection, with the injection event initiated at 40°BTDC on compression, fuel is injected into relatively motionless air and impinges directly on the piston crown. It was found that only 50% of the injected fuel was evaporated by 20°BTDC on compression. This indicates that the particular geometric configuration that was analyzed would not be able to operate in the stratified-charge mode.

The interaction of the fuel spray with the in-cylinder air flow was investigated by Kono *et al.* (1990). Figure 16(a) shows the combustion chamber geometry that was used, and illustrates the locations of the injector nozzle and spark plug. A single-hole nozzle was used to inject fuel into the piston bowl of the engine. Three spark plug locations were selected, all having the same distance from the nozzle tip. As a result, three injection directions were used to direct the fuel towards the spark plug, including one with swirl, which is denoted as the forward direction (F), one radial, denoted as the central injection (C), and one against the swirl, denoted as the reverse direction (R). The measured engine performance and emissions obtained using these three different injection directions with different quantities of injected fuel are shown in Fig. 16(b). The KIVA calculations of the spray-dispersion characteristics inside the piston bowl indicate that significant bowl wall wetting occurs for injection in the forward direction. A rich fuel zone appears in the vicinity of the cavity wall and the spark gap. The fuel is dispersed by the swirl, and a lean mixture zone is formed downstream of the injector. As a result, the measured fuel consumption and UBHC emissions are higher for the forward direction than for the other injection directions. Large fluctuations in IMEP are obtained, indicating that ignition and combustion are not stable for this injection direction. For the case of reverse-direction injection, the tip of the spray is rapidly decelerated and the mixture cloud is formed in the vicinity of the spark gap. As a consequence, good fuel economy and stable combustion are obtained. For the central injection direction, the spray development and penetration are not significantly influenced by the swirl, and the tip of the spray penetrates across the cylinder bowl and impinges on the far wall. Therefore, a combustible mixture is not formed around the spark gap. Interestingly, the UBHC emissions for central injection were found to be equal to that for the reverse direction for all of the fuel amounts that were evaluated. This work demonstrates the important relationship between the air flow motion and the spray orientation.

The effects of the in-cylinder swirl and nozzle type on the resulting penetration of a gasoline, direct-injection spray were reported by Harrington (1984a). Six nozzle types of one, two and three-hole configurations were used to obtain spray penetration data for three swirl ratios and four in-cylinder pressure levels. It was found that the spray-tip penetration and trajectory was strongly dependent on both swirl ratio and the nozzle geometry. The variation in the spray-tip velocity with distance and time for in-cylinder injection of gasoline was also measured by Harrington (1984b). It was found that penetration histories based upon individual droplet drag were not accurate, as the spray tip is not comprised of a single collection of droplets during the injection event and does not experience the same drag force as an individual droplet. A three-regime method of penetration and velocity correlation was found to be necessary; an early-time regime that is determined by the injector

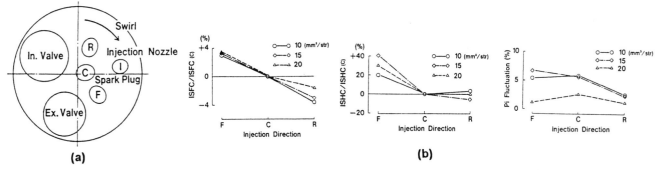

Fig. 16 The effect of fuel injection directions on engine performance and emissions (Kono *et al*, 1990); (a) configuration of the cylinder head; (b) measured engine performance and emissions (F : forward; R : reverse; C : central)

opening characteristics, a middle-time region that is controlled by spray-tip drag, and a late-time region that is dominated by the in-cylinder flow field.

In-Cylinder Charge Cooling - As a result of the thermodynamic effects of in-cylinder charge cooling, the GDI engine exhibits the advantages of higher torque and a higher knock-limited compression ratio as compared to the PFI engine. A low-pressure spray having fuel droplets of 120 to 200 μm SMD is injected into the intake port of the PFI engine, and a liquid film is formed on the intake valve and port wall. The liquid fuel vaporizes under the influence of concentration gradients, port vacuum, and by absorbing thermal energy from the valve and wall. For a GDI engine, in contrast, fuel is injected into the cylinder as a well-atomized spray that is vaporized mainly by absorbing thermal energy from the air. This is an efficient mechanism for cooling the charge air inside the cylinder. The decrease in the charge temperature can result in an increased mass of air in the cylinder during the induction process, as well as a slightly reduced loss of mass that is associated with the backflow near the end of the intake stroke. As a result, the in-cylinder trapped mass can be increased, thus increasing the volumetric efficiency. The cooling of the induction air due to the direct injection of a vaporizing fuel spray has been found to provide the following advantages :

- increased volumetric efficiency
- enhanced full-load torque
- reduced compression temperature
- reduced autoignition tendency at the same compression ratio
- higher knock-limited compression ratio
- reduced heat loss during compression at high load.

The decreased charge temperature that is associated with injection during the induction process was found to provide an increase of 2% in the trapped cylinder mass over the non-injection case; however, as would be expected, the gain in volumetric efficiency disappears when injection is retarded to the end of the intake stroke (Han *et al*., 1997a). At the other extreme, if the injection timing is advanced to the early stages of the intake stroke, fuel impaction on the piston can occur. The increased wall-wetting reduces the fuel vaporization rate, resulting in a smaller temperature-decrease and a smaller gain in the trapped mass. Therefore, for injection circumstances that result in significant wall wetting, the benefits of charge cooling will be diminished. The cooling of the intake charge also modifies and improves the engine heat transfer process, especially for early injection for high-load operation. The charge density is increased during the compression stroke, and the charge temperature is elevated to values higher than the wall temperature, with heat being transferred to the walls. The decreased charge temperature results in a smaller temperature differential late in the compression stroke, which decreases the heat loss. This heat transfer advantage diminishes rapidly as the injection timing is retarded towards BDC of the intake stroke.

Anderson *et al*. (1996a) estimated the magnitude of the charge cooling effect on volumetric efficiency for two extreme cases. One case is similar to that of the PFI engine where the fuel is vaporized only by heat transfer from the intake port and valve surfaces. The other case corresponds to the basic GDI engine concept where the fuel is vaporized only by absorbing thermal energy from the air. Assuming an initial intake air temperature of 100°C and a fuel temperature of 50°C, it was found that for direct injection the volume of the mixture after vaporization is about 5% smaller than the volume of intake air. Under the same condition, however, if fuel is vaporized and heated to the intake air temperature by heat transfer from the wall only, the mixture volume will increase by 2% due to the volume of the fuel vapor. Thus, the total difference in the mixture volume of the two extreme cases can be as large as 7%. However, it should be noted that these extreme cases are not totally descriptive of GDI and PFI processes, because some vaporization of fuel in air occurs in PFI engines, and some wall film vaporization of fuel occurs in GDI engines. Also, the compressed charge in a PFI engine is an air-gasoline mixture that has a lower specific heat ratio than air alone. This yields a lower temperature at the end of compression. Moreover, for the cold-start case, the fuel, air and engine cylinder wall have similar temperatures. As a result, the actual difference in the engine volumetric efficiency between the PFI and GDI engines can be significantly less than that computed from the ideal limiting cases, and is quite dependent on the specific engine design, fuel characteristics and operating conditions. At a constant pressure, the difference in the calculated charge temperature for the two extreme cases can be as large as 30°C, depending on the assumed intake air and fuel temperatures. The charge temperature for the case of injection during intake was found to decrease 15°C by the end of the induction process due to fuel evaporative cooling (Anderson et al., 1996a). This translates into a significant decrease in the gas temperature at the end of the compression stroke. For example, the temperature is reduced 116°C from 539°C for the non-injection case to 423°C when the fuel is injected at 120°ATDC on intake (Han et al., 1997a). Therefore, the GDI engine utilizing a mid-induction injection of fuel will exhibit a lower knocking tendency, and it is found that the knock-limited spark timing may be substantially advanced. Alternatively, the knock-limited compression ratio can be increased by as much as 2 full ratios for 91 RON fuels, thus easily achieving a significant gain in thermal efficiency. Lake et al. (1996) reported the benefit of the charge cooling resulting from early fuel injection to be an octane number improvement of 4 to 6, allowing an increase of compression ratio of up to 1.5. The reason for this improvement was thought to be a combination of in-cylinder charge cooling and changes in the end-gas air/fuel ratio which suppress detonation. The use of this allowable increase in the compression ratio will directly enhance the power and torque characteristics of the GDI engine, and will result in an immediate and substantial improvement in engine BSFC.

Takagi (1996a) reported that with a stoichiometric air/fuel ratio, intake charge cooling has the effect of lowering the air temperature by approximately 20°C. As a result, even under WOT operation at low engine speed, the GDI engine exhibits an improvement in the power output of 6% as compared to that of the PFI engine. This improvement is attributed in part to the higher intake manifold absolute pressure (MAP) resulting from the improved charging efficiency. Another contributing factor is the ignition timing advance made possible by the reduced knock tendency. The effect of injection timing on the GDI engine volumetric efficiency was compared with that obtained for the baseline PFI engine by Anderson et al. (1996a). As shown in Fig. 17, the volumetric efficiency improvement was found to be about 1/3 of the theoretical maximum difference, or about 2.5%, and exhibited a strong dependence on the injection timing. When injection occurs at the beginning of the intake stroke, the piston is close to the injector and is moving at a low velocity. For this case, fuel impaction on the piston crown occurs, and the amount of energy supplied by the

induction air to vaporize the fuel is reduced. Retarding the injection timing to later in the induction stroke, when the piston is moving away with a much higher velocity, reduces the spray impaction and results in an increase in the volumetric efficiency. If injection occurs too close to the time of intake valve closure, the fuel droplets will have insufficient time to vaporize before the intake valve closes. For this timing the charge-cooling effect is negligible. Therefore, the benefit of in-cylinder charge cooling is only fully realizable within a relatively narrow window of injection timing.

COMBUSTION CONTROL STRATEGIES

Over the past twenty years, a number of concepts have been proposed to exploit the potential benefits of direct gasoline injection for passenger car applications, but until recently none were incorporated into a production GDI engine. The primary reason for this was the lack of controllability in the fuel injection system. Systems that were based upon diesel fuel systems and pressure-activated poppet nozzles experienced significant limitations on performance and control. Although the engines used in the previous attempts were reasonably successful in producing improvements in BSFC, the mechanical pumps that were utilized had limited speed and timing ranges, and the power of these engines were often less than that of the diesel engine. The reexamination of the GDI engine over the last five years has benefited from the application of electronic, common-rail, injection systems to four-stroke GDI engine requirements. It should be noted that this type of injection system has also been utilized on two-stroke DI gasoline engines and has long been the production PFI hardware. Such injection systems provide fully flexible timing over the entire speed range of current gasoline engines by allowing a strategy of optimum injection timing at both full and part load. According to Lake et al. (1996), the GDI engine equipped with an electronic common-rail system has the ultimate potential of achieving a fuel economy equivalent to that of the diesel engine and of achieving a specific power output that is equivalent to that of the PFI engine.

Injection Timing and Control - The objective of charge stratification in the GDI engine is to operate the engine unthrottled at part load at an air/fuel ratio that is leaner than is possible with the conventional lean-burn or homogeneous mixture. This is achieved by creating and maintaining charge stratification in the cylinder such that the air/fuel ratio at the spark gap is compatible with stable ignition and flame propagation, whereas areas further from the point of ignition are either very lean or devoid of fuel. In general, air/fuel mixture stratification is realized by injecting the fuel into the cylinder during the compression stroke; however, it may also be possible to achieve stratification with early injection, and some success towards this goal has been obtained with an air-assisted fuel system (Ghandhi and Bracco, 1995). The use of charge stratification with an overall lean mixture can provide a significant improvement in engine BSFC (Kume et al., 1996). This is obtained primarily by significantly reducing the pumping losses that are associated with throttling, but there are also additional benefits such as reduced heat loss, reduced chemical

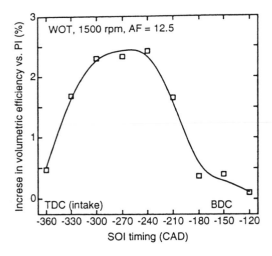

Fig. 17 Increase in volumetric efficiency of the GDI engine over the PFI engine for the early-injection mode (Anderson et al., 1996b)

dissociation from lower cycle temperatures, and an increased specific heat ratio for the process.

There is a consensus that achieving stable, stratified-charge combustion while controlling the engine-out UBHC emissions to a very low level is a difficult task. The interrelationships of injector location, spray characteristics, combustion chamber geometry, EGR rate, injection timing, and spark timing are quite complex, and must be optimized for each system. The GDI engine operating range in which charge stratification can be effectively utilized to obtain the available thermodynamic benefits should be chosen to be wide enough to cover the most frequent engine operating conditions (Karl et al., 1997). In general, it is exceptionally difficult to achieve full air utilization with a late injection strategy due to incomplete mixing, which may lead to excessive smoke emissions. For most prototype GDI combustion systems, part-load, stratified operation with acceptable values of COV of IMEP has proved to be quite difficult to achieve. It should be noted that no stratified-charge GDI engine has yet demonstrated a sufficiently low level of engine-out NOx for the attainment of ULEV emissions without a lean-NOx catalyst.

If the stratified charge mode is extended to high-load operation, smoke is likely to be produced due to over-rich regions near the spark gap, even though the amount of air is maximized by the absence of throttling. In this case the smoke problem can be alleviated by operating the engine in the homogeneous-charge mode using early injection. The injection timing has been found to be an influential parameter in suppressing the soot formation inside the combustion chamber of GDI engines. A carefully designed transition between the low-load stratified and high-load homogeneous mixture formation is important for realizing particulate-free combustion (Pischinger and Walzer, 1996). It was reported by Kume et al. (1996) that the stratified-charge mode using late injection can be extended to 50% of full load without an apparent increase of the soot for the Mitsubishi GDI engine. It was theorized that sufficient excess air exists in the cylinder for the soot to oxidize rapidly. When the load exceeds 50%, however, soot particulates begin to be emitted from the GDI engine. For the DI diesel engine the amount of fuel that can be injected must be restricted so as to limit the soot emissions to permissible levels, resulting in lower performance. In the GDI engine, however, this problem can be circumvented over the entire engine operating range without degrading the engine performance by adjusting the fuel injection timing (Matsushita et al., 1996).

It has been demonstrated that the strategy of early injection in the GDI engine can meet the requirements for higher loads. In combination with a complementary in-cylinder charge motion, this strategy has the potential of providing full-load performance that is comparable to that of conventional PFI engines (Anderson et al., 1996b). Early injection also makes it possible for the engine to operate at a slightly higher compression ratio due to an improved octane requirement, which can provide an additional incremental improvement in fuel economy. An early injection strategy with an essentially homogeneous charge can also achieve a number of benefits in the areas of cold-start and transient emissions.

The amount of fuel/wall impingement is known to vary significantly with the injection timing and engine speed, thus the injection timing must be optimized in order to avoid

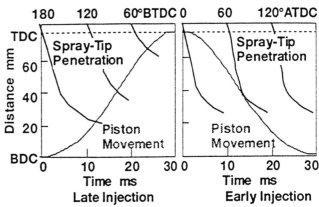

Fig. 18 Spray-tip penetration and piston trajectory for an engine speed of 1000 rpm (Kume et al., 1996)

spray overpenetration and wall-wetting (Anderson et al., 1996a; Kume et al., 1996). There is a general consensus that the timing for early injection should be adjusted so that the spray tip chases the receding piston without impacting it. Figure 18 shows a comparison of spray tip penetration and piston surface trajectory at 1000 rpm (Kume et al., 1996). To enhance the fuel evaporation and the fuel/air mixing process, it is necessary to set a minimum time interval between the end of fuel injection and the occurrence of the spark in order to avoid an over-rich mixture near the spark plug. As a result, the injection timing should be advanced as the engine speed increases. According to Matsushita et al. (1996), fuel injection must occur between TDC on intake and 160°BTDC on compression in order to provide enough time for the fuel to vaporize. For the avoidance of spray impaction, start-of-injection timing is the most useful, whereas for mixture preparation an end-of-injection timing is the most meaningful parameter. Both should be recorded during GDI engine development programs.

For the early injection mode, a later injection timing reduces the ignition delay and advances the 50% heat release point due to the presence of a rich mixture zone around the spark gap. However, retarding the injection timing does not significantly change the combustion duration. In the late-injection, stratified-charge mode, an excessively rich mixture near the spark gap must be avoided in order to maintain a stable ignition and to minimize smoke. Therefore, it is imperative that a mixture preparation strategy that will accommodate a wide range of engine operating conditions be developed for each specific GDI design (Ando, 1996a). A moderate crank-angle window exists for the timing of early injection for homogeneous charge combustion, but only a very narrow operating window is available for the timing of late injection for stratified-charge combustion. At higher loads, a longer injection duration with the accompanying richer mixture results in a decreased sensitivity to injection timing (Jackson et al., 1997).

Two fuel injection strategies that have been utilized are illustrated in Fig. 19. Both the early injection mode for homogeneous-charge operation and the late injection mode for stratified-charge operation are widely utilized in current GDI combustion systems; however, misfire or partial burns may occur during the transition between these modes due

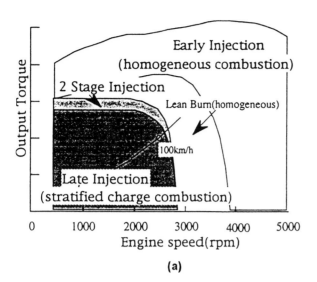

(a)

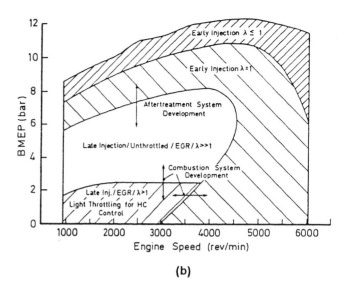

(b)

Fig. 19 Fuel injection strategies that have been utilized; (a) Toyota GDI engine (Matsushita *et al.*, 1996); (b) Ricardo GDI engine (Jackson *et al.*, 1996)

to the overall lean mixture. For the transition region, Matsushita *et al.* (1996) proposed a two-stage or split-injection strategy that injects portions of the fuel into the cylinder during both the intake and the compression strokes in order to form a weak stratification. As a result, excessively rich or lean mixtures can be avoided and stable combustion can be obtained from an overall lean mixture. It was reported that this two-stage injection permits a smooth transition from lean-stratified combustion to full-load, homogeneous combustion.

Miyamoto *et al.* (1994) and Hattori *et al.* (1995) conducted an extensive study of the potential of two-stage injection. It was reported by Miyamoto *et al.* (1994) that stable combustion could be realized in this two-stage-injection engine over a wide range of operation without detonation. Optimized combinations of spark timing, secondary fuel injection timing, fuel fraction between the primary and secondary injections, and spray geometry were found to be essential for improved ignition and combustion. It was also found that less volatile fuels and lower octane numbers could be used with this two-stage injection process, and efficient combustion without knock was possible even with gasoline containing 50% diesel fuel. According to Hattori *et al.* (1995), the optimum fraction of fuel in the secondary injection had to be controlled carefully in order to improve both the lean limit and the BSFC. If the fraction of fuel in the secondary injection is too high, the main pre-mixed homogeneous mixture becomes too lean and the normal flame propagation of the lean homogeneous mixture will be hindered.

Considerations in Locating the Injector and Spark Plug - The location and orientation of the fuel injector relative to the ignition source are very critical geometric parameters in the design and optimization of a GDI combustion system. During high-load operation the selected injector and spark plug positions must promote good mixing with the induction air in order to maximize the air utilization. For late injection, the chosen configuration should ideally provide an ignitable mixture at the spark gap at the ignition timing that will yield the maximum work from the cycle for a range of engine speeds. In general, no single

set of positions is optimum for all speed-load combinations, thus the positioning of the injector and spark plug is nearly always a compromise.

In converting successful designs of PFI engine combustion chambers to GDI applications, a number of additional requirements must also be met. To minimize the flame travel distance and to increase the knock-limited power for a specified octane requirement, a single spark plug is generally positioned in a near-central location. This location usually provides the lowest heat losses during combustion for open-chamber designs. Once a near-central location has been specified for the spark plug, numerous factors must be taken into account in positioning and orienting the fuel injector. The location selection process can be aided by laser diagnostics (Felton, 1996) and CFD analyses (Kono *et al.*, 1990) that include spray and combustion models; however, spray models and wall-film submodels are considered to be in the development stage. This means that much of the verification work on geometric positioning will consist of prototype hardware evaluation on an engine dynamometer. The basic considerations are that the injector should be located in a position that can provide a stable stratified charge at light load, a homogeneous mixture with good air utilization at high load, and avoid the impingement of fuel on the cylinder wall. Other important factors include the injector tip temperature and the fouling tendencies of the spark plug and injector, the trade-off between intake valve size and injector location, and the design constraints of injector access and service. The effective area available for the intake valves is noted because most proposed injector locations generally reduce the space available for the engine valves. There are six important parameters that will influence the final selected location of the spark plug and injector. These are :

- injector spray characteristics
- structure and strength of the in-cylinder flow field
- injector body and tip geometry
- spark plug design and the extent of electrode extension
- combustion chamber geometry
- piston crown geometry.

Generally the GDI injector should not be located on the exhaust side of the chamber, as injector tip temperatures of

124

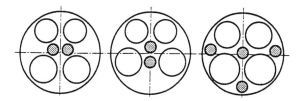

Fig. 20 Valve size limitation for optional injector and spark plug locations (Fraidl *et al.*, 1996)

more than 185°C may be obtained, causing injector deposits and durability problems as discussed in detail in the section of injector deposits. Both external and internal deposits can occur at these temperature levels, resulting in fuel spray distortion and fuel metering errors. The fuel injector was positioned near the center of the chamber in tests of the Isuzu GDI engine (Shimotani *et al*, 1995, 1996), and in the Ford GDI engine (Anderson *et al.*, 1996a, 1996b), with the spark plug being located near the spray cone. This location is preferable in many chamber designs as it assures the presence of a rich mixture near the spark gap at the time of ignition. In addition, the vertical, centrally-mounted injector location has the advantages of distributing the atomized fuel symmetrically in the cylindrical geometry, thus promoting good air utilization (Zhao *et al.*, 1997a, 1997b). For late injection, however, the centrally-mounted, vertical location directs the fuel spray at the piston and is likely to yield higher UBHC emissions than an optimum inclined orientation. Also, it should be noted that the combustion stability of a GDI engine that uses a centrally-mounted, vertical injector for stratified mode operation is very sensitive to variations in the fuel spray characteristics. For example, variations in spray symmetry, skew or cone angle due to production variations or injector deposits can result in excessive values of COV of IMEP. For significant distortions of the expected baseline spray geometry, misfires and partial burns can occur. Thus, for GDI engine designs that incorporate a centrally-mounted, vertical fuel injector, maintaining the designed spray geometry and quality is critical. The requirement for very close spacing of the injector and spark plug in some designs results in an additional reduction of valve sizes. It also introduces considerations of ignition fouling from the impingement of the spray periphery on the electrodes. The formation of injector deposits is always a concern, but placing the injector tip closer to the ignition source may aggravate the problem.

The valve size limitation for optional injector and spark plug locations is shown in Fig. 20 (Fraidl *et al.*, 1996). The significant design constraint is evident, and necessitates the use of auxiliary methods for improving the volumetric efficiency, such as optimizing the intake port design. According to Iwamoto *et al.* (1997), the centrally-mounted injector location also increases the possibility of smoke emissions. Some cases of spark-plug fouling have been noted for spark gaps positioned within the periphery of the fuel spray from a centrally-mounted injector. Spark plug fouling resulting from the impingement of liquid fuel can be reduced by using a very high-energy ignition system, which, however, may affect the durability of the spark plug.

Two important ways to increase the probability of a stable stratified mixture are to decrease the time interval between injection and ignition and to decrease the distance between the injector tip and the spark gap (Tomoda *et al.*, 1997). However, decreasing the distance between the injector and the spark gap decreases the time available for mixture preparation, which generally has a negative effect on UBHC emissions and soot formation. Conversely, improved fuel/air mixture formation at the sacrifice of some combustion stability can be obtained by increasing the separation between the spark gap and the injector tip, or by increasing the time delay between injection and ignition. For open chamber designs in which the stratification is supported mainly by charge motion, a stable stratification can be obtained with directed spray impingement on the piston crown or bowl cavity, with subsequent transport of the fuel vapor towards the spark plug by the charge motion (Ando, 1996a). With the use of a specially-contoured piston and an optimized balance of tumble and swirl, the transport of the fuel that impinges on the piston can be controlled to obtain either stratified or homogeneous combustion.

Miok *et al.* (1997) analyzed the mixture formation process that is associated with an injector located at the periphery of the cylinder between the intake and exhaust valves. For this study, the spray was directed towards the center of the combustion chamber at an angle of 70° down from the horizontal plane. It was found that this injector location and orientation produced a poor mixture distribution around the spark plug with a flat piston; however, this may be improved by modifying the piston crown geometry. Lake *et al.* (1996) also verified that an optimized piston-bowl position is very important in directing the fuel spray towards the spark plug when using an injector mounted at the periphery of the cylinder.

The concept of locating the fuel injector underneath the intake port and between the two intake valves and positioning the spark plug at the center of the cylinder has been invoked by a number of research groups (Kume *et al.*, 1996; Takagi, 1996a, 1996b; Lake *et al.*, 1994). This geometric configuration is considered to be a key element of a compact design for a multi-cylinder GDI engine, particularly for smaller bore diameters (Jackson *et al.*, 1997). It provides an improved entrainment of the injected fuel into the induction air, as well as enhanced cooling of the injector tip (Lake *et al.*, 1994). As opposed to the configuration using a centrally-mounted injector, systems using spray impingement on solid surfaces to create a stratified mixture generally have reduced sensitivity to variations in spray characteristics. Hence, these systems tend to be more robust regarding spray degradation from deposits, pulse-to-pulse variability or part-to-part production variances in the spray envelope. Shimotani *et al.*, (1995) reported that locating an injector at the periphery of the cylinder on an Isuzu single-cylinder GDI engine produced higher UBHC emissions and fuel consumption than was obtained with the injector located at the center of the combustion chamber. This was theorized to be due to fuel penetrating to the cylinder wall on the exhaust valve side, with an increased absorption of liquid fuel and desorption of fuel vapor by the lubricating oil film (Zhao *et al.*, 1997a). The impingement of fuel on the piston cavity forms a film of liquid fuel on the solid surface, which can result in some pool-burning and an increase in UBHC emissions. An additional consideration in using a spray-impingement system is the necessary compromise that must be made in shaping the piston crown in order to obtain part-load stratification. This compromise will most likely be somewhat detrimental to the air utilization at the high-load conditions for which a homogeneous mixture is required. The piston curvature may even increase the heat losses from the combustion gases.

Some additional combinations of fuel injector and spark plug locations proposed by Fraidl *et al.* (1996) are illustrated in Fig. 13. Some of these proposals have been evaluated and found to be lacking in one or more key areas. For example, predictions of the equivalence ratio at the spark gap for a four-valve configuration with a central spark plug and the injector positioned at the bore edge between the intake valves showed that the mixture at the spark gap is too lean at the time of ignition with a flat-top piston and late injection timings (Jackson *et al.*, 1997).

The spark plug electrodes must always be positioned in a zone of ignitable and combustible mixture when the spark occurs, and this position is affected by the swirl and tumble ratios, spray cone angle, mean droplet size, and injection and spark timings. Guidelines have been reported on the recommended position of the spark gap for open chamber GDI systems by many researchers, such as Fraidl *et al.* (1996) and Ando (1996b). For instance, for systems that do not use wall impingement, the spark plug electrodes must generally project into the periphery of the spray cone in order to ignite the combustible mixture, and should also be positioned to shield the gap from being affected by the bulk flow. It is known that the optimum spark plug location changes with engine load, thus any one configuration installation represents a compromise location for the entire operating map. For low-load operation, it is suggested that the spark gap be near the injector tip; however, too short a distance will cause the system to be too sensitive to variations in the fuel injection system.

Combustion Characteristics of Gasoline Direct-Injection Systems - The combustion characteristics of the GDI engine change significantly with the combustion control strategy that is used. Using flame luminosity analysis, Kume *et al.* (1996) reported that combustion associated with early injection is characterized by flames that are typical of premixed lean or stoichiometric mixtures. The flame luminescence that is observed is attributed to OH and CH chemiluminescence, as well as to CO-O recombination emission. Luminescence at the longer wavelengths normally associated with soot radiation was not observed. For the case of late injection, it was found that the major component of the flame luminosity consists of continuous blackbody radiation emitted from soot particles formed inside the cylinder, which is typical of a stratified-charge combustion process. The soot radiation was found to decrease abruptly after sufficient air has been entrained into the reaction zone. For heavy-load operation, soot may be generated for some injection timings due to the formation of a liquid film on the piston crown. This can occur for injection early in the intake stroke. Another cause of soot generation is an insufficient fuel/air mixing time, which can occur for injection very late in the intake stroke. Many design parameters such as the piston crown geometry, spray cone angle, injection timing, and the in-cylinder air motion must be optimized to minimize soot formation.

When the air/fuel mixture is successfully stratified for an idle or low-load condition, the mixture around the spark gap is designed to be slightly rich. If this is achieved, the reaction rate will be high enough to sustain efficient and stable combustion. For the PFI engine operating at the idle condition, the combustion rate is low, and the combustion stability is marginal, primarily because of the large fraction of residual gas. For the GDI engine at idle, the initial combustion rate was reported to be approximately the same as that for the full-load condition (Toyota, 1996). According to Jackson *et al.* (1996), the GDI engine demonstrates a significant reduction in both the ignition delay and the burn duration when compared to a PFI engine of equivalent geometry. After ignition the flame develops rapidly in the rich mixture region near the spark gap; however, the rate of flame propagation is reduced in the lean outer region of the stratified charge. The significantly reduced combustion rate near the end of the combustion process is one of the causes for the observed increase in UBHC emissions (Karl *et al.*, 1997). The combustion rate and stability are enhanced rather than degraded with a reduction in idle speed. As a result, the idle fuel consumption of a GDI engine can be less than half of that required for a PFI engine if ultra-lean air/fuel ratios are utilized (Toyota, 1996). The maximum brake-torque (MBT) ignition timing of the stratified-charge GDI engine at idle and part load is generally more advanced than that of the conventional PFI engine. Fraidl *et al.* (1996) reported that the main part of the stratified GDI combustion occurs before TDC on compression for MBT timing, which is quite advanced. For the GDI engine operating in the homogeneous mode at full load, a heat release curve is obtained that is nearly identical to the PFI engine. A slightly reduced heat release rate may be observed for the GDI engine, which is indicative of some charge non-homogeneity.

The throttling losses of the gasoline SI engine are relatively small for high-load operation. For this mode, the efficiency of both the PFI engine and the GDI engine is determined primarily by the compression ratio and the combustion characteristics; however, increases in the compression ratio and advances in the ignition timing for best efficiency are limited by mixture autoignition. Improvements in combustion-chamber geometry, in piston and charge cooling, and in residual gas control to modify the flame propagation at high load have proven to be effective means for knock reduction. Autoignition generally occurs in the end gas region where flame arrival is delayed. Modification of the charge motion in order to obtain symmetric flame propagation is an effective way to improve the knock behavior (Wojik and Fraidl, 1996). In general, the best tradeoff among UBHC, NOx, BSFC, and COV of IMEP can be obtained by a combustion process that offers a fast and stable initial phase, a moderate main combustion rate and a locally uniform end of combustion. Anderson *et al.* (1996a) studied the effect of injection timing on the knock-limited spark advance in a GDI engine and compared the value to that obtained for a PFI engine. As shown in Fig. 21, the knock-limited spark advance continues to increase as the injection timing is retarded for the heavy-load condition. Retarding the injection timing not only reduces wall wetting by the fuel spray but also reduces charge heating due to heat transfer from the walls. This results in a lower mixture temperature near the end of compression, and permits a more advanced spark timing; however, retarding the injection timing does reduce the time available for mixing, which generally results in a higher cycle-by-cycle variation

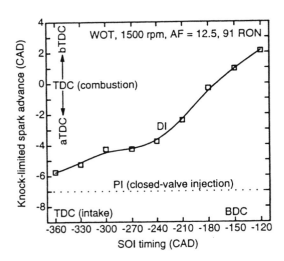

Fig. 21 Comparison of knock-limited-spark-advance between the GDI and PFI engines (Anderson *et al.*, 1996b)

in the IMEP.

It is necessary to carefully phase the injection timing with the ignition timing in order to achieve accurate control of mixture stratification. Jackson *et al.* (1997) reported that the combustion stability of a tumble-dominated, stratified-charge, GDI engine can be improved significantly when the ignition is advanced even beyond the MBT timing; however, when ignition is advanced further the combustion stability deteriorates very rapidly, which is attributed to insufficient time for evaporation and mixing. UBHC emissions were found to decrease with ignition advance, whereas fuel consumption initially decreases before increasing. The optimized ignition timing for the lowest COV of IMEP was found to be in the range of 13 to 23°BTDC on compression. It was reported that the optimum ignition timing for the best trade-off between UBHC emissions and COV of IMEP is around 23°BTDC on compression. At this timing, however, combustion phasing is too far advanced for best fuel economy, as 50% of the charge is burned at 5°BTDC on compression, as compared to 7°ATDC on compression for best fuel economy. NOx emissions are also relatively high due to the high local gas temperature. By using a local rich mixture near the spark plug, while maintaining a constant overall lean air/fuel ratio, Arcoumanis *et al.* (1996) reduced the cyclic variations of IMEP by more than 60% in a modified PFI engine. It was reported that 40% of the observed improvement can be attributed to the presence of the rich mixture around the spark gap, with the remaining 60% attributed to the enhancement of mixing due to the injection event. Ronald *et al.* (1996) reported that the limiting compression ratio for the GDI engine can be as high as 15:1 for part-load operation. At high load, however, detonation when operating with a compression ratio higher than 12 is a problem with gasolines currently available in the field. It was noted that an expensive and complex solution could be a variable compression ratio, with a higher compression ratio utilized at part load.

The combustion characteristics associated with early fuel injection under cold-start conditions were investigated in detail by Shimotani *et al.* (1996). A homogeneous, stoichiometric mixture was used for all of the cold-start tests. Figure

22(a) shows the measured in-cylinder pressures and mass burning rates for fuel injection timings of 30°ATDC on intake (-300° ATDC in the figure) and 110° ATDC on intake (-250°ATDC in the figure) at a coolant temperature of 20°C. For the very early injection timing of 30°ATDC on intake, the heat release rate during the last half of combustion is quite low due to the slow evaporation of liquid fuel from the piston crown. As a result, the UBHC emissions and BSFC are degraded. The effect of injection timing on the combustion delay and the 90-percent-burn duration at a coolant temperature of 20°C is shown in Fig. 22(b). Early injection timing delays the initial combustion and extends the main combustion period due to the lean mixture resulting from the slow vaporization. Visualization of the spray impinge-ment on the piston shows that for an injection timing of 30°ATDC on intake the spray impingement velocity is higher, and the impingement footprint is smaller, than for later timings. Due to the brief time available before impingement, the contribution of the in-cylinder air flow field to spray evaporation is limited and, as a result, a substantial liquid film is formed on the piston crown. For a later injection timing, with the piston moving away from the injector more rapidly, the impingement velocity of the spray is reduced and the impingement footprint is larger. As a consequence, the thickness of the fuel film on the piston crown is reduced and the evaporation of fuel droplets in the ambient air flow field is enhanced due to the increased time available prior to impingement. All of these factors contribute directly to a reduction in the amount of fuel on the piston.

The visualization of combustion indicated that for the case of early injection (30°ATDC on intake), the entire chamber is initially filled with a blue flame, after which a yellow flame is observed at the center of the chamber. This yellow flame persists until the beginning of the exhaust stroke, and is attributed to pool burning of the film of liquid fuel on the piston surface. The presence of a yellow flame is quite limited for an injection timing of 110°ATDC on intake, and it occurs at the periphery of the chamber rather than at the center. It was noted that the yellow flame observed at the periphery of the chamber results from a lower film evaporation rate due to reduced air velocities in this region. If high velocity air can be directed to the area of spray impingement, the wall film evaporation and fuel transport to the ignition area can be enhanced. The time histories of the exhaust temperature of both a GDI and a PFI engine for a cold start using a coolant temperature of 20°C are shown in Fig. 22(c). The improved A/F control capability of the GDI system results in a more rapid increase in the GDI exhaust temperature than is obtained for the PFI engine.

Matsushita *et al.* (1996) investigated the combustion characteristics of the Toyota GDI engine for lean, stratified mixtures. It was reported that the initial burning rate at light load is higher than that observed in the PFI engine, even though the GDI engine is operated on a mixture that is overall twice as lean. Both the in-cylinder pressure at the beginning of the compression stroke and the peak compression pressure are higher as a result of the unthrottled, stratified operation. High-speed photography of the combustion event shows that a luminous yellow flame surrounded by a blue flame is formed initially. Later in the combustion event, the blue flame predominates, but the yellow flame is still visible. The rapid flame development

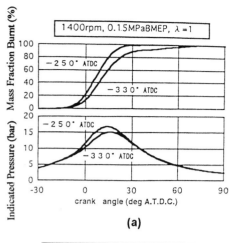

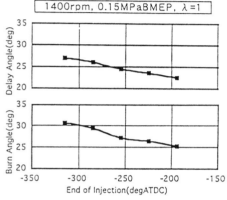

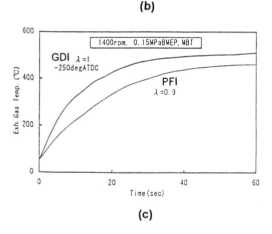

Fig. 22 Combustion characteristics of the Isuzu GDI engine (Shimotani *et al.*, 1996); (a) indicated pressure and mass fraction burnt; (b) ignition delay and burn duration; (c) comparison of exhaust gas temperature histories for the GDI and PFI engines for a cold-start

associated with stratified combustion and the higher O_2 and low CO_2 level in the exhaust gas are the reasons for the enhanced EGR tolerance of the GDI engine. It was reported by Jackson *et al.* (1997) that the cyclic variability of combustion becomes excessive at an EGR rate of 20% for the baseline PFI engine, whereas excellent stability is possible for stratified operation in a GDI engine for an EGR rate of up to 40%. EGR has little effect on the burn duration for stratified operation, while delay and burn durations are

significantly increased for the PFI engine as the EGR rate is increased. At an EGR rate of 40% the GDI engine exhibits improvements of 3% in BSFC, 81% in NOx and 35% in UBHC emissions as compared to operation without EGR. The improvement in UBHC is most likely the result of a richer mixture core, resulting in less bulk quenching. A similar PFI combustion system in the comparative test configuration exhibited poor EGR tolerance, with an associated increase in fuel consumption. As expected, the GDI engine requires more EGR for the same NOx reduction when compared to the PFI configuration. Another favorable characteristic of the GDI engine is the improved tolerance to spark retard at higher EGR rates. This is principally due to a lengthening of the delay period from the time of ignition to the time at which 10% of the mass is burned, as the burn period from 10% to 90% burned was found to be relatively insensitive to EGR rate. The result for the GDI engine without any EGR shows an early combustion phasing with a higher peak cylinder pressure. The introduction of 40% EGR delays combustion, and improves the fuel economy while substantially reducing the peak cylinder pressure. The research GDI engine with a top-entry, bowl-in-piston configuration was run at MBT with 40% EGR, and it was found that the point of 50% mass burned always occurs at about 15°ATDC on compression rather than at the typical 7°ATDC on compression for the conventional open-chamber PFI engine. At this condition the burn rates are similar for both the PFI and the GDI engines, although the GDI engine exhibits a mass-burn profile having a rapid initial rate and a slower rate near the end of combustion that is typical of stratified-charge combustion.

Injector Deposit Issues - Deposit formation is a significant concern with nearly all injector designs for GDI application and should be accorded sufficient time and resources in any development program. To neglect this engine issue is to lengthen development time and decrease the required service interval in the field. The operating environment of the fuel injector is much more harsh for in-cylinder injection than for port injection, and the formation of deposits on the injector tip can be much more rapid than are experienced with PFI systems. Although the operating environment of diesel injectors is even more extreme, deposits tend to form much more gradually on the tips of diesel injection nozzles than on GDI injector tips. One important factor is the fuel, but another major influence is the 50 to 140 MPa injection pressure level. The corresponding pressure of 4 to 7 MPa in GDI system is not sufficient to mechanically affect the continued deposition of carbonaceous material on the injector tip. Thus, the design of the injector tip for inherent resistance to deposit formation is particularly critical for GDI applications.

The effects of injector deposits are manifested in two distinct areas. The first is a degradation in the spray quality delivered by the injector, and the second is a reduction in the static flow capacity of the injector, with less fuel mass per injection being delivered at the same fuel pulse width. For most GDI injectors, the early stages of deposit formation do not result in significant flow reduction, but *can* result in substantial changes in spray skew, cone angle and spray symmetry, and in some changes in droplet-size distribution. The spray geometry, particularly the symmetry, is generally perturbed long before the cumulative deposit blockage reduces the flow by even 10% from the clean condition. As the operation of the GDI engine is much more sensitive to spray parameter variations than the PFI engine, the GDI engine, in general, will exhibit a much greater sensitivity to low levels of deposits. For GDI engines that have critical positioning of the spark gap relative to the spray periphery for light-load, stratified-charge operation, shifts in the spray geometry may be expected to result in a significant degradation of combustion, including misfires. Combustion systems using controlled vaporization from fuel-impact surfaces, such as that in the Mitsubishi GDI engine, would be expected to show less sensitivity to changes in spray characteristics, and thus less sensitivity to injector deposit formation. It should be emphasized that flow reductions due to deposits do not occur uniformly along a bank of cylinders, thus one ramification of deposit formation is to cause cylinder-to-cylinder variations in the air/fuel ratio. All cylinders on the bank will receive the same fuel pulse width command, but those injectors with more deposits will inject less fuel.

In comparison with the fuel spray from a clean injector, the spray from the injector with deposits can exhibit a distorted, non-symmetric spray envelope. The phase-Doppler data for most GDI injectors with deposits shows that the distribution of droplet sizes is affected very little, whereas the droplet velocity distribution is significantly changed. It is interesting to note that the atomization level of the injector may not be significantly degraded by deposit formation, even by a deposit accumulation that gives an 8% flow rate reduction. The velocity distribution and spray shape are generally affected, and the spray cross-section may be changed from near-circular to a narrow ellipse that is similar to the cross-section of a fan spray.

The observed deposits on the critical tip surfaces of GDI injectors generally fall into two categories. One is the thin, brittle-coating type that contains sulfides, and the other is the softer carbonaceous type. The latter forms more rapidly, and is more easily removed with common solvents and cleaners. Deposits can form internally, externally, or in both locations, with internal deposits at or near the lapped seat of the needle or pintle being the most undesirable. The external deposits may form on the downstream face of an isolating director plate, or on the downstream face of the pintle. Internal deposits near the minimum metering area of the fuel flow path cause fuel flow rate reductions, whereas external deposits generally result in a degradation in the spray geometry. The rate at which deposits form on the tip surfaces of a GDI injector depends upon both the inherent resistance of the injector design to deposit formation and the specific operating configuration in the combustion chamber. As with a port fuel injector, the inherent resistance of a GDI injector to deposit formation would ideally be determined by a standardized bench test using either an oven or a dynamometer in conjunction with a standard fuel, test configuration and test conditions. The lower the level of deposits that are formed in such a standard test, the higher the inherent resistance of the particular injector design. Examples of design parameters that would affect the inherent deposit resistance of an injector include a director plate for isolating the needle seat from direct contact with combustion gases, a more heat-conductive path from the tip to the injector threads or mounting boss, an increased fuel volume extending closer to the needle seat and special plating for the tip surfaces. Unfortunately, no proven standardized test for deposit resistance as yet exists for

GDI injectors, which is a situation that will have to be quickly resolved if rapid progress in direct-injection gasoline engines is to continue.

The standard deposit tests for pintle-type and director-plate type port fuel injectors are discussed by Caracciolo and Stebar (1987) and by Harrington et al. (1989). These tests were developed over a number of years, and were proven to correlate with field data from PFI engines; however, it is considered unlikely that the standard PFI tests can be effectively utilized for GDI injectors. It is well established that the hot soak time and tip temperature history are of critical importance for PFI deposit formation, and that deposit formation rates are very low for continuous operation. The initial indications for GDI applications, however, are that deposits *do* form under continuous operation, thus indicating that the formation mechanism differs from that of the PFI injector. The tip temperature is generally considered to be an important parameter for both the PFI and GDI injector, but the contribution of tip temperature windows and history is not known for GDI applications. The role of the hot soak interval is a complete unknown in the current GDI literature, making this an important research topic. Without an accepted industry test, the inherent resistance of a particular GDI design can only be evaluated by an ad hoc test within each company.

Knowledge of the inherent resistance of a GDI injector is necessary, but not sufficient, for interpreting the observed rate of deposit formation in a particular application. The inherent resistance of a specific injector design may be either enhanced or degraded by the configuration of the injector in the combustion chamber, and by the specific fuel that is being used. The tip temperature will be affected by the protrusion of the tip into the cylinder, the conductive path from the injector mounting boss to the coolant passage, and by the in-cylinder air velocity history at the tip location. It is recommended that the tip temperature be considered to be an important variable, and that it be measured and logged during any development program. Micrographs of the injector tip from a scanning electron microscope should also be obtained periodically during engine down periods.

The centrally-mounted location for the injector is known to be subject to a higher thermal loading than is experienced at a location under the intake port. GDI combustion systems such as those of Mercedes-Benz, Isuzu and Ford that use a centrally-mounted injector would be expected to yield 10 to 15°C higher injector tip temperatures than would be obtained with the Mitsubishi, Nissan, Toyota, or Ricardo systems, which locate the injector far from the exhaust valves and derive additional tip cooling from the intake air. A measured GDI tip temperature of 126°C was reported by Engineers at Mitsubishi (Iwamoto et al., 1997) for a steady, road-load cruise condition, with 140°C obtained for full-load operation. It was claimed that no deposit problems were encountered with these tip temperatures for steady running conditions; however, no controlled hot-soak periods were included. Some deposit formation has been noted for combustion systems that use a centrally-mounted injector, where full-load tip temperatures of more than 150°C may be achieved.

The following items represent important considerations in minimizing the formation of GDI injector deposits :

- injector tip temperature (maintained at less than 145°C)

- injector protrusion distance into chamber
- heat path from injector body to engine coolant passages
- air velocity variation at the injector tip during an engine cycle
- proximity of bulk fuel in injector to tip area
- gasoline additives to inhibit cumulative deposit formation
- special coating or plating of tip surfaces
- operating cycle, including hot-soak intervals (to be verified by future research).

The discussion above has dealt mainly with preventing or minimizing the formation of deposits, but there is another philosophy that can be invoked in parallel, although only by injector manufacturers. An injector can be designed with the important goal of minimizing the influence of deposits on the resulting flow rate and spray. This is known as an inherent tolerance to deposit. It is desirable, of course, to reduce the basic rate of deposition on the injector during the operation of the GDI engine. But, bearing in mind the inevitability of some deposition occurring with time, it is also desirable to have an injector flow-path design that exhibits a good tolerance for deposits that form. For example, 100 μg of deposits may form in 100 hours of operation on two different injector designs, which is the same rate of deposition. However, one injector design may exhibit a 3% flow reductions for this weight of deposits, while the flow in the second is reduced by only 1%. Some injector designs using swirl have proven to be quite sensitive to small amounts of deposits, particularly in terms of spray-symmetry degradation.

Finally, it is fair to note that a reasonable portion of the GDI deposit problem results from the lack of proven and effective GDI anti-deposit additives in the fuel. The additives that are currently in the gasoline supply in North America were developed and improved over the time period from 1984 to 1993 to minimize the effect of the port deposit formation mechanism. In the early development of PFI systems, injector deposits were a significant problem that had to be alleviated (Caracciolo and Stebar, 1987; Zhao *et al.*, 1995a), and this will have to be done for GDI systems. The problem is that with no GDI vehicle currently in production in North America, there is no strong incentive to develop and blend new additives for this application. Even if such additives were available, it would have to be carefully verified that adding them to the fuel supply would not adversely affect the formation of injector deposits in the 100 million PFI vehicles now in operation.

ATTAINABLE BSFC IMPROVEMENTS FOR GASOLINE DIRECT-INJECTION

A current strategic objective for the automotive application of the 4-stroke, gasoline engine is a substantial improvement in BSFC while meeting the required levels of pollutant emissions and engine durability. The improvement of passenger car fuel economy represents a very important goal that will determine the future development and use of SI engines relative to the diesel engine (Pischinger and Walzer, 1996). As outlined conceptually by Karl *et al.* (1997), the thermal efficiency can be enhanced by increasing the compression ratio, and by reducing the throttling losses, wall heat losses and mechanical losses.

The GDI engine using charge stratification offers the potential for reducing the part-load fuel consumption by 20~25% by incorporating enhancements in three of the above areas when the gas cycle, heat transfer and geometric configuration are optimized. The potential for improvements in the fuel economy of PFI and GDI engines was discussed by Seiffert (1996) and it was concluded that a GDI engine that can achieve load control without throttling can exhibit significant improvements in fuel economy that results mainly from the decrease of pumping work. Indeed, thermodynamic analyses do indicate that the reduction of pumping loss is the principal factor contributing to the improvement of the fuel economy of a GDI system that is able to achieve part-load operation by using an overall-lean, but stratified, mixture. Also contributing, but to a smaller degree, are reductions in the heat loss and in the specific heat ratio (Kume *et al.*, 1996). The gas near the cylinder wall is cooler for part-load, stratified operation, thus there is a smaller heat differential between the wall and the burning gas, and less energy is lost to the wall (Toyota, 1996). As discussed in the section on charge cooling, early injection reduces the octane number requirement (Anderson *et al.*, 1996a, 1996b) and, as a result, the knock-limited engine compression ratio for the GDI engine can generally be increased above 11 : 1. As is well known, the effect of this higher compression ratio on fuel economy is significant. The major factors contributing to the improved BSFC of a GDI engine over that of a conventional PFI engine are :

- decreased pumping losses due to unthrottled part-load operation using overall lean mixtures
- increased cooling of the intake charge due to early injection
- increased knock-limited compression ratio due to lower end-gas temperatures
- decreased wall heat loss due to stratified combustion.

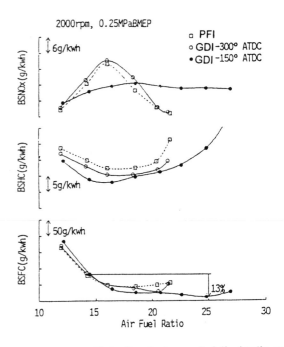

Fig. 23 The effect of injection timing and air/fuel ratio on the fuel consumption and emissions of the Isuzu GDI engine (Shimotani *et al.*, 1995)

130

Shimotani *et al.* (1995) compared the fuel consumption of a PFI engine with that of the Isuzu GDI engine that was operated in the early-injection, homogeneous mode. As shown in Fig. 23, the lean limit achieved by the use of an injection timing of 60° ATDC on intake (-300° ATDC in the figure) on the GDI engine was basically the same as that obtained for the PFI engine. However, retarding the fuel injection timing to 210°ATDC on intake (-150° ATDC in the figure) significantly extended the lean combustion limit for the GDI engine. It was found that the fuel economy is generally improved when lean air/fuel ratios are used, although the effect becomes marginal for air/fuel ratios leaner than 20. The fuel economy for the homogeneous mode was improved about 13% at an air/fuel ratio of 25 over that obtained for a stoichiometric mixture. This air/fuel ratio yields the lowest BSFC for the engine operating condition tested.

A significant improvement in fuel economy due to the use of a stratified charge in a GDI engine was also reported by Takagi (1996b). The results of this study are shown in Fig. 24. It was found that stable combustion with a stratified-charge can be achieved for mixtures leaner than 40 with an acceptable COV of IMEP. The fuel economy was improved about 20% as compared to the PFI engine. Wojik and Fraidl (1996) noted that the GDI engine offers a fuel economy improvement potential of more than 25% when compared to current PFI engine technology. According to Fraidl *et al.* (1996), the theoretical potential for improving the fuel economy by using direct gasoline injection is about 20% at part load, with an associated potential reduction of 35% in idle fuel consumption. Kume *et al.* (1996) and Iwamoto *et al.* (1997) reported that the BSFC may be improved by 30%, based upon tests of the Mitsubishi GDI engine. A Mitsubishi "Galant", outfitted with a Mitsubishi GDI engine operating on the Japanese 10-15 mode cycle, demonstrated that the fuel economy can be improved up to 35% over a conventional vehicle outfitted with a PFI engine. Idle fuel consumption was reduced by 40% by utilizing the benefit of more stable combustion, which permitted a lower idle speed. It is very important to note that an unknown portion of the fuel economy improvement came from vehicle component changes, such as a reduction in the vehicle weight. Ronald *et al.* (1996) reported that an improvement of 25% in the fuel economy over the PFI engine is possible at an engine speed of 2000 rpm, and 30% may be achieved at 1200 rpm. Anderson *et al.* (1996) reported that improvements in fuel consumption of up to 5% at part load and 10% at idle with stoichiometric operation were obtained for the steady-state operation of the Ford GDI combustion system. The best operation at part load was found to offer a fuel economy improvement of up to 12% as compared to the baseline PFI engine. The engine-out UBHC emissions, however, are comparable to those obtained with PFI operation using the same combustion system. Regarding the net effect of two parameters that give opposite trends, it was found that the NOx emissions were less than those of the baseline PFI engine as a result of both charge cooling and the higher residual content of the charge, thus more than offsetting the effects of the maximum increase in compression ratio that could be used on the GDI engine.

Lake *et al.* (1996) reported that the conventional PFI, homogeneous-charge, lean-burn engine can, when operated at an air/fuel ratio of 25, provide an overall fuel economy improvement of about 10 to 12% when compared

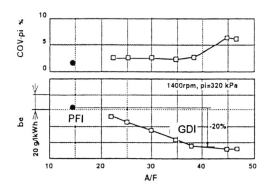

Fig. 24 The improvement of fuel consumption by the use of stratified-charge combustion for the Nissan GDI engine (Takagi, 1996b)

to stoichiometric operation without EGR. A stratified GDI engine operating at an overall air/fuel ratio exceeding 40 was found to provide an overall fuel consumption improvement of up to 25% when compared to the same baseline. It was found that the fuel economy is improved slightly by adding uncooled EGR at light load, even though the engine operates unthrottled. This was assumed to be due to the additional enhancement of mixing by EGR heating. The use of EGR, however, had little effect on fuel economy at intermediate load, and ultimately degrades the obtainable fuel economy at high load. It was reported by Jackson *et al.* (1996, 1997) that for the best trade-off between fuel consumption and emissions characteristics the system should employ maximum EGR at part load, limiting the overall air/fuel ratio to richer than 30. The combustion stability for a GDI system using a curved-crown piston was found to be improved relative to that of a flat-crown system over the entire load range, even at 0.15 MPa brake mean effective pressure (BMEP) at an air/fuel ratio of 50, yielding a 17~20% reduction in the fuel consumption at part load. This was considered to be due to the enhanced stratification that was achieved using the curved-crown piston. At a constant engine operating condition of 2000 rpm and a BMEP of 0.2 MPa, a reduction in fuel consumption of 23% was achieved for the unthrottled Mercedes-Benz GDI combustion system when compared to homogeneous, stoichiometric operation of the same engine. It was noted that this was achieved not only due to the reduction in throttling losses, but also due to the decrease in the peak cycle temperature (Karl *et al.*, 1997).

Even though the GDI engine offers a significant potential for vehicle fuel economy improvement, some mitigation in the fuel economy gain will occur when meeting strict emission limits. For most automotive GDI applications, this is the dominate optimization criterion. The required emission-related compromises in GDI system operating strategy will depend upon the combustion process, the EGR strategy, and the exhaust aftertreatment adopted (Karl *et al.*, 1997). For example, it is difficult to realize ultra-lean combustion while maintaining an effective catalyst operating temperature during steady, warm engine operation and, as a result, extended lean operation may be limited by the exhaust temperature. Kagawa *et al.* (1993) investigated the rotary DISC engine and concluded that it is necessary to utilize some manifold vacuum in order to maintain the exhaust gas temperature above the effective operating temperature of the catalyst for extended lean operation. With such a limitation, the stratified GDI engine can achieve a 10% improvement in fuel economy when compared with the conventional PFI combustion system. Although un-throttling the engine is a very effective way to improve the fuel economy of the GDI engine, slight throttling has been found to provide emission improvements at the expense of some fuel economy (Lake et al., 1996).

Using a two-zone model, Iiyama and Muranaka (1994) calculated the fuel consumption of a GDI engine for a range of operating conditions. It was found that an improvement in fuel economy up to 20% could be achieved when compared with current PFI engines that use a three-way catalyst if the GDI engine were operated under stratified conditions over the entire operating range. The active operating temperature of the catalyst was found to be a very important parameter, and the overall mixture cannot be ultra-lean if the catalyst inlet temperature is to be maintained above a critical threshhold. The analysis also indicates that the operating range of the stratified engine is narrowed somewhat when the emission constraint is applied, with the result that the fuel economy improvement is reduced to the order of 10%. It was noted that improvements in catalyst technology will be required in order to take full advantage of the benefits of highly-stratified combustion using overall ultra-lean mixtures.

EMISSIONS FROM GASOLINE DIRECT-INJECTION SYSTEMS

The concepts and significant potential of the GDI engine are quite evident, and have been addressed in previous sections of this paper. The realization of a viable production engine, however, requires the successful implementation of emission control strategies, hardware and control algo-rithms in order to achieve certifiable emission levels and acceptable driveability levels during load transients. This is one of the barriers to simply converting current PFI engines to GDI engines. Experience has shown that some fraction of the potential fuel economy gains will have to be com-promised in order to achieve increasingly stringent emission standards such as the U.S. low-emission-vehicle (LEV) and ULEV requirements. The key question for original equip-ment manufacturers (OEMs) in relation to GDI engines is whether the final margin in fuel economy for a particular engine system is sufficient to offset the required direct-injection hardware. Another important consideration is that the fuel economy margin must be measured relative to the increasingly sophisticated PFI engine, which, of course, is a target that moves annually. It is the specific details of the trade-off between BSFC and test-cycle emissions, coupled with the required hardware complexity, which will be the most important factor in determining the production feasibility of a GDI engine application.

UBHC Emissions - One of the significant emission advantages of the GDI engine is a potential reduction in UBHC emissions during a cold start. Direct in-cylinder injection of gasoline can completely eliminate the formation of a liquid fuel film on the intake port walls, which has the benefit of reducing the subsequent fuel transport delay that is associated with the PFI engine. This significantly improves the engine response under both cold start and transient conditions. As a result, it is very likely that

enrichment compensation may be substantially reduced in the engine calibration, further reducing the total UBHC emissions.

The engine performance and emission levels of a GDI engine and a PFI engine under cold start conditions were compared by Shimotani et al. (1995). The GDI engine exhibits a rapid rise in the IMEP following the first injection event, whereas the PFI engine requires about 10 cycles for the engine to attain stable combustion. This is attributed to the formation and growth of a fuel film on the intake valve and port wall of the PFI engine under these conditions. The fuel mass entering the cylinder on each cycle is not necessarily what is being metered by the injector. As a result, misfires and partial burns occur and the UBHC emissions for the PFI engine are quite high during these cycles. To compensate for the delay in reaching a steady-oscillatory state for the wall film, significant additional fuel must be added at cold start for the PFI engine. The GDI engine however, can be started cold using a stoichiometric or even slightly lean mixture.

Anderson et al. (1996a) compared the cold start performance of GDI and PFI engines. The IMEP traces in Fig. 25(a) verify that the Ford GDI engine fired on the second cycle, whereas the PFI engine failed to fire until the seventh cycle, after which the misfire and partial-burn events continued sporadically. In comparison, the GDI engine exhibited relatively good combustion with little cycle-to-cycle variation following the first combusting cycle. The GDI engine might be expected to fire on the very first cycle if a slightly greater amount of fuel were injected to account for the fact that there may be no residual gas in the combustion chamber when the first fuel injection occurs. The high IMEP that is obtained for the second cycle is due to the fact that this is the first and the only firing cycle in which combustion occurs without residual. The IMEP produced is therefore larger than is obtained in any later cycle in spite of the fact that the combustion chamber surfaces are the coldest on the first firing cycle. The lack of combustion on the first cycle with the GDI engine however, does result in high UBHC emissions as shown by the fast FID measurements in Fig. 25(b). By the fifth cycle, the UBHC emissions attain a steady-state level, whereas the UBHC level builds steadily in the PFI engine until the first fire in the seventh cycle. The pattern of high UBHC emissions continues through as many as 35 cycles as misfire or partial burns continue for the PFI engine.

Takagi (1996b) reported results from a cold starting test of the Nissan, 1.8 liter, 4-cylinder GDI engine. As shown in Fig. 26, it was found that the engine exhibited improved transient response and a significant reduction in UBHC emissions when compared with a standard PFI engine. Four cycles were required for the GDI engine to attain a stable IMEP when operating at an air/fuel ratio of 14.5, whereas the PFI engine required 12 cycles to reach stable operation at an air/fuel ratio of 13.

A vehicle test of the Isuzu prototype GDI engine was conducted by Shimotani et al. (1996). The injection timing was set at the end of the intake stroke for cranking, and at 110°ATDC on intake following the first firing and synchro-nization. A stoichiometric air/fuel ratio was used. The UBHC and CO emissions upstream of the catalysts were measured for the first cycle in the LA-4 mode. The emissions data that were obtained during the first 60 seconds after the cold start are shown in Fig. 27(a). By

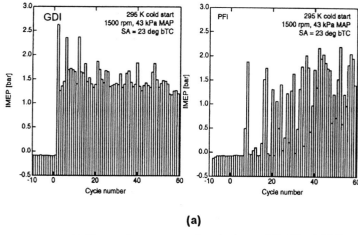

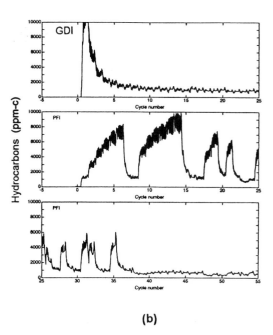

(a)

(b)

Fig. 25 The performance and emissions of the Ford GDI engine (Anderson *et al.*, 1996a); (a) measured cold-start IMEP variations for a single-cylinder research engine operating with direct and port injection; (b) measured cold-start UBHC variations for a single-cylinder research engine under the conditions of 295 K, 1500 rpm, 43 kPa MAP, and spark advance of 23°BTDC

means of an optimized injection timing and the use of a leaner mixture, the GDI engine exhibits a clear reduction in both UBHC and CO. The catalyst temperature, UBHC conversion efficiency, and the integrated UBHC concentration after the catalyst are shown in Fig. 27(b), which is for the same engine operating condition as in Fig. 27(a). It was noted that the GDI engine provides a more rapid catalyst response, and that the catalyzed UBHC emission level is lower than that obtained with the PFI engine. The rapid increase of the catalyst temperature after a cold start results from the improved A/F control capability afforded by direct injection. The time history of the air/fuel ratio from the 11th cycle of the LA-4 mode is shown in Fig. 27(c). For the conventional PFI engine, numerous calibration compensations are required to minimize the effect of fuel transport lag caused by the fuel/wall wetting. During speed and load transients the effectiveness of such compensation is limited, and tip-in and tip-out spikes of UBHC emissions occur. The more accurate control of the transient air/fuel ratio for the GDI engine is clearly illustrated in Fig. 27(c), and the associated catalyst efficiency is shown in Fig. 27(d). It may be seen that the PFI engine shows a large fluctuation in the catalyst efficiency during engine transients, indicating a

decrease in the UBHC conversion efficiency. This was not observed during the operation of the GDI engine, which implies that the accurate A/F control of the GDI engine can provide an associated improvement in the catalyst conversion efficiency. As a result, the UBHC emissions can be reduced, even during steady-state operation. Both the emissions and the BSFC are improved during the entire LA-4 mode for the GDI engine. It was reported that the total reduction ratio of the UBHC emissions can reach 45% for the integrated LA-4 mode, showing the potential of the GDI engine for meeting the LEV and ULEV emission standards.

The UBHC emissions for low-load operation represent a significant problem that is historically associated with gasoline DISC engines (Giovanetti *et al.*, 1983). In the case of gasoline direct-injection, the flame is eventually quenched in the extra-lean area at the boundary of the stratified region where the mixture is in transition from slightly rich to no fuel content (Takagi *et al.*, 1996b). If the boundary of the stratified charge is not sharp, a significant amount of UBHC remains in the total charge. The inability of the flame front to propagate from the rich mixture at the spark gap through all of the very lean mixture in the outer portion of the combustion chamber is a key factor contributing to the part-load UBHC emissions in GDI engines, and a substantial research effort is being directed towards the resolution of this problem. In general, it has been found that direct gasoline injection leads to a slight increase in UBHC emissions at idle, and to a substantial increase at part load. For part-load operation at higher engine speeds, another cause of higher UBHC levels is the decrease in the time available for mixture preparation, which is limited for direct injection, and can lead to liquid droplet combustion. The application of EGR at low load can also increase the UBHC emissions substantially, and the use of a higher compression ratio for the GDI application can result in additional amounts of UBHC being forced into crevices, resulting in an incremental increase in the UBHC emissions (Anderson *et al.*, 1996a; Drake *et al.*, 1995).

A range of possible reasons for the increase in the UBHC emissions for DISC systems at low-load were listed by

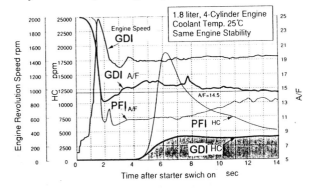

Fig. 26 Reduction of UBHC emissions for the Nissan GDI engine under cold-start condition (Takagi, 1996b)

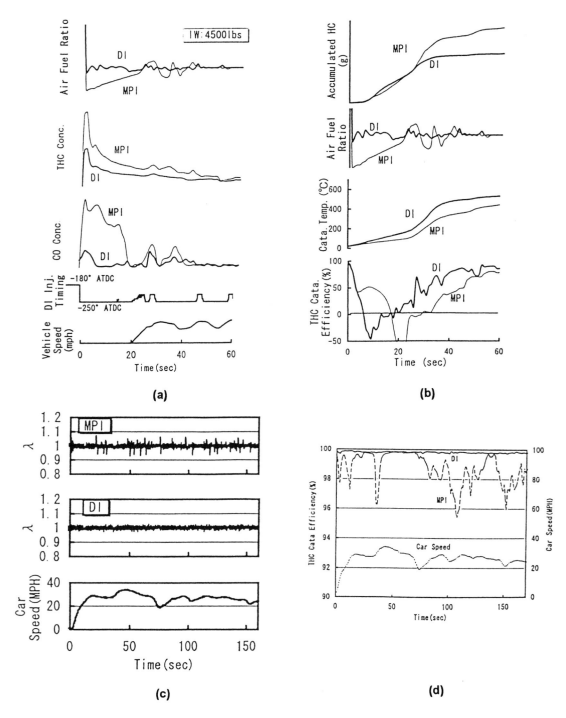

Fig. 27 Performance and emissions of the Isuzu GDI engine for the LA-4 mode vehicle test (Shimotani *et al.*, 1996); (a) engine-out UBHC emissions during the first cycle; (b) tailpipe UBHC emissions during the first cycle; (c) variation of air/fuel ratio during the 11th cycle; (d) catalyst efficiency during the 11th cycle

Balles *et al.* (1984), and by Frank and Heywood (1989) for the TCCS combustion system. It was noted that at low load the mixture becomes too lean prior to ignition due to the mixing and diffusion of the injected fuel. As a result, the combustion efficiency is low for many cycles, and the UBHC emissions are excessive. Generally, at low load an overall ultra-lean mixture is provided to the TCCS engine, which may cause overmixing. Due to the small quantity of

fuel injected into the combustion chamber, the cyclic variation of the evaporation rate becomes a significant factor. The UBHC emissions were traced mainly to cycles with the lowest combustion efficiency. Frank and Heywood (1990a, 1990b, 1991) reported that the UBHC emissions exhibit only limited dependence on the piston temperature, which would indicate that the adherence of liquid fuel on the piston surface is not the main source of the UBHC emissions for the combustion system tested. Any delay in the flame kernel development around the spark gap, or in the propagation rate of the flame through the mixture, is a contributing factor that increases the UBHC emissions. Frank and Heywood (1989) pointed out that engine combustion events with large combustion delays yield a marked increase in the UBHC emissions. This would indicate that for the stratified combustion of lean mixtures, the resident time of the fuel inside the combustion chamber should be as short as possible. Thus, the spray characteristics at the end of the fuel injection event are important in determining the UBHC emissions. It was also noted that the fuel injected as the pintle is closing may not be sufficiently well atomized, and may penetrate separately from the main spray. This will result in an increase in the UBHC emissions.

Iiyama and Muranaka (1994) concluded from calculations that pockets of mixture within the combustible range for air/fuel ratios richer than 30 are distributed randomly inside the flame. As a result, the flame does not propagate uniformly through these regions, yielding an increase in UBHC emissions. It was reported that the high UBHC emissions at low load is an inherent feature of the stratified combustion concept itself. To reduce this type of UBHC emissions, it is necessary to generate a highly stratified flow field. This tends to suppress the formation of rich pockets inside the combustion zone. One of the technologies proposed for accomplishing this is the two-stage injection system that is discussed by Miyamoto et al. (1994) and by Hattori et al. (1995). Ronald et al. (1996) studied the effect of air/fuel ratio on the UBHC emissions of a single-cylinder research GDI engine. The UBHC emissions were found to increase continuously with an increase in the air fuel ratio for the stratified-charge mode. This was attributed to increasingly prevalent pockets of excessively lean mixture which leads to incomplete combustion. A further contributing factor may be a reduced reaction rate along the exhaust pipe due to the lower exhaust gas temperature.

Throttling is generally found to provide emission improvements at the expense of some fuel economy, although the GDI engine offers the potential for improving the BSFC by operating at part load without throttling. Theoretically, non-throttling operation is possible at part load by using a stratified mixture; however, some throttle control is usually required in the production GDI engine (Harada et al., 1997). This is because the GDI engine, even with stratified operation, must also operate for conditions such as cold starting which require a homogeneous charge. For such an operating condition it has been found that some throttling is advantageous. Throttling can also be quite effective in achieving smoother transitions in engine operation. The application of some throttling at low load has been demonstrated to be effective in increasing the very low exhaust temperatures that result from the extended periods of stratified operation, which can improve the catalyst conversion efficiency. In addition, the flow of EGR for essential NOx reduction may require a moderate amount of intake manifold vacuum that can be obtained by some throttling. A conventional vacuum-assisted power brake system also requires some vacuum which may be most expediently provided by throttling. It has been found that some GDI engines demonstrate an improvement in the positioning of the line representing the trade-off between emissions and fuel economy with some degree of throttling.

An electronically-controlled throttling system is generally used for GDI applications because of the difficulty of achieving proper control with mechanically-linked systems, and it was effectively employed in the Toyota GDI engine system (Harada et al., 1997). It was found that less throttling is required for warm engine operation than for cold engine operation.

In general, it is possible to use moderate throttling of the engine at low load to reduce the UBHC emissions without significantly degrading the fuel economy. In addition, throttling may be successfully combined with the use of EGR to improve the overall compromises among fuel economy, NOx and UBHC emissions. Lake et al. (1996) investigated the effect of throttling on fuel economy and emissions as a function of EGR, and reported that a 20% reduction in UBHC emissions can be obtained by the use of light throttling at an EGR rate of 20%, with an associated fuel economy penalty of only 2.5%. Halving the UBHC emissions by using some throttling results in an increase of approximately 8% in fuel consumption, whereas throttling the engine without any EGR yields the lowest UBHC emissions at any fuel consumption level. Moreover, throttling does not cause any apparent change in NOx emissions at a fixed EGR rate. As throttling is increased to enrich the A/F ratio beyond 26, NOx decreases at a fixed EGR rate, but to the detriment of fuel economy. A 55% reduction in NOx emissions can be achieved with a 2% loss of fuel economy by applying EGR with no throttling. An 80% reduction in NOx emissions can be achieved by the use of throttling and EGR, but with an associated 15% loss in fuel economy. However, since over half of the total GDI fuel economy benefit derives from the reduction of pumping losses (Kume et al., 1996), throttled operation may be expected to result in some loss of fuel economy, thus the degree of throttling should be restricted to the minimum necessary.

NOx Emissions - The conventional, homogeneous, lean-burn, PFI engine exhibits decreasing NOx formation and engine-out emissions as the mixture becomes leaner, which occurs due to reductions in the reaction zone temperature. However, in the case of the GDI engine that operates with a stratified charge, the local temperature of the reaction zone remains high because ideally some areas with a stoichiometric or slightly rich mixture exist in the stratified charge. NOx production is high in these areas even though the cylinder thermodynamic temperature is reduced due to the overall lean operation. Also, the GDI engine can achieve important fuel economy benefits because of the possible operation with a higher knock-limited compression ratio, but this tends to elevate the NOx emission levels. As a result, the NOx level of the GDI engine without EGR is similar to that for the PFI engine, even though some GDI engines can operate at an air/fuel ratios leaner than 50 at low load. As reported by Ronald et al. (1996), despite an enleanment of the overall air/fuel ratio, a continuous *increase* of NOx is observed for a GDI engine that is operating in the stratified-charge mode, whereas the NOx emission level *decreases* for a PFI engine that is operating with an air/fuel ratio leaner than 15. The GDI engine also exhibits significantly elevated NOx emissions at idle as compared to the PFI engine. This is the result of locally stoichiometric combustion and the attendant high heat release rate for GDI engines, as compared to the slower homogeneous combustion of PFI engines at a lower level of pressure and temperature.

As illustrated earlier in Fig. 23 (Shimotani *et al.*, 1995), the use of an early fuel injection timing of 30° ATDC on intake (-300°ATDC in the figure) for a GDI engine produces NOx at approximately the same level as that of the PFI engine. Retarding the GDI injection timing to 210°ATDC on intake (-150°ATDC in the figure) results in a significant decrease in the peak NOx. However, for a mixture with an air/fuel ratio leaner than 20 the NOx level is higher than is obtained with a later injection timing. A rich mixture formed around the spark gap due to the fuel stratification is the cause of this elevation of NOx level.

For an engine such as the GDI that operates with lean or even ultra-lean mixtures, a conventional three-way catalyst cannot be used to remove NOx, therefore other techniques for in-cylinder NOx reduction or exhaust aftertreatment are necessary. There is a consensus that NOx reduction after-treatment for lean burn engines is a challenging task, especially in view of the lower exhaust gas temperature. Furthermore, part-load, lean operation of the GDI engine is quite frequent, and contributes about half of the NOx emissions for the total test (Iwamoto *et al.*, 1997).

EGR is widely used for in-cylinder NOx reduction, and it functions primarily as a diluent to the fuel/air mixture. The dilution of the mixture by EGR is a very straightforward method of reducing the peak combustion temperature. As compared to dilution using air, however, dilution using exhaust gases decreases the polytropic index due to the presence of the CO_2 and H_2O molecules, with an associated higher specific heat ratio. Therefore, the effect of EGR dilution on thermal efficiency was found to be inferior to that of air dilution. Furthermore, the amount of EGR that can be introduced is limited because it tends to degrade the combustion stability. Consequently, it may be appropriate to use EGR in combination with another technique for NOx reduction (Ando, 1996b). In a conventional PFI engine, NOx reduction using EGR fails near the lean combustion limit because the mixture in proximity to the spark gap is diluted by EGR, which results in misfires. For a GDI engine using charge stratification, the mixture near the spark gap is ideally either stoichiometric or slightly rich; therefore a stable combustion is possible with a much higher level of EGR. However, there is an associated trade-off between any NOx reduction and a counter-productive increase in both UBHC emissions and fuel consumption.

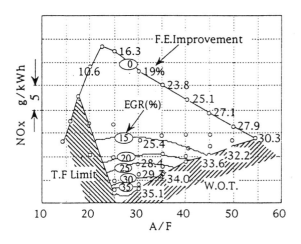

Fig. 29 The effect of EGR rate on the reduction of NOx for the Mitsubishi GDI engine (2000 rpm, 15 mm³/st.) (Kume *et al.*, 1996)

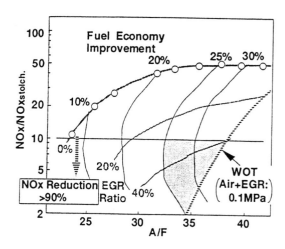

Fig. 30 The effect of EGR rate on the reduction of NOx for the Toyota GDI engine (1200 rpm, 12 mm³/st.) (Matsushita *et al.*, 1996)

A comparison of the NOx reduction performance as a function of EGR rate for PFI, GDI, and diesel engines is illustrated in Fig. 28 (Kuwahara *et al.*, 1996). It is evident that a larger NOx reduction can be realized with EGR in the GDI engine than may be obtained with either the PFI engine or the diesel engine. For the GDI engine the fuel/air mixing time is comparatively longer than that of the diesel engine and, as a result, the NOx emissions can be reduced significantly. Kume *et al.* (1996) also reported the effect of EGR on NOx reduction, and these data are shown in Fig. 29. When compared with the engine-out NOx for a PFI engine operating with a stoichiometric mixture, the NOx reduction for the Mitsubishi GDI engine with EGR can exceed 90% while maintaining improved fuel economy. It was noted that the flame temperature was found to be reduced by about 200 K with the use of 30% EGR. Lake *et al.* (1996) reported that stable combustion can be maintained in the Ricardo GDI engine operating in the early injection mode for EGR rates of up to 30%. A reduction of

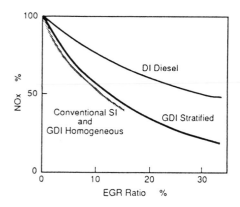

Fig. 28 The effect of EGR on NOx reduction for different types of internal combustion engines (Kuwahara *et al.*, 1996)

88% to 95% in NOx was achieved for the stratified-charge, late-injection mode when EGR was applied. Both the ignition delay and the combustion period were found to be extended with EGR due to the decrease in the laminar flame speed. The high EGR tolerance is attributed to the improvement in fuel/air mixing due to EGR heating effects, and to small-scale variations in the local air/fuel ratio. It was claimed that a high EGR tolerance yields an additional fuel economy benefit when compared to a PFI engine. The effect of EGR on the lean limit and NOx emissions of a Toyota stratified-charge GDI engine is shown in Fig. 30 (Matsushita *et al.*, 1996). It was reported that a stable combustion was obtained even when the air/fuel ratio was increased to 55 without throttling. For such an operating condition a 30% improvement in the fuel economy was achieved, but NOx production was deemed to be a problem when compared to conventional homogeneous combustion. Without the use of EGR, the engine-out NOx emissions for the Toyota GDI engine increase as the mixture is enriched from an air/fuel ratio of 55, and reach a maximum at an air/fuel ratio of 22. Although the NOx emissions decrease when the amount of EGR increases, it was noted that when the EGR rate is increased for the stratified-charge case, misfire may occur. A stable combustion is obtained at an EGR rate of 40%, which yields both a 90% reduction in NOx, when compared with engine operation without EGR, and a 35% improvement in fuel economy when compared with the conventional PFI engine.

Even though EGR is widely used for reducing the NOx emissions in SI engines, there is a general view in the literature that the progress of GDI engines is strongly coupled to the development of lean-NOx catalysts. This is because EGR cannot reduce the NOx emissions over the entire engine speed-load map to meet the scheduled emission limits in North America and has little or no margin for the emission limits in Europe and Japan. For example, within the emission test cycles the GDI engine generally operates with a lean homogeneous mixture at a stable lean limit. In such an operating condition, large amounts of EGR cannot be used as the engine combustion stability would be degraded. Also, the stratified-charge operating map is narrowed when restricted to low NOx levels, thus reducing the achievable fuel economy improvement. The lean-NOx-catalyst technology would definitely be a welcome tool in order to make optimum use of stratified combustion technology. The conversion efficiencies and the durability of lean-NOx catalysts are still lacking when compared with the current three-way catalyst; however, a significant current research effort is being directed toward the desired goal. A stoichiometric homogeneous mixture could be used in a GDI engine in combination with a three-way catalyst, but the fuel economy would be significantly degraded. If a reasonable fuel economy improvement is to be achieved for a GDI engine using a homogeneous mixture, a strategy using a lean homogeneous mixture in conjunction with a lean-NOx catalyst should be investigated.

A number of promising technologies are currently being explored for NOx reduction in lean-burn and stratified-charge engines. These include NOx reducing catalysts of zeolite and precious metal for oxygen-rich conditions, and an NOx storage catalyst that is capable of trapping NOx when the exhaust is oxygen-rich, then converting the stored NOx during short periods of over-fueling. One of the problems that hinder the application of NOx storage catalyst

technology to stratified lean combustion is the generation of a precisely-controlled rich spike in the exhaust flow during extended periods of part-load operation. During homogeneous lean mixture operation in the Toyota GDI system, where fuel is injected during the intake stroke, the required rich mixture spike is obtained by the synchronized control of both fuel pulse width and spark retard. For stratified-charge operation this spike is obtained not only by mixture enrichment and spark control, but also by a complex system that briefly modifies the injection timing, throttle opening, and the positions of other actuators such as the swirl control valve (SCV) and EGR valves. It was reported that the NOx emission level without either EGR or an NOx catalyst is 1.85g/km for the Japanese 10-15 mode test. With EGR control, NOx is reduced to 0.6g/km, which is a 67% reduction. This can then can be further reduced to 0.1 g/km with a stabilized NOx catalyst. With the use of an NOx storage-reduction catalyst, the introduction of an inter- mittent rich mixture results in a 2% loss of fuel economy (Harada *et al.*, 1997). Iwamoto *et al.* (1997) proposed using a de-NOx catalyst for the Mitsubishi GDI engine to further reduce the NOx emissions not only during the urban driving cycle, but also for the lean-burn condition associated with the highway driving cycle. This de-NOx catalyst is a selective-reduction-type with three-way catalytic functions, a high conversion efficiency and a high resistance against sulfur contamination; however, it currently has a relatively narrow working temperature that limits its performance. Another new technology for potential application to GDI engines is a plasma system that provides a simultaneous conversion of NOx, UBHC and CO (Jackson *et al.*, 1997). The system features a surface plasma having a low tem- perature, a low pressure and a low energy that could be packaged in a volume similar to the conventional catalyst. Conversion efficiencies of up to 50% for CO and 70% for UBHC and NOx were claimed. The advantages of this system are a reduced light-off temperature and an expanded range of operating temperature.

In summary, it is apparent from the current literature that the attainment of regulatory levels of NOx emissions is one of the major challenges facing GDI engine developers in the 1999-2009 time frame.

SPECIFICATIONS AND PERFORMANCE OF PRODUCTION AND PROTOTYPE ENGINES

Operating with a single, fixed ignition position imposes very stringent requirements on the mixture preparation process in GDI engines, as it is very difficult to provide a combustible mixture at the spark gap over the entire engine operating map. That is why most of the combustion control strategies are primarily directed towards the specifics of preparing the air/fuel mixture. Generally, homogeneous combustion for high-load operation can be fairly readily achieved by means of injection during the induction stroke. Obtaining a stable, stratified mixture that is required in order to achieve the best fuel economy at part load is more difficult, and additional developmental problems must be overcome to achieve a stable, stratified combustion during idle or light-load operations. In order to minimize the heat transfer to the cylinder liners, and to avoid fuel en- croachment into the crevice volumes, radial stratification with the mixture cloud stabilized in the cylinder center is preferred. In the case of GDI homogeneous operation with

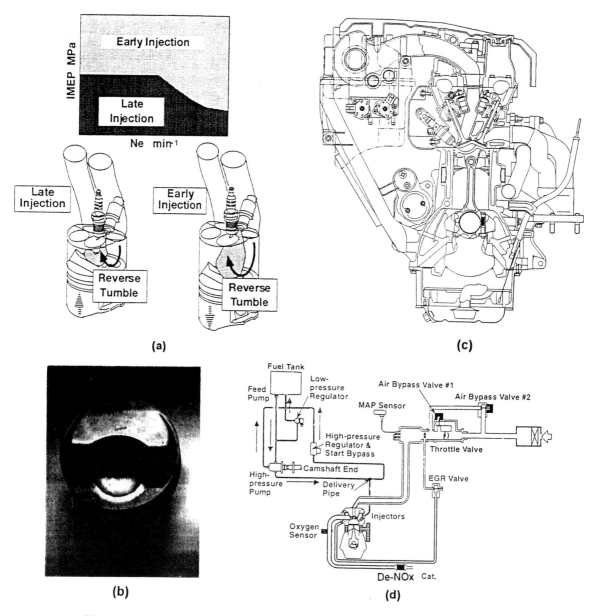

Fig. 31 The Mitsubishi GDI combustion system; (a) basic concept of mixture preparation (Kume *et al.*, 1996); (b) piston photography (Yamaguchi, 1996b); (c) engine cutaway (Iwamoto *et al.*, 1997); (d) system layout (Iwamoto *et al.*, 1997)

early injection, wetting of the cylinder liner by the fuel spray must be avoided. For full-load operation, a homogeneous mixture with good air utilization should be produced by a combination of injection strategy and in-cylinder charge motion. It is a significant advantage to have the capability of operating the GDI engine in both the stratified mode and homogeneous lean mode, although there are GDI prototype and research engines that operate only in the homogeneous mode. It is indeed difficult to design a GDI combustion system that will optimize the mixture formation process over the entire engine operating map, particularly when transitions must be made between stratified and homogeneous combustion. The operating map conditions range from cold start to hot-soak start, from cranking speed to 6000 rpm, from idle to full load, and cover both steady-state and transient operation. Optimizing the GDI combustion for this

range of operation requires advanced hardware and sophisticated control systems.

In recent years, new mixture preparation and combustion control strategies for GDI engines have been proposed and developed by a number of automotive companies, fuel system manufacturers and research institutions. Two have even become available in production automobiles. The key features of the several specific GDI combustion systems that have been developed during the past few years are discussed and summarized in this section.

Mitsubishi GDI Engine - The Mitsubishi GDI combustion system was discussed in detail by Kume *et al.* (1996), Ando (1996a, 1996b), Kuwahara *et al.* (1996), Iwamoto *et al.* (1997), Yamaguchi (1995, 1996b), and McCann (1995). The schematic of the engine layout, the cutaway of the

Mitsubishi engine, and the piston photo are shown in Fig. 31. The GDI engine is based on the lower half of the production Type-4G93, dual overhead camshaft, four-valve-per-cylinder, inline, four-cylinder engine. The main differences from the 4G93 are the cylinder head and piston design, and the high-pressure fuel pump and injectors. According to Yamaguchi (1996b), Mitsubishi's new Galant and Legnum, the station wagon variant, are powered by this GDI engine. It is claimed that the fuel economy provided by the Mitsubishi 49G93 GDI engine is ~35% better than that obtained using the comparable-displacement, conventional, PFI engine, and is markedly superior to that of Mitsubishi's own turbocharged, intercooled, 2.0-L, inline, four-cylinder, diesel engine in the Japanese urban test cycle, with the 0-to-100-km/h acceleration time also being reduced by 5%. Due to the in-cylinder charge cooling and the improved intake port design, the volumetric efficiency of the Mitsubishi GDI engine is increased by 5% over the entire engine operating range. In combination with an increased compression ratio, the total power output is increased by 10%. The principal features of the system are :

MITSUBISHI GDI ENGINE
- four-valve DOHC pentroof combustion chamber
- four cylinders
- bore x stroke : 81 x 89 mm
- displacement : 1864 cm^3
- compression ratio : 12 : 1
- central spark plug
- a spherical compact piston cavity to redirect the mixture plume to maintain charge stratification
- intake port : up right straight intake ports to generate an intense tumble with a direction of rotation opposite that of the conventional horizontal intake ports of PFI engines
- injector located underneath intake port and between two intake valves
- high-pressure, swirl injector
  ◊ a sinusoidally-shaped spray for stratified-charge operation
  ◊ a hollow-cone spray for homogeneous-charge operation
- fuel pressure : 5 MPa
- part-load mode : late injection at the air/fuel ratio of up to 40
- high-load mode : early injection
- NOx reduction : EGR at a rate of 30% and De-NOx catalyst
- EGR control : achieved by an EGR valve actuated by a stepping motor.
- octane requirement : 98 RON
- engine idle speed : 600 rpm
- power : 112 kW JIS net at 6500 rpm
- torque : 128 N•m at 5000 rpm.

Toyota GDI Engine - The Toyota direct-injection combustion system has been described in several recent publications (Matsushita et al., 1996, Harada et al., 1997; Toyota, 1996; Tomoda et al., 1997; Gualtieri and Sawyer, 1995; Yamada, 1996; Yamaguchi, 1996a). The principal components and stratification concepts that are used in the GDI engine, which is designated as the D-4, are illustrated in Fig. 32. As shown in Fig. 32(a), a uniquely shaped cavity in the piston is employed. The narrower zone (A) of the

cavity is designed to be the mixture formation area, and is positioned upstream of the spark plug. The wider zone (B) is designed to be the combustion space and is effective in promoting rapid mixing. It is claimed that the increased width in the swirl flow direction enhances the flame propagation after the stratified mixture is ignited. The involute shape (C), is designed to direct the vaporized fuel towards the spark plug. The intake system consists of both a helical port and a straight port, which are fully independent. An electronically-activated SCV of the butterfly-type is located upstream of the straight port. When the SCV is closed the resulting swirl ratio is reported to be 2.1. The helical port utilizes a variable valve timing intelligent (VVT-I) cam-phasing system on the intake camshaft. For light-load operation the SCV is closed, which forces the induction air to enter through the helical port, thus creating a swirling flow. Atomized fuel is injected into the swirling flow near the end of the compression stroke. The air flow moves the rich mixture to the center of the chamber around the spark plug, while part of the fuel disperses into the air in the combustion chamber, forming a stratified air/fuel mixture. For heavy-load operation the SCV is opened, and the intake air is inducted into the cylinder with a lower pressure loss. The fuel is injected during the intake stroke, resulting in a homogeneous mixture. The main features of this GDI system are summarized as follows :

TOYOTA GDI ENGINE
- four-valve pentroof combustion chamber
- four cylinders
- bore x stroke : 86 x 86 mm
- displacement : 1998 cm^3
- compression ratio : 10 : 1
- spark plug located slightly away from the center of the combustion chamber
- piston and bowl : deep-bowl-in-asymmetrical-crown
- intake port : one helical port with VVT-I system, and another straight port with a SCV valve
- injector located under the straight port
- fuel injector : high-pressure swirl; two-stage injection
- fuel pressure : 12 MPa
- four-mode operation : stratified, semi-stratified, lean, and stoichiometric
- part-load mode : late injection at the air/fuel ratio of up to 55
- high-load mode : early injection
- torque transition between stratified-charge and homogeneous-charge operation : two-stage injection and electronic throttle control system
- NOx reduction : electronically controlled EGR at a rate of 40%, NOx storage-reduction catalyst, and standard three-way catalyst
- octane requirement : 91 RON.

As illustrated earlier in Fig. 19(a), a special two-stage injection process is used for the transition between light-load and heavy-load operation. This process creates a weakly-stratified mixture in order to achieve a smooth transition of torque from the ultra-lean operation to either lean or stoichiometric operation. The D-4 engine operates using four operating regimes having distinctly different mixture strategies and/or distributions, which are described as follows. The first is stratified-charge operation using an air/fuel ratio of 25 to 50. The vehicle operates in this ultra-lean zone under steady light-load for road-loads of up to

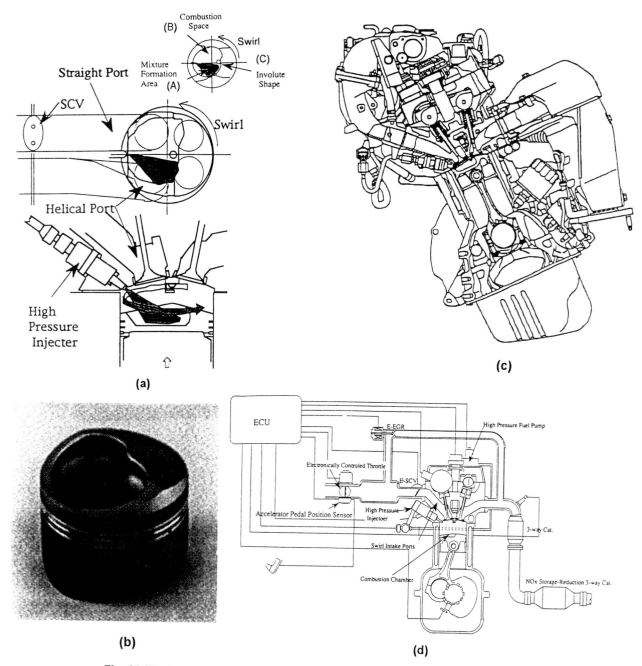

Fig. 32 The Toyota GDI combustion system; (a) basic concept of mixture preparation (Matsushita *et al.*, 1996); (b) piston photography (Yamaguchi, 1996a); (c) engine cutaway (Toyota, 1996); (d) system layout (Harada *et al.*, 1997)

100 km/h. The second operating regime is a transitional, semi-stratified zone with the air/fuel ratio ranging from 20 to 30. Within this zone, fuel is injected twice within a cycle, once during the intake stroke and once during the compression stroke. The third operating regime utilizes injection during the intake stroke to operate the engine in the homogeneous combustion mode, using a lean air/fuel ratio of 15 to 23. The fourth regime is denoted as the stoichiometric power zone, and is similar to the third, except that the air/fuel ratio range is from 12 to 15.

The Toyota D-4 engine incorporates an electronic-throttle-control system that is purported to reduce the harshness of torque transitions, thus improving the driveability. The VVT-

I valve-phasing technology is used to maximize torque in the low to mid-speed ranges, and to maximize power at high speed. At low engine loads the intake valves are opened earlier, increasing the overlap between the intake and exhaust valves. This results in internal EGR, which increases the total EGR ratio while requiring less from the external EGR system. As a result, the VVT-I system contributes to reducing NOx emissions. With the VVT-I system and an increase in the compression ratio to 10:1, about 10% more torque is obtained from the D-4 engine in the low to mid-rpm ranges than is obtainable with the conventional PFI engine.

The emission levels of oxides of nitrogen are claimed to

be reduced by as much as 95% on the Japanese test cycle. The fuel economy of the D-4 engine, when utilized with a four-speed automatic transmission, was reported to be about 30% better than that of the conventional PFI engine. For operation on the Japanese urban 10/15-mode cycle, a fuel consumption of 17.4 km/liter was achieved, which compares to 13 km/liter for the conventional PFI engine. The vehicle acceleration was also reported to be improved by ~10%, reaching 100 km/h from a standstill in less than 10 seconds.

Nissan GDI Engine - The Nissan prototype GDI engine is designed with a centrally-mounted spark plug, whereas the injector is located underneath the intake port and between the two intake valves (Takagi, 1996a, 1996b; Iiyama, 1996). A schematic of the combustion chamber is shown in Fig. 33(a). The engine can operate in both the stratified-charge mode and the homogeneous-charge mode. A 30% reduction of the UBHC emissions for a cold start is obtained when compared to the baseline PFI engine. The engine can be operated with stable combustion using a mixture leaner than an air/fuel ratio of 40. This results in a 20% improvement in fuel economy when compared with the baseline PFI engine, which operates with a stoichiometric air/fuel ratio. As shown in Fig. 33(b), the GDI engine also shows an improvement in power output of about 6% when

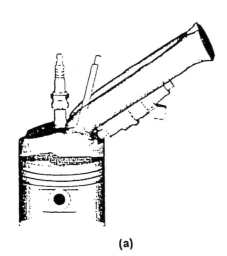

(a)

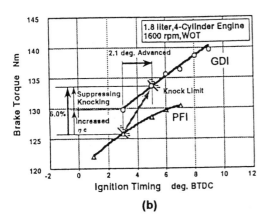

(b)

Fig. 33 The Nissan GDI combustion system (Takagi, 1996b); (a) configuration of combustion chamber; (b) improvement of output torque at low engine speeds

compared to the PFI engine, even for full-power operation at low engine speed. The principal features of the combustion system are summarized as follows :

NISSAN GDI PROTOTYPE ENGINE
- four-valve pentroof combustion chamber
- four cylinders
- bore and stroke : 82.5 x 86.0 mm
- displacement : 1838 cm^3
- compression ratio : 10 : 1
- centrally-mounted spark plug
- curved piston
- conventional intake port design
- reverse tumble obtained by shrouding the upper side of the intake valve
- injector located between the two intake valves and underneath the intake port, at an angle of 36° down from the horizontal plane
- high-pressure swirl injector to generate a hollow-cone spray with a cone angle of 70° and a SMD of 30 μm
- fuel pressure : 5 MPa
- part-load mode : late injection for stratified operation at an air/fuel ratio of up to 40
- high-load mode : early injection for homogeneous operation.

Ford GDI Engine - The prototype Ford research GDI engine is designed to operate in the homogeneous, stoichiometric mode, and utilizes a centrally-located fuel injector and spark plug (Anderson et al., 1996a, 1996b). A specially-fabricated, four-valve, single-cylinder head was used in order to compensate for the loss of breathing efficiency resulting from the decrease in the intake valve flow area. The main features of the Ford GDI combustion system are summarized as follows :

FORD GDI RESEARCH ENGINE
- four-valve, DOHC pentroof combustion chamber
- single cylinder
- bore x stroke : 90.2 x 90.0 mm
- displacement : 575 cm^3
- exhaust /intake valve area ratio : 0.77
- compression ratio : 11.5 : 1
- in-cylinder flow : no net swirl, but small tumble component
- near-centrally-mounted spark plug
- vertical, near-centrally-mounted injector
- Zexel high-pressure swirl injector
- fuel pressure : 5 MPa
- early injection mode only
- octane requirement : 91 RON.

The steady-state operation of the Ford GDI engine was reported to result in improvements in fuel consumption of up to 5% at part load and 10% at idle with stoichiometric operation. It was claimed that the best lean operation at part load can offer a potential reduction of up to 12% when compared to their baseline PFI engine, whereas emissions are comparable to the PFI values.

Isuzu GDI Engine - A pentroof, four-valve, SI engine was converted to operate as a direct-injection gasoline engine (Shimotani et al., 1995). Two engine configurations are

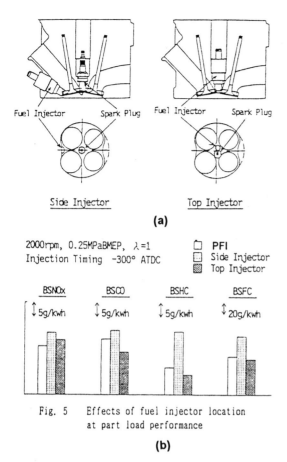

(a)

2000rpm, 0.25MPaBMEP, λ=1
Injection Timing -300° ATDC

☐ PFI
▦ Side Injector
▨ Top Injector

BSNOx   BSCO   BSHC   BSFC

↕5g/kwh   ↕5g/kwh   ↕5g/kwh   ↕20g/kwh

Fig. 5   Effects of fuel injector location
at part load performance

(b)

Fig. 34 The Isuzu GDI combustion system (Shimotani *et al.*, 1995); (a) configuration of combustion chamber; (b) effect of injector location on engine emissions

used; a single-cylinder research engine and a six-cylinder prototype engine. The research engine has the provision for either a centrally-mounted or a side-mounted fuel injector. Figure 34(a) shows the combustion chamber geometry and the injector locations. A comparison of the engine performance variation with injector location was made between the center mounting and the side mounting. Figure 34(b) shows the effect of injector location on the engine performance at part load. The center-mount location shows a level of exhaust emissions (UBHC and NOx) that is identical to that of the equivalent PFI engine when injection is completed by 60°ATDC on intake (-300°ATDC in the figure), when operating at an engine speed of 2000rpm and an air/fuel ratio of 14.7. The side-mount location exhibits an apparent increase in the UBHC emissions and fuel consumption as compared to the baseline PFI engine. This was considered to be the result of fuel impingement on the cylinder liner near the exhaust valve, with some of the fuel being absorbed in the oil film. For the center-mount location the emissions and engine performance are quite similar to that of the PFI engine when the fuel is injected during the early portion of the intake stroke. As a result, a center-mounting location was selected for the Isuzu GDI combustion system. The prototype V-6 engine has a centrally-mounted injector, and operates in the homogeneous mode. Vehicle tests with the six-cylinder GDI engine under cold-start condition have been conducted (Shimotani *et al.*,

1996). It was reported that the total reduction ratio of the UBHC emissions can reach 45% for the integrated LA-4 mode, showing the potential of the this GDI engine for meeting the LEV and ULEV emission standards. The principal features of the Isuzu GDI combustion system are summarized as follows :

ISUZU GDI RESEARCH AND PROTOTYPE ENGINES
- four-valve, DOHC pentroof combustion chamber
- research engine : single cylinder
- prototype engine : six cylinder V-type
- bore x stroke : 93.4 x 77.0 mm
- displacement (single cylinder) : 528 cm³
- compression ratio : 10.7 : 1
- near-centrally-mounted spark plug
- flat piston
- standard intake port : at a tumble ratio of 0.63 and a swirl ratio of 0
- vertical, near-centrally-mounted injector
- high-pressure swirl injector to produce a spray with a cone angle of 40° ~ 60°
- fuel injection pressure : 5 MPa
- early injection mode only.

Ricardo GDI Engine - Engineers at Ricardo (Lake *et al.*, 1996; Jackson *et al.*, 1996, 1997) investigated the combustion and charge motion requirements of a GDI engine that incorporates a top-entry port head in combination with a curved piston, as depicted in Fig. 35. This engine was designed to operate using both the early-injection and late-injection modes. For stratified-charge operation mode, the minimum values of UBHC and COV of IMEP were obtained with a narrow cone spray having a relatively low injection rate for such a top-entry configuration. The combustion characteristics were found to be more sensitive to the fuel injection rate than to the spray cone angle. For a given fuel injection pressure, the use of an injector with a reduced fuel injection rate lengthens the fuel pulse width, thus reducing the spray penetration into the combustion chamber. This, in turn, reduces the spray impingement on the piston crown and cylinder walls, which is a primary source of UBHC emissions. It was noted, however, that the injector flow capacity must be compatible with rated power operation, which requires an acceptable injection pulse width at

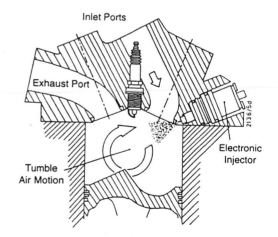

Fig. 35 The Ricardo GDI combustion system (Jackson *et al.*, 1996)

142

maximum speed and load. From the engine speed-load map that was shown earlier in Fig. 19(b) it may be seen that this GDI engine operates in the homogeneous-charge mode above 50% rated load. Early injection with either a stoichiometric or a slightly rich mixture is used above 70% load. Stratified-charge operation is thus restricted to the lower load and speed areas where fuel consumption is critical. The system runs fully unthrottled from 70% to 20% of full-load, and with light throttling below 20% of full-load to control UBHC emissions. The typical features of the Ricardo GDI engine are as follows :

RICARDO GDI RESEARCH ENGINE
- four-valve pentroof combustion chamber
- single cylinder
- bore x stroke : 74 x 75.5 mm
- displacement : 325 cm^3
- compression ratio : 10.4 : 1 to 12.7 : 1
- centrally-mounted spark plug
- curved piston crown
- hydra head with a top-entry port to generate reverse tumble
- fuel injection pressure range : 5~10 MPa
- part-load mode : late injection for stratified operation
- high-load mode : early injection for homogeneous operation.

Mercedes-Benz GDI Engine - The Mercedes-Benz GDI combustion system is a basic bowl-in-piston type using a vertical, centrally-mounted, fuel injector (Karl *et al.*, 1997). The schematic of the combustion chamber is illustrated in Fig. 36. Dynamometer tests of the Mercedes-Benz GDI combustion system for a range of injection pressures from 4 to 12 MPa indicate that the fuel consumption, UBHC emissions and COV of IMEP are minimized at 8 MPa. The NOx emission level was found to be at a maximum at this fuel pressure. The data reveal that the combustion duration measured at a fuel injection pressure of 8 MPa is the shortest, and show that the corresponding 50% heat release point occurs before TDC on compression. Parametric tests of the effect of spray cone angle indicate that the optimum spray cone angle for this system configuration is 90° for minimum indicated specific fuel consumption (ISFC), 105° for minimum UBHC emissions, and 75° for minimum NOx emissions. Both the ISFC and the UBHC emissions are reduced by utilizing an open piston-bowl configuration, but this increases the NOx emissions. It was found that in addition to the fuel economy improvement, engine-out NOx emissions are reduced by approximately 35% as compared to a conventional PFI engine; however, the engine-out UBHC emissions are significantly higher. For the load range of 0.2 to 0.5 MPa IMEP, the minimum ISFC is obtained at an engine speed of 2000 rpm. It was claimed that at lower or higher engine speeds, the fuel consumption increases as a result of a degradation in the mixture uniformity. The specifications of the GDI combustion system are summarized as follows :

MERCEDES-BENZ GDI ENGINE
- single-cylinder, four-valve engine
- most of the chamber is formed by a bowl in the piston, in conjunction with the flat valve angle
- single cylinder
- bore x stroke : 89 x 86.6 mm
- stroke to bore ratio : 0.973

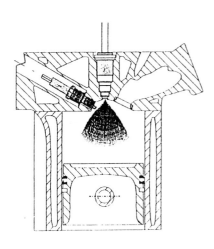

Fig. 36 The Mercedes-Benz GDI combustion system (Karl *et al.*, 1997)

- displacement volume : 538.5 cm^3
- compression ratio : 10.5 : 1
- valve inclination to deck : 19.5°
- spark plug located between the intake valves and in the immediate vicinity of the injector tip
- axisymmetric-reentrant-bowl-in-piston centered on cylinder axis
- vertical, centrally-mounted injector
- high-pressure, common-rail fuel system
- fuel pressure : 4 to 12 MPa
- cone spray with a geometric cone angle between 75° and 105°
- part-load mode : late injection for stratified operation
- high-load mode : early injection for homogeneous operation.

Other Research Engines - Other GDI research engines and concepts to be discussed in this section include :
- Ford PROCO
- Texaco TCCS
- MAN-FM
- OSKA
- Volkswagen
- FEV
- AVL/DMI.

A stable stratified-mixture region is important for achieving the fuel-economy potential of a GDI combustion system design. Many control strategies were proposed to attain such a stratified region during the early days of DISC engine development. Stable stratification can be most easily achieved by using a divided chamber in which a subvolume and the main combustion chambers provide well-separated mixture regions. One of the better-known production applications of this technique is the combustion chamber geometry of the Honda CVCC engine (Date and Yagi, 1974). For this combustion system, an auxiliary fuel preparation system feeds a fuel-rich mixture through an extra intake valve into a prechamber containing the spark plug. A very lean mixture is supplied to the main combustion chamber through the main fueling system. After combustion is initiated in the prechamber, a rich burning mixture issues as a jet through an orifice into the main chamber, both entraining and igniting the lean main-chamber charge. The disadvantage is that increased wall heat losses and throttling losses between the prechamber and the main combustion chamber reduce engine combustion efficiency.

Other systems have utilized jet/wall interaction and film evaporation to control the fuel distribution and the combustion process in an open chamber, such as the MAN-FM system (Meurer and Urlaub, 1969; Urlaub and Chmela, 1974]. Extensive testing led to the conclusion that the MAN-FM system has the drawback of increased wall heat losses, as well as increased UBHC and soot emissions. Also, as is the case for the CVCC divided chamber, only stratified operation is possible. Even if the higher compression ratio of the MAN-FM engine is taken into account, the engine power is still limited when compared to homogeneous, stoichiometric operation of the conventional PFI engine. Some of these early engines exhibit an improved tolerance for fuels of lower octane number, and a significant segment of the early work on prototype engines focused on the inherent multifuel capability of direct injection (Mitchell *et al.*, 1972; Lewis, 1986; Kim and Foster, 1985).

Another basic concept locates the fuel injector and the spark gap in close proximity in order to place the ignition source directly in the periphery of the fuel spray. The well known systems that use this method are the Ford PROCO (Simko *et al.*, 1972; Scussei *et al.*, 1978) and the Texaco TCCS (Alperstein *et al.*, 1974). A central injector provides a fuel spray with a hollow-cone shape that is stabilized by a strong air swirl (Ford PROCO) or provides a narrow jet injected tangentially into a piston bowl with swirl motion (TCCS). In both cases ignition stability is achieved primarily by the close spatial and temporal spacing of injection and ignition, which can tolerate only small cycle-to-cycle variations of the ignition area. The PROCO stratified-charge engines employed an injection system with an operating fuel pressure of 2 MPa with injection occurring early in the compression stroke, while the TCCS engines utilized a high-pressure, diesel-type injection system. Although the BSFC of both of these engines was quite good, the control of UBHC emissions was extremely difficult for light-load operation. Even with very high rates of EGR, the NOx emissions for both the PROCO and TCCS systems significantly exceeded the current emissions standards. The output of the late-injection TCCS system was also soot-limited.

Kato *et al.* (1987, 1990) proposed a direct-injection combustion system for both GDI and DI diesel engines that operate with a central pedestal in a cavity of the piston. This was denoted as the OSKA combustion system. The fuel is injected by a single-hole nozzle against the flat surface of the pedestal. The fuel deflects symmetrically in a disk shape and forms the air/fuel mixture. Since a comparatively rich mixture exists in the vicinity of the pedestal, a stable combustion may be attained by locating the spark plug in this area. As the mixture is always formed near the pedestal, there is little fuel in the squish area. Therefore, it is possible to prevent end-gas detonation in GDI applications and this type of engine can be operated at compression ratios of up to 14.5. However, the UBHC emissions are found to be substantially higher with this system, and represents a significant problem that must be solved.

A prototype engine incorporating the Volkswagen GDI combustion system was developed in 1991 (Schapertons *et al.*, 1991). This system is based on the MAN-FM concept, and utilizes a relatively compact injector that sprays fuel tangentially into a piston bowl. The electrodes of a spark plug project into the periphery of the fuel spray envelope, and mixture formation under high-load operation is promoted by a strong swirl induced by the intake port configuration. The swirling air is forced into a small-piston-bowl during compression, producing an intense movement of air which enhances the evaporation of the fuel film on the bowl wall. It was reported that injection is restricted to a short time window prior to ignition to ensure that a combustible mixture is present at the spark gap. Even at full load, however, injection cannot commence much earlier with the high compression ratio of 16:1 because detonation would occur.

The prototype FEV GDI engine utilizes a roof-shaped combustion chamber with a flat piston and a fuel injection nozzle that is centrally-mounted (Spiegel and Spicher, 1992). The effects of different injector nozzle types and configurations on the combustion characteristics and operating performance of this prototype engine were evaluated. It was found that for early injection timing with either a hollow-cone or hole-type nozzle, engine combustion characteristics comparable to PFI engines are achieved. Moving the spark plug location relative to the injected spray had little or no effect because of the early injection timing. For injection during the compression stroke, no stable engine combustion could be obtained with a six-hole nozzle. It was concluded that the performance of the hollow-cone nozzle with a wider cone angle is superior to the other types of injection nozzles tested for achieving a stable stratification by means of late injection.

All of the GDI engines exhibit the principal trade-off between the requirement of a minimum mixture formation time and the maximum time interval between injection and combustion if stratification is to be maintained. A separation of these two opposing requirements should provide the highest flexibility for a GDI system. Based upon this con-

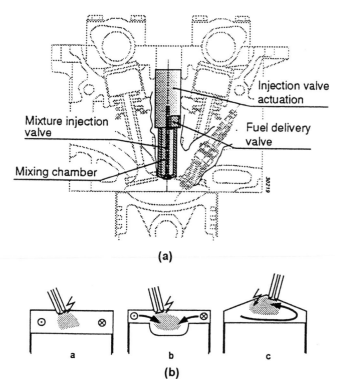

(a)

(b)

Fig. 37 The direct mixture injection concept (Fraidl *et al.*, 1996); (a) layout of the combustion chamber for direct mixture injection; (b) mixture preparation strategies based on direct mixture injection

sideration, engineers at AVL (Fraidl *et al.*, 1996) proposed the direct mixture injection (DMI) concept. The goal of this injection strategy is to combine the advantages of air-assisted injection and pre-vaporization of the fuel without the need for an external pressurized air supply. Figure 37(a) shows the schematic of the DMI system. The DMI valve incorporating a standard poppet valve geometry with an electrically-controlled actuation is responsible for the injection of the mixture and the recharging of the prechamber. The DMI injector recovers the gas pressure required for injection by withdrawing a small amount of compressed charge from the cylinder in the preceding engine cycle. To avoid combustion in the prechamber, the DMI valve is closed at the time of the spark. However, the mixture can be injected into the main chamber at any time prior to ignition. After the DMI valve is closed, liquid fuel is injected into the prechamber and is evaporated for injection during the next engine cycle. The mixture preparation quality is substantially enhanced by the almost complete vaporization of the fuel. The DMI system effectively decreases the engine crankangle interval that is required for fuel vaporization when compared to the conventional GDI injection. It is claimed that fuel metering and injection into the prechamber can be achieved by a constant displacement method using a low fuel pressure that exceeds the maximum prechamber pressure only slightly, as a high degree of fuel atomization is not required. The DMI system offers additional advantages regarding mixture stratification strategies. These advantages are due to the reduced mixture penetration velocity that results from the comparably low pressure difference between the pre-chamber and the combustion chamber at the start of fuel injection, and from the low momentum associated with the fuel vapor as compared to fuel droplets. Several DMI-based mixture preparation strategies are illustrated in Fig. 37(b).

The DMI system was found to have some problem areas that need addressing. Since the mixture is injected into the main chamber by the small pressure difference between the prechamber and the main chamber, all of the evaporated fuel may not be injected. This metering error is commonly designated as the fuel hang-up, and is associated with injection from intermediate cavities. At high engine speeds or under cold-start conditions, the fuel inside the prechamber may not evaporate completely, and cavity wetting is likely to occur. These conditions result in fuel metering errors. Moreover, the pressure inside the prechamber, as sampled from the previous cycle may not be appropriate for the metering for the current cycle, particularly if some throttling is used to smooth GDI mode transients. This will result in UBHC spikes during engine transient operation.

CLOSURE

The theoretical potential and the current research status of the spark-ignition, four-stroke, gasoline direct-injection engine has been discussed in detail in each of the sections of this paper. It is quite evident that significant incremental gains in engine performance and emission parameters are indicated; however, there are numerous practical concerns that will have to be successfully addressed if this type of engine is to realize its potential and become the primary automotive powerplant for the 2003-2013 time frame. Some of these key considerations are : Can a sufficient advantage in operating BSFC be achieved to offset the additional complexity as compared to a PFI engine? Can the applicable US, European, and Japanese emission standards be achieved and maintained for the required durability interval? Is a production-feasible trade-off between the required control systems and the overall system reliability obtainable; bearing in mind that the required GDI hardware may incorporate multi-stage injection, variable swirl-and-tumble control hardware and variable fuel pressure? Can component performance for fuel injectors and high pressure gasoline fuel pumps be sustained for reasonable time periods? Is it possible to minimize injector fouling due to deposit formation for the wide range of fuel quality and composition in the field to such a degree that reasonable service intervals can be achieved? And, finally, can reliable control-system strategies and algorithms be developed and implemented such that sufficiently smooth transitions from stratified-charge, late-injection operation to mid-range to homogeneous-charge, early-injection operation may be obtained, thus yielding driveability levels that are comparable to current sequential PFI systems? The field / warranty experience of the production Mitsubishi and Toyota GDI systems will have to be reviewed, and the actual emission indices and BSFC values will have to be evaluated over the entire operating map. This will permit a more knowledgeable evaluation of the relative merits of the two contrasting GDI concepts.

Design engineers, managers and researchers who must evaluate and prioritize the published, measured advantages of GDI engines over PFI engines should be aware of one area of data reduction and reporting that is somewhat disconcerting. This is the intermingling of practical test limitations, and in some cases even marketing, with the reporting of technical data on GDI systems and subsystems. Thus, it is sometimes very difficult for the reader to make a direct engineering comparison between GDI and PFI performance. One somewhat extreme, but true, example is the comparison of GDI and PFI fuel economy data that was obtained using two different vehicles with two different inertial weights. An example of a more subtle difference is the evaluation of the BSFC reduction resulting from the complete elimination of throttling, but not noting or subtracting the parasitic loss of a vacuum pump that would have to be added for braking and other functions. Most published comparisons lie between these two extremes. The readers are cautioned to review all claims of comparative GDI/PFI data carefully as to the precise test conditions for each, and the degree to which the systems were tested under different conditions or constraints.

It is likely that the development of future GDI engines will follow the concept of dual-mode stratified/homogeneous operation to equal the power output level of current PFI engines, and the engine will make inroads into the market in two stages. First, prior to the emergence of the proven lean-NOx-catalyst, the homogeneous-mode GDI will be prevalent. Increasing the engine specific power at high load will be a key goal. These engines will be operated in the homogeneous mode using both lean and stoichiometric mixtures, and will achieve BSFC values that are similar to current lean-burn engines. The exhaust emissions will be controlled by the combination of the three-way catalyst and EGR. A marked reduction in UBHC emissions during the cold start, coupled with improved transient response, will be the major improvements when compared to the conventional PFI engine. The availability of an efficient and durable lean-NOx-catalyst will likely trigger the second stage of GDI development, and engines incorporating ultra-lean stratified operation for load control will become available. These engines will provide both the specific power of the homogeneous mode and a 20% increase in vehicle fuel economy. For the same reasons that port fuel injection gradually replaced carburetion and throttle-body injection, a GDI combustion configuration that is an enhancement of one of the concepts outlined in this paper will emerge as the predominate engine system, and will gradually displace the sequential PFI applications.

REFERENCES

1. Alperstein, M., Schafer, G. H., and Villforth, III, F. J., Texaco's stratified charge engine - multifuel, efficient, clean, and practical, *SAE Technical Paper*, No. 740563 (1974).
2. Anderson, R. W., Brehob, D. D., Yang, J., Vallance, J. K., and Whiteaker, R. M., A new direct injection spark ignition (DISI) combustion system for low emissions, *FISITA-96 Technical Paper*, No. P0201 (1996a).
3. Anderson, R. W., Yang, J., Brehob, D. D., Vallance, J. K., and Whiteaker, R. M., Understanding the thermodynamics of direct injection spark ignition (DISI) combustion systems : an analytical and experimental investigation, *SAE Technical Paper*, No. 962018 (1996b).
4. Ando, H., Combustion control technologies for direct-injection gasoline engines, *Proceedings of the 73rd JSME Annual Meeting (V) (in Japanese)*, No. WS 11-(4), pp.319-320 (1996a).
5. Ando, H., Combustion control technologies for gasoline engines, *I.MechE. Seminar of Lean Burn Combustion Engines*, S433, December 3~4, 1996 (1996b).
6. Arcoumanis, C., Gold, M. R., Whitelaw, J. H., and Xu, H. M., Local charge stratification in spark-ignition engines, *I.MechE. Seminar of Lean Burn Combustion Engines*, S433, December 3~4, 1996 (1996).
7. Balles, E. N., Ekchian, J. A., and Heywood, J. B., Fuel injection characteristics and combustion behavior of a direct-injection stratified-charge engine, *SAE Technical Paper*, No. 841379 (1984).
8. Baranescu, G. S., Some characteristics of spark assisted direct injection engine, *SAE Technical Paper*, No. 830589 (1983).
9. Buchheim, R. and Quissek, F., Ecological and economical aspects of future passenger car powertrains, *FISITA Technical Paper*, No. P1404 (1996).
10. Caracciolo, F. and Stebar, R. F., An engine dynamometer test for evaluating port fuel injector plugging, *SAE Technical Paper*, No. 872111 (1987).
11. Daisho, Y., Shimizu, A., Saito, T., and Choi, K. H., A fundamental study on charge stratification, *Proceedings of COMODIA-85*, pp.423-432 (1985).
12. Das, S. and Dent, J. C., A study of air-assisted fuel injection into a cylinder, *SAE Technical Paper*, No. 941876 (1994).
13. Diwakar, R., Fansler, T. D., French, D. T., Ghandhi, J. B., Dasch, C. J., and Heffelfinger, D. M., Liquid and vapor fuel distributions from an air-assist injector - an experimental and computational study, *SAE Technical Paper*, No. 920422 (1992).
14. Dodge, L. G., Fuel preparation requirements for direct-injected spark ignition engines, *Proceedings of ILASS-America*, pp.120-124 (1996a).
15. Dodge, L. G., Fuel preparation requirements for direct-injected spark ignition engines, *SAE Technical Paper*, No. 962015 (1996b).
16. Douaud, A., Tomorrow's efficient and clean engines and fuels, *FISITA Technical Paper*, No. K0006 (1996).
17. Drake, M. C., Fansler, T. D., and French, D. T., Crevice flow and combustion visualization in a direct-injection spark-ignition engine using laser imaging techniques, *SAE Technical Paper*, No. 952454 (1995).
18. Duggal, V. K., Kuo, T.-W., and Lux, F. B., Review of multi-fuel engine concepts and numerical modeling of in-cylinder flow processes in direct injection engines, *SAE Technical Paper*, No. 840005 (1984).
19. Emerson, J., Felton, P. G., and Bracco, F. V., Structure of sprays from fuel injectors part III : the Ford air-assisted fuel injector, *SAE Technical Paper*, No. 900478 (1990).
20. Enright, B., Borman, G. L., and Myers, P. P., A Critical review of spark ignited diesel combustion, *SAE Technical Paper*, No. 881317 (1988).
21. Evers, L. W., Characterization of the transient spray from a high pressure swirl injector, *SAE Technical Paper*, No. 940188 (1994).
22. Fansler, T. D., French, D. T., and Drake, M. C., Fuel distribution in a firing direct-injection spark-ignition engine using laser-induced fluorescence imaging, *SAE Technical Paper*, No. 950110 (1995).
23. Felton, P. G., Laser diagnostics for direct-injection gasoline engines, *I.MechE. Seminar of Lean Burn Combustion Engines*, S433, December 3~4, 1996 (1996).
24. Fraidl, G. K., Piock, W. F., and Wirth, M., Gasoline direct injection : actual trends and future strategies for injection and combustion systems, *SAE Technical Paper*, No.960465 (1996).
25. Frank, R. M. and Heywood, J. B., Combustion characterization in a direct-injection stratified-charge engine and implications on hydrocarbon emissions, *SAE Technical Paper*, No. 892058 (1989).
26. Frank, R. M. and Heywood, J. B., The importance of injection system characteristics on hydrocarbon emissions from a direct-injection stratified-charge engine, *SAE Technical Paper*, No. 900609 (1990a).
27. Frank, R. M. and Heywood, J. B., The effect of fuel characteristics on combustion in a spark-ignited direct-injection engine, *SAE Technical Paper*, No. 902063 (1990b).
28. Frank, R. M. and Heywood, J. B., The effect of piston temperature on hydrocarbon emissions from a spark-ignited direct-injection engine, *SAE Technical Paper*, No. 910558 (1991).
29. Fujieda, M., Siraisi, T., and Oosuga, M., Influence of the spray pattern on combustion characteristics of the direct injection SI engine, *Proceedings of ILASS-Japan (in Japanese)*, pp.173-177 (1995).
30. Furuno, S. et al., The effects of inclination angle of swirl axis on turbulence characteristics in a 4-valve lean burn engine with SCV, *SAE Technical Paper*, No. 902139 (1990).
31. Ghandhi, J. B. and Bracco, F. V., Fuel distribution effects on the combustion of a direct-injection stratified-charge engine, *SAE Technical Paper*, No. 950460 (1995).
32. Giovanetti, A. J., Ekchian, J. A., and Heywood, J. B., Analysis of hydrocarbon emission mechanisms in a direct injection spark-ignition engine, *SAE Technical Paper*, No. 830587 (1983).
33. Gualtieri, J. and Sawyer, C. A., Toyota, Mitsubishi DI gas engines debut, *Automotive Industries*, p.30 August, 1995 (1995).
34. Han, Z., Reitz, R. D., Claybaker, P. J., Rutland, C., Yang, J., and Anderson, R. W., Modeling the effects of intake flow structures on fuel/air mixing in a direct-injected spark ignition engine, *SAE Technical Paper*, No. 961192 (1996a).
35. Han, Z., Parrish, S. E., Farrell, V., and Reitz, R. D., Modeling atomization processes of pressure-swirl hollow-cone fuel sprays, *Paper Submitted to Atomization and Sprays*, (1996b).
36. Han, Z., Reitz, R. D., Yang, J., and Anderson, R. W., Effects of injection timing on air/fuel mixing in a direct-injection spark-ignition engine, *SAE Technical Paper*, No. 970625 (1997a).
37. Han, Z., Fan, L., and Reitz, R. D., Multi-dimensional modeling of spray atomization and air/fuel mixing in a direct-injection spark-ignition engine characteristics, *SAE Technical Paper*, No. 970625 (1997b).
38. Harada, J., Tomita, T., Mizuno, H., Mashiki, Z., and Ito, Y., Development of a direct injection gasoline engine, *SAE Technical Paper*, No. 970540 (1997).
39. Harrington, D. L., Interactions of direct injection fuel sprays with in-cylinder air motions, *SAE Transition*, Vol. 93, (1984a).

40. Harrington, D. L., Analysis of spray penetration and velocity dissipation for non-steady fuel injection, *ASME Technical Paper* 04-DGP-13 (1984b).

41. Harrington, D. L., Stebar, R. F., and Caracciolo, F., Deposit-induced fuel flow reduction in multiport fuel injectors, *SAE Technical Paper*, No. 892123 (1989).

42. Hauser, G., Heitland, H., and Wislocki, K., Can the no man's land between diesel- and otto-engines be bridged?, *SAE Technical Paper*, No. 951962 (1995).

43. Hattori, H., Ota, M., Sato, E., and Kadota, T., Fundamental study on DISC engine with two-stage fuel injection, *JSME International J.*, Series B., Vol. 38, No. 1, pp.129-135 (1995).

44. Heisler, H., Advanced engine technology, *SAE* (1995).

45. Hiroyasu, H., Experimental and theoretical studies on the structure of fuel sprays in diesel engines, *Proceedings of ICLASS-91*, Keynote Lecture, pp.17-32 (1991).

46. Hoffman, J., Martin, J. K., and Coates, S. W., Spray photographs and preliminary spray mass flux distribution measurements of a pulsed pressure atomizer, *Proceedings of ILASS-America*, pp.288-291 (1996).

47. Hoffman, J., Eberhardt, E., and Martin, J. K., Comparison between air-assisted and single-fluid pressure atomizers for direct-injection SI engines via spatial and temporal mass flux measurements, *SAE Technical Paper*, No. 970630 (1997).

48. Iida, Y., The current status and future trend of DISC engines, *Preprint of JSME Seminar* (in Japanese), No. 920-48, pp.72-76 (1992).

49. Iiyama, A. And Muranaka, S., Current status and future perspective of DISC engine, *Proceedings of JSAE* (in Japanese), No. 9431030, pp.23-29 (1994).

50. Iiyama, A., Direct injection gasoline engines, *Technical Seminar at Wayne State University*, February 26, 1996 (1996).

51. Ikeda, Y., Nakajima, T., and Kurihara, N., Spray formation of air-assist injection for two-stroke engine, *SAE Technical Paper*, No. 950271(1995).

52. Iwamoto, Y., Noma, K., Yamauchi, T., and Nakayama, O., Development of gasoline direct injection engine, *SAE Technical Paper*, No. 970541 (1997).

53. Iwata, M., Furuhashi, M., and Ujihashi, M., The spray characteristics and engine performance of EFI injector, *Proceedings of the Technical Conference of JSAE* (in Japanese), No. 861, pp.29-32 (1986).

54. Jackson, N. S., Stokes, J., Whitaker, P. A., and Lake, T. H., A direct injection stratified charge gasoline combustion system for future European passenger cars, *I.MechE. Seminar of Lean Burn Combustion Engines*, S433, December 3~4, 1996 (1996).

55. Jackson, N. S., Stokes, J., and Lake, T. H., Stratified and homogeneous charge operation for the direct injection gasoline engine - high power with low fuel consumption and emissions, *SAE Technical Paper*, No. 970543 (1997).

56. Kagawa, L., Okazaki, S., Somyo, N., and Akagi, Y., A study of a direct-injection stratified charge rotary engine for motor vehicle application, *SAE Technical Paper*, No. 930677 (1993).

57. Karl, G., Kemmler, R., and Bargende, M., Analysis of a direct injected gasoline engine, *SAE Technical Paper*, No. 970624 (1997).

58. Kato, S. and Onishi, S., New mixture formation technology of direct fuel injection stratified combustion SI engine (OSKA), *SAE Technical Paper*, No. 871689 (1987).

59. Kato, S. and Onishi, S., Direct fuel injection stratified charge engine by impingement of fuel jet (OSKA) - performance and combustion characteristics, *SAE Technical Paper*, No. 900608 (1990).

60. Kim, C. and Foster, D. E., Aldehyde and unburned fuel emission measurements from a methanol-fueled Texaco stratified charge engine, *SAE Technical Paper*, No. 852120 (1985).

61. Kim, K. S. and Kim, S. S., Spray characteristics of an air-assisted fuel injector for two-stroke direct-injection gasoline engines, *Atomization and Sprays*, Vol. 4, pp.501-521 (1994).

62. Kono, S., Kudo, H., and Terashita, A Study of spray Direction Against Swirl in D.I. Engines, *Proceedings of COMODIA-90*, pp.269-274 (1990).

63. Kono, S., Study of the stratified charge and stable combustion in DI gasoline engines, *SAE Technical Paper*, No. 950688 (1995).

64. Kume, T., Iwamoto, Y., Iida, K., Murakami, M., Akishino, K., and Ando, H., Combustion control technologies for direct injection SI engine, *SAE Technical Paper*, No. 960600 (1996).

65. Kuwahara, K., Watanabe, T., Shudo, T., and Ando, H., A study of combustion characteristics in a direct injection gasoline engine by high-speed spectroscopic measurement, *Proceedings of the Internal Combustion Engine Symposium - Japan* (in Japanese), pp.145-150 (1996).

66. Laforgia, D., Chehroudi, B., and Bracco, F. V., Structure of sprays from fuel injectors - part II, the Ford DFI - 3 fuel injector, *SAE Technical Paper*, No. 890313 (1989).

67. Lake, T. H., Christie, M. J., Stokes, J., Horada, O., and Shimotani, K., Preliminary investigation of solenoid activated in-cylinder injection in stoichiometric S.I. engine, *SAE Technical Paper*, No. 940483 (1994).

68. Lake, T. H., Sapsford, S. M., Stokes, J., and Jackson, N. S., Simulation and development experience of a stratified charge gasoline direct injection engine, *SAE Technical Paper*, No. 962014 (1996).

69. Lee, C. F. and Bracco, F. V., Initial comparisons of computed and measured hollow-cone sprays in an engine, *SAE Technical Paper*, No. 940398 (1994).

70. Lewis, J. M., UPS multifuel stratified charge engine development program - field test, *SAE Technical Paper*, No. 860067 (1986).

71. MacInnes, J. M. and Bracco, F. V., Computation of the Spray from an Air-Assisted Fuel Injector," *SAE Technical Paper*, No. 902079 (1990).

72. Matsushita, S., Nakanishi, K., Gohno, T., and Sawada, D., Mixture formation process and combustion process of direct injection S.I. engine, *Proceedings of JSAE* (in Japanese), No. 965, Oct. 10, 1996, pp. 101-104 (1996).

73. Mccann, K. A., MMC ready with first DI gasoline engine, *WARD's Engine and Vehicle Technology Update*, Vol. 21, No. 11, June 1, 1995, pp.1-2 (1995).

74. Meyer, J., Kiefer, K., Von Issendorff, F., Thiemann, J., Haug, M., Schreiber, M., and Klein, R., Spray visualization of air-assisted fuel injection nozzles for direct injection SI-engines, *SAE Technical Paper*, No. 970623 (1997).

75. Meurer, S. and Urlaub, A., Development and operational results of the MAN FM combustion system, *SAE Technical Paper*, No. 690255 (1969).

76. Miok, J., Huh, K. Y., and Noh, S. H., Numerical prediction of charge distribution in a lean burn direct-injection spark-ignition engine, *SAE Technical Paper*, No. 970626 (1997).

77. Mitchell, E., Alperstein, M., Cobb, J. M., and Faist, C. H., A stratified charge multifuel military engine - a progress report, *SAE Technical Paper*, No. 720051 (1972).

78. Miyamoto, N., Ogawa, H., Shudo, T., and Takeyama, F., Combustion and emissions in a new concept DI stratified charge engine with two-stage fuel injection, *SAE Technical Paper*, No. 940675 (1994).

79. Miyamoto, T., Kobayashi, T., and Matsumoto, Y., Structure of sprays from an air-assist hollow-cone injector, *SAE Technical Paper*, No. 960771 (1996).

80. Naitoh, K. and Takagi, Y., Synthesized spheroid particle (SSP) method for calculating spray phenomena in direct-injection SI engines, *SAE Technical Paper*, No. 962017 (1996).

81. Ohyama, Y., Nogi, T., and Ohsuga, M., Effects of fuel/air mixture preparation on fuel consumption and exhaust emission in a spark ignition engine, *IMechE Paper*, No. 925023, C389/232, pp.59-64 (1992).

82. Parrish, S. and Farrel, P. V., Transient spray characteristics of a direct-injection spark-ignited fuel injector, *SAE Technical Paper*, No. 970629 (1997).

83. Pischinger, F. and Walzer, P., Future trends in automotive engine technology, *FISITA Technical Paper*, No. P1303 (1996).

84. Pontoppidan, M., Gaviani, G., and Marelli, M., Direct fuel-injection - a study of injector requirements for different mixture preparation concepts, *SAE Technical Paper*, No. 970628 (1997).

85. Ronald, B., Helmut, T., and Hans, K., Direct fuel injection - a necessary step of development of the Si engine, *FISITA Technical Paper*, No. P1613 (1996).

86. Saito, A., Kawamura, K. and Tanasawa, Y., Improvement of fuel atomization electronic fuel injector by air flow, *Proceedings of ICLASS-88*, pp.263-270 (1988).

87. Salters, D., Williams, P., and Greig, A., Fuel spray characterization within an optically accessed gasoline direct injection engine using a CCD imaging system, *SAE Technical Paper*, No. 961149 (1996).

88. Schapertons, H., Emmenthal, K.-D., Grabe, H.-J., and Oppermann, W., VW's gasoline direct injection (GDI) research engine, *SAE Technical Paper*, No.910054 (1991).

89. Schechter, M. M. and Levin, M. B., Air-forced fuel injection system for 2-stroke D.I. gasoline engine, *SAE Technical Paper*, No. 910664 (1991).

90. Scussei, A. J., Simko, A. O., and Wade, W. R., The Ford PROCO engine update, *SAE Technical Paper*, No. 780699 (1978).

91. Seiffert, U., The automobile in the next century, *FISITA Technical Paper*, No. K0011 (1996).

92. Shimotani, K., Oikawa, K., Horada, O., and Kagawa, Y., Characteristics of gasoline in-cylinder direct injection engine, *Proceedings of the Internal Combustion Engine Symposium - Japan* (in Japanese), pp.289-294 (1995).

93. Shimotani, K., Oikawa, K., Tashiro, Y., and Horada, O., Characteristics of exhaust emission on gasoline in-cylinder direct injection engine, *Proceedings of the Internal Combustion Engine Symposium - Japan* (in Japanese), pp.115-120 (1996).

94. Simko, A., Choma, M. A., and Repko, L. L., Exhaust emissions control by the Ford Programmed combustion process - PROCO, *SAE Technical Paper*, No. 720052 (1972).

95. Spiegel, L. and Spicher, U., Mixture Formation and Combustion in a Spark Ignition Engine with Direct Fuel Injection, *SAE Technical Paper*, No.920521(1992).

96. Sugimoto, T., Takeda, K., and Yoshizaki, H., Toyota air-mix type two-hole injector for 4-valve engines, *SAE Technical Paper*, No. 912351 (1991).

97. Takagi, Y., The role of mixture formation in improving fuel economy and reducing emissions of automotive S.I. engines, *FISITA Technical Paper*, No. P0109 (1996a).

98. Takagi, Y., Combustion characteristics and research topics of in-cylinder direct-injection gasoline engines, *Proceedings of the 73rd JSME Annual Meeting* (V) (in Japanese), No. WS 11-(3), pp.317-318 (1996b).

99. Tomoda, T., Sasaki, S., Sawada, D., Saito, A., and Sami, H., Development of direct injection gasoline engine - study of stratified mixture formation, *SAE Technical Paper*, No. 970539 (1997).

100. Toyota, Direct-injection 4-stroke gasoline engine, *TOYOTA Press Information'96*, August 1996 (1996).

101. Urlaub, A. G. and Chmela, F. G., High-speed, multifuel engine : L9204 FMV, *SAE Technical Paper*, No. 740122 (1974).

102. Wirth, M., Piock, W. F., and Fraidl, Actual trends and future strategies for gasoline direct injection, *I.MechE. Seminar of Lean Burn Combustion Engines*, S433, December 3~4, 1996 (1996).

103. Wojik, K. and Fraidl, G. K., Engine and vehicle concepts for low consumption and low-emission passenger cars, *FISITA Technical Paper*, No. P1302 (1996).

104. Wood, C. D., Unthrottled open-chamber stratified charge engines, *SAE Technical Paper*, No. 780341(1978).

105. Yamada, T., Trends of S.I. engine technologies in Japan, *FISITA-96 Technical Paper*, No. P0204 (1996).

106. Yamaguchi, J., Mitsubishi DI gasoline engine prototype, *Automotive Engineering*, pp.25-29, September, 1995 (1995).

107. Yamaguchi, J., Toyota readies direct-injection gasoline engine for production, *Automotive Engineering*, pp.74-76, November, 1996 (1996a).

108. Yamaguchi, J., Mitsubishi Galant sedan and Legnum wagon, *Automotive Engineering*, pp.26-29, November, 1996 (1996b).

109. Yamauchi, T. and Wakisaka, T., Computation of the hollow-cone sprays from high-pressure swirl injector from a gasoline direct-injection SI engine, *SAE Technical Paper*, No. 962016 (1996).

110. Zhao, F.-Q., Taketomi, M., Nishida, K. and Hiroyasu, H., Quantitative imaging of the fuel concentration in a SI engine with laser Rayleigh scattering, *SAE Technical Paper*, No.932641 (1993).

111. Zhao, F.-Q., Taketomi, M., Nishida, K. and Hiroyasu, H., PLIF measurements of the cyclic variation of mixture concentration in a SI engine, *SAE Technical Paper*, No.940988 (1994).

112. Zhao, F.-Q., Lai, M.-C., and Harrington, D. L., The spray characteristics of automotive port fuel injection - a critical review, *SAE Technical Paper*, No. 950506 (1995a).

113. Zhao, F.-Q., Yoo, J.-H., and Lai, M.-C., The spray characteristics of dual-stream port fuel injectors for applications to 4-valve gasoline engines, *SAE Technical Paper*, No. 952487 (1995b).

114. Zhao, F.-Q., Lai, M.-C., Liu, Y., Yoo, J.-H., Zhang, L., and Yoshida, Y., Spray characteristics of direct-injection gasoline engines, *Proceedings of ILASS-America*, pp.150-154 (1996a).

115. Zhao, F.-Q., Yoo, J.-H., Liu, Y., and Lai, M.-C., Spray dynamics of high pressure fuel injectors for DI gasoline engines, *SAE Technical Paper*, No. 961925 (1996b).

116. Zhao, F.-Q., Yoo, J.-H. and Lai, M.-C., The spray structure of air-shrouded dual-stream port fuel injectors with different air mixing mechanisms, *Proceedings of the 1996 Spring Technical Conference of the ASME Internal Combustion Engine Division*, ICE-Vol. 26-2, pp.21-29 (1996c).

117. Zhao, F.-Q., Yoo, J.-H., and Lai, M.-C., Mixture formation and combustion characteristics of a direct-injection gasoline engine, *Paper submitted to ILASS-America* (1997a).

118. Zhao, F.-Q., Yoo, J.-H., and Lai, M.-C., Mixture formation process of a direct-injection gasoline engine, *Paper submitted to 1997 SAE Spring Fuels and Lubricants meeting*, (1997b).

119. Zhao, F.-Q., Yoo, J.-H., Liu, Y., and Lai, M.-C., Characterization of direct-injection gasoline sprays under different ambient and fuel injection conditions, *Paper submitted to ICLASS-97* (1997c).

148

Computation of the Hollow-Cone Sprays from a High-Pressure Swirl Injector for a Gasoline Direct-Injection SI Engine

Toyosei Yamauchi
Subaru Research Center Company, Ltd.

Tomoyuki Wakisaka
Kyoto Univ.

ABSTRACT

The hollow-cone sprays generated by a high-pressure swirl injector are numerically analyzed using the author's Generalized Tank and Tube (GTT) code with spray models. The effects of the prescription of fuel injection conditions on the spray behavior are investigated and discussed in detail. The calculated results are compared with the results of measurement by Phase Doppler Anemometer (PDA), etc. A reasonable method to numerically reproduce the structure of a hollow-cone spray including the coarse droplet phenomenon is presented. As a result, the structure of a hollow-cone spray and the temporal transition of the structure are clarified in terms of fluid dynamics.

INTRODUCTION

Much efforts have been concentrated on the investigation of spray behavior injected from various types of injector nozzles for gasoline direct-injection spark ignition engines. Reitz & Diwakar[1][2], and Lee & Bracco[3][4] investigated the behavior of the hollow-cone sprays generated by an injector having an oscillating conical poppet valve and a circular slit nozzle. Recently, the hollow-cone sprays generated by a single-fluid (liquid) high-pressure swirl injector, which is investigated in this paper, have been put into research. This type of injector has a circular hole nozzle as described later in detail. For this type of injector the atomization process of injected liquid is thought to be rather different from that for a poppet-valve type injector, which was discussed in Reference[4]. In numerical simulation of the spray behavior, it seems that a submodel for atomization phenomenon and also the prescription of fuel injection conditions are closely related to the reproducibility of such an atomization process. Unfortunately, for the type of injector treated here, there are very few published papers reporting detailed knowledges of the primary atomization process or the breakup process of

liquid sheet. Therefore, special consideration is required in treating the initial conditions of droplets injected at each computational time step.

In this paper, as the first step of a numerical study on the hollow-cone sprays injected from a single-fluid (liquid) high-pressure swirl injector, the effects of the prescription of fuel injection conditions on the behavior of the sprays are investigated systematically and discussed in detail. The special feature of this study is that not only the spray behavior until the end of injection but also the spray behavior after the end of injection (in the period about three times longer than the injection period) is discussed. This will give us a useful knowledge regarding the structure of a fully-developed spray, which will be prerequisite in numerically simulating the spray behavior under actual engine operating conditions. In regard to the spray behavior in the downstream region from an injector nozzle, the results of calculation are compared with those of experiments. Then a reasonable method to reproduce the structure of a hollow-cone spray including the coarse droplet phenomenon is developed. The calculated droplet size distribution is verified by PDA measurement. As a result of this study, the temporal transition of the spray structure is clarified in terms of fluid dynamics.

METHOD OF NUMERICAL ANALYSIS

GAS FLOW - In this study, gas flows are calculated by the author's GTT code[5] which is based on the finite volume and fully-implicit discretization methods with generalized curvilinear coordinates. The details of the calculation method are described in Reference[5]. As a differencing scheme for the convection terms of the Navier-Stokes equations, the third-order Chakravarthy-Osher TVD scheme[6] is used. The Hybrid scheme is used for the convection terms of the conservation equations of enthalpy, turbulence energy k and its dissipation rate ε. Pressure-velocity coupling is accomplished by means of the

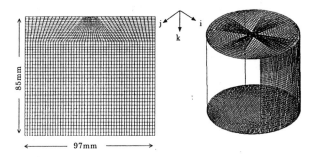

Figure 1: Partial view of the computational grid for free sprays

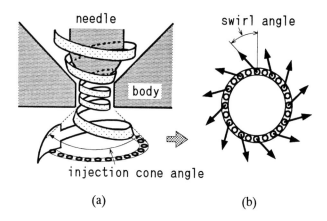

(a)   (b)

Figure 2: Schematics of (a) the nozzle outlet of a high-pressure swirl injector, and (b) initial droplets located on a circle with a swirl component (in computation)

SIMPLEC algorithm[7]. The concentration field of fuel vapor is calculated by the CIP (Cubic Interpolated pseudo-Particle) method[8], whose numerical diffusion is very small.

In this study, the k-ε turbulence model is used for calculating the spray behavior. For calculating gas velocity, however, the turbulent viscosity computed with this turbulence model is not used but the laminar viscosity is used as viscosity in the Navier-Stokes equations, because small vortices to be generated by a spray are not reproduced if the turbulent viscosity, which is over-estimated in some cases, is employed. As the wall boundary conditions for velocity, enthalpy, k and ε, the law of the wall is used. The validity of the above-mentioned method of calculating gas velocity was confirmed by comparing the calculated results of the velocity field in an engine cylinder with the experimental ones[9].

SPRAY MODEL - Liquid fuel sprays are numerically analyzed using the Discrete Droplet Model (DDM) in KIVA code[10]. The bag and stripping breakup model by Reitz, et al.[2] is employed as a droplet breakup model, and the models of droplet coalescence and evaporation in KIVA code are used. Concerning the interaction between the gas phase and the fuel droplets, the vaporized mass, momentum and enthalpy lost by the droplets are given to the source terms in their respective conservation equations as to the gas phase. At present, quantitative information on the primary breakup process of the spray, that is, the breakup process of liquid sheet is not provided for the injector treated here; therefore in this study, the liquid sheet is ignored, and as a first attempt, initial droplets to be injected are placed near the nozzle outlet according to the prescribed fuel injection conditions. Consequently, the spray behavior in the downstream is determined through the secondary atomization process.

COMPUTATIONAL CONDITIONS - Figure 1 shows the partial view of the grid used in the free spray computations. The number of grid points is $51 \times 51 \times 46$, and the minimum grid size is about 0.2 mm at the injector

nozzle outlet. The time increment Δt for calculating gas velocity is set at 2.5 μsec during the injection period, and set at 5 μsec after the end of injection. The behavior of droplets and fuel vapor concentration are calculated explicitly in the subcycles with a smaller time increment Δt_s ($= 0.1\, \Delta t$). Schmidt number is set at 1.0. The number of injected droplet parcels (representative droplets) is 9000.

HIGH-PRESSURE SWIRL INJECTOR

The schematic drawing of the injector nozzle outlet is shown in Fig.2(a). As shown in this figure, the injection nozzle has a circular hole and the injection velocity of a spray has a swirl component which is given in the upstream region of the injector. The injection cone angle (see Fig.2(a)) of the injector treated here is about 70 degrees.

The injector operating conditions mainly used in this investigation are as follows: The injection pressure of liquid fuel (octane) is 7 MPa; injection period (pulse width for opening an injector needle valve in experiment) is 0.8 msec; amount of injection fuel mass is 8 mg; ambient gas is air at atmospheric pressure and room temperature.

EFFECTS OF THE PRESCRIPTION OF FUEL INJECTION CONDITIONS ON THE HOLLOW-CONE SPRAY BEHAVIOR

There are various fuel injection conditions which influence the behavior of a hollow cone spray. As a fundamental study, investigating the effect of each of these conditions on the spray behavior separately in computation will give useful information, even if the condition prescribed in computation may be imaginary. In this study, among various conditions, the effects of injection mass flow rate, initial droplet diameter, and swirl component on the spray behavior are investigated systematically.

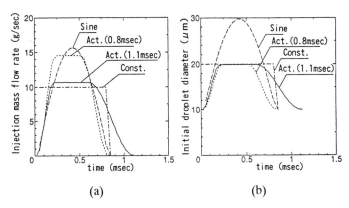

(a)   (b)

Figure 3: Temporal variations in (a) injection fuel flow rate, and (b) injection droplet size used in computation

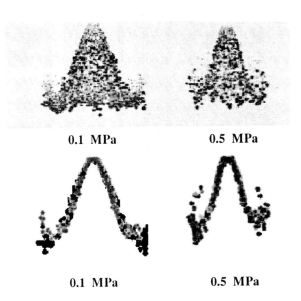

0.1 MPa   0.5 MPa

0.1 MPa   0.5 MPa

Figure 4 : Effects of ambient gas pressure on the spray cone angle reproduced by computation with the initial droplet diameter of 20 μ m. Top row: spray shape , Bottom row: spray structure.

In this section, firstly, the fuel injection conditions to be prescribed in computation are described,
and then the effects of these conditions on the structure and behavior of hollow cone sprays are discussed in detail. Furthermore, the calculated results under the injection condition which is prescribed most reasonably are compared with the experimental results.

PRESCRIPTION OF FUEL INJECTION CONDITIONS -

Injection Position of Droplets - There are two possible choices concerning the location of injecting droplets. One is to place the droplets just at the nozzle outlet[2]. The other is to place them somewhat downstream from the outlet. In the former case, reproduction of swirling motion is difficult unless very fine meshes are provided at the nozzle outlet. In the latter case, initial droplets to be injected are located on a circle, and a swirl component is given to the injection velocity of each droplet as shown in Fig.2(b) so that swirling motion may be appropriately reproduced by computation with not so many meshes. For this reason, the latter case, namely placing the droplets somewhat downstream from the outlet is selected in this study.

Injection Fuel Flow Rate and Droplet Size - The main purpose of this study is to investigate how the profiles of the temporal variations in injection fuel flow rate and injection droplet size affect the spray behavior. As a first step, the profiles of injection fuel flow rate shown in Fig.3(a) and those of injection droplet size variation shown in Fig.3(b) are considered. Figure 3(a) shows four profiles of fuel flow rate; "Const.(Constant)", "Sine", and two "Act.(Actual)"s. "Act." denotes an approximated profile of actual fuel flow rate derived from a curve of injector needle lift. "Act.(0.8msec)" and "Act.(1.1msec)" denote that the opening period of the needle valve are 0.8 msec and 1.1 msec, respectively. As for the injection droplet size variation, each profile shown in Fig.3(b) corresponds to each profile in Fig.3(a). The maximum droplet sizes for "Act." and "Const." are set at the same size. These profiles are chosen so that the droplet size may vary with the needle

lift. This way of determining the injection droplet size is similar to the way used by Reitz, et al. for a poppet-valve type injector[2].

Initial Droplet Size - The injection pressure (7 MPa) is much higher than the ambient gas pressures (0.1 MPa - 0.5 MPa) in this study. Therefore, the injection velocity of fuel can be treated as constant, and it is assumed that the initial droplet size does not vary with the ambient gas pressure[2]. The initial droplet diameter during the constant fuel flow rate (see Fig.3(a)) is selected in such a way that the calculated spray behavior can be reasonable in the full range of the above-mentioned ambient gas pressures. Here, the diameter is set at 20 μ m because the well known feature that the spray cone angle decreases with an increase in ambient gas pressure is reproduced reasonably well with this diameter as shown in Fig.4.

Injection Velocity and Direction of Droplets - The injection velocity and injection cone angle are fixed at 60 m/s and 70 degrees, respectively, during the injection period. These values are determined from spray shape photographs taken under the standard atmospheric condition. As for the swirl component of the velocity vector, a swirl angle of 40 degrees is selected (see Fig.2(b)). This swirl angle is determined by a rough estimation for the injector treated here. The accurate swirl angle has to be determined in future.

STRUCTURE OF THE HOLLOW-CONE SPRAYS - The structure of a hollow-cone spray is represented clearly by the distribution of spray droplets on a vertical cross section. Figure 5 shows the calculated results of spray structure under various injection conditions presented in Figs.3(a) and (b). The schematic representations of the spray structures are also shown in the last column of Fig.5.

As shown in Fig.5, the temporal transition of the hollow-cone spray structure is divided into three typical stages as follows:

1. first stage: The droplets form a hollow-cone shape at the end of injection.

2. second stage: The droplets which form the top part of the cone are catching up with the previously injected droplets. A vortex ring, namely, toroidal vortex is generated around the lower part of the cone, and is developing. Due to this vortex ring, droplets form a ring, which appears as circles on a vertical cross section. Under some injection conditions, the ring of droplets is not formed. At this stage, the spray structure changes transiently.

3. third stage: The droplets which formed the top of the cone have already caught up with the previously injected droplets and the vortex ring has been completely formed. The vortex ring grows bigger and is slowly descending. At this stage, the spray structure is almost in a steady state.

By this classification the spray structure in each stage and its sequence of change through the three stages are made clear. The typical shape appearing in each stage is produced by the droplets injected during each specified period in the whole period of fuel injection shown in Fig.3. The injection period can be divided into the following three periods:

1. opening period
2. plateau period
3. closing period

The droplets rolling-up from the lower fringe of the cone in the first stage are those injected during the opening period, and the droplets which form the top part of the cone

in the second stage are those injected during the closing period.

EFFECTS OF THE PRESCRIPTION OF FUEL INJECTION CONDITIONS - The effects of the prescription of fuel injection conditions on the spray structure are discussed on the basis of the above-mentioned three stages.

The first stage - The differences of injection conditions among "Const.(1), (2)", "Sine (3)", and "Act.(4), (5), (6)" in Fig.5 do not noteworthily influence the spray structure.

The second and third stages - Only in the Const.-40 μm case (Fig.5 (1)), the droplets do not form a ring. They do go straight forward in the shape of a cone. In the Const.-20 μm case (Fig.5 (2)), part of injected droplets roll up, but these droplets do not form a complete ring, while in the Sine case (Fig.5 (3)) and in all of the Act. cases (Fig.5 (4), (5), (6)) the rolling-up droplets form a complete ring. In all these cases, a vortex ring is generated as a consequence of gas motion induced by a spray, but the interaction between gas flow and droplets varies with the prescription of injection conditions; the interaction is stronger for smaller droplets. In the two Const. cases, droplets are not small enough to be influenced by gas flow, which moves from outside to inside across the cone surface. In the Act. cases, the droplets in the upper part of the spray cone concentrate towards the cone axis due to the influence of the vortex ring. In the Sine case (Fig.5 (3)), the majority of the droplets go straight and the small part of the droplets are involved in the vortex ring, while in all of the Act. cases, the droplets are distributed almost equally in the vortex ring and in the cone shape. The diameter of the circle formed by droplets on a vertical cross section in the Sine case is smaller than that in the Act. cases.

Comparing the two Act. cases (Fig.5 (4), (5)), through all of the three stages there is no outstanding difference in hollow-cone spray structure, although their mass flow rates

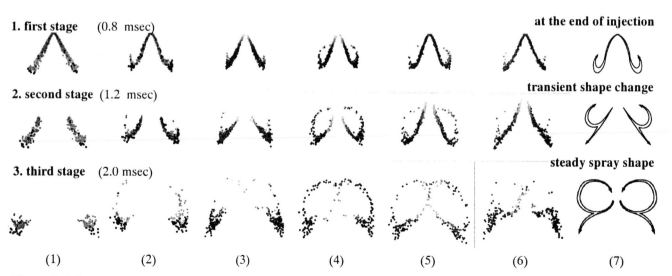

Figure 5: Effects of the prescrition of fuel injection conditions on hollow-cone spray structures : (1) Const. droplet diameter 40 μm, (2) Const. droplet diameter 20 μm, (3) Sine , (4) Act. injection period 0.8 msec, (5) Act. injection period 1.1 msec, (6) Act. injection period 1.1 msec with swirl, and (7) schematic representation of hollow-cone spray structure at each stage.

in the plateau period are different (see Fig.3(a)). Comparing the Sine and Act.(0.8msec) cases, the spray structures are considerably different, although their mass flow rates are about the same. It seems that this difference comes from the difference between the maximum droplet diameter in the Sine case and that in the Act.(0.8msec) case.

As to swirl component, the hollow-cone spray structure is considerably influenced by it. Comparing the Act.(1.1msec) (Fig.5 (5)) and Act.(1.1msec) with swirl (Fig.5 (6)) cases, the effects of swirl component are recognized as follows: The hollow-cone spray shapes near the branching point from going-straight part to rolling-up part are very different between the cases with and without swirl component; in the case with swirl component, many droplets are distributed horizontally just below the vortex ring in the third stage. It seems that this typical shape in the case with swirl component results from the interaction between gas flow and droplets, e.g., the effect of centrifugal force.

As a summary, it is found that the pattern of injection mass flow rate has little influence on the hollow-cone spray structure but the initial droplet diameter has a serious influence on it. Initial swirl component has also a significant influence.

COMPARISON WITH EXPERIMENTAL RESULTS - The calculated results in the Act.-1.1 msec with swirl case (Fig.5 (6)) are compared with the results of experiments using the high-pressure swirl injector.

Shape and Structure of a Spray - Figure 6 shows the calculated results and the measured results at four specified times after the start of injection. The first one is 0.2 msec from the beginning of injection - the end of opening period. Second one is 0.8 msec - the first stage described earlier. The remaining two (1.2 msec and 2.0 msec) correspond to the typical points in the second and third stages.

The calculated spray shapes in Fig.6(b) are generally in good agreement, except at 0.2 msec, with spray pictures in Fig.6(a) which were obtained experimentally by strobe light.

Furthermore, the calculated results of the hollow-cone spray structures in Fig.6(d) are in reasonable agreement, except at 0.2 msec, with the measured results in Fig.6(c) which were taken by YAG laser light sheet.

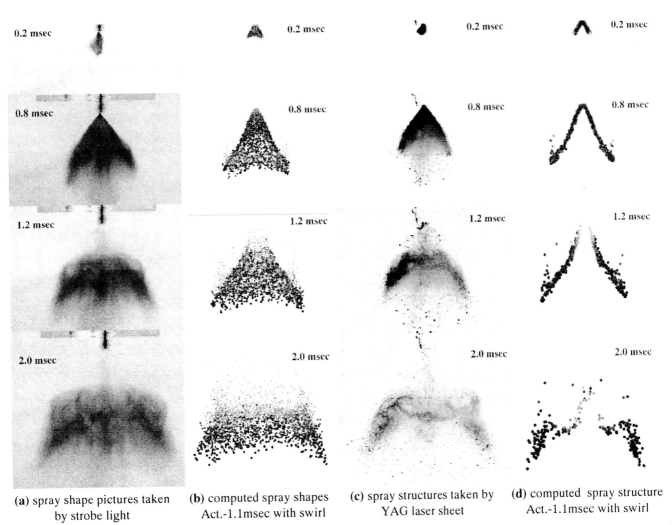

(a) spray shape pictures taken by strobe light

(b) computed spray shapes Act.-1.1msec with swirl

(c) spray structures taken by YAG laser sheet

(d) computed spray structure Act.-1.1msec with swirl

Figure 6: Calculated results by means of the prescription of fuel injection conditions of Act. injection period 1.1 msec with swirl and the measured results at four specified times after the start of injection (ambient gas pressure 0.1 MPa)

In the measured results (Fig.6(c)), many coarse droplets are clearly seen in the spray tip and even near the cone axis at 0.8 msec and 1.2 msec. This phenomenon is called the "coarse droplet phenomenon" in this paper. This phenomenon will be discussed in the next section.

<u>Distribution of Droplet Size</u> - In order to validate the calculated distribution of droplet size, the PDA measurement data was provided. Figure 7 shows the trapezoidal measurement area of PDA. The upper border line of the measured area locates 10 mm below the injector nozzle. The pointwise PDA measurements were carried out at each of 25 × 9 grid points on a vertical cross section. By these measurements, the time histories of the 2-dimensional distribution of physical quantities such as volume flux, droplet number density, SMD (Sauter mean diameter), etc. are obtained.

The distributions of droplet number density on a vertical cross section in the third stage are shown in Figs.8(a), (b), (c) and (d), which correspond to the four different ranges of droplet diameter. In Fig.8(c), which shows the distribution of droplets in the diameter rage of 10 to 25 μm, it can be seen that a clear vortex ring is formed. In contrast to this, in the figures of other diameter ranges, such a clear vortex ring is not seen. Besides, the total number of the droplets in the diameter rage of 10 to 25 μm is larger than that in the other droplet diameter ranges. As shown in Fig.8(d), the droplets in the diameter range of 25 to 58 μm are distributed below the vortex ring, but the total number of the droplets in this range is very small. These noteworthy distributions of droplets play an important role in forming the hollow-cone spray structure. In the calculated results shown in Fig.6, these features are reproduced very well.

In Fig.9(a), which shows the variation of the calculated droplet size (Sauter mean diameter) with a distance from the nozzle outlet, the size of droplets forming a ring (distance = 20~40mm) is about 20 μm and that of the droplets going straight forward (distance > 40mm) is about 50 μm. These calculated results coincide with the above-mentioned experimental results.

Figures 8(a) and (b) show the distributions of the droplets in the diameter rages of 0 to 5 μm and 5 to 10 μm, respectively. As shown in these figures, smaller droplets have a stronger tendency to concentrate toward the center of the vortex ring. In calculation, however, such a tendency with respect to small-sized droplets are not reproduced well. Details of this poor reproducibility will be discussed later.

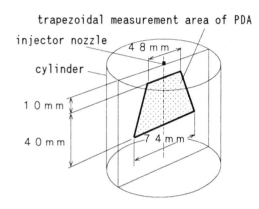

Figure 7: Schema of the trapezoidal measurement area of PDA

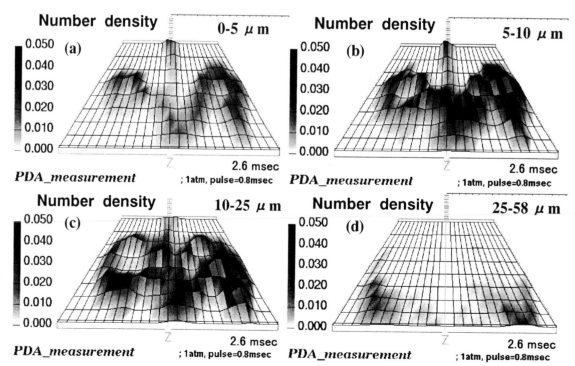

Figure 8: Distributions of droplet number density on a vertical cross section in four different ranges of droplet diameter in the third stage (measured by PDA)

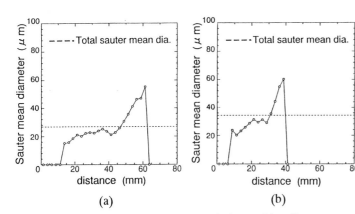

(a)   (b)

Figure 9: Calculated droplet size variations with a distance from the nozzle outlet at ambient gas pressure (a) 0.1 MPa, and (b) 0.5 MPa.

In the case of a high-pressure swirl injector, it is a well known characteristic that the droplet size in downstream region increases with an increase in ambient gas pressure. This characteristic is also reproduced by calculation as is obvious from Figs.9(a) and (b). In these figures, the dotted lines express the total Sauter mean diameter (SMD), and SMD at an ambient gas pressure of 0.5MPa (Fig.9(b)) is larger than that at an ambient gas pressure of 0.1MPa (Fig.9(a)).

COARSE DROPLET PHENOMENON

REPRODUCING THE COARSE DROPLET PHENOMENON - The coarse droplet phenomenon is observed in the pictures taken by strobe light (Fig.6(a)) or YAG laser light sheet (Fig.6(c)). Here, through the investigation of the spray structure in an earlier period of injection process, a method to numerically simulate the coarse droplet phenomenon is presented.

The distributions of droplet number density measured by PDA at 0.9 msec (fuel injection has just ended at this time in the experiment) are shown in Figs.10(a), (b), (c) and (d) (the divisions of diameter ranges are the same as in Fig.8.). From these figures the following features can be seen:

(i) As shown in Fig.10(d), the coarse droplets are distributed in the injected hollow-cone and also in the front cone which caps the injected hollow-cone (see Fig.11).

(ii) The middle-sized droplets are distributed inside the coarse droplet region (Fig.10(c)).

(iii) The small-sized droplets ($< 10 \mu$m) do not form a hollow-cone but have a strong tendency to concentrate toward the center of the front cone (Figs.10(a) and (b)).

From the above-mentioned observations, the hollow-cone spray seems to have a schematic structure shown in Fig.11.

Figure 12 shows the number density distributions of coarse (25 to 58 μm) droplets at two different times.

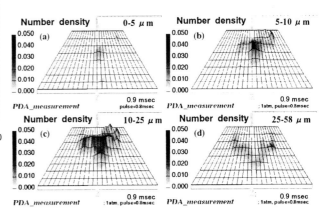

Figure 10 : Distributions of droplet number density on a vertical cross section in four different ranges of droplet diameter at the end of injection (measured by PDA)

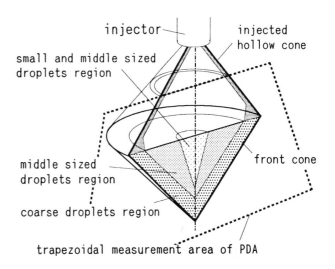

Figure 11: Schema of the hollow-cone spray structure at 0.9 msec (fuel injection has just ended at this time in experiment) obtained by PDA measurement.

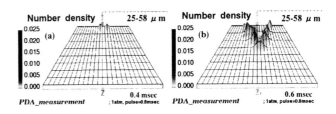

Figure 12 : Distributions of droplet number density on a vertical cross section at two different times (measured by PDA)

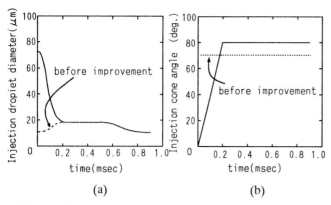

Figure 13: Modified temporal variations in (a) injection droplet size and (b) injection cone angle

Figure 12(a) is an image obtained at 0.4msec after the start of injection. The droplets on the upper border line are concentrated in a small region around the injection axis. This implies that the cone angle at the start of injection is very small. Figure 12(b) is an image obtained at 0.6 msec after the start of injection. In this image, the droplets on the upper border line concentrate at two far-separated points. This implies that the cone angle is much widened at 0.6 msec after the start of injection. As a result, the injection cone angle increases from nearly zero to the steady value in proportion to elapsed time during the valve opening period. On the basis of these considerations, the prescription of fuel injection conditions is modified as shown in Figs.13(a) and (b). Here, the maximum injection cone angle is increased up to 80 degrees as shown in Fig.13(b). As to the variation of injection droplet diameter with time (Fig.13(a)), further investigation may be required.

The calculated results by means of the modified prescription of fuel injection conditions are shown in Figs.14(b), (c) and (d). For comparison, the PDA measurement results are shown in Fig.14(a). The calculated results in these figures, especially at 0.9 msec, show that the coarse droplets form the front cone which caps the injected hollow-cone. Furthermore, the spray shape at 0.2 msec is reproduced reasonably well, although it has not been reproduced in Fig.6. Eventually, the behavior and structure of a hollow-cone spray are reproduced successfully by computation.

SMALL-SIZED DROPLET - The modified injection conditions mentioned in the above subsection do not influence the behavior of small droplets. In order to predict the distribution of small-sized droplets accurately, the droplet breakup model has to be improved on the basis of more detailed information on the atomization process. At present, however, such detailed information is not available. Here, instead of improving the breakup model, small-sized droplets are introduced to the injection conditions by assuming a $\chi 2$-distribution as the distribution of initial droplet diameter. The mean droplet diameter for this $\chi 2$-distribution is the same as that shown in Fig.13(a). By calculating the spray behavior under this modified injection

condition, the distribution of small-sized droplets is improved considerably as shown in Fig.15. This fact implies that the behavior of hollow-cone sprays will be reproduced more accurately if such an appropriate breakup model is employed as produces small child droplets in the injection period in addition to larger parent droplets. For this purpose, the modified version[11] of the wave breakup model by Reitz[12] may be useful.

CONCLUSIONS

The hollow-cone sprays generated by a single-fluid (liquid) high-pressure swirl injector are numerically analyzed using the author's GTT code which employs the Discrete Droplet Model in KIVA code along with the bag and stripping breakup model by Reitz, et al. The effects of the prescription of fuel injection conditions on the spray behavior are investigated systematically. Injection fuel flow rate, injection droplet diameter, injection cone angle and swirl component are examined, and the basic effects of these conditions on the spray behavior are clarified.

Through these computational studies the spray structure for this type of injector is clearly elucidated in terms of fluid dynamics. Namely, it is found that the temporal transition of the spray structure can be divided into three stages, and the fully developed hollow-cone spray structure is characterized by the droplets forming a ring and by the droplets going straight forward in the shape of a cone.

The calculated results show a good agreement with the experimental results with respect to the spray structure, especially the distributions of relatively large-sized droplets, which play an important role in determining the structure of the hollow-cone spray. On the basis of the PDA measurement data, a reasonable method to numerically simulate the coarse droplet phenomenon is presented. In contrast to this, for predicting the distribution of small-sized droplets accurately, more detailed information on the atomization process is necessary by means of experimental and numerical studies.

REFERENCES

[1] Reitz, R. D. and Diwakar, R., "Effect of Drop Breakup on Fuel Spray", SAE paper 860469, 1986.

[2] Reitz, R. D. and Diwakar, R., "Structure of High-Pressure Fuel Spray", SAE paper 870598, 1987.

[3] Lee, C. F. and Bracco, F. V., "Initial Comparisons of Computed and Measured Hollow-Cone Sprays in an Engine", SAE paper 94398, 1994.

[4] Lee, C. F. and Bracco, F. V., "Comparisons of Computed and Measured Hollow-Cone Sprays in an Engine", SAE paper 950284, 1995.

[5] Wakisaka, T., Shimamoto, Y., Isshiki, Y., Sumi, N., Tamura, K. and Modien, R. M., "Analysis of the Effects of In-Cylinder Flows during Intake Stroke on the Flow

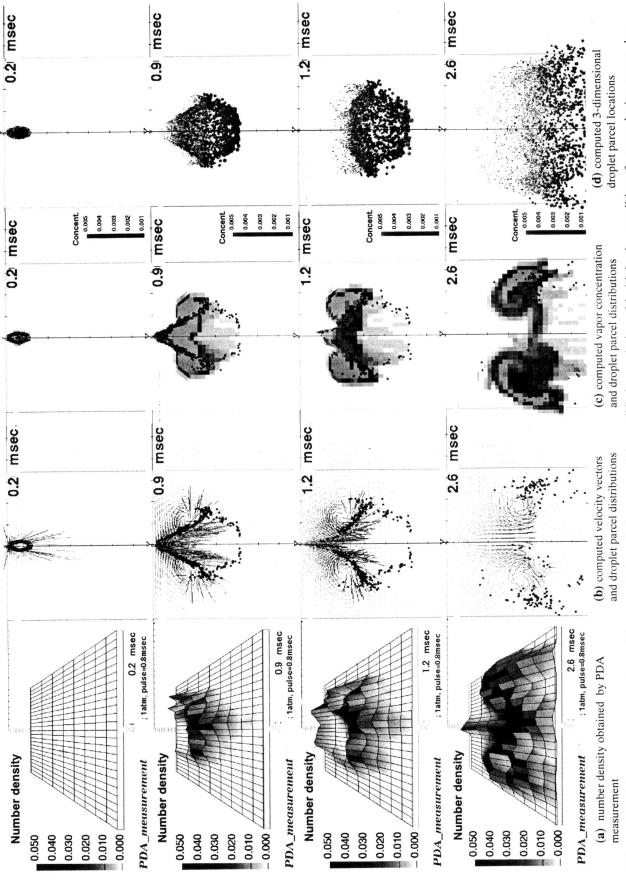

Figure 14: PDA measurement results and the computed results by means of the modified prescription of fuel injection conditions for reproducing coarse drop phenomenon : (a) (b)(c) on a vertical cross section, (d) projected view

(a) number density obtained by PDA measurement

(b) computed velocity vectors and droplet parcel distributions

(c) computed vapor concentration and droplet parcel distributions

(d) computed 3-dimensional droplet parcel locations

157

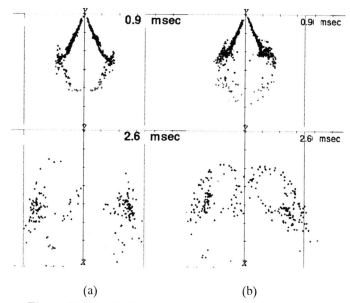

(a)   (b)

Figure 15: Distributions of droplets computed under the conditions of initial droplet diameter of (a) mono-size , and (b) χ^2-distribution

Characteristics near Compression TDC in Four-Stroke Cycle Engines", Proc. Int. Symp. COMODIA 90, pp.487-492, 1990.

[6] Chakravarthy, S. R. and Osher, S., "A New Class of High Accuracy TVD Schemes for Hyperbolic Conservation Laws", AIAA paper 85-0363, 1985.

[7] Van Doormaal, J. P., and Raithby G. D., "Enhancements of the SIMPLE Method for Predicting Incompressible Fluid Flows", Numerical Heat Transfer, Vol.7, pp.147-163, 1984.

[8] Yabe, T. and Takei, E., "A New Higher-Order Godunov Method for General Hyperbolic Equations", Journal of the Physical Society of Japan, Vol.57, No.8, pp.2598-2601, 1988.

[9] Wakisaka, T., Takeuchi, S. and Shimamoto, Y., "Numerical Analysis of the Behavior of Tumbling Vortices in Engine Cylinders Using Higher-Order Schemes", Proc. 1st Asian Computational Fluid Dynamics Conference, Vol.1. pp.227-233, 1995.

[10] Amsden, A. A., Ramshaw, J. D., O'Rourke, P. J. and Dukowicz, J. K., "KIVA: A Computer Program for Two- and Three-Dimensional Fluid Flows with Chemical Reactions and Fuel Sprays", Los Alamos Report No. LA-10245-MS, 1985.

[11] Chung, J. H., Wakisaka, T. and Ibaraki, K., "An Improved Droplet Breakup Model for Three-Dimensional Diesel Spray Simulation", Proc. 3rd KSME/JSME Thermal Engineering Conference, (to appear), 1996.

[12] Reitz, R. D., "Modeling Atomization Processes in High-Pressure Vaporizing Sprays", Atomisation and Spray Technology, Vol.3, pp.309-337, 1987.

Cycle-Resolved PDA Measurement of Size-Classified Spray Structure of Air-Assist Injector

Yuji Ikeda, Shigeo Hosokawa, Fukashi Sekihara, and Tsuyoshi Nakajima

Kobe Univ.

ABSTRACT

Very high data rate Phase Doppler Measurements were carried out in order to demonstrate the spray characteristics at each cycle and how each injection differed from each other. Conventional time-averaged data analysis can hardly provide information to analysis cyclic variation of spray formation and droplet dynamics so that a cycle-resolved PDA system was developed in the study. A direct gasoline injector for two-stroke marine engine was used for the experiment. For data analysis, droplet dynamics and characteristics of different droplet diameter were examined. The results show that cycle variation of injector was remarkable, the maximum spray tip velocity differed from 63 m/s to 93 m/s even for the consecutive injection. The data rate obtained was over 40 kHz (Max : 85 kHz) and bin width was carefully examined to show the spray collision to air and entrained air motion. The high speed cycle formed large velocity fluctuation and small vortex time scale, while the low speed cycle was not so clear for the vortex entrainment.

INTRODUCTION

Understandings of mixture formation [1-4] in a cylinder have been investigated in order to reduce NOx and optimize combustion characteristics for power and exhaust gas. Recently, direct injection methods have been implemented for gasoline engines [5-8]. Many spray measurements [9-11] and numerical simulations [12-14] have been carried out for better control of spray and combustion characteristics. But it is very difficult to understand the spray characteristics quantitatively so far because the conventional measurement techniques can hardly provide three-dimensional image in time series with high temporal resolution.

When a gasoline fuel is injected in the cylinder, the pressure and droplet momentum and its characteristics change dramatically, and these facts are yielded; large influence from residual hot gas, large shear stress, collision at a spray tip, droplet break up, droplet separation, cluster, large entrainment

and so on. The values which we can measure and use for discussion are almost time-averaged one. Once we want to understand the practical spray characteristics, time series information is highly required, which can hardly be obtained by conventional measurement techniques. For injector spray, an ensemble-averaged data was used for engine performance valuation, which is not so useful for cyclic variation analysis [15-18]. The data for cyclic variation analysis should have very high temporal resolution and long time for cycles. The measurement system and experimental conditions have to be optimized.

The two image measurement methods can show the spray shape [19], vapor area [20], entrained vortex [21] and coherent vortex scale [22], but these methods can not demonstrate time series information with high temporal resolution. Further developments of high repetition laser and CCD camera are needed for this analysis.

In order to understand cyclic variation of gas flows in engines, several experiments have been tried [15, 23-27]. But, droplets information such as diameter, velocity, drag coefficient, relative Reynolds number which a phase Doppler technique (PDA/PDPA) can provide should be demonstrated with high temporal resolution.

We have been measuring gas flow characteristics in a two-stroke engine [15, 28-33] and evaluating cyclic variation. In this study, a direct injector for a two-stroke marine engine was used and the measurement techniques obtained by these two-stroke gas flow experiments were applied for high data rate analysis, that is, cycle-resolved method. The spray structure of this air assisted injector was reported [33] in detail. In this report, the droplet characteristics over cycle are investigated.

EXPERIMENTAL APPARATUS

An air-assisted injector used in this experiment was a practical injector for the two-stroke marine engine of over 22 kW (30 ps) as shown in Fig. 1 [33]. A fuel was injected in a cavity and then an air injector was operated by opening a

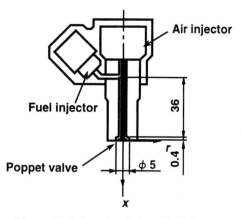

Figure 1 Schmatic of air-assist injector

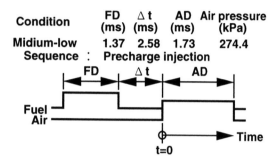

Figure 2 Experimental conditions

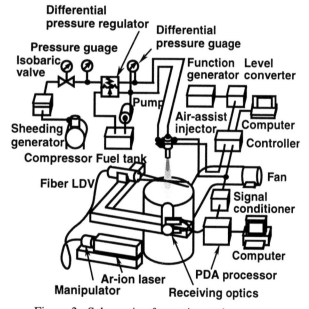

Figure 3 Schematic of experimental apparatus

Table 1 Performance of FLDV probe for PDA measurement

	Type A	Type B	Type C	Type D
Forcal length (mm)	300.0	600.0	310.0	600.0
Spot diameter (μm)	63.0	126.0	228.0	440.0
Fringe spacing (μm)	2.6	5.2	4.3	8.3
Fringe number	24	24	54	53
Maximum diameter (μm)	50.7	101.4	81.7	157.5
Maximum velocity (m/s)	31.6	47.1	128.2	248.0
Minimum velocity (m/s)	-7.9	-15.7	-25.6	-49.6
Wavelength (nm)	632.8	632.8	514.5	514.5

where,

Focal length of receiving optics : 310 mm
Band widht of signal processor : 36 MHz
Frequency shift   : 40 MHz

done under open air condition and surrounding air was sucked out by a blower to prevent from adhesion of fuel on to the optics. An argon-ion laser was used and the laser power at the measurement volume was set up over 300 mW [34]. For optimizing the measurement systems, four types of optics have been tried to optimize this measurement system having high data rate and less measurement uncertainly as shown in Table 1. Since a bandwidth of a processor was fixed, measurable velocity and diameter were determined by the size and the fringe spacing of measurement volume. Long focal length is desirable, but the resolution in velocity and diameter decrease, furthermore, the light scattered intensity from large measurement volume causes low SNR (Signal-to-Noise ratio) signal [29]. The small measurement volume is the best way to increase SNR. On the contrary, the maximum velocity and diameter were limited due to small fringe spacing and its measurement volume size. In this way, the optimization process is a very important factor in actual measurements to enhance reliability and data rate.

A drysolvent which has a refractive index of 1.427 was used as a fuel instead of gasoline. The specific density of the drysolvent of 0.77 g/cm^3 is very similar to that of gasoline (0.7-0.8g/cm^3). The scattering angle of 68 degree was determined by first order refraction angle [34]. An ensemble averaged technique using a phase locking method of one degree resolution was implemented [34].

MEASUREMENT RESULTS

SPRAY IMAGE - A direct picture and laser sheet pictures are compared in Fig. 4. The spray tip velocity calculated from the direct picture was about 64 m/s [33], which was almost same to the laser sheet images. This laser sheet images were measured by YAG Laser (l : 532 nm, 400 mJ, 6 ns) and conventional 35 mm still camera. Those pictures show a development of rolling vortex (mushroom vortex) at 2 ms after injection. Direct pictures cannot demonstrate spray characteristics such as diameter, concentrations, and number density. Even for the laser sheet images, these data could not be obtained. It was found more clearly by the laser sheet image that there were strong illuminated regime like regament up to x = 25 mm and spray angle started to be wide from that location.

poppet valve.

Figure 2 shows an experimental conditions, and an experimental apparatus is shown in Fig. 3. All experiments were carried out under non-combustion and open air condition. The pressure difference of air and fuel was set up to 100 kPa and the air pressure was varied for different engine load conditions. A crankangle was set up and its timing signal was used to demonstrate an actual operating condition. The injector drive voltage was used as a starting trigger signal and the set-up timing was used as reference signal. The experiment was

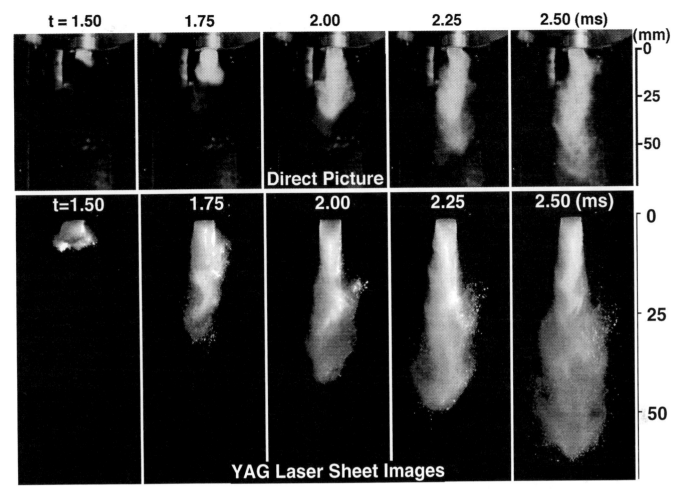

t = 1.50　　　1.75　　　2.00　　　2.25　　　2.50 (ms)　(mm)

Direct Picture

t=1.50　　　1.75　　　2.00　　　2.25　　　2.50 (ms)

YAG Laser Sheet Images

Figure 4　Images of spray

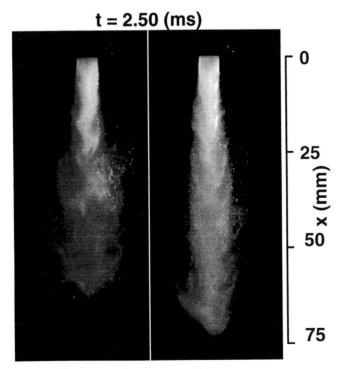

t = 2.50 (ms)

Figure 5　Cyclic valiation

Figure 5 shows same spray images at different cycle. It was also obvious that there was some variation or cyclic variation of the spray development. In order to understand the physics, the data of high temporal resolution is indispensable.

PDA FOR CYCLIC VARIATION - The PDA can measure the droplet velocity and its diameter. The classical data analysis used Sauter mean diameter [10], but its value could not be defined well in the acceleration droplet [33]. Furthermore, size-classified techniques [35-39] can be improved and applied to show high data rate analysis for each classes. Many improvements were taken into account in the optics and data analysis for cycle-resolved method [40-44]. First, let's see the uncertainty of the ensemble-averaged data analysis in Fig. 6. At six locations on and off the center axis, axial droplet velocity at each classes are shown in Fig. 6. As explained in the previous report [33], there are some no data period which may be formed by a poppet valve and negative pressure field behind the valve. This fact could not be explained well just by this ensemble averaged data and laser sheet image. The spray at each cycle was totally different one.

As shown in Fig. 5, spray formation is not the same, there is some variation or large fluctuation. These two pictures were taken at the same timing but different cycles. From the

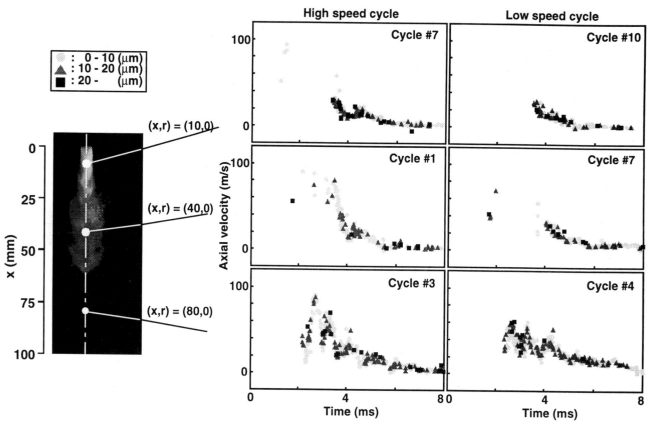

Figure 6 Size-classified droplet axial velocity

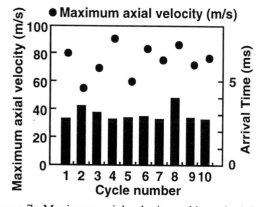

Figure 7 Maximum axial velocity and its arrival time

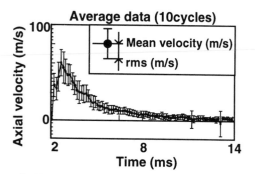

Figure 8 Ensemble averaged velocity profile of 10 cycles

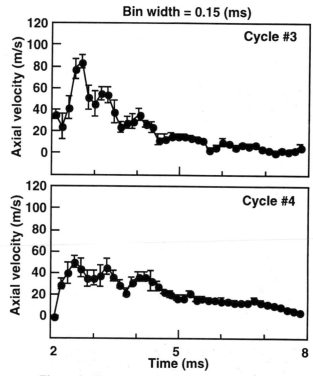

Figure 9 Two velocity variations at x = 80 mm

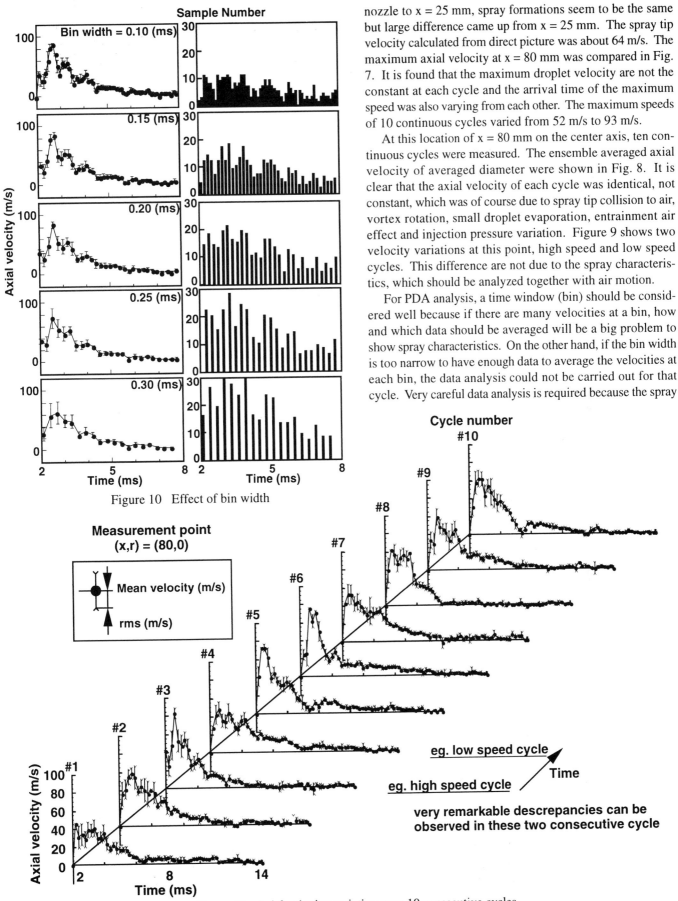

Figure 10   Effect of bin width

nozzle to x = 25 mm, spray formations seem to be the same but large difference came up from x = 25 mm. The spray tip velocity calculated from direct picture was about 64 m/s. The maximum axial velocity at x = 80 mm was compared in Fig. 7. It is found that the maximum droplet velocity are not the constant at each cycle and the arrival time of the maximum speed was also varying from each other. The maximum speeds of 10 continuous cycles varied from 52 m/s to 93 m/s.

At this location of x = 80 mm on the center axis, ten continuous cycles were measured. The ensemble averaged axial velocity of averaged diameter were shown in Fig. 8. It is clear that the axial velocity of each cycle was identical, not constant, which was of course due to spray tip collision to air, vortex rotation, small droplet evaporation, entrainment air effect and injection pressure variation. Figure 9 shows two velocity variations at this point, high speed and low speed cycles. This difference are not due to the spray characteristics, which should be analyzed together with air motion.

For PDA analysis, a time window (bin) should be considered well because if there are many velocities at a bin, how and which data should be averaged will be a big problem to show spray characteristics. On the other hand, if the bin width is too narrow to have enough data to average the velocities at each bin, the data analysis could not be carried out for that cycle. Very careful data analysis is required because the spray

Figure 11   Axial velocity variation over 10 consecutive cycles

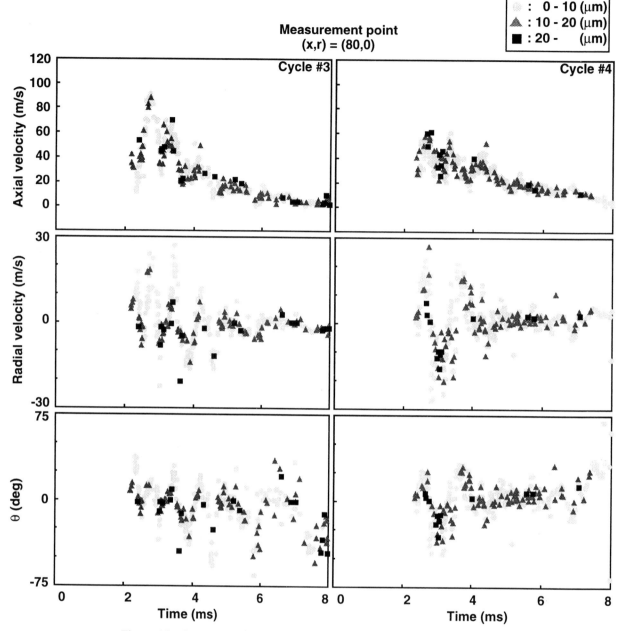

: 0 - 10 (μm)
: 10 - 20 (μm)
: 20 - (μm)

Figure 12 Comparison of axial, radial velocity and velocity vector angle

at different cycle is the very different spray. This bin is the key point in the PDA application for cycle variation analysis. In order to determine the time window (bin), each cycle was examined as shown in Fig. 10. Five bins were examined for a certain cycle in order to determine the optimum bin width and keep data reliability high level. It is found that the clear and sharp velocity variation can be obtained with small time window (bin) but the data sample at each bin decreases, that is, the data reliability decreases. We chose the bin width of 0.2 ms because of the data sample number over 10 at each bin at least and velocity variation smoothness. This criteria is the very empirical. For better analysis, the most important parameter to be considered was how to obtain high data rate and long data record. In this cycle data rate over 75 kHz was archived.

The axial velocity variation over cycles (10 continuous cycles) are shown in Fig. 11. The measurements were performed at the same location of x = 80 mm on the axis. Each cycle has continuous velocity variation, that is, acceleration region, maximum velocity, large fluctuation and spray tail. These data was measured at over 40 kHz. In this experiment, these 11 cycles were the maximum for data memory. Further measurement will be reported in the next paper soon.

CYCLIC RESOLVED ANALYSIS - As shown in Fig. 9, the axial velocity of each classified droplets were not the constant for cycles; let's see #3 for high speed cycle and #4 low speed cycle are the consecutive one. These velocity and vector angle are shown in Fig. 12. The x-axis is an arrival time. The cycle #3 was the high speed cycle and the

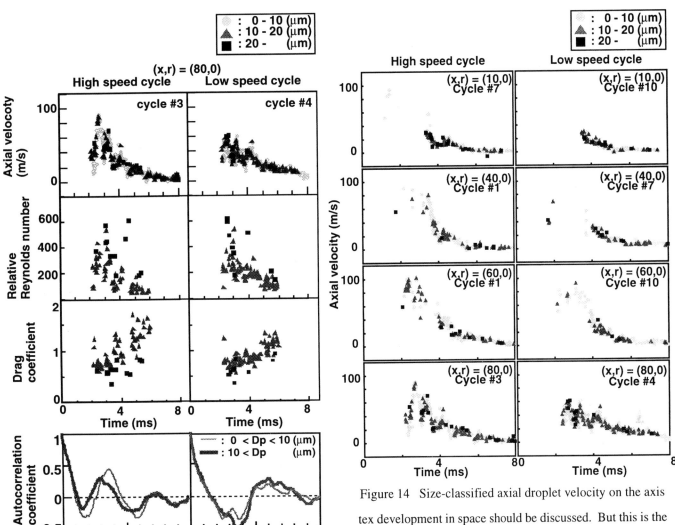

Figure 13 droplet characteristics at x = 80 mm

Figure 14 Size-classified axial droplet velocity on the axis

maximum axial velocity was about 93 m/s. The axial velocity of #4 cycle was about 63 m/s.

Once a strong injection spray reached at a certain point (here : x = 80 mm), there were obvious velocity fluctuations both in axial and radial direction. The fluctuation period of high speed cycle was shorter than that of low speed cycle. This fact can be easily seen both in axial and radial direction. The velocity vector angle was also shown the same tendency, that is, the high speed cycle has large velocity fluctuation and short variation period. The droplet dynamics of each droplet size classes were seen in this figure. In order to understand the further droplet dynamics, relative Reynolds number and its drag coefficient [33] were calculated together with auto-correlation coefficient of axial velocity as shown in Fig. 13. There are large discrepancies in drag coefficient between high and low speed cycles in small droplet classes but less in large droplet classes.

The auto-correlation coefficient shows that the faster speed cycles had short time scale of variation and the lower was larger. Because this was the only one point data so that vor-

tex development in space should be discussed. But this is the measurement technique limitation so far. As introduced, PDA can only provide data in time series at a certain point. Imaging methods can demonstrate spatial spray structure but not so detail as that of PDA. This is trade-off relation. Then, this report can contribute to demonstrate how the spray formation and its variation occurs over cycles by cycle-resolved PDA method.

CYCLIC VARIATION AT DIFFERENT LOCATION
- Detailed data analysis at one point was done in the pervious section. Since PDA is a one point measurement method, it is not possible to obtain two point data at the same time, that is, this is the limitation of the PDA for cyclic variation analysis. It is possible to measure high data rate measurement at several points but those data could hardly be compared in order to understand spatial development of droplet behavior and its dynamics. Then, let's see how the velocity variation occurs and how large and short the variation period are.

Four point measurements on the center axis were carried out at very high data rate from 40 - 80 kHz and all bin widths were examined to identify the velocity variations as shown in Figs. 14 and 15. The high speed cycle data shows large velocity fluctuation both in axial and radial, especially x > 60 mm. As shown in Fig. 5, the spray formation show the same

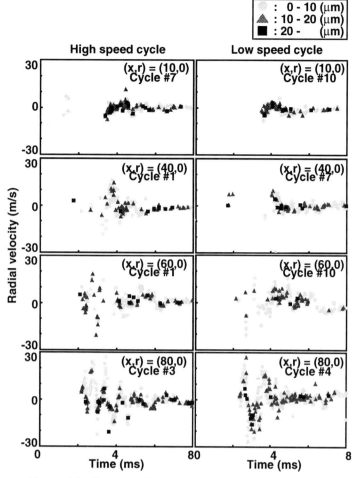

Figure 15 Size-classified radial droplet velocity on the axis

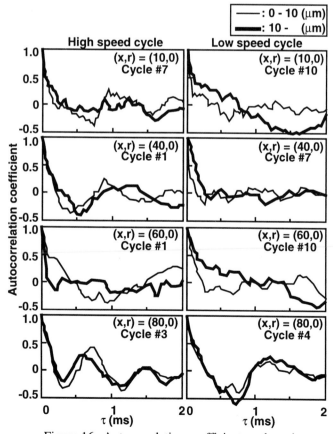

Figure 16 Autocorrelation coefficient on the axis

tendency at x < 25 mm. At x < 40 mm, there are some spray shell and empty region in which no data could be detected. As seen in the previous section at x = 80 mm, the fast speed cycle has short variation period than that of the low speed. This was due to the fact that the fast speed cycle means large momentum collision to air and strong shear stress at the spray tip so as to form some vortex in the spray tip and shell. While, the low speed cycle spray can not form such strong wake and vortex. This formed vortex shows the short period of velocity fluctuations both in axial and radial direction, which can be said, the high speed cycle has strong vortex of small time scale.

The auto-correlation at these points were calculated and shown in Fig. 16. Here, an air motion was denoted with the droplet less than 10 µm [33] and the droplet motions were compared to those both in high and low speed cycles. The air motion seems to be faster than droplet momentum somehow. But it is difficult to compare the characteristics at different points because these data was measured at different time.

CONCLUSION

In order to investigate a cyclic variation of gasoline injector, a cycle-resolved PDA system was demonstrated. The data evaluation was done by its data rate over 40 kHz and reasonable bin width to show obvious velocity changes over cycles. These velocity variations over cycles was very large both in axial and radial direction. The high speed cycle and the low speed cycle were obviously different droplet characteristics from each other. This means that the ensemble-averaged mean data show different nature of gasoline spray formation. The high speed cycle spray has strong momentum so as to have large velocity fluctuation and short variation period, which means strong vortex of small timescale. This fact could be observed at different locations. It was found that the cycle-resolved PDA could provide very detail spray information of time series and very useful to identify the spray characteristics at each cycle.

REFERENCES

[1] G. P. Blair, "The Basic Design of Two-Stroke Engines", SAE Publication R-104, 1990.

[2] J. B. Heywood, "Internal Combustion Engine Fundamentals", McGraw-Hill Book CO., 1988.

[3] H. P. Lenz, "Mixture Formation in Spark-Ignition Engine", SAE Publications, 1992.

[4] C. Arcoumanis, "Internal Combustion Engines", Academic Press, 1988.

[5] N. J. Beck, W. P. Johnson, R. L. Barlhimer and S. H. Patterson, "Electric Fuel Injection for Two-Stroke Cycle Gasoline Engines", SAE Paper No. 861242, 1986.

[6] M. M. Schechter, E. H. Jary and M. B. Levin, "High Speed Fuel Injection System for 2-Stroke D. I. Gasoline Engine", SAE Transaction, Paper No. 910666, 1991.

[7] T. Sato and M. Nakayama, "Gasoline Direct Injection for a Loop-Scavenged Two-Stroke Cycle Engine", SAE Paper No. 871690, 1987.

[8] R. G. Kenny, R. J. Kee, C. E. Carson and G. P. Blair, "Application of Direct Air-Assisted Fuel Injection to a SI Cross-Scavenged Two-Stroke Engine", SAE Paper No. 932396, 1993.

[9] K. K. Kuo, "Recent Advances in Spray Combustion: Spray Combustion Measurements and Model Simulation", AIAA Progress in Astronautics and Aeronautics, Volume 171, 1996.

[10] H. Lefebvre, "Atomization and Sprays", Hemisphere Publishing Corporation, 1989.

[11] H. Lefebvre, "Gas Turbine Combustion", Hemisphere Publishing Corporation, 1983.

[12] A. A. Amsden, P. J. O'Rourke, and T. D. Butler, "KIVA-II : A Computer Program for Chemically Reactive Flows with Sprays", Los Aramos National Laboratory Report, LA-11560-MS, 1989.

[13] A. A. Amsden, P. J. O'Roueke, T. D. Butler, K. Meintjes and T. D. Fansler, "Comparisons of Computed and Measured Three-Dimensional Velocity Fields in a Motored Two-Stroke Engine", SAE Paper No. 920418, 1992.

[14] T. L. McKinley, R. J. Primus, P. J. O'Rourke and T. Dan Butler, "Comparison Computed and Measured Air Motion in Circular and Square Piston Cups", SAE Paper No. 881612.

[15] T. Ohira, Y. Ikeda, K. Kakemizu and T. Nakajima, "In-Cylinder Flow Measurement and Its Application for Cyclic Variation Analysis in a Two-Stroke Engine", SAE Transaction, Paper No. 950224, 1995.

[16] B. Johansson, "Influence of the Velocity Near the Spark Plug on Early Flame Development", SAE Paper No. 930481, 1993.

[17] P. C. Miles, R. M. Green, and P. O. Witze, "In-Cylinder Gas Velocity Measurements Comparing Crankcase and Blower Scavenging in a Fired Two-Stroke Cycle Engine", SAE Paper No. 940401, 1994.

[18] E. Nino, B. F. Gajdeczko, and P. G. Felton, "Two-Color Particle Image Velocimetry in an Engine with Combustion", SAE Paper No. 930872, 1993.

[19] K. Kuwahara , T. Kawai and H. Ando, "Influence of Flow Field Structure after the Distortion of Tumble on Laser-Burn Flame Structure", International Symposium, COMODIA94, pp. 89-94, 1994.

[20] P. G. Felton, "Fluorescence Imaging of Engine Sprays", Res. Paper ASI on Unsteady Combustion, L38., 1993.

[21] E. Tomita, Y. Hamamoto and H. Tsutsumi, "A Study on Ambient Gas Entrainment into Transient Gas Jet (Continued Report)",, The 11th Internal Combustion Engine Symposium, Japan, No. 92, pp. 541-546, 1993, (in Japanese).

[22] M. Shioji, K. Yamane, H. Isogami and M. Ikegami, "Turbulent Eddies in a Jet Flame as Visualized by a Laser Sheet Method", Transaction of JSME, Vol. 57, No. 542, pp. 3562-3568, 1991, (in Japanese).

[23] M. Abraham and S. Parakash, "Cyclic Variations in a Small Two-Stroke Cycle Spark-Ignited Engine - An Experimental Study", SAE Paper No. 920427, 1992.

[24] R. E. Winsor, and D. J. Patterson, "Mixture Turbulence - A Key of Cyclic Variation", SAE Transaction, Paper No. 730086, 1973.

[25] M. B. young, "Cyclic Dispersion - Some Quantitative Cause and Effect Relationship", SAE Paper No. 800459, 1980.

[26] B. D. Papers and G. S. Borman, "Cyclic Variations and Average Burning Rates in a S. I. Engine", SAE Paper No. 700064, 1970.

[27] K. Tsuchiya, Y. Nagai and T. Gotoh, "A Study of Irregular Combustion in 2-Stroke Cycle Gasoline Engines", SAE Paper No. 830091, 1983.

[28] Y. Ikeda, M. Hikosaka, T. Nakajima, "Scavenging flow measurements in a motored two-stroke engine by fiber LDV", SAE Paper No. 910669, 1991.

[29] Y. Ikeda, M. Hikosaka, T. Nakajima and T. Ohira, "Scavenging Flow Measurements in a Fired Two-Stroke Engine by Fiber LDV", SAE Transaction, Vol. 101, Section 3, pp. 990-998, Paper No. 910670, 1991.

[30] Y. Ikeda, T. Ohira, T. Takahashi and T. Nakajima, "Flow Vector Measurements at the Scavenging Port in a Fired Two-Stroke Engine", SAE Transaction, Vol. 102, Section 3, pp. 635-645, Paper No. 920420, 1992.

[31] T. Ohira, Y. Ikeda, T. Takahashi, T. Ito and T. Nakajima, "Exhaust Gas Flow Behavior on a Two-Stroke Engine", SAE Transaction, Vol. 103, Section 3, pp. 672-680, Paper No. 930502, 1993.

[32] Y. Ikeda, T. Ohira, T. Takahashi and T. Nakajima, "Misfiring Effects on Scavenging Flow at Scavenging Port and Exhaust Pipe in a Small Two-Stroke-Engine", SAE Paper No. 930498, 1993.

[33] Y. Ikeda, T. Nakajima, and N. Kurihara, "Size-classified Droplet Dynamics and Its Slip Velocity Variation of Air-Assist Injector Spray", SAE Transaction, Paper No. 970632, 1997.

[34] Y. Ikeda, T. Nakajima and N. Kurihara, "Spray Formation of Air-Assist Injection for Two-Stroke Engine", SAE Transaction, Paper No. 950271, 1995.

[35] C. F. Edwards, "Measurement of Correlated Droplet Size and Velocity Statics, Size Distribution, and Volume Flux in a Steady Spray Flame", 5th International symposium on Applications of Laser Techniques to Fluid Mechanics, 1990.

[36] J. Seay, V. McDonell and G. Samuelsen, "Atomization and Dispersion from a Radial Airblast Injector in a Subsonic Crossflow", AIAA Paper 95-3001.

[37] C. Presser, A. K. Gupta, J. T. Hodges and C. T. Avedisian, "Interpretation of Size-Classified Droplet Velocity Data in Swirling Spray Flames", AIAA Paper, 95-0283, 1995.

[38] N. Kawahara, Y. Ikeda and T. Nakajima, "Droplet Followability and Slip Velocity Analysis of Evaporating Spray on Gun-Type Oil Burner", PARTEC95, pp. 593-602, 1995.

[39] N. Kawahara, Y. Ikeda and T. Nakajima, "Droplet Dispersion and Turbulent Structure in a Pressure-Atomized Spray Frame", AIAA Paper, 97-0125, 1997.

[40] T. M. Liou and D. A. Santavicca, "Cyclic Resolved LDA Measurements in a Motored IC Engine", Trans. ASME : J. Fluids Eng., Vol. 107, pp. 232-240, 1985.

[41] R. A. Fraser and F. V. Bracco, "Cycle-Resolved LDV Integral Length Scale Measurements in an I.C. Engine",, SAE Paper No. 880381, 1988.

[42] M. Lorenz and K. Rescher, "Cycle resolved LDV Measurements on a Fired SI-Engine at High Data Rates Using Conventional Modular LDV-System", SAE Paper No. 900054, 1990.

[43] T. D. Fansler, "Turbulence Production and Relaxation in Bowl-in-Piston Engines", SAE Paper No. 920479, 1993.

[44] K. Kobashi, et al., "Measurement of Fuel Injector Spray Flow of I.C. Engine by FFT Based Phase Doppler Anemometer - An Approach to the Time Series Measurement of Size and Velocity", Application of Laser Techniques to Fluid Mechanics, Springer-Verlag, pp. 268-687, 1991.

1999-01-0500

Modeling of Pressure-Swirl Atomizers for GDI Engines

C. Arcoumanis and M. Gavaises
Imperial College of Science, Technology & Medicine

B. Argueyrolles and F. Galzin
Renault

ABSTRACT

A new simulation approach to the modeling of the whole fuel injection process within a common-rail fuel injection system for direct-injection gasoline engines, including the pressure-swirl atomizer and the conical hollow-cone spray formed at the nozzle exit, is presented.

The flow development in the common-rail fuel injection system is simulated using an 1-D model which accounts for the wave dynamics within the system and predicts the actual injection pressure and injection rate throughout the nozzle. The details of the flow inside its various flow passages and the discharge hole of the pressure-swirl atomizer are investigated using a two-phase CFD model which calculates the location of the liquid-gas interface using the VOF method and estimates the transient formation of the liquid film developing on the walls of the discharge hole due to the centrifugal forces acting on the swirling fluid. Parametric studies reveal the effect of various nozzle operating and design parameters, such as the injection and back pressure, the needle lift and the radius of curvature of the discharge hole both at its inlet and exit, on the development of the liquid film. The nozzle CFD calculations are extended outside the injection hole in order to predict the initial development of the cone angle of the hollow-cone spray formed by the pressure-swirl atomizer. The nozzle flow exit characteristics are then used as inputs to a liquid sheet atomization model which estimates the size of the droplets formed after the disintegration of the injected liquid. Images obtained with a CCD camera of the spray structure, as a function of the injection pressure, close to the injection hole confirm that the proposed computational approach can simulate the near-nozzle spray development as a function of the geometric and operating characteristics of the fuel injection system and the pressure-swirl atomizer itself, thus providing the necessary initial conditions for spray predictions in the engine cylinder.

INTRODUCTION

It is expected that sooner or later automobile manufacturers world-wide will be faced with the very difficult task of simultaneous reduction of exhaust emissions and fuel consumption to very low regulated levels satisfying a number of urban and global environmental concerns. This will require a joint effort by the automotive companies, the oil and lubricant industries and the manufacturers of exhaust aftertreatment systems. In the heart of the problem is the mixture preparation strategy which is intimately related to the fuel injection system. The recent development of direct-injection gasoline engines (GDI) represents a promising approach in achieving low fuel consumption and emission levels and, as such, it requires further investigation.

GDI engines have been introduced over the last few years into the Japanese market [1-6] and more recently into Europe, offering fuel consumption levels comparable to those of indirect-injection Diesels and equivalent to an air/fuel ratio of 50 and above at part-load conditions. This is achieved by means of charge stratification, where a fuel-rich mixture is present in the vicinity of the spark plug at the time of ignition and a very lean mixture or pure air in the rest of the cylinder.

There are two main approaches to the mixture preparation and combustion strategy in GDI engines depending on the relative position of the injector and spark plug which are better known as close-spacing (spray-guided) and wide-spacing (wall or air-guided) approach [7-8]. Each one has advantages and disadvantages which have been investigated in more detail by consulting companies [9-14]. In brief, the close spacing approach allows at idle and part-load conditions better control of charge stratification, combustion of extremely lean overall mixtures and use of conventional piston design but suffers from spark plug electrode wetting and packaging difficulties of the injector and spark plug in the central part of the cylinder head. On the other hand, in the wide-spacing approach the injector is conveniently positioned at the side which prevents electrode wetting but the required control of charge stratification can only be achieved by means of a special piston cavity and strong tumble or swirl motion generated by modified cylinder heads.

To achieve high degree of stratification requires very accurate control of the quantity and timing of fuel injection which can only be achieved by electronic common-rail systems. To satisfy the conflicting requirements of mixture preparation during high-load (homogeneous stoichiometric) and part-load (stratified overall lean) conditions, high pressure swirl injectors have been used in GDI engines to inject the well-atomized fuel directly into the cylinder during the induction stroke (early injection) or during the compression stroke (late injection), respectively.

Swirl atomizers are capable of generating different spray patterns during induction and compression due to the different back pressures present in the cylinder at the time of injection. These spray patterns are also affected by the design of the swirl chamber and the exit angle/radius of curvature of the discharge hole. These details determine to a large extent the characteristics of the injected sprays and, as a result, they require special attention.

Due to the limited number of studies published in the open literature, the characteristics of the flow in the injection system of GDI engines and its effect on the spray formation and mixing are relatively unknown. Most of the recent studies, both experimental and computational, are focusing on the characterization of the spray by assuming a-priori known injection conditions and conventional CFD models widely used in diesel sprays, and its subsequent mixing process both under atmospheric conditions and in real engine configurations [15-26]. These CFD models usually represent the spray in the vicinity of the nozzle using the conventional approach of spherical blobs of computational parcels instead of the continuous liquid film exiting the nozzle. The widely used Eulerian-Lagrangian approximation is then adopted since it can be easily applied in order to simulate the spray development away from the nozzle. However, emerging experimental evidence indicates that the flow processes in pressure swirl atomizers represent a more complex physical problem since, due to the swirling motion, a liquid film is formed inside the discharge hole of the nozzle which leads to the formation of a hollow cone spray. Such nozzle flow phenomena have not been investigated in detail up to now, and, furthermore, it is not clear how far from the nozzle the spray pattern is affected by the nozzle exit flow characteristics.

It thus becomes clear that the conventional Eulerian-Lagrangian approximation cannot be justified in the vicinity of the nozzle exit hole prior to the disintegration of the liquid film into smaller droplets and ligaments. Furthermore, this method is associated with the inherent problem of using arbitrary rather than experimentally verified assumptions for the size and velocity characteristics of the initially injected parcels which are directly affected by the flow conditions in the discharge hole of the pressure-swirl atomizer.

In the following sections, an alternative approach to the modeling of GDI sprays injected from a pressure-swirl atomizer connected to a common-rail fuel injection system is presented. Initially, a description of the fuel injection system and the nozzle itself is given, followed by a general discussion of the computational methods that have been used up to now for the modeling of GDI sprays. Then, the flow models used in the present investigation are described. These include a 1-D flow model simulating the pressure waves developing in the pipes of the common-rail that cause variable injection pressure conditions, a two-phase CFD model that resolves the gas-liquid surface interface within the pressure-swirl atomizer and predicts the characteristics of the liquid film emerging from the discharge hole and finally a liquid sheet atomization model which estimates the size of the initially formed droplets.

FUEL INJECTION SYSTEM

Figure 1 shows a schematic representation of the common-rail GDI system used in the present investigation. It consists of the low pressure pump transferring the liquid from the fuel tank to the high pressure pump, the high pressure pump that pressurizes the fuel using a conical rotating plate, the connecting high pressure pipe, the common-rail, the four injection nozzles connected to the rail with flexible pipes and the pressure regulator which determines the nominal high pressure level in the rail.

The injection nozzle of the fuel injection system is the most critical part since its design controls the following spray formation. In the present system investigated, this is a pressure-swirl atomizer, comprised by the nozzle gallery, the needle and the nozzle tip. The needle is forced to be closed due to the action of a spring and it is electromagnetically forced to open during a specified injection timing which is flexibly programmed. The swirling motion inside the discharge hole is created by the off-center tangential slots which allow the fuel to flow from the high-pressure nozzle gallery to the nozzle exit discharge hole through a conical 360 degrees slot. Their number allows faster flow mixing and uniform swirl flow development in the conical slot, while their angle relative to the tangent at the periphery of the conical slot determines the angular momentum of the incoming liquid and, as a result, the swirl velocity at the beginning of the conical slot. A schematic representation of the nozzle can be seen in Figure 2 where the maximum needle lift is approximately half of the width of the conical slot. The radius of curvature at the inlet of the discharge hole is expected to affect the film development process while the radius of curvature at the nozzle exit is expected to be one of the most important parameters affecting the near-nozzle spray angle and the initial spray formation.

MODELLING APPROACH OF PRESSURE-SWIRL ATOMIZER

Conventional Approach

For the modeling of the gasoline sprays injected from pressure-swirl atomizers, it seems that different assumptions have to be employed than those used in diesel engine sprays in

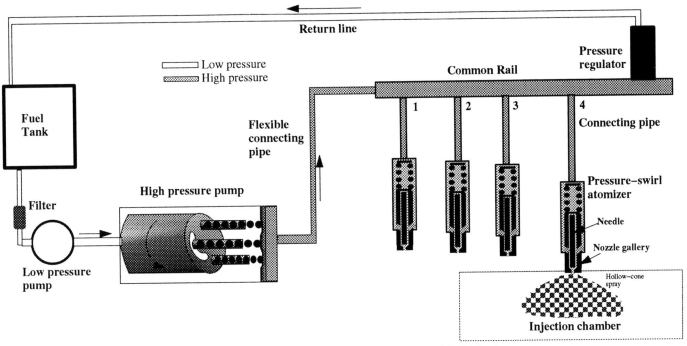

Figure 1 : Schematic representation of the GDI common–rail fuel injection system

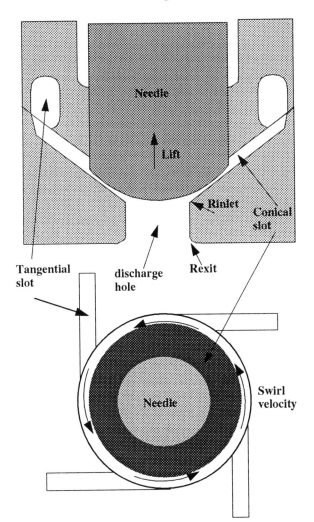

Figure 2 : Schematic representation of the pressure–swirl atomizer

order to estimate the liquid-gas mixing process. On the basis of previous experience from pintle or multi-hole gasoline and diesel sprays, the easiest approach to the modeling of GDI sprays would be to employ the conventional Eulerian-Largangian approximation, according to which the spray is represented by a discrete number of computational parcels. In this methodology, which is suitable for perfectly atomized dilute sprays, the droplet initial conditions represent one of the most important factors affecting the following spray development. The injection rate is usually taken from experimental data or from FIE flow models while arbitrary assumptions have to be employed for the initial size of the injected parcels. Either an initial droplet size distribution is assumed and then the size of the droplets in each parcel is randomly selected from this distribution function (Monte-Carlo approximation), or initial blobs are injected at the nozzle exit and then a liquid fragmentation model is employed in order to predict the mean size of the formed droplets. The initial blob size can be assumed to be that of the liquid film thickness at the nozzle exit, which necessitates the use of a nozzle flow model for the estimation of the nozzle flow exit conditions. The parameters required as input to the spray model are the liquid film thickness, and the mean axial, radial and swirl velocity components. Relevant studies employing this conventional approach to the modeling of GDI sprays have been reported in [27,28]; this approximation is presently used in most of the commercial CFD codes.

However, a number of discrepancies may arise from the employment of the Eulerian-Lagrangian methodology to the prediction of hollow-cone swirling sprays. The first is related to the effect of the nozzle flow exit conditions on the predicted spray characteristics. In order to demonstrate their importance, a set of calculations were performed using the KIVA CFD code; the aim of this study was to estimate the effect of small variations in the initial film thickness and liquid axial velocity on the predicted spray characteristics.

The test cases investigated included spray injection into a quiescent constant-volume chamber at a pressure of 4.5 bar and temperature of 300 and 500 K, in order to examine the effect of both non-evaporating and evaporating droplet conditions on the predicted spray structure. The injection pressure was assumed constant and equal to 70 bar while typical values for the injection rate and period were selected to be 11.9 g/s and 3 ms, respectively. The numerical grid was consisted of approximately 40×40 cells with 2×2.25 mm grid spacing in the radial and axial directions, respectively; 6000 parcels were assumed to represent the spray while the fuel physical properties were represented by those of iso-octane.

For each of the two chamber pressures, two different injection conditions were selected. Since recent studies on the liquid film thickness in GDI injectors [46,47] have indicated that its value is approximately 1/3 of the radius of the discharge hole, a value of 100μm was initially selected which corresponds to a mean injection velocity of 79m/s. In the second set of calculations, the injection velocity was arbitrary selected as 20% smaller than the previous case (63 m/s) which, from mass balance considerations for the assumed flow rate, it gives rise to a film thickness of 131.4μm. The far spray angle estimated from the spray images was approximately 30 deg.

The spray sub-models employed to resolve the spray development include the TAB model [29] for calculating the blob fragmentation and the subsequent droplet secondary break-up, a model for estimating droplet collision and coalescence [30], a model for estimating droplet turbulent dispersion assuming isotropic turbulence [31] and, finally, when injected into a high temperature environment, a droplet evaporation model where the droplet drag coefficient is taken as that of spherical solid spheres. These are the standard spray sub-models incorporated into the KIVA code [32].

Figures 3a and 3b show a cross section of the droplet parcels on the plane of symmetry as calculated at 1.5 ms after the start of injection for the two different injection conditions discussed previously. Although the spray penetration presented in Figure 4 for the two cases, is almost the same, the spray structure is different; apart from the spray dispersion, considerable differences exist in the droplet size and velocity. These differences are more important when injecting into a high pressure and temperature environment where droplet deceleration is faster and droplet evaporation is the most important physical process affecting the fuel vapor to gas ratio in the computational domain. Figure 5, which presents the calculated results for injection into high pressure and temperature gas for the two different inlet conditions, confirms the above statement. As can be seen in Figures 5a and 5b, despite the fact that the calculated spray tip penetration was identical between the two cases, the calculated spray pattern varies considerably but also large differences exist in the calculated fuel vapor volume fraction presented along the same cross section in Figures 5c and 5d, respectively. As can be

seen, for the spray injected with higher velocity and smaller initial blob size, the evaporation process is faster, which, according to Figure 6, results in almost double total fuel evaporated mass than in the other injection case. It is thus clear that the initial droplet parcel conditions represent one of the most important uncertainties in spray calculations using the conventional approach.

Apart from the need for accurate estimation of the injection conditions, four more issues centered on the validity of the Eulerian-Lagrangian methodology, as applied to GDI hollow-cone sprays, require further consideration.

The first one refers to the numerical treatment of the liquid sheet before its disintegration into smaller ligaments and droplets. Assuming that a hollow cone spray is emerging from the injection nozzle, and since the sheet travels a small distance in the surrounding gas before its disintegration into droplets, it is obvious that the only possible way of air entrainment in the inner zone of the sheet is from below the sheet rather than through it [33,34]; thus representation of the liquid sheet using parcels leads to unavoidable gas motion through the conical spray.

The second point that requires further attention is related to the transient phenomena associated with the early spray development before the formation of a liquid film. Following the opening of the needle, it takes some time until the formation of the liquid film on the walls of the discharge hole which is associated with the injection of a fuel quantity at the center of the hole with relatively small velocity; present droplet parcel models cannot account for such transient flow phenomena related to the opening and closing of the needle and its effect on the hole exit flow characteristics.

Furthermore, the angle of the spray and its initial motion cannot be accurately resolved by the parcel representation. This is due to the gradual film movement from the center of the hole towards the walls and the resulting isolation of the gas phase at the two sides of the film, which creates a depression of the conical sheet resulting in a gradual contraction of the liquid as it travels downstream of the point of injection. This phenomenon is more pronounced when the surrounding pressure is higher. As a result, the calculation of the motion of the continuous film cannot be accurately estimated from that of spherical droplets moving along the trajectories in the direction of their initial injection.

The last point that requires special attention is related to the conical expansion of the liquid sheet which leads to a significant decrease of its thickness even within a very small distance downstream of the injection nozzle, due to the relatively wide angle of the sprays injected from swirl atomizers. This reduction in thickness has a strong effect on the spray break-up length and the size of the formed droplets and introduces complications to the modeling of the atomization process of the liquid. From existing liquid sheet atomization models, the linear instability analysis is presently the only reliable theory for estimating the atomization process of the liquid sheet; however, since the sheet thickness is one of the most important parameters involved in the relevant equations, the effect of its conical expansion should be taken

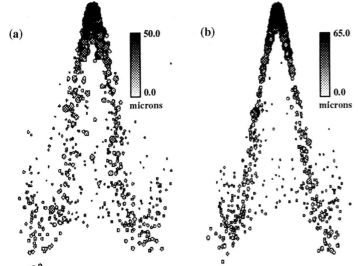

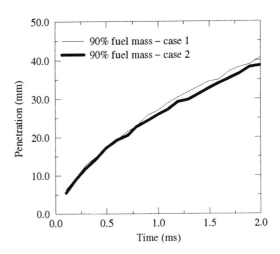

Figure 3 : CFD (KIVA) calculations of droplet parcel location on a plane of symmetry at 1.5 ms after the start for injection under 4.5 bar and 300 K (non–evaporating spray). The two different plots corrspond to two different initial velocity and film thickness inputs (a) Case 1 :Uinj=79 m/s, film thickness= 100 μm, and (b) Case 2 : Uinj=63 m/s, film thickness=131 μm

Figure 4 : Calculated spray penetration (based on 90% fuel mass in a computational cell) for the two injection cases of Figure 3

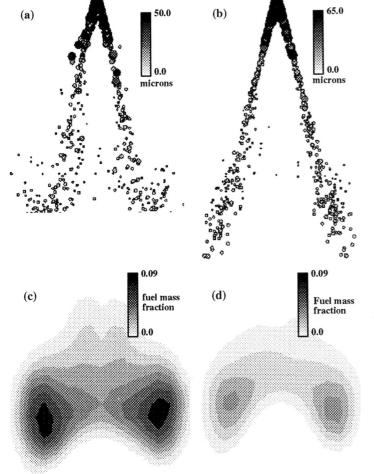

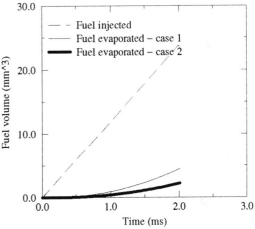

Figure 6 : Calculated quantity of total fuel evaporated for the two injection cases of Figure 5

Figure 5 : CFD (KIVA) calculations of droplet parcel location and fuel vapor mass fraction on a plane of symmetry at1.5 ms after the start for injection under 4.5 bar and 500 K (evaporating spray).The two different plots corrspond to two different initial velocity and film thickness inputs (a) Case 1 :Uinj=79 m/s, film thickness= 100 μm, (b) Case 2 : Uinj=63 m/s, film thickness=131 μm, (c) Case 1 and (d) Case 2

into account. Unfortunately, the film thickness reduction resulting from this conical expansion of the liquid sheet is of the same order of magnitude as that of the amplitude of the surface waves developing on its surface, thus raising questions about the validity of the relevant atomization theories.

Alternative Approach

In this section, the alternative methodology employed here in order to predict the initial development of sprays injected from GDI pressure-swirl atomizers is briefly described. This approach includes the simulation of the pressure waves developing in the common-rail system using a 1-D model that estimates the actual injection pressure in the nozzle gallery and the flow rate through the discharge hole, and the simulation of the liquid film development within the injection nozzle and just after its exit using a recently developed two-phase CFD code

[35]; liquid sheet atomization models based on a linear instability analysis of the waves developing on the surface of the jet are also described since they can be combined with existing spray models [36] for the estimation of the subsequent spray development.

1-D pressure wave dynamics in the common-rail system

This model accounts for the pressure wave dynamics in the injection system assuming 1-D flow which is justified in view of the length of the pipes being much longer than their diameter. Experimental studies on the pressure variations in such injection systems have confirmed that although the volume of the rail is very large compared to the injection quantity, pressure pulses can develop through the system with an amplitude approximately 20% of the nominal pressure level in the system.

The 1-D model uses the mass and momentum conservation equations for estimating the pressure and flow rate in the pipes of the system, the mass balance for the calculation of the mean pressure in the volumes of the system, such as the nozzle gallery, and the equilibrium of forces for estimating the mechanical movement of the needle. Variable discharge coefficients, pressure losses, leakage and variable liquid physical properties, have been implemented into the 1-D model in order to enhance its predictive capability [36].

Liquid film flow simulation

Despite the significant importance of the flow characteristics inside the injection nozzle in obtaining better estimates for the nozzle exit conditions, only a few studies usually employing single-phase CFD codes, have been published up to now in this area; for example, see [41-47]. More recently, commercial CFD models have been employed for the simulation of the film thickness inside the injection hole by applying a two-step approach. In the first step, the single-phase is solved assuming that the hole is filled with liquid and in the second step it is assumed that the liquid-gas interface is located in the area where the axial velocity changes direction due to the recirculation zone formed adjacent to the hole axis [45,47]; apparently, large differences for the liquid axial velocity have been calculated using the two steps which implies that the actual film thickness plays an important role on the exit velocity of the injected liquid. Furthermore, such models cannot be used to account for any transient effects related to the initial development of the liquid film inside the injection hole.

The nozzle flow model employed in the present investigation is based on a recent two-phase extension of a previously developed CFD code (GFS) [35], which accounts for the tracking of liquid-gas interface surfaces using a VOF methodology [37]. The model solves numerically the full Navier-Stokes equations describing the turbulent motion of the moving fluids. The time-averaged form of the continuity, momentum and conservation equations for scalar variables are numerically solved using collocated Cartesian velocity components on a Cartesian non-uniform, curvilinear, non-orthogonal numerical grid. Turbulence is simulated by the two equation k-ε model. The discretization method is based on the finite volume approach; the spatial discretization scheme used here was a second order one while the time discretization was based on a fully implicit Crank-Nicholson scheme. Although these numerical schemes have an effect on the computational time, they were selected in order to minimize numerical diffusion which can become the most important source of numerical error in the estimation of the accurate location of the liquid-gas interface. The general form of these equations written in a non-orthogonal curvilinear coordinate system and extended to include the second phase is given in [35,36,38-40]. Although the presence of the tangential slots creates a 3-D flow configuration, the computational time required for prediction of the transient film development was prohibitively long which necessitated the use of a 2-D approximation in the present investigation. The relevant assumption that had to be made refers to the creation of an axisymmetrically swirling motion at an early stage in the conical slot of the pressure-swirl atomizer. Since the scope of the present investigation is to describe the methodology employed in the simulation of such systems and to identify the parameters affecting the film development inside the nozzle, the use of a 2-D instead of the 3-D version which is required for the accurate estimation of the swirl velocity component as a function of the geometry of the tangential slots can be justified. Thus, a simple expression for the conservation of the angular momentum can give at present a reasonable estimate of the swirl velocity component at the inlet boundary of the computational domain while the inlet axial velocity is determined by the flow rate calculated from the 1-D model. Since the simplified configuration represents an axisymmetric geometry, the angular momentum conservation equation is solved in addition to the other flow equations in order to account for the swirl velocity component inside the nozzle and the related centrifugal forces responsible for the creation of the liquid film. The computational domain has been extended approximately two nozzle diameters away from the nozzle exit in order to allow the air entrainment process at the nozzle exit to be considered and also to account for the initial spray angle as a function of the geometric details of the nozzle itself and the centrifugal forces acting on the liquid.

Liquid sheet atomization and spray model

Having identified the characteristics of the liquid flow inside the injection nozzle and at its exit using the two-phase CFD model, the computational domain can be extended in order to solve for the following spray development using the conventional droplet parcel approximation of the liquid phase. It is important to note that the developed code [40] is able to identify different computational domains through the use of appropriate conditions at the common cell boundaries. The Eulerian-Largangian approximation requires the use of a liquid sheet atomization model which should predict the size distribution of the droplets formed during the disintegration of the liquid film.

174

It is accepted that the linear instability theory can be employed in order to account for the liquid sheet atomization process, and a number of investigations have been already reported on this issue; for example, see [48-54]. Important improvements to the original method include the conical expansion of the film and the corresponding reduction of its thickness as well as the presence of the swirling motion which, not only increases the relative velocity between the liquid and the surrounding gas and enhances the disintegration process, but also allows for the development of surface waves in the direction of the swirling motion; details can be found in [55]. Since such models predict only an average droplet size, a droplet size distribution sub-model is required for the calculation of all possible droplet sizes present in the spray. Instead of using an arbitrarily selected distribution function, a model based on the maximum entropy formalism which is suitable for such applications [56] has been employed here. In addition to the droplet size distribution, this type of model can predict a droplet velocity distribution function resulting from the non-uniform velocity profile within the bulk of the liquid film [57].

Following the formation of droplets with size and velocities calculated from the liquid sheet atomization model, a previously developed CFD spray model [36] can then be employed in order to predict the subsequent spray development. The sub-models implemented into the code account for the droplet secondary break-up which estimates the droplet deformation process during its life time, the droplet turbulent dispersion, the droplet collision and coalescence and the droplet evaporation while the droplet drag coefficient can be based on that of the deformed droplets.

RESULTS AND DISCUSSION

In this section, the results obtained with the above simulation models are described. These include the 1-D pressure wave simulation in the common-rail GDI system and the CFD prediction of the liquid film developing inside the pressure-swirl atomizer; estimates for the initial spray characteristics from the phenomenological liquid sheet atomization model are also presented. Finally, spray images obtained with a fast CCD camera at various times during the spray development under atmospheric conditions and for different injection pressures are included and briefly discussed.

Common-Rail Injection System

In this section the results obtained from the 1-D flow model simulating the flow development within the common-rail fuel injection system are described. Since the operation of the high pressure pump has not been modeled at present, the pressure in the rail was given as input to the model. Two different types of input data have been selected in order to examine the effect of the operation of one or more injectors on the injection characteristics. Initially the rail pressure is assumed to be constant and equal to the nominal value specified by the pressure regulator while in the second set of input data the measured rail pressure with all four injectors operating has

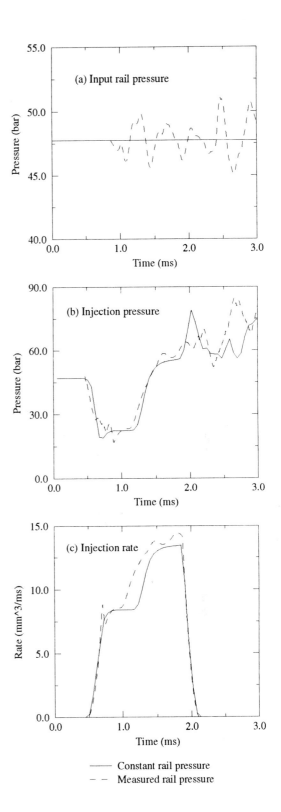

Figure 7 : 1–D simulation of the flow in the common–rail GDI injection system for full load conditions (a) input rail pressure (b) predicted injection pressure in the nozzle gallery and (c) predicted injection rate.

been used; in this case the model calculates the pressure wave dynamics within the connecting flexible pipe and the nozzle. The two different input pressure signals for a typical rail pressure of around 50 bar are given in Figure 7a.

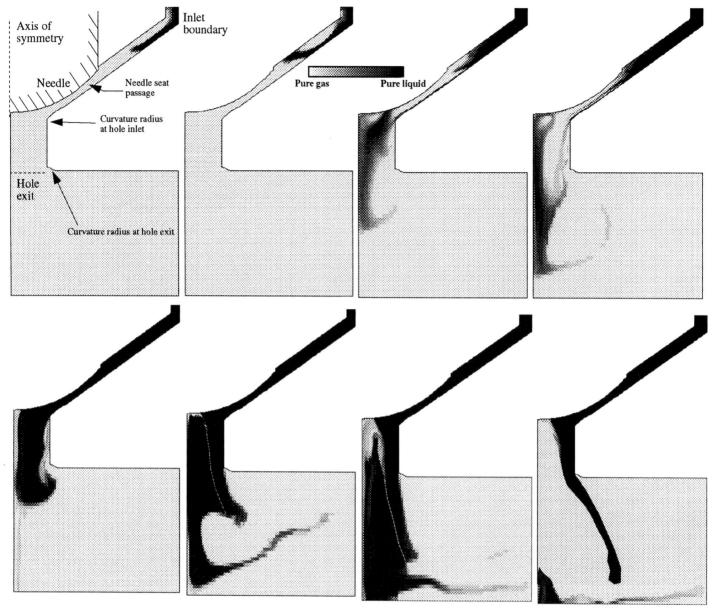

Figure 8 : Calculated liquid flow development inside and close to the hole exit of the pressure–swirl atomizer for injection under atmospheric conditions. Injection pressure 50 bar, full needle lift.

Figure 7b shows the calculated injection pressure at the gallery of the nozzle just upstream of the tangential slots. As can be seen, following the opening of the needle, the pressure in the nozzle gallery decreases substantially, which implies that the assumption of constant injection pressure usually adopted in commercial CFD codes for the modeling of common-rail GDI sprays may not be valid. In fact, the pressure drop can be as high as half of the nominal rail pressure. The effect of the operation of all four injectors connected to the rail is to induce pressure fluctuations to the actual injection pressure which are expected to affect the formation of the liquid film inside the nozzle and the subsequent spray characteristics.

Figure 7c shows the predicted fuel injection rate assuming a variable discharge coefficient of the nozzle as a function of the

transient needle lift, with a maximum value of 0.66 at full lift. As can be seen, the injection rate varies considerably during the injection period which is expected to have a strong effect on the transient spray characteristics and the fuel evaporation process inside the engine cylinder.

From the analysis presented above, it becomes obvious that FIE flow models can be used to provide estimates of the injection pressure and injection rate in pressure-swirl atomizers; the common assumption adopted in standard CFD codes, as applied to GDI spray simulations, of constant injection pressure and injection rate cannot be justified even in common-rail systems and may lead to unrealistic spray structure, air-entrainment and fuel/air mixing predictions.

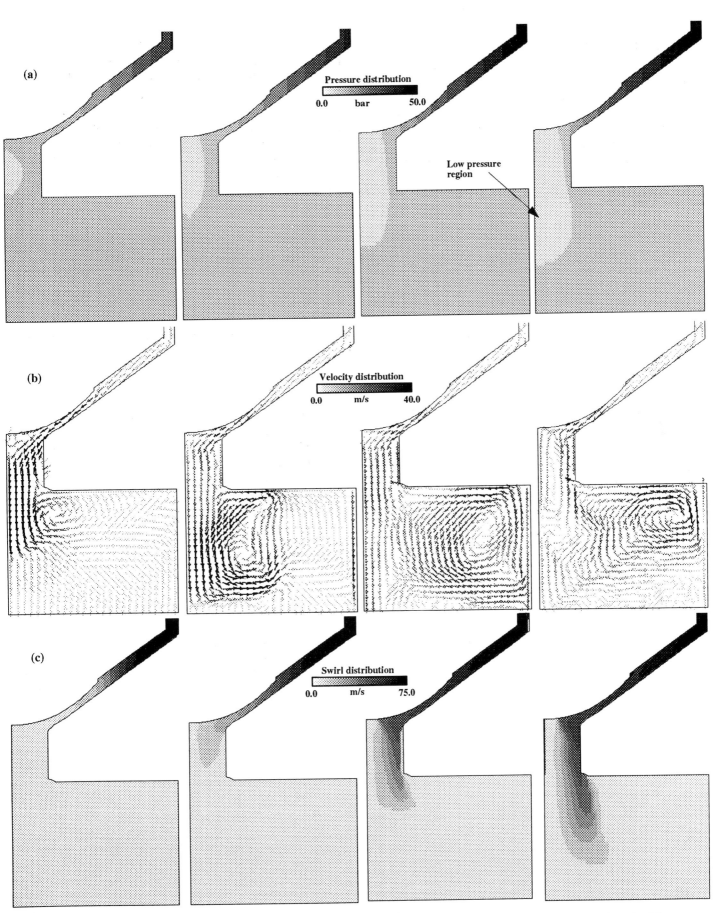

Figure 9 : Calculated development of (a) pressure distribution, (b) velocity flow field and (c) swirl flow inside and close to the exit of the pressure−swirl atomizer for the same conditions as in Figure 8

Internal Nozzle Flow Simulation

Having identified the flow conditions within the common-rail system, the two-phase CFD model was used in order to investigate the film formation process inside the discharge hole of the pressure-swirl atomizer. The aim of the present investigation has been to identify the mechanisms leading to the formation of the film and the parameters affecting its development as well as their influence on the nozzle exit characteristics. The numerical grid used in the present 2-D solution comprises approximately 8000 computational cells with a grid spacing inside the discharge hole of approximately 9μm. This grid resolution was considered adequate to allow estimation of the relative effect of the various geometric and operating parameters on the development of the liquid film inside the discharge hole; however, it is not adequate to predict accurately the liquid film thickness that should be used as input to the liquid sheet atomization model. Figure 8 presents a sequence of events describing the development of the liquid-gas surface interface for a constant injection pressure of 50 bar, the needle placed at its full lift position and injection taking place under atmospheric chamber conditions. Liquid with uniform axial and swirl velocity is assumed to enter into the nozzle from the upper corner of the conical slot, which is gradually moving inside the discharge hole through the needle seat passage. As can be seen, the liquid that was initially identified not to occupy the whole conical slot falls into the center of the hole due to its momentum; this early moving liquid seems to lose most of its angular momentum and exits the hole with nearly zero swirl velocity, thus creating a poorly developed spray. Following this first stage, the main bulk of the liquid which has filled completely the conical slot is entering into the hole and is gradually moving towards the wall of the discharge hole due to the action of the centrifugal forces resulting from its swirling motion. Interestingly, two main liquid streams can be observed to exist in the injection hole. The first comes from the main flow entering through the needle seat passage and moving towards the exit, while the second one is formed from the liquid trapped into the low pressure recirculation region developed at the center of the hole and shown clearly in Figure 9a. As can be seen in Figures 9b and 9c, this bulk of liquid has almost zero swirl velocity and much smaller axial velocity compared to the liquid forming the film on the walls of the discharge hole. This liquid forms a lamela at its front as it moves downstream parallel to the axis of symmetry of the nozzle. There is a considerable time delay until the formation of the liquid film which depends on the inlet velocity and the opening of the needle. It is expected that this flow would create a different spray pattern at the early stages of injection relative to that later under steady-state conditions. As discussed in a following section, this bulk of liquid can be clearly observed in the images obtained during the development of the spray. At the exit of the hole, a vortex structure is induced to the gas by the moving liquid which interacts with the injected liquid assisting in its disintegration. Clearly, the initial development of the liquid in the injection hole until the formation of the liquid film on the hole walls, as identified by the present two-phase CFD model, cannot be described in terms of parcels. Points that require special attention are related to the pressure and swirl velocity distributions in the injection nozzle, presented in Figures 9a and 9c respectively. The pressure drops gradually within the conical slot and becomes equal to the back pressure in the area of the discharge hole occupied by the liquid film, while a low pressure region is developing at the center of the injection hole. As can be seen, only this area occupied by the liquid film exhibits a high-swirling motion. Unfortunately, even within a very small distance from the hole exit, the film thickness becomes comparable to the cell size which precludes any realistic predictions to be obtained further downstream.

Having identified the processes leading to the formation of the film at the early injection period, it was considered important to demonstrate the effect of different nozzle geometric and operating parameters on the flow exit characteristics. The operating parameters investigated here include the initial swirl velocity component, and the flow rate through the nozzle which is equivalent to the combined effect of the injection and back pressures, while the geometric parameters include the needle position and the radius of curvature of the discharge hole both at its inlet and exit.

Figure 10 shows the effect of a smaller initial swirl velocity on the pressure distribution, swirl distribution, velocity flow field and liquid film thickness for the same flow rate through the nozzle as in the previous case presented in Figures 8 and 9. For the lower swirl value selected here, the film is developing at a much slower pace while the spray exhibits a relatively smaller angle; it is also highly possible that under certain conditions a liquid core or a cylindrical sheet may form, instead of a liquid film, on the walls of the discharge hole. Since the level of the swirl velocity is a function of the geometry of the tangential slots, it is expected that nozzles with the same discharge hole but different geometric configuration inducing swirl, will produce considerably different sprays.

The rail pressure is one of the most important parameters affecting the following spray characteristics. As can be seen in Figure 11, increasing the flow rate through the nozzle not only increases the injection velocity but also creates a thicker liquid film. In addition, the time delay until the formation of the steady-state film becomes shorter and the spray angle smaller. It is thus expected that during the injection period when the actual flow rate through the discharge hole varies, the spray pattern (spray angle, air-fuel ratio) should also vary considerably.

Another important parameter affecting the formation of the film inside the nozzle is the back (chamber) pressure. Increasing the back pressure causes a reduction to the spray angle for the same flow rate and swirl velocity inside the nozzle, due to the increased drag forces acting on the moving liquid stream inside the nozzle before it reaches the walls of the discharge hole and the lower pressure region formed at the inner side of the emerging from the hole liquid; this flow development can be seen in Figure 12. As already known from a number of investigations on the spray characteristics of pressure-swirl atomizers, at higher back pressure levels the

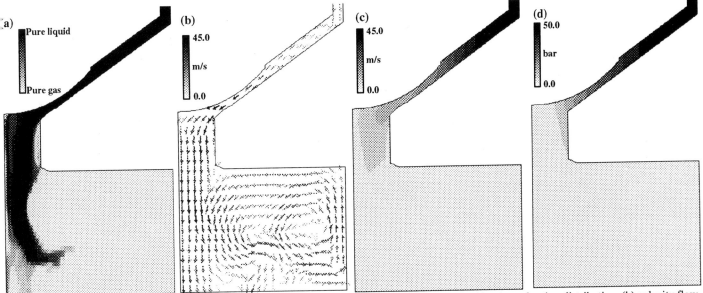

Figure 10 : Effect of initial swirl velocity on the flow structure in the pressure–swirl atomizer (a) liquid volume fraction distribution, (b) velocity flow field, (c) swirl velocity distribution and (d) pressure distribution. Injection under atmospheric conditions, full needle lift, injection pressure 50 bar

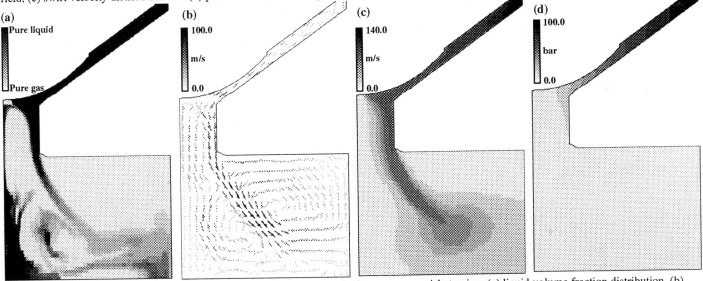

Figure 11 : Effect of injection pressure of 100 bar on the flow structure in the pressure–swirl atomizer (a) liquid volume fraction distribution, (b) velocity flow field, (c) swirl velocity distribution and (d) pressure distribution. Injection under atmospheric conditions, full needle lift

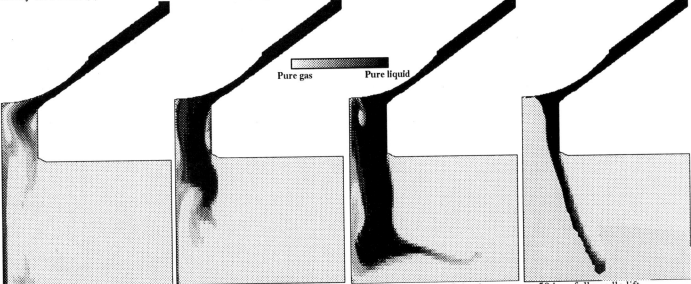

Figure 12 : Calculated liquid flow development for injection against a back pressure of 10 bar. Injection pressure 50 bar, full needle lift

spray angle collapses which may even lead to a solid cone rather than a hollow cone spray.

Having identified the effect of the major operating conditions of the nozzle on the liquid film flow structure inside the discharge hole, we can now proceed to the investigation of the effect of various nozzle geometric parameters on the nozzle flow exit characteristics. The first parameter that required investigation was the needle lift. In swirl-pressure atomizers the needle stop position is such that, the lift is usually smaller than the width of the conical slot, resulting to a sudden contraction at the needle seat area just before the turn of the flow inside the discharge hole. As can be seen in Figure 13, for a much smaller needle lift than that presented in previous figures, the injection pressure required to inject the same amount of fuel increases and the liquid velocity at the needle seat area also increases considerably, which initially results in formation of a liquid core at the nozzle center instead of a film on the walls of the discharge hole; however, as can be seen in Figure 13c, there is still a significant swirling motion which may lead to the formation of a liquid film as the flow develops. It is thus expected that under transient engine operation, the flow structure in the discharge hole will be strongly affected not only by the variable flow rate and swirl velocity but also by the needle movement.

Another geometric parameter that is related to the film formation process is the radius of curvature at the inlet to the discharge hole. As can be seen in Figure 14, for a sharper inlet corner it becomes more difficult for the liquid to turn inside the hole which results in the accumulation of a larger fuel quantity in the recirculation zone formed at the center of the discharge hole. Compared with Figure 8, it is clear that it takes more time for a relatively larger amount of liquid mass to form the poorly developed spray until the steady-state conditions are achieved. Interestingly, a larger spray cone angle is initially formed due to the smaller mass (less inertia) of the liquid contributing to the conical sheet at the exit of the nozzle. However, at subsequent time steps, the spray angle is expected to become equal to that of a nozzle with a smoother inlet shape of the discharge hole.

Finally, the last parameter investigated was the radius of curvature at the nozzle exit. As can be seen in Figure 15, this geometric detail of the nozzle may have a strong effect on the spray cone angle since it alters the actual direction of injection. This parameter, together with the centrifugal forces acting on the liquid stream, determine the spray angle of the hollow cone spray formed at the nozzle exit under atmospheric conditions.

Liquid Sheet Atomization Process

Since it is expected that the previously described two-phase CFD model can provide realistic estimates for the initial conditions of the liquid film forming on the walls of the discharge hole, an atomization model employing the linear instability approximation for the waves developing on the surface of the conical liquid sheet, has been used in order to examine how these nozzle exit characteristics are expected to influence the size of the formed droplets.

For the model employed here [55], the break-up length has been estimated from an empirical correlation initially reported in [27] and modified in [47]. This correlation leads to an estimate of the break-up length of the order of 1 to 2 mm, depending on the initial film thickness, the mean exit velocity, the spray cone angle and the chamber pressure and density conditions. Figure 16 shows the calculated droplet size as a function of the film exit thickness for two different values of the swirl velocity and gas chamber density. The selected values of the gas density correspond to injection under atmospheric conditions and during compression, respectively, while the swirl velocity values correspond to different configurations of the tangential slots. The selected range of values for the film thickness correspond to fuel injection quantities from low to full load in a typical passenger car GDI engine. Two different values of near-nozzle spray angle of 35 and 70 deg. have been selected and the corresponding results are presented in Figures 16a and 16b, respectively; these values would correspond to different geometric configurations at the exit of the discharge hole of the injection nozzle.

It is evident that the mean droplet size, as calculated here from the linear instability analysis, is much smaller than the film thickness. Apart from the initial spray angle which affects not only the size of the formed droplets but also the break-up length, it can be seen that by increasing the swirl velocity the droplet size decreases considerably due to the increased relative velocity between the gas and the injected liquid which leads to higher Weber numbers. A more dramatic effect can be observed for the case of the gas pressure, where a higher chamber pressure (i.e. higher gas density) leads to the formation of smaller droplets. It should be noted that the predicted mean droplet size presented here corresponds to a very small distance from the nozzle exit where experimental data cannot be easily obtained. In addition, phenomena related to droplet secondary break-up have not been considered although it is expected that their relative effect will be more important for injection under atmospheric conditions where larger droplets are formed that are traveling at higher velocities due to the reduced drag effects associated with the lower gas densities.

From this analysis it can be concluded that the existing atomization model is sensitive to a number of parameters related to the flow distribution inside the injection nozzle. It is thus expected that computer models, such as the one proposed here, that simulate the whole injection process from the pressure inside the common-rail to the emerging from the hole spray can be used as research and design tools while assisting at the same time in the assignment of input data to existing CFD codes for predicting the mixture distribution in direct-injection gasoline engines operating under part- and full-load conditions.

Global Spray Characteristics

In order to validate the previously described computer models, a GDI common-rail injection system has been

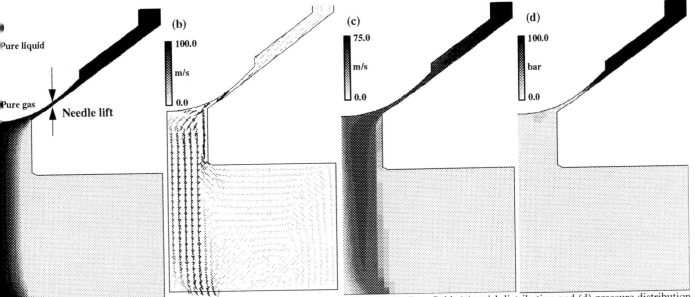

Figure 13 : Effect of needle lift of 20 µm on (a) liquid volume fraction, (b) velocity flow field, (c) swirl distribution and (d) pressure distribution. Injection under atmospheric conditions, Injection pressure100 bar

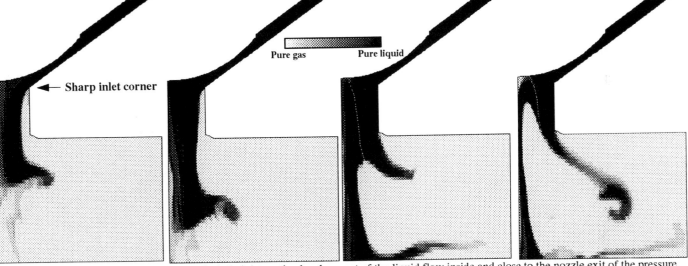

Figure 14 : Effect of curvature radius at the hole inlet on the development of the liquid flow inside and close to the nozzle exit of the pressure swirl atomizer. Injection under atmospheric conditions, full needle lift, Injection pressure 50 bar

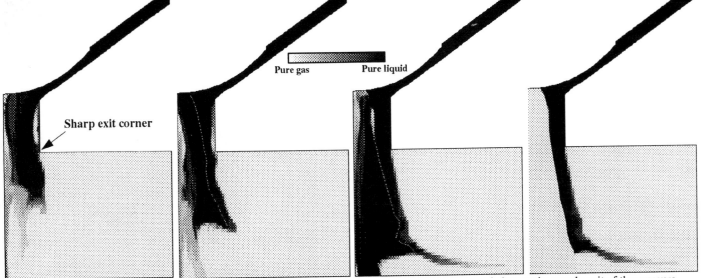

Figure 15 : Effect of curvature radius at the hole exit on the development of the liquid flow inside and close to the nozzle exit of the pressure swirl atomizer. Injection under atmospheric conditions, full needle lift, Injection pressure 50 bar

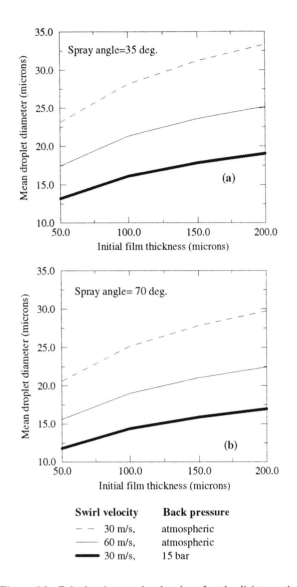

Figure 16 : Calculated mean droplet size after the disintegration of the liquid sheet as a function of the film thickness at the hole exit for two different values of the swirl velocity component and the back pressure. (a) spray angle 35 deg., (b) spray angle 70 deg.

instrumented and it is currently used to obtain detailed measurements of the spray structure and droplet characteristics under both atmospheric and high pressure/temperature conditions.

The first part of this investigation is imaging of the atmospheric spray using a high-resolution CCD camera. Figure 17 shows three pictures obtained close to the injection hole at 0.05ms after the start of injection for three different nominal values of the rail pressure (20, 50 and 70 bar). The images reveal that at this early stage of injection, the spray is not developed into a swirling conical spray. Instead, a bulk of liquid consisting of large droplets and ligaments is emerging from the hole close to the axis of symmetry while, at the same time, smaller droplets are found at the periphery of the spray in areas where, at subsequent times during the spray development, the fully developed conical hollow-cone spray

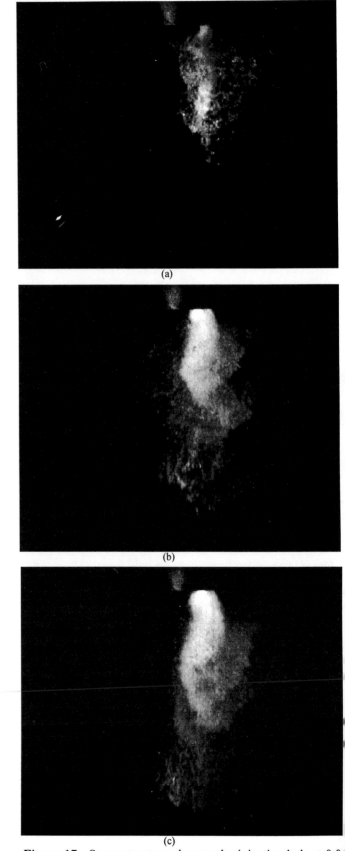

Figure 17 : Spray structure close to the injection hole at 0.0: ms after the start of injection for different nominal rai pressure values (a) 20 bar, (b) 50 bar and (c) 70 bar

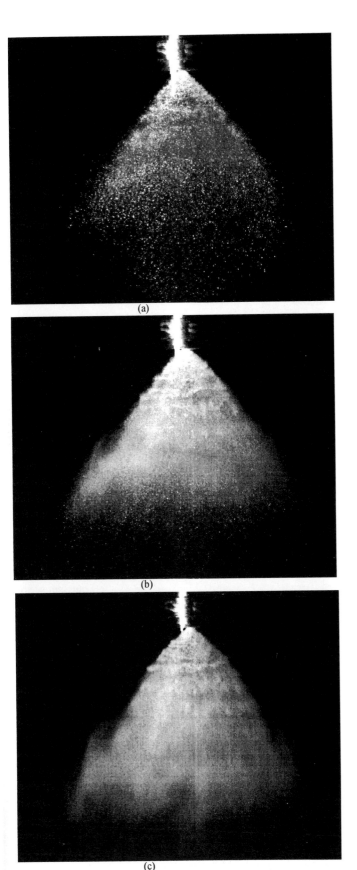

will be formed. These images can be indirectly linked to the liquid structures identified with the CFD model during the development of the liquid film within the injection hole.

Figure 18 shows three images for the same injection conditions as in Figure 17, but obtained at approximately 1.0 ms after the start of injection. At this later time a hollow-cone spray is formed with recirculation zones evident at its front edge. For the higher injection pressure case (70 bar), a more dense and better atomized spray is observed with small droplets trapped in the recirculation zones formed at the periphery of the spray which, as they move backwards, create a mushroom-shape spray. For the lower injection pressure case (20 bar) presented here, a substantial amount of droplets can be seen in the central part of the spray which are over-penetrating relative to the rest of the conical spray. It is expected that these droplets have been formed during the initial development of the conical spray which takes place at relatively slower pace as the injection pressure decreases. Experimental work is in progress at Imperial College to identify the detailed transient spray characteristics under both atmospheric and high pressure/ temperature conditions and the degree to which the linear instability analysis can represent the atomization process in pressure-swirl atomizers.

CONCLUSIONS

A new approach to the modeling of GDI sprays injected from a pressure-swirl atomizer connected to a common-rail fuel injection system, has been presented. The model takes into account the pressure wave dynamics within the lines and the rail of the injection system, caused by the operation of the electronically controlled needle in all four injectors connected to the common-rail by using a 1-D compressible flow sub-model. Considerable pressure drop and pressure fluctuations in the actual injection pressure in the nozzle gallery were predicted resulting to a variable injection rate.

Following the calculations of the injection pressure and injection rate through the nozzle, a CFD model that employs the VOF methodology and accounts for the liquid-gas surface interface developing within the injection nozzle, was used to identify the formation mechanism of the liquid film inside the slots and the discharge hole of the pressure-swirl atomizer. This calculation was extended outside the nozzle in order to predict the spray cone angle as a function of the swirling motion of the injected liquid and the geometric details of the nozzle itself. The mechanism of formation of the liquid film was presented as a function of the operating conditions of the nozzle and the internal geometric details of the pressure-swirl atomizer. Images of the liquid emerging from the nozzle, obtained with a CCD camera, have confirmed the flow types identified by the model during the early flow development within the injection hole. Finally, estimates of the nozzle exit characteristics such as the liquid film thickness, the injection velocity profile and the spray angle were then used as inputs to a liquid sheet atomization model which, based on the linear instability analysis of the surface waves developing on the

Figure 18 : Effect of injection pressure on the structure of the spray under atmospheric conditions at t=1.5 ms after the start of injection (a) 20 bar, (b) 50 bar (c) 70 bar

liquid sheet emerging from the nozzle, predicts the average size of the droplets formed after the disintegration of the sheet.

The proposed methodology which accounts for the two-phase flow development inside and just outside the nozzle hole using the Eulerian VOF method may be combined with the conventional Eulerian-Lagrangian approximation to allow representation of the spray development after the disintegration of the conical liquid sheet. This approach offers promise for predicting the transient spray characteristics in direct-injection gasoline engines as a function of the geometric and operating characteristics of the fuel injection system and the nozzle itself and the thermodynamic conditions in the engine cylinder.

ACKNOWLEDGEMENTS

The authors would like to thank Renault SA for providing the fuel injection system and financial support. In particular they would like to acknowledge the contributions of Dr. Thierry Mantel to the early stage of the project and Dr. A. Ahmed for permission to publish the results. Also, Mr. E. Abo-Serie from Imperial College for his contribution to the development of the liquid sheet atomization model and the experimental rig. Finally Dr. A. Theodorakakos from the National Technical University of Athens for his assistance in the implementation of the VOF method into the CFD code used in the presence investigation.

REFERENCES

1. Harada J., Tomita T, Mizuno H., Mashiki Z. and Ito Y. "Development of direct injection gasoline engine", SAE Paper 970540, 1997

2. Iwamoto Y., Noma K., Nakayama O., Yamauchi T. and Ando H. "Development of direct injection gasoline engine", SAE Paper 970541, 1997

3. Tomoda T., Sasaki S., Sawada D., Saito A. and Sami H. "Development of direct injection gasoline engine - study of stratified mixture formation", SAE Paper 970539, 1997

4. Kume T., Iwamoto Y., Iida K., Murakami M., Akishino K. and Ando H. "Combustion control technologies for direct injection SI engine", SAE Paper 960600, 1996

5. Ando H. "Mitsubishi GDI engine - Strategies to meet European requirements", AVL Conference, 4-5 September, 1997

6. Ando H. "Combustion control technologies for gasoline engines", IMechE Intl. Seminar on Lean Burn Engines, 3-4 December, 1996

7. Zhao F-Q, Lai M-C and Harrington D.L "A review of mixture preparation and combustion control strategies for spark-ignited direct-injection gasoline engines", SAE Paper 970627, 1997

8. Arcoumanis C. "Research issues in passenger car engines", The 4th Int. Symp. On Diagnostics and Modeling of Combustion in Internal Combustion Engines, COMODIA 98, July 20-23, Kyoto, Japan, 1998

9. Lake T.H., Sapsford S.M., Stokes J. and Jackon N.S. "Simulation and development experience of a stratified charge gasoline direct-injection engine", SAE Paper 962014, 1996

10. Jackson N.S., Stokes J., Whitaker P.A. and Lake T.H. "A direct injection stratified charge gasoline combustion system for future european passenger cars", IMechE Intl. Seminar on Lean Burn Engines, 3-4 December, 1996

11. Wirth M., Piock W.F. and Fraidl G.K. "Actual trends and future strategies for gasoline direct injection", IMechE Intl. Seminar on Lean Burn Engines, 3-4 December, 1996

12. Wirth M., Piock W.F., Fraidl G.K., Schoeggl P. and Winklhofer E. "Gasoline DI engines : The complete system approach by interaction of advanced development tools", SAE Paper 980492, 1998

13. Whitaker P.A., Stokes J. and Lake T.H. "Comparison of top-entry and side-entry direct injection gasoline combustion systems", Paper presented at IMechE Intl. Conference on Combustion Engines and Hybrid Vehicles, 28-30 April, 1998

14. Fraidl G., Piock W.F. and Wirth M. "The potential of direct injection gasoline engine", Paper presented at the 18th Intl. Vienna Motor Symposium, 24-25 April, 1997

15. Lai M-C., Zhao F-Q., Amer A.A. and Chue T-H. "An experimental and analytical investigation of the spray structure from automotive port injectors", SAE Paper 941873, 1994

16. Zhao F-Q, Lai M-C and Harrington D.L. "The spray characteristics of automotive port fuel injection - A critical review", SAE Paper 950506, 1995

17. Meyer J., Kiefer K., von Issendorff F., Thiemann J., Haug M., Schreiber M and Klein R. "Spray visualization of air-assisted fuel injection nozzles for direct injection SI engines", SAE Paper 970623, 1997

18. Miok J., Huh K.Y., Noh S.H. and Choi K.H. "Numerical prediction of charge distribution in a lean burn direct-injection spark ignition engine", SAE Paper 970626, 1997

19. Parrish S.E. and Farrell P.V. "Transient spray characteristics of a direct-injection spark-ignited fuel injector", SAE Paper 970629, 1997

20. Hoffman J.A., Eberhardt E. and Martin J.K. "Comparison between air-assisted and single-fluid pressure atomizers for direct-injection SI engines via spatial and temporal mass flux measurements", SAE Paper 970630, 1997

21. Han Z., Reitz R.D., Yang J. and Anderson R.W. "Effects of injection timing on air-fuel mixing in a direct-injection spark-ignition engine", SAE Paper 970625, 1997

22. Stanglmaier R.H., Hall M.J. and Matthews R.D. "Fuel-spray/charge-motion interaction within the cylinder of a direct-injected 4-valve SI engine", SAE 980155, 1998

23. Shin Y., Cheng W.K. and Heywood J.B. "Liquid gasoline behavior in the engine cylinder of a SI engine", SAE Paper 941872, 1994

24. Papageorgakis G. and Assanis D.N. "Optimizing gaseous fuel-air mixing in direct injection engines using an RNG based k-ε model", SAE Paper 980135, 1998

25. Shelby M., VanDerWege B.A. and Hochgreb S. "Early spray development in gasoline direct-injected spark ignition engines", SAE Paper 980160, 1998

26. Kano M., Saito K., Basaki M., Matsushita S. and Gohno T. "Analysis of mixture formation of direct injection gasoline engine", SAE Paper 980157, 1998

27. Han Z., Fan Li and Reitz D. "Multidimensional modelling of spray atomization and air-fuel mixing in a direct-injection spark-ignition engine", SAE Paper 970884, 1997

28. Yamauchi T. and Wakisaka T. "Computation of the hollow-cone sprays from a high pressure injector for a gasoline direct-injection SI engine" SAE Paper 962016, 1996

29. O'Rourke P.J. and Amsden A.A. "The TAB method for numerical calculation of spray drop break-up", SAE Paper 872089, 1987

30. O'Rourke P.J. "Collective drop effects in vaporizing liquid sprays", Ph.D. Thesis, Prinston University, 1981

31. O'Rourke P.J. "Statistical properties and numerical implementation of a model for turbulent dispersion in a turbulent gas", J. Comp. Physics, vol.83, pp. 345-360, 1989

32. Amsden A.A., O'Rourke P.J. and Butler T.D. "KIVA II - A computer program for chemically reactive flows with sprays", Los Alamos Labs, report LS 11560 MS, 1989

33. Lee C. and Bracco F.V. "Initial comparison of computed and measured hollow-cone sprays in an engine", SAE Paper 940398, 1994

34. Lee C. and Bracco F.V. "Comparisons of computed and measured hollow-cone sprays in an engine", SAE Paper 950284, 1995

35. Theodorakakos A. 'Numerical simulation of the flow in IC engines', Ph.D. Thesis, National Technical University of Athens, 1997

36. Gavaises M. 'Modeling of diesel fuel injection processes', Ph.D. Thesis, Imperial College, University of London, 1997

37. Ubbink O. "Numerical prediction of two fluid systems with sharp interfaces", Ph.D. Thesis, Imperial College, University of London, 1997

38. Glekas J.P. and Bergeles G. "A numerical method for recirculating flows on generalised coordinates: Application to environmental flows", Appl. Math. Modelling, vol. 17, p. 605, 1993

39. Demerdzic I. And Peric M. "Finite volume method for prediction of fluid flow in arbitrary shaped domains with moving boundaries", Int. J. Numer. Meth. Fluids, vol. 10, p. 771, 1990

40. Gavaises M. and Theodorakakos A. "Development of a 3-D flow model for prediction of two-phase flows with sharp interfaces", in preparation

41. Preussner C., Doring C., Fehler S. and Kampmann S."GDI: Interaction between mixture preparation, combustion system and injector performance", SAE Paper 980498, 1998

42. Xu M. and Markle L.E. "CFD-aided development of spray for an outwardly opening direct injection gasoline injector", SAE Paper 980493, 1998

43. Dumouchel C., Bloor M.I.G., Dombrowski N., Ingham D.B. and Ledoux M. "Viscous flow in a swirl atomizer", Chemical Eng. Sci., Vol. 48, No. 1, pp.81-87, 1993

44. Ren W.M., Shen J. and Nally Jr. "Geometrical effects, flow characteristics of a gasoline high pressure swirl injector", SAE Paper 971641, 1997

45. Chinn and Yule A. "Computational analysis of swirl atomizer internal flow', ICLASS-97, Seoul, pp. 868-875, 1997

46. Cousin J., Ren W.M. and Nally S. "Transient flows in high pressure swirl injectors", SAE Paper 980499, 1998

47. Ren W.M. and Nally J.F. "Computation of hollow-cone sprays from a pressure-swirl atomizer", SAE Paper 980499, 1998

48. Jeng S.M., Jog M.A. and Benjamin M.A. "Computational and experimental study of liquid sheet emanating from simplex fuel nozzle", AIAA JOURNAL, vol. 36, no. 2, February 1998

49. Panchagnula M.V., Sojka P.E. and Santagelo P.J. "On the three-dimensional instability of a swirling, annular, inviscid liquid sheet subject to unequal gas velocities", Phys. Fluids 8 (12), 1996

50. Ponstein J. "Instability of rotating cylindrical jets", Appl. Sci. Res. A 8, pp. 425-456, 1959

51. Crapper G.D., Dombrowski N. and Pyott G.A.D. "Kelvin-Helmoltz wave growth on cylindrical sheets", J. Fluid Mech., vol. 68, part 3, pp. 497-502, 1975

52. Hopfinger E.J. and Lasheras J.C. "Explosive breakup of a liquid jet by a swirling coaxial gas jet", Phys. Fluids 8 (7), July 1996

53. Dombrowski N. and Foumeny E.A. "On the stability of liquid shets in hot atmospheres", Atomization and Sprays, vol.8, pp. 235-240, 1998

54. Jeandel X. and Dumouchel C."Influence of the viscocity on the linear stability of an annular liquid sheet", ILASS-Europe, Manchester UK, 1998

55. Abo-Serie E., Arcoumanis C. and Gavaises M. "Modeling of atomization processes in swirl-pressure atomized sprays", in preparation

56. Boyaval S. and Dumouchel C. The maximum entropy formalism and the determination of spray drop size distribution", ILASS-Europe, Manchester UK, 1998

57. Van Der Geld C.W.M. and Vermeer H. "Prediction of drop size distribution in sprays using the maximum entropy formalism: formation of satelite droplets", Int. J. Multiphase Flow, Vol.20, No 2, pp. 363-381, 1994

1999-01-0498

Spray Formation of High Pressure Swirl Gasoline Injectors Investigated by Two-Dimensional Mie and LIEF Techniques

Wolfgang Ipp, Volker Wagner, Hanno Krämer, Michael Wensing and Alfred Leipertz
Lehrstuhl für Technische Thermodynamik, Universität Erlangen-Nürnberg, Erlangen, Germany

Stefan Arndt
Robert Bosch GmbH, Stuttgart, Germany

Amar K. Jain
Indian Institute of Petroleum, Dehradun, India

ABSTRACT

Two-dimensional Mie and LIEF techniques were applied to investigate the spray formation of a high pressure gasoline swirl injector in a constant volume chamber. The results obtained provide information on the propagation of liquid fuel and fuel vapor for different fuel pressures and ambient conditions. Spray parameters like tip penetration, cone angles and two new defined parameters describing the radial fuel distribution were used to quantify the fuel distributions measured. Simultaneous detection of liquid and vapor fuel was applied to study the influence of ambient temperature, injector temperature and ambient pressure on the evaporating spray.

INTRODUCTION

In the last years a variety of direct injection strategies for SI engines has been presented and some gasoline direct injection (GDI) engines are already available on the market [1-8]. Nearly all of the strategies and the GDI engines presented so far, use high pressure swirl injectors. High pressure swirl injectors offer the advantage to combine a well atomized spray with relatively low penetration at moderate injection pressure levels [9]. But, at the same time the spray formation of this type of injectors is to a high extent dependent on the ambient conditions, injection parameters and details of the injectors design [9-12]. Most direct injection strategies require a wide operating range of the injectors from low ambient temperatures and pressures in case of early injection to high temperatures and pressures at late injection, different rail pressure levels and even secondary injection. All this has contributed to an increasing demand of measurement techniques for the investigation of GDI sprays. Especially optical techniques offer the required temporal and spatial resolution and are non-intrusive as well. Consequently, a large number of the publications on gasoline direct injection include optical spray investigations: The optical measurement techniques commonly used are integral and two-dimensional Mie scattering techniques [13,14], Laser and Phase Doppler Analysis (LDV/PDA) [15, 16] and laser-induced (exciplex) fluorescence [17-19]. Examples of fundamental spray investigations of high pressure swirl injection can be found in [9,11,12,20,21], measurements inside injection chambers and optical accessible engines in [1-3,5,7,8,21-27].

This work presents two-dimensional Mie scattering and LIEF techniques optimized for the investigation of spray structures and spray formation (liquid and vapor phase) of high pressure swirl injectors. The results obtained show the spray formation and spray evaporation at different ambient conditions providing very high resolution in space, time and measured intensity. The two-dimensional Mie measurements give information on the distribution of liquid fuel in an axial and in a radial cut of the sprays. Image processing routines were developed to automatically extract spray parameters out of the Mie images. By means of LIEF measurements the liquid fuel phase and the fuel vapor were simultaneously acquired onto two separate images, so that the influence of ambient conditions on the fuel vapor phase and spray evaporation was visualized.

EXPERIMENTAL SETUP

In spite of the fact that laser sheet investigations of fuel sprays and mixture formation in combustion engines have become quite common in the last years, the experimental setup and details of the technique applied have to be carefully chosen for each spray type not only to get the best possible result, but also to ensure a realistic image of the spray process. Fig. 1 illustrates the experimental setup used for the investigation of the GDI sprays

The injector tested in a constant volume injection chamber is a typical swirl injector used for gasoline direct injection with nominal 90° cone angle and a static flow of 15 cm³/s. The injector was fitted to a high pressure/high temperature injection chamber. The chamber is electrically heated by means of a continuous air flow through the chamber. The velocity of the air flow is less than 0.1m/s inside the spray region. The maximum possible temperature is 800K, the maximum air pressure 5.0MPa. The injector is cooled by water. A constant pressure diaphragm system supplied by pressurized nitrogen was used to create the fuel pressure in order to avoid pressure fluctuations in the fuel rail.

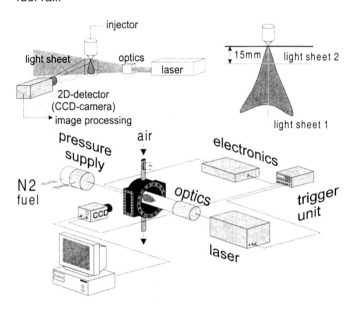

Fig. 1: Experimental setup

SETUP FOR MIE MEASUREMENTS

The laser beam of a frequency doubled Nd:YAG laser was formed to a thin light sheet of 80mm height and less than 200µm thickness (FWHM). Two different light sheet positions were used. First the light sheet illuminated one central plane of the spray including the spray axis. Secondly, a plane normal to the first one and in 15mm distance to the injector orifice was used. For the second position the laser beam was separated into two beams and the measurement plane was illuminated from two sides to minimize extinction effects in the dense spray. Due to the relatively large cone angle of 90°, it was not necessary to illuminate the spray from two sides, when the first light sheet position was used. In the present study light sheet 1 was introduced from the spray tip and directed towards the injector orifice. The images given e.g. in Fig. 3 demonstrate, that the attenuation of the laser light is small. However, for sprays with narrower cone angles as well as for light sheet 2, it is necessary to use illumination from two sides. The signal strength of

the Mie scattering process makes it possible to use non-intensified CCD cameras which offer a better dynamic range than intensified ones. The physical dynamic range of the detector is essential for spray investigations since strong signals from dense spray regions coexist in the same spray image with weak signals in other parts of the spray. The images were stored in a 12bit format (which should not be mixed up with the physical dynamic range) which is sufficient to represent spray structures at high signal levels in dense spray regions as well as structures at low signal levels in thinner parts of the spray. An extension tube was put in between camera body and lens to adjust the image magnification and reduce the focal depth. Additionally, an interference filter was used to reduce the influence of daylight. The exposure time of the camera was set to be 0.1µs while the time resolution of the measurement is defined by the laser pulse duration of about 10ns. A dimension of 0.05mm was imaged on each pixel due to the size of the measurement plane of 60mm x 50mm and the 1280 x 1024 camera pixels.

SETUP FOR LIEF MEASUREMENTS AND CHOICE OF SEED

LIEF measurements are a challenging task in dense sprays like GDI sprays with the aim to distinguish between liquid fuel and fuel vapor. For a couple of reasons no multi-component fuels like customized or standard gasoline can be used for such investigations. Major reasons are the fluorescence spectra of the liquid fuel phase and that the vapor phase of a multi-component fuel may change during the evaporation process, each component is affected by quenching in a different way, and absorbtion of the irradiating laser occurs in the liquid and in the vapor phase. Additionally, it is nearly impossible to correct for absorption in a completely instantaneous spray and (since the fluorescence spectra from fuel vapor and the by far stronger signal from the liquid fuel at least overlap) it is very hard to distinguish between the two phases. Therefore, a non-fluorescing base fuel was doped with a suitable combination of seeds. The choice of seeds is an important point for the technique. For spray investigations the physical properties of the seeds have to be close to that of the base fuel since only the seeds are detected and both, base fuel and seeds should represent a mid-range component of gasoline. Secondly, the fluorescence of the seeds has to be sufficiently characterized even for qualitative investigations in order to make sure that a strong signal always indicates a comparatively high fuel concentration. Thirdly, since fluorescence is a mass proportional signal the fluorescence of the liquid phase is by far stronger than the signal from the fuel vapor due to the higher densities and therefore, a wide spectral shift between the fluorescence of the two phases is required or at lest very helpful. We have chosen the exciplex seed system of benzene and triethylamine found and characterized by Fröba et. al. [18] for the non-fluorescing base fuel isooctane. The fluorescence spectra of the seed combination that forms an exciplex with a red shifted fluorescence spectrum in the liquid phase is shown in Fig. 2.

The experimental setup used in the LIEF measurements is similar to that one shown in Fig. 1 while only the first light sheet position (axial) was investigated. A KrF-excimer laser at 248nm and two intensified CCD cameras (752 x 580 Pixel, images stored in 8 bit format) were used. The filters applied to separate liquid fuel and fuel vapor are 307±25nm (vapor phase) and 360-400nm (liquid).

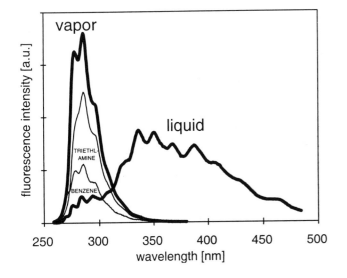

Fig. 2: Fluorescence spectra of the mixture benzene-TEA [18]

RESULTS – MIE LIGHT SHEET MEASUREMENTS

The first series of spray images in Fig. 3 represent the spray formation process at atmospheric conditions. The different parts of the spray (leading edge, trailing edge, body, vortex region) described by Evers [21] can clearly be identified in the image series. The swirl injector produces a hollow cone spray structure, which is preceded by a „pre-jet". The acquisition times given in the figure represent the time delay to the electrical opening signal of the injector and not to actual injection start. This start of injection is 0.34ms delayed to the electrical signal. After start of injection first a pre-jet is formed and afterwards the main spray cone builds up (0.55ms). The pre-jet structure is compact near injection start at 0.55ms and breaks up at 0.75ms while the intensity distribution indicates relatively large drops in the pre-jet. The main spray body forms a hollow cone (0.75ms). At 1.15ms a vortex is visible resulting from the interaction between the injected fuel and the quiescent air. The vortex grows by moving downstream. It is clearly visible that liquid fuel is transported upstream again into the spray region by the vortex. This is believed to enhance the transport of mainly small droplets into the spray center evoked by a lower pressure in the center of the spray cone. A large number of small droplets in the spray center can simulade a filled cone structure in Mie measurements because the intensity of Mie scattered light is – depending on the polarization [28] – in a first approximation proportional to the liquid surface. Later in this paper a comparison between Mie and LIF images of the same sprays will be shown. However, in the image series shown, clearly a

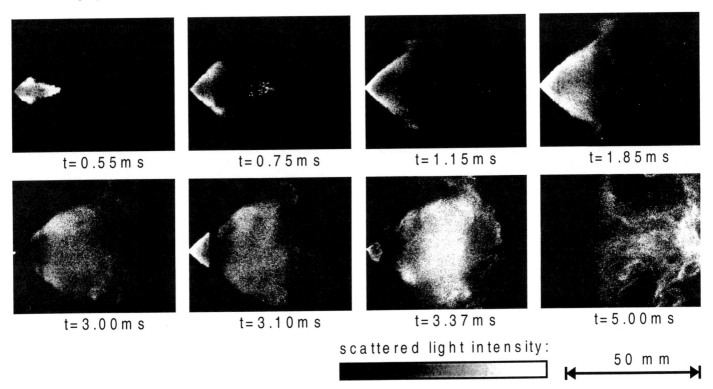

Fig. 3: Spray formation at atmospheric conditions (axial light sheet position; $p_{Chamber}$=0.1MPa; $T_{Chamber}$=293K; p_{Rail}=10MPa; $t_{injection}$=2.5ms)

189

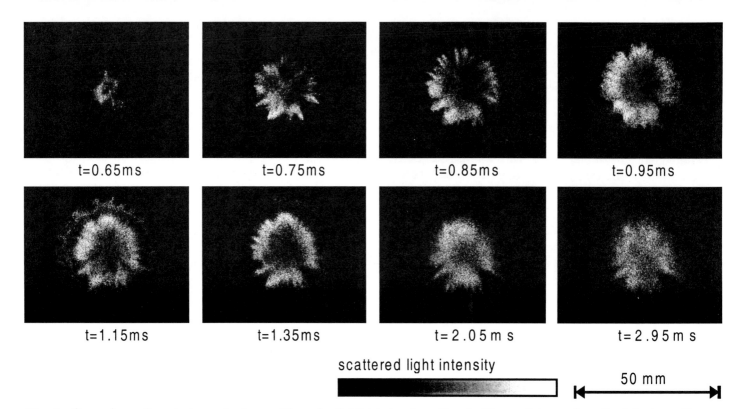

t=0.65ms	t=0.75ms	t=0.85ms	t=0.95ms
t=1.15ms	t=1.35ms	t=2.05ms	t=2.95ms

scattered light intensity

50 mm

Fig. 4: Spray formation at atmospheric conditions (radial light sheet position in 15mm distance from the injector orifice; $p_{Chamber}=0.1MPa$; $T_{Chamber}=293K$; $p_{Rail}=10MPa$; $t_{injection}=2.5ms$)

hollow cone structure can be recognized by the Mie measurements.

After the end of the main injection three bouncers of the needle were detected and two are shown in the figure. The first of these needle bouncers creates a second hollow cone spray. At about 5.0ms the injection generated flow in the chamber is visible in the liquid fuel distribution. The highest droplet concentration is at that time found at the spray tip. Please note again, that the intensity distribution measured by means of the Mie-technique in a first approximation corresponds to the distribution of the total surface area and not to the mass distribution.

The images plotted in Fig. 4 represent the evolution of the liquid fuel distribution in the second light sheet plane, which is perpendicular to the propagation direction. Angular irregularities are visible in the spray patterns and these irregularities explain that, depending on the position (orientation), different results could be obtained when the first light sheet position is used. At 0.65ms (Fig. 4) parts of the pre-jet can be observed followed by the hollow cone of the main spray body. The contour of the hollow cone is highly irregular (0.75ms and 0.85ms) with steep gradients. When the vortex starts to grow up the contour gets blurred, and afterwards a second ring is visible covering the hollow cone spray. This ring grows while the diameter of the hollow cone stays approximately constant. After injection end the spray pattern shows a nearly regularly filled cone. Please note that the position of the angular irregularities stays constant although the images shown were acquired in

different injection events. Later in this paper we will present a definition for two new spray parameters that describe the measured angular and radial fuel distributions.

AMBIENT PRESSURE The influence of the ambient pressure on the spray formation is given in Fig. 5. With increasing pressure the spray becomes increasingly compact. All images have been taken at 1.15ms. At very high pressures pre-jet and main spray body do not separate. For pressure levels of and above 0.65MPa (not included in the figure) vortex formations were found even in the pre-jet region. The lower signal intensities near the injector orifice occur due to extinction of the laser light. But the dynamic range of the measurement is sufficient to represent the fuel distribution in that part of the spray, too. Note that the rail-pressure and not the pressure difference was kept constant in this variation. The influence of the ambient pressure is quantified by the spray parameters plotted in the next section.

AMBIENT TEMPERATURE The influence of the ambient temperature on the liquid fuel distribution is given in Fig. 6 in the range from 293K to 523K. Due to the decrease in ambient density the spray tip penetration increases up to temperatures of about 373K. For a further increase the reduced density is compensated by an enhanced evaporation so that a constant tip penetration was measured. At times later in the injection process an increase of the tip penetration was found in the LIEF investigations up to about 450K. The spray cone angle stays approximately constant in this variation. The intensity distribution inside the main spray body is only

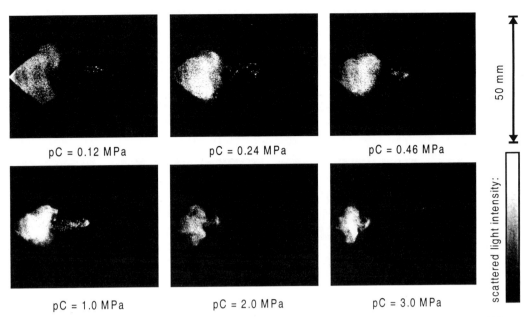

pC = 0.12 MPa   pC = 0.24 MPa   pC = 0.46 MPa

pC = 1.0 MPa   pC = 2.0 MPa   pC = 3.0 MPa

Fig. 5: Spray formation under ambient pressure variation (axial light sheet position; $T_{Chamber}=393K$; $p_{Rail}=10MPa$; $t_{injection}=1.0ms$; $t_{acquisition}=1.15ms$)

slightly changed while the significant changes of the fuel distribution in the pre-jet region are cyclic fluctuations. The effect of the ambient temperature on the fuel vapor phase is visible in the LIEF results.

RAIL PRESSURE A change of the fuel pressure in the rail significantly affects the spray formation. The effect of the variation of the rail pressure is shown for an elevated ambient pressure of 0.46MPa in Fig. 7. For high rail pressure a regular filled cone structure of the spray is found which changes to a hollow cone structure with decreasing fuel pressure. It is obvious in the measurements, that the spray is well atomized at high rail pressures while the atomization is significantly reduced at the lower pressure levels.

RESULTS – SPRAY PARAMETERS EXTRACTED FROM THE MIE LIGHT SHEET MEASUREMENTS

For a more quantitative description of the spray formation and also for a reduction of the enormous amount of data the spray parameters tip penetration, cone angle and two newly defined parameters describing the radial fuel distribution (e.g. that ones given in Fig. 4) were extracted out of the acquired spray images. Every single shot was evaluated while each measurement was repeated 15 times. Consequently, mean values and standard deviation from 15 measurements are given. 15 repetitions is a very small sample for estimating mean values and even less for quantifying fluctuations. On the other hand it was not possible for us to increase the number of repetitions in this investigation since the total amount of data already exceeded 60GB using the high resolution in space and intensity, which we wanted to keep. Therefore, we ask the reader to keep the small sample size in mind. The spray parameters were evaluated automatically using image processing routines. The spray contour is automatically found by means of a threshold in the background corrected images. The tip penetration is separately evaluated for pre-jet and main spray. The cone angle is derived from the measurements using the

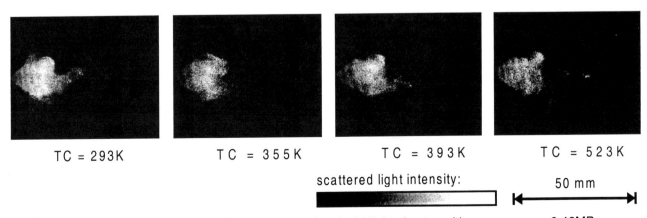

TC = 293K   TC = 355K   TC = 393K   TC = 523K

scattered light intensity: 50 mm

Fig. 6: Spray formation under ambient temperature variation (axial light sheet position; $p_{Chamber}=0.46MPa$; $p_{Rail}=10MPa$; $t_{injection}=1.0ms$; $t_{acquisition}=1.15ms$)

191

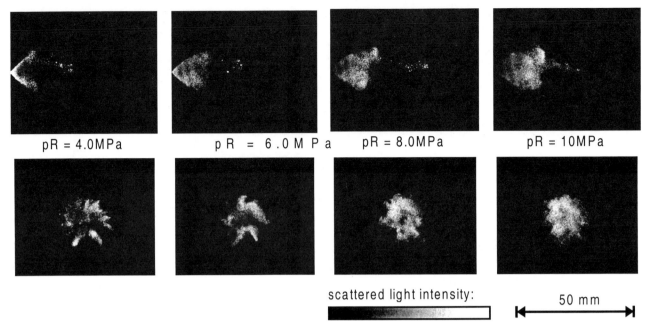

pR = 4.0MPa   pR = 6.0MPa   pR = 8.0MPa   pR = 10MPa

scattered light intensity:

50 mm

Fig. 7: Spray formation for rail pressure variations (radial light sheet position; $p_{Chamber}$=0.46MPa; $T_{Chamber}$=393K; $t_{injection}$=1.0ms; $t_{acquisition}$=1.15ms)

axial light sheet position including 20% of the main spray from the orifice to 20% of the main body penetration in the determination. Two regressions are fitted on this part of the spray and the cone angle is defined as the angle between the two regressions on the spray contour.

Fig. 8 gives the time history of tip penetration and cone angle evaluated from the spray images in Fig. 3. The spray tip penetration of the main spray can well be represented by a square-root fit. This tip penetration was evaluated separately for the upper and lower half of the spray. The values measured for the two half's are nearly the same and only one was plotted. The pre-jet shows a faster and larger penetration with significantly higher cyclic fluctuations than the main spray. The spray cone angle shows a decrease vs. time which is also known from PFI and Diesel injection processes. For the swirl injector tested here, the decrease is superimposed by fluctuations which can be interpreted to be caused by pressure fluctuations inside the injector (although a constant pressure supply was used).

$t_{injection}$=2.5ms)

The influence of ambient pressure and of ambient temperature on the spray parameters is given in Fig. 9 and Fig. 10, respectively. An increase of the ambient pressure naturally reduces tip penetration and cone angle. The measured spray parameters are only slightly effected by the ambient temperature. The tip penetration shows a very weak increase in the region up to ca. 373K and is constant afterwards. For the cone angle a significant change was only found for the highest temperature tested (523K).

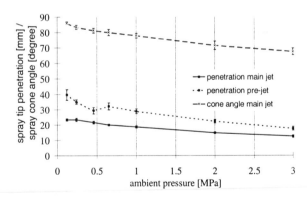

Fig. 9: Influence of ambient pressure on spray tip penetration and spray cone angle (images in Fig. 5; $T_{Chamber}$=393K; p_{Rail}=10MPa; $t_{injection}$=1.0ms; $t_{acquisition}$=1.15ms)

The rail pressure variation that was performed for an ambient pressure of 0.46MPa has not only a strong effect on the spray pattern (Fig. 7), but also changes the spray parameters extracted (Fig. 11). As it could be expected the tip penetration increases with the fuel injection pressure, while the cone angle is reduced with increasing rail-pressure.

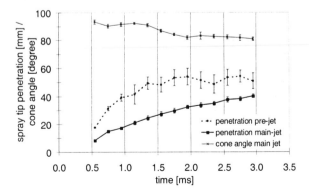

Fig. 8: Time history of spray tip penetration and spray cone angle (images in Fig. 3; axial light sheet position; $p_{Chamber}$=0.1MPa; $T_{Chamber}$=293K; p_{Rail}=10MPa;

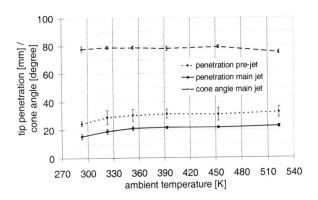

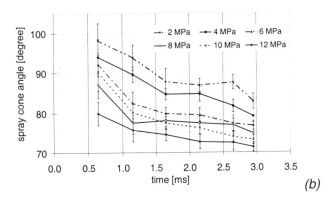

Fig. 10: Influence of ambient temperature on spray tip penetration and spray cone angle (images in Fig. 6; $p_{Chamber}$=0.46MPa; p_{Rail}=10MPa; $t_{injection}$=1.0ms; $t_{acquisition}$=1.15ms)

Fig. 11: Influence of rail pressure on (a) spray tip penetration and (b) spray cone angle (images in Fig. 7; $p_{Chamber}$=0.46MPa; $T_{Chamber}$=393K; $t_{injection}$=1.0ms; $t_{acquisition}$=1.15ms)

PARAMETERS DESCRIBING THE RADIAL FUEL DISTRIBUTION

The images given in Fig. 4 can very well be used to get a quick impression on the spray structure, but it is difficult to quantify differences in the spray pattern from these images. Therefore, it is difficult to compare different injectors. We have tried to represent the distributions in polar coordinates by two curves, the function R giving the fuel distribution versus the distance to the spray axis and the function Φ which gives the angular distribution of the spray. These functions are given by:

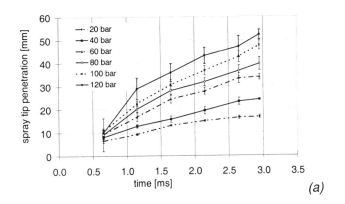

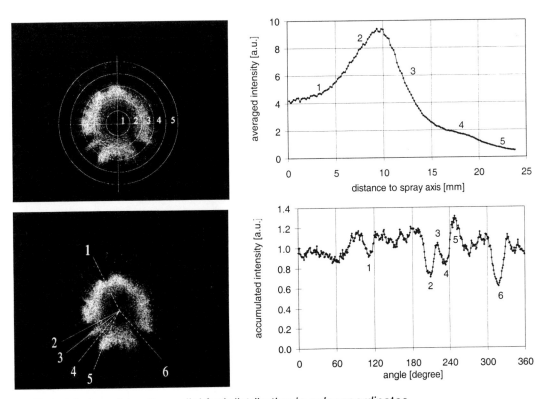

Fig.12: Functions R and Φ describing the radial fuel distribution in polar coordinates

193

$$R(r) = \frac{1}{360} \sum_{\varphi=1°}^{360} F(\varphi, r) = \overline{F}(r) \quad \text{and}$$

$$\Phi(\varphi) = \sum_{r=1}^{\infty} F(\varphi, r) = sum(F(\varphi))$$

F: intensity of the measured light signal

Fig. 12 illustrates these functions for a typical hollow cone spray. The function R represents the average intensity measured on a ring with radius r. The function Φ gives the accumulated intensity acquired in each direction φ. The functions are given in arbitrary units and depend on the experimental setup and the laser intensity used. Therefore, comparisons are valid only for constant measurement conditions.

Fig. 13 represents the temporal development of the radial fuel distribution. The curves show the building up of the hollow cone structure and the vortex formation. Slight differences in the distributions can easier be identified in the curves than in the images in Fig. 4. Fig. 14 exhibits the radial fuel distribution for the variation of the ambient pressure. The curves illustrate the change from the hollow cone spray with the visible vortex ring to a filled cone structure at larger pressures. In Fig. 15 and Fig. 16 the angular fuel distribution is given for five identical measurements and in its' temporal evolution, respectively. Strong oscillations of the curves can be seen in the plot. Both diagrams indicate by the uniformity of the angular structures measured, that the angular fuel distribution is a characteristic of the (probably individual) injector. The structures in the angular distribution are a consequence of the injector`s internal geometry and surface roughness. For the high pressure swirl injector investigated here, the angular fuel distribution is (in spite of the skeinness detected) more uniform than e.g., distributions produced by pintle type injectors used in port fuel injection SI engines [14].

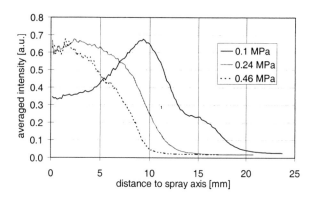

Fig.14: Change of the radial distribution function R for a ambient pressure variation (images in Fig. 7; p_{rail}=10MPa; $t_{injection}$=1.0ms; $t_{acquisition}$=1.15ms)

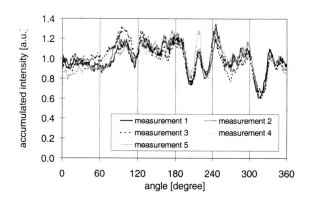

Fig.15: Angular fuel distribution ϕ ($p_{Chamber}$=0.1MPa; $T_{Chamber}$=293K; p_{Rail}=10MPa; $t_{injection}$=2.5ms; $t_{acquisition}$=1.15ms)

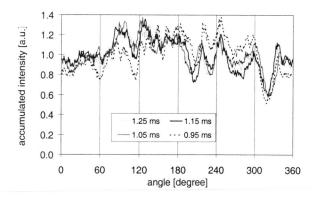

Fig.16: Temporal development of the angular distribution ϕ ($p_{Chamber}$=0.1MPa; $T_{Chamber}$=293K; p_{Rail}=10MPa; $t_{injection}$=2.5ms)

SIMULTANEOUS AND SEPARATE IMAGING OF THE LIQUID AND THE VAPOR FUEL PHASE

The LIEF technique explained above was used to simultaneously acquire the liquid and the vapor fuel phase onto two separate images for three different conditions. At first, a time history of the evaporation

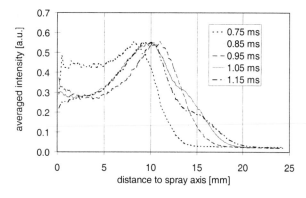

Fig.13: Temporal development of the radial distribution function R (images in Fig. 4; radial light sheet position in 15mm distance to the injector orifice; $p_{Chamber}$=0.1MPa; $T_{Chamber}$=293K; p_{Rail}=10MPa; $t_{injection}$=2.5ms)

194

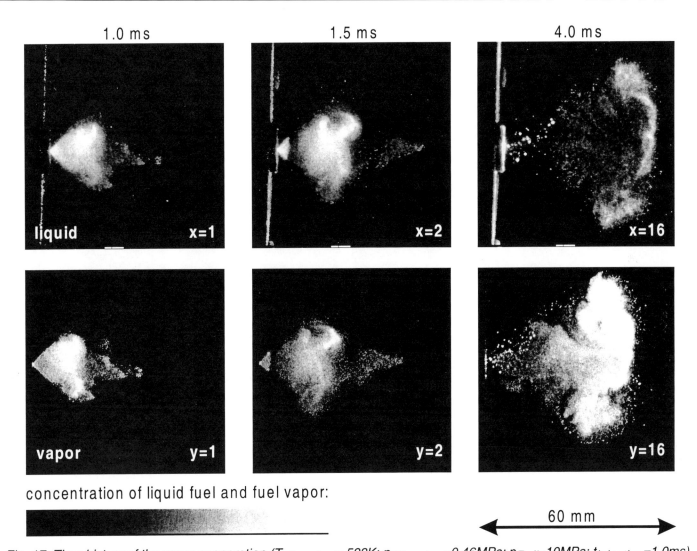

1.0 ms	1.5 ms	4.0 ms
liquid ... x=1	x=2	x=16
vapor ... y=1	y=2	y=16

concentration of liquid fuel and fuel vapor:

60 mm

Fig. 17: Time history of the spray evaporation ($T_{Chamber}$= 523K; $p_{Chamber}$=0.46MPa; p_{Rail}=10MPa; $t_{injection}$=1.0ms)

process is shown for relatively high temperature (Fig. 17). Secondly, a variation of the ambient temperature has been used comparing the resulting fuel distribution after end of the injection process (Fig. 18). In the last image series a comparison between measurements performed with and without injector cooling has been performed (Fig.19).

It can be seen in the images that the separation of the two phases, liquid and vapor phase, is not complete. The influence of the vapor phase on the image of the liquid phase can be neglected, but the vapor phase image is superimposed by the signal from the liquid phase. This can well be seen in the second pair of the images in Fig. 17. The fuel injected because of the needle bounce can be assumed to consist nearly completely of liquid fuel in this case. But this fuel is, although strongly reduced in its' signal intensity, also visible on the vapor phase image.

However, strong signals in the vapor phase image are to a significant part caused by fuel vapor. Additionally, weak signals in parts of the spray with weak liquid phase signals (like the pre-jet region in the same image) are also caused from the fuel vapor phase. Therefore, it can be concluded that the vapor phase image is interfered

by the liquid phase signal, but with respect to the degree of that interference, the vapor phase images can well be interpreted, when the superimposition is kept in mind. The factors x and y given in the images are magnification factors of the intensity scaling. This means the plotted intensity in every series of images was raised from the first image to the last image of the series by the given factors and with respect to the first image of the series while the image of the liquid series and the vapor series were amplified in the same way.

In the time history plotted in Fig. 17 it can be seen that the fuel vapor follows the path of the evaporating liquid fuel. Both phases are found at the leading edge of the spray. At 1ms vaporized fuel was already detected. From the raise of the vapor signal, compared to the nearly constant signal level of the liquid phase, it can be derived that evaporation starts in a distance of approximately 10mm to the orifice. But further investigations are necessary to get a better separation between the two phases to verify this result. Even 4ms after the electrical start of the injection no clear separation in the flow path of the two phases was found. Due to the continuous evaporation process, the liquid signal decreases more rapidly than the vapor phase.

195

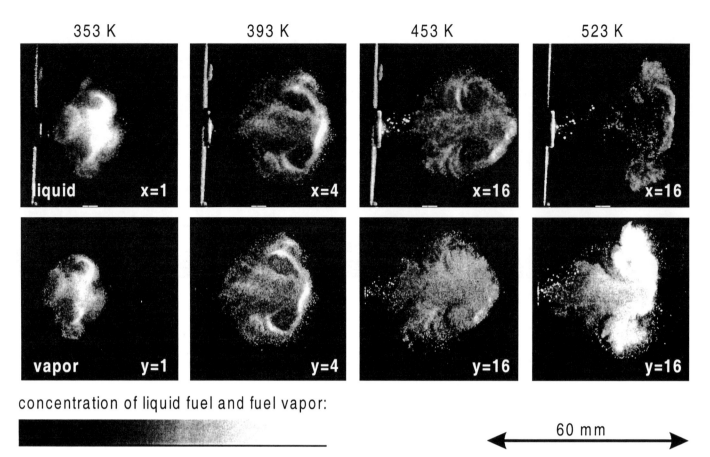

concentration of liquid fuel and fuel vapor:

60 mm

Fig. 18: Spray evaporation for different ambient temperatures ($p_{Chamber}$=0.46MPa; p_{Rail}=10MPa; $t_{injection}$=1.0ms; $t_{acquisition}$=4.0ms)

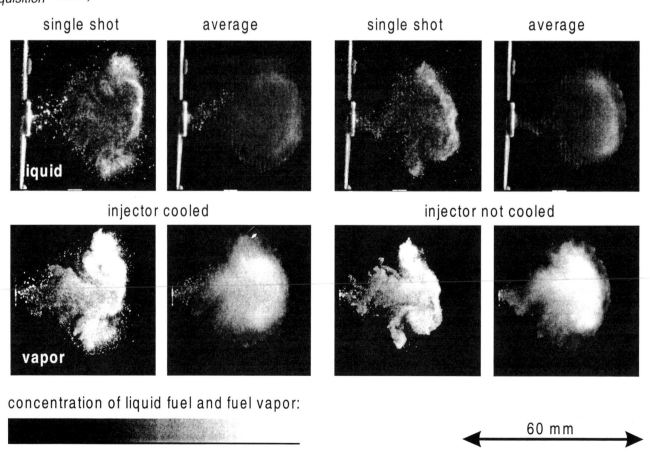

concentration of liquid fuel and fuel vapor:

60 mm

Fig. 19: Effect of injector body and fuel temperature ($T_{Chamber}$= 523K; $p_{Chamber}$=0.46MPa; p_{Rail}=10MPa; $t_{injection}$=1.0ms; $t_{acquisition}$=4.0ms

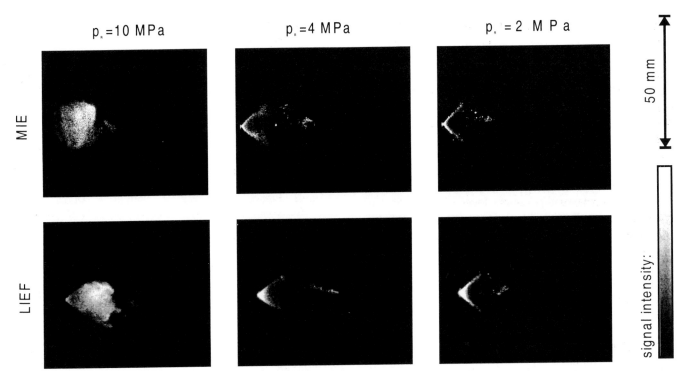

Fig. 20: Comparison of Mie and LIEF images of the liquid fuel phase $T_{Chamber}= 453K$; $p_{Chamber}=0.46MPa$; $t_{injection}=1.0ms$; $t_{acquisition}=1.0ms$

Fig. 18 shows that the temperature enhancement in the range from 353K to 523K enhances the spray evaporation while spray structure is changed mainly by the varied ambient density.

Fig. 19 shows the influence of the an elevated injector body and fuel temperature on spray formation and spray evaporation by switching of the injector cooling. For this test the water cooling of the injector which was used in all other tests was switched off. The effects measured can be compared to the flash boiling described by Hochgreb et. al. [29]. Hochgreb et. al. found strong changes in the spray formation of high pressure swirl injectors when the fuel was superheated relative to the boiling point about 20K and injected into subatmospheric pressure. Due to the higher ambient pressure in our test no dramatic change of the spray evaporation was found and due to the already compact spray structure caused by the elevated air pressure, no strong change in the spray structure is detected. But, both the liquid fuel and the fuel vapor show a slightly more compact structure. This can reduce the fuel air mixing, and although the evaporation process is enhanced by the fhigher fuel temperature the overall mixture formation can appear to be slower. On the other hand much less single large drops are detected, when the injector cooling is switched off.

COMPARISON OF MIE AND LIEF MEASUREMENTS OF THE LIQUID FUEL DISTRIBUTION

As already mentioned above it is not easy to decide which of the two experimental techniques applied is best suited to visualize the formation of the liquid fuel phase in a GDI spray process. The LIEF technique offers the advantage to have a mass proportional signal. On the other hand for Mie scattering experiments it is possible to use non-intensified cameras with a significant advantage in the dynamic range. A comparison between images of the liquid fuel phase acquired using the LIEF and the Mie technique is given in Fig. 20.

CONCLUSIONS

Two-dimensional Mie and LIEF techniques were used to investigate the spray formation (liquid and vapor phase) of a high pressure swirl gasoline injector in a constant volume injection chamber.

- The formation of the liquid fuel phase was visualized in two measurement planes by means of a Mie technique with high spatial and temporal resolution. The results show a strong influence of the ambient conditions and of the fuel rail pressure on the spray formation. The influence of ambient pressure, temperature and rail pressure was quantified in spray parameters describing the axial (cone angle, tip penetration) and radial (radial and angular distribution curves) fuel distribution (Mie-intensity distribution). The measurements using the radial light sheet position visualize the fuel distribution in the spray in a better way than the distribution measured in the axial plane. Large angular irregularities were found in the fuel distributions. These irregularities stayed constant in repetitions of the measurements and throughout the spray process. We conclude, that the irregularities in the fuel distribution are a peculiarity of the (individual) injector. Further investigation will have to show to what extend these irregularities vary for different injectors especially of the same type.

- The LIEF technique was used to simultaneously acquire the liquid fuel phase and fuel vapor onto two separate images while the separation at present is not 100%, but sufficient for the evaluation. The results show that the fuel vapor phase follows the liquid fuel phase throughout the injection process and even more than 2ms after injection end. A higher injector body and fuel temperature causes for an elevated ambient pressure level of 0.46MPa a more compact distribution of both, liquid and vapor fuel, accompanied by a visible reduction of large drops.

- The comparison of the Mie and LIEF technique has shown that both techniques are well suitable to visualize the liquid fuel phase in GDI sprays.

ACKNOWLEDGMENTS

The authors gratefully acknowledge financial support for parts of the work by the German Federal Ministry of Education, Science, Research and Technology (BMBF) under conduct of the VDI-TZ Physical Technologies in the frame of project 13N7179.

REFERENCES

1. Y.Iwamoto, K. Noma, O. Nakayama, T. Yamauchi, H. Ando; Development of Gasoline Direct Injection Engine, SAE Paper 970541 (1997)
2. T. Kume, Y. Iwamoto, K. Lida, M. Murakami, K. Akishino, H. Ando; Combustion Control Strategies for Direct Injection SI Engine; SAE Paper 960600 (1996)
3. D. Andriesse, A. Ferrari, R. Imarisio, Experimental Investigation on Fuel Injection Systems for Gasoline Direct Injection Engines; Direkteinspritzung im Ottomotor, HAUS DER TECHNIK, Essen, March, 1997
4. R. W. Anderson, J. Yang, D.D. Brehob, J.K. Vallance, R.M. Whiteaker; Understanding the Thermodynamics of Direct Injection Sorak Ignition (DISI) Combustion Systems: An Analytical and Experimental Investigation, SAE Paper 962018 (1996)
5. G. K. Fraidl, W.F. Piok, M. Wirth, Gasoline Direct Injection: Trends and Future Strategies for Injection and Combustion Systems, SAE Paper 960465 (1996)
6. G. Karl, R. Kemmler, M. Bargende, J. Abthoff, Analysis of a Direct Injection Gasoline Engine, SAE Paper 970624 (1997)
7. J. Harada, T. Tomita, H. Mizuno, Z. Mashiki, Y Ito Development of Direct Injection Gasoline Engine, SAE Paper 970540 (1997)
8. M. Pontoppidan, G. Gaviani, G. Bella, V. Rocco, Direct Fuel Injection – A Study of Injector Requirements for Different Mixture Preparation Concepts, SAE Paper 970628 (1997)
9. A. H. Lefebvre, Atomization and Sprays, Hemisphere Publishing Corporation, 1989
10. F. Ruiz, N. Chigier, The Effects of Design and Operating Conditions of Fuel Injectors on Flow and Atomization, SAE Paper 870100 (1987)
11. H. Gebhardt, Zerstäubung mit Dralldüsen, Wiisenschaftliche Zeitschrift der Technischen Hochschule Dresden, 7 Heft 2, S. 249-273, 1957
12. S. Beer, O.J. Haidn, Sprayerzeugung mit Hohlkegeldüsen, Proc SPRAY ´94, 1994
13. K.-U. Münch, Zweidimensionale Mie-Streulichttechniken zur Spraydiagnose in Dieselmotoren, PhD Thesis, Universität Erlangen-Nürnberg, BEV Berichte zur Energie- und Verfahrenstechnik, Heft 93.1, ESYTEC GmbH, Erlangen, 1993
14. M. Wensing, H. Krämer, K.-U. Münch, A. Leipertz; Mixture Formation and Combustion of a Four-Valve SI Engine Investigated by Advanced Two-Dimensional Laser Measurement Techniques; Proc. COMODIA 98, pp. 379-385, 1998
15. G. Brenn, J. Domnick, V. Dorfner, F. Durst; Unsteady Gasoline Injection Experiments: Comparison of Measurements in Quiescent Air and in a Model Intake Port, SAE Paper 950512 (1995)
16. F. Vannobel, J.B. Dementhon, D. Robart,; Phase Doppler Anemometry Measurements on a Gasoline Spray inside the Inlet Port and Downstream of the Induction Valve; Steady Flow Conditions; in Laser Techniques and Applications in Fluid Mechanics, Proc. 6th Intern. Symp. M.V. Heitor, M. Maeda, J.H. Whitelaw, Springer Verlag, 1992
17. L.A. Melton, J.F. Verdiek; Vapor/Liquid Visualization in Fuel Sprays; 20th Symposium on Combustion, pp. 1283-1290, 1984
18. A. Fröba, F. Rabenstein, K.-U. Münch, A. Leipertz; Mixture of Triethyamine (TEA) and Benzene as a New Seeding Material for Quantitative Two-Dimensional Laser-Induced Exciplex Flourescence Imaging of Vaopr and Liquid Fuel Inside SI Engines, Combust. Flame 112, 1998, 199-209
19. H. Krämer, K.-U. Münch, A. Leipertz, Investigation of Fuel Evaporation Inside the Intake of a SI Engine Using Laser Induced Exciplex Fluorescence with a NEW Seed, SAE Paper 961930 (1996)
20. T. Yamauchi, T. Wakisaka, Computation of the Hollow-Cone Sprays from High-Pressure Swirl Injector for a Gasoline Direct Injection SI Engine, SAE Paper 962016 (1996)
21. L.W. Evers, Characterization of the Transient Spray from a High Pressure Swirl Injector, SAE Paper 940188, (1994)
22. S.E. Parrish, P.V. Farell, Transient Spray Characteristics of a Direct Injection Spark-Ignited Fuel Injector, SAE Paper 970629 (1997)
23. M. Ohsuga, T. Shiraishi, T. Nogi, Y. Nakayama, Y. Sukegawa, Mixture Preparation for Direct-Injection SI Engines, SAE Paper 970542 (1997)
24. Observation on In-Cylinder Sprays in a Firing Single-Cylinder Direct-Injection Gasoline Engine, Direkteinspritzung im Ottomotor, U. Spicher (Hrsg.), S. 117-131, Expert-Verlag, 1998
25. M.J. Hall, R:D: Matthews, In-Cylinder Flow and Fuel Transport in a 4-Valve GDI Engine: Diagnostics and Measurements, Direkteinspritzung im Ottomotor, U. Spicher (Hrsg.), S. 132-146, Expert-Verlag, 1998
26. A. Kakuhou, T. Urushihara, T. Itoh, Y. Takagi, LIF Visualization of In-Cylinder Mixture Formation in a Direct-Injection SI Engine, Proc. COMODIA 98, pp 305-310, 1998
27. T. Fujikawa, Y. Hattori, K. Akihama, M. Koike, T. Kobayashi, S. Matsushita, Quantitative 2-D Fuel Distribution Measurements in a Direct Injection Gasoline Engine Using Laser-Induced Fluorescence Technique, Proc. COMODIA 98, pp. 317-322, 1998
28. M. Wensing, T. Stach, K.-U. Münch, A. Leipertz, Two-Dimensional Droplet Size Distribution Measurements in Two-Phase Flows, Proc. PARTEC ´98, pp. 645-654, 1998
29. S. Hochgreb, B. A. Van der Wege, The Effect of Fuel Volatility on Early Spray Development from High-Pressure Swirl Injectors, Direkteinspritzung im Ottomotor, U. Spicher (Hrsg.), S. 107-116, Expert-Verlag, 1998

CFD-Aided Development of Spray for an Outwardly Opening Direct Injection Gasoline Injector

Min Xu and Lee E. Markle
Delphi Automotive Systems

ABSTRACT

A high pressure outwardly opening fuel injector has been developed to produce sprays that meet the stringent requirements of gasoline direct injection (DI) combustion systems. Predictions of spray characteristics have been made using KIVA-3 in conjunction with Star-CD injector flow modeling. After some modeling iterations, the nozzle design has been optimized for the required flow, injector performance, and spray characteristics. The hardware test results of flow and spray have confirmed the numerical modeling accuracy and the spray quality.

The spray's average Sauter mean diameter (SMD) is less than 15 microns at 30 mm distance from the nozzle. The DV90, defined as the drop diameter such that 90% of the total liquid volume is in drops of smaller diameter, is less than 40 microns. The maximum penetration is about 70 mm into air at atmospheric pressure. An initial spray slug is not created due to the absence of a sac volume.

INTRODUCTION

Improved fuel economy, reduced engine-out emissions, quick starting and fast transient response have been considered as major advantages of gasoline DI spark-ignition (SI) engines over port fuel injection (PFI) SI engines. Although DI's great theoretical potential has been recognized for a long time, the technology realization and implementation has been delayed due to many technical difficulties [1-3]. The current development has demonstrated some of the DI advantages, but it has also unveiled many problems and new challenges. In particular, excessive unburned hydrocarbon (UHC) emissions, injector fouling, low conversion efficiency and durability of the lean NOx catalysts remain as the major concerns [1]. New technology breakthroughs and in-depth understanding of the fundamentals are necessary for further improvement and development of gasoline DI engines.

In gasoline DI engines, fuel is directly injected into the cylinder, therefore the spray characteristics and mixture formation are of primary importance. Depending on the engine operating condition, the injection time window and the available cylinder volume for the injected fuel are often very constrained. Small droplet size, rapid evaporation, stable ignition, appropriate mixing (or charge stratification), and minimal wall impingement are generally required for DI engines [4].

For high engine load conditions, injection during the intake stroke (early injection) requires a large spray angle and sufficient penetration for maximum air utilization. However, excessive penetration aided by intake air flow and low cylinder pressure can result in significant cylinder wall and/or piston wetting [5]. A well atomized spray is required for homogeneous mixture formation prior to ignition.

For part load conditions, injection occurs during the compression stroke (late injection) to achieve adequate charge stratification for lean burn combustion, which is desired for high thermal efficiency. During the compression stroke, the cylinder pressure and temperature increase with crank angle. The charge turbulence intensity increases substantially near TDC due to the collapse of the large scale flow structures. This phenomenon tends to improve the fuel droplet evaporation and air-fuel mixing. However, the increased air drag causes rapid deceleration of fuel droplets and increases the probability of droplet coalescence and the formation of large droplets. The prompt atomization of a spray with a short intact liquid length must take place in such severe environments. The excessive UHC emissions experienced in many gasoline DI engines have been attributed to the flame quenching in the overly lean region and to wall wetting. To avoid piston impingement and to create an appropriate degree of spatial charge stratification for stable ignition and combustion, the spray should be compact with low penetration [6].

Consequently, to ensure a high combustion efficiency and low emissions levels, a well-atomized spray is always required over the entire engine operating range. An SMD of 15 microns or smaller with Rosin-Rammler N values greater than 1.9 (equivalent to DV90 smaller than 43 microns) was set as the target for the DI sprays [7]. DI combustion systems are very sensitive to spray quality. A few large droplets could significantly degrade the emissions and combustion efficiency. Therefore, a spray with a maximum droplet size larger than 50 microns cannot be accepted. For typical automotive engines, the spray penetration distance should be less than 80 mm in the early injection mode and less than 30 mm in the late injection stratified charge mode. Fortunately, the cylinder pressure variation as a function of the crank angle affects the penetration and angle of the spray favorably for the DI requirements.

In order to meet the severe DI requirements, significant effort has been placed on fuel injector development. Among many design concepts, the inwardly opening high pressure swirl type injector is predominant [1]. High liquid velocity resulting from high fuel pressure is needed for rapid breakup of the bulk liquid and secondary atomization. Swirl motion imparted on the fuel droplets has been effectively utilized for spray penetration reduction while maintaining adequate atomization. In high cylinder pressure environments, angular momentum was found to be preserved for a longer time than the momentum in other directions [8]. Thus, smaller droplet sizes can be expected for a swirling spray compared to a non-swirling spray in the late injection regime. A hollow cone spray can be generated from a swirl type injector. Swirl also affects the spray angle due to the centrifugal forces exerted on the individual droplets. Various spray angles can be obtained by varying the swirl intensity.

Although the inwardly opening high pressure swirl injector has many advantages, there are also drawbacks associated with the design. Because the spray characteristics such as drop size, spray angle, and penetration are all closely related to the swirl intensity, the customized tailoring of a spray for a specific combustion system requirement is difficult. Trade-offs and compromises have to be made case by case, depending on which characteristics are most critical to the application. For example, combustion chamber geometry constraints may require a small spray angle. A small spray angle may be accompanied by relatively large droplets and penetration. Therefore, a lower fuel supply pressure may be required for penetration control, but a lower pressure may increase the droplet size.

In general, excessive spray penetration and droplet size are still problems associated with most DI sprays. Wall-controlled charge stratification combustion systems have been developed and realized in production engines, partially because of their insensitivity to spray quality [2,3]. These spray-wall impingement systems, however, risk producing UHC emissions. An initial fuel slug, created by most inwardly opening DI injectors, travels ahead of the main spray body [9]. Compared to the main spray body, the slug has a larger speed, and a longer penetration distance. The slug possesses large droplets which contain a notable percentage of the total fuel injected per pulse. Also, combustion product deposits on the inwardly opening injector tip may degrade spray quality and reduce fuel flow.

A new design concept which has the potential to produce an improved spray meeting the critical gasoline DI requirements, has been implemented into a Delphi DI gasoline injector, by successfully utilizing computational fluid dynamics (CFD). The Delphi DI injector has an outwardly opening nozzle. A hollow cone spray is formed from the nozzle. Different from the inwardly opening pressure-swirl injectors, the Delphi DI injector has design flexibility which may allow the spray angle, penetration and droplet size to be controlled separately. Experimental spray analysis has shown that the Delphi DI injector's spray meets the stringent DI combustion system requirements. The nozzle design procedure, numerical modeling and test results will be discussed in this paper.

NOZZLE DESIGN

Of the hundreds of different nozzle concepts for spray generation, two main classifications exist: round jet and liquid sheet disintegration. For a fixed flow rate, an annular liquid sheet has a much larger surface area than a cylindrical liquid jet. Therefore, liquid sheets are more easily disintegrated into small droplets. Annular liquid sheets can be generated by either pressure-swirl inwardly opening nozzles or outwardly opening nozzles. However, outwardly opening nozzles have the following advantages over inwardly opening nozzles. A liquid sheet is formed by extrusion of the liquid through the conical seat annulus, rather than by centrifugalization of swirling liquid as in inwardly opening nozzles. As the sheet moves outwardly from the nozzle axis, the thickness of the sheet will decrease because of mass conservation. The initial sheet thickness is directly controlled by the pintle valve stroke regardless of swirl intensity. In general, the spray droplet size, characterized by SMD, is affected by the combination of the sheet thickness, the sheet velocity (i.e., the fuel supply pressure), and the initial disturbances. Outwardly opening nozzles, therefore, allow for additional atomization control by adjusting the initial liquid sheet thickness. The spray angle is also predetermined geometrically by the cone angle of the seat passage. Although swirl is not necessary for the liquid sheet generation and spray angle determination, it can be used for reducing penetration, and also for extending the spray angle and evenly distributing the fuel. The control of the droplet size, spray angle and penetration can be decoupled, to some extent, by varying the initial sheet thickness, the valve and seat angle, and the swirl intensity, respectively. Therefore, it is possible to create a well-atomized and compact spray for late injection stratified charge combustion, even though small drop size, small penetration and small spray angle in a single spray have been considered as contradictory requirements [6].

The outwardly opening nozzle has other benefits as well. During the valve opening and closing, the initial sheet thickness is small. The upstream pressure of the discharge orifice is greater than the fully opened condition due to decreased pressure drop at the swirler. Therefore, the initial spray and the spray tail have even smaller droplets than the main spray. There is no sac volume between the valve seal surface and the discharge orifice. The potentially detrimental initial fuel slug is not created. The outwardly opening valve continuously cleans the conical valve seat. There are no nozzle holes exposed to the combustion environment. Therefore, this design should be robust to combustion product deposits, and nozzle clogging should not be a problem.

Another problem associated with the inwardly opening swirl nozzle is flow control. The minimum flow area varies from the valve seat to final nozzle orifice during opening. The discharge coefficient also varies depending on the instantaneous flow velocity. However, in outwardly opening nozzles, the flow area is a simple function of the valve stroke, and the discharge coefficient is nearly constant and almost independent of the flow velocity. Hence, precise flow control is easily realized. Based on the above-mentioned potentials, the design concept of the outwardly opening nozzle with an internal swirl generating feature was selected for the Delphi DI injector.

Recently, computational fluid dynamics techniques have been used increasingly in industry and research facilities for improved accuracy, fast turn-around and relatively low cost, as compared to experiments. In the Delphi DI injector development process, CFD modeling was used extensively prior to the hardware design and build. Star-CD (Adapco Co.) was chosen for the injector internal flow modeling. The angles of the valve and seat were determined from the spray angle target. The valve stroke was derived by considering the required spray drop size, the internal contamination sensitivity and the actuator constraints. The critical dimensions of the injector were optimized through the flow modeling until the specific requirements, such as the steady-state flow rate, swirl strength, pressure distribution inside the injector, and force balance for reliable seal and fast actuation, were met. Once the flow modeling was accomplished, its output

was input to the multidimensional CFD program, KIVA-3, for spray simulation and assessment [10,11]. Ambient pressures of 100 kPa absolute and 1.5 MPa absolute were chosen for representing the cylinder pressures of the early injection and late injection modes, respectively. Special treatments and modifications were incorporated into the program for simulation of the hollow cone spray emanating from the annular orifice of the outwardly opening Delphi DI injector. The predicted spray characteristics were used to modify the injector spray controlling features as needed. This procedure was repeated several times until the resulting spray met the targets. During the design process, the limitations and constraints of actual injector design and manufacturability were taken into consideration. Finally, the nozzle design was completed and implemented into the Delphi DI injector by efforts of the injector design and manufacturing teams. Figure 1 shows a sketch of the Delphi DI injector.

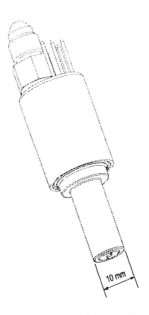

Figure 1: Sketch of Delphi DI gasoline injector.

CFD MODELING RESULTS

The flow passages inside the Delphi DI injector are complicated. However, the most restriction is in the lower part of the injector including the nozzle. The pressure drops occurring in all upstream passages are assumed to be negligible. Therefore, only the flow domain in the lower part of the injector was meshed and modeled with Star-CD. Pressure boundary conditions were imposed on the inlet and outlet of the model. The standard k-ε turbulence model was used for the turbulence simulation. Flow was assumed to be incompressible. Typical gasoline fluid properties (e.g., density of 736 kg/m^3 and viscosity of 4.25×10^{-4} Pa·s) were used in the model. A steady-state flow solution was obtained for the fully opened injector during the main injection period.

Figure 2 is a contour plot of the pressure distribution in the nozzle model. The main pressure drops occur at the swirler holes and the annular discharge orifice. Negative pressure in a portion of the valve seat passage indicates the presence of cavitation, but the actual cavitation phenomena is not simulated because Star-CD does not have a cavitation model. However, the flow model is still valid. Figure 3 shows the flow velocity vectors of the flow model. Large rotational velocities are generated in the swirl chamber. They are sustained downstream until exiting the injector. The modeling results at the injector exit for the two ambient pressures are summarized in Table 1. A slightly reduced flow is predicted in the elevated ambient pressure case due to decreased pressure differential across the nozzle.

Table 1: Summary of the Star-CD flow modeling of the Delphi DI injector

Characteristic	P_a=100 kPa	P_a=1.5 MPa
Initial swirl angle*	17.8°	17.7°
Initial cone angle	60°	60°
Steady-state flow rate	11.59 g/s	10.74 g/s
Turbulence intensity	7.1%	7.1%

* Initial swirl angle is defined as the arctangent of the ratio of swirl velocity to metering velocity.

KIVA-3 is a multidimensional modeling code developed for spray dynamics and combustion simulation particularly in internal combustion engines. In this paper, only the work of modeling the DI sprays injected into a constant volume, initially still chamber is reported. To accommodate the annular opening orifice of the Delphi DI injector, the KIVA-3 code was appropriately modified. The Star-CD modeled injector flow was directly input into the KIVA-3 model. The initial swirling velocity component and the turbulence were incorporated into the KIVA-3 model with some modifications [12,13]. A square wave form was assumed for the injection pulse. A reference flow rate of 14 mg per injection was chosen for the spray analysis. The injected gasoline was divided into 5000 droplet parcels. The initial drop size was set as the initial liquid sheet thickness. Chamber pressure was varied from 100 kPa absolute to 1.5 MPa absolute at room temperature (20°C).

Many of the complex processes related to spray including: liquid breakup, secondary atomization of the droplets, liquid droplet vaporization, air entrainment into the spray, mixture formation, gas and liquid droplet dynamics, droplet collision and coalescence, as well as air and liquid droplet interactions are modeled in the KIVA simulation [10]. By analyzing the spatially and temporally resolved spray simulation, characteristics of interest, such as the spray angle, penetration, SMD, DV90 and the Rosin-Rammler N value, are obtained. In addition, the spray-induced air flow, vapor concentration

distribution, pressure and temperature distribution, and turbulence intensity, etc., are also available.

The snapshot of the spray modeled at 1 ms after the start of injection (SOI) into air at 100 kPa absolute pressure is illustrated in Figure 4 (a), and its sectional view is shown in Figure 4 (b). The spray induced air flow field is illustrated in Figure 4 (c). The penetration of the spray leading edge is 34 mm from the injector exit. In the model, instead of an intact liquid sheet, discrete but closely packed liquid droplet parcels are assumed. A very narrow high flux region is predicted along the hollow cone sheet path. The initial spray has an angle of approximately 60°, following the injector conical seat. Beyond the distance of 15 mm, the spray region becomes much thicker. Many small droplets are produced as a result of the secondary atomization and the continuous evaporation of the droplets. The small droplets lose their momentum quickly against the air drag force, and therefore are easily carried by the air flow. Figure 4 (c) depicts a large vortex in the air flow near the outer edge of the spray's main flux cone. This was confirmed numerically and experimentally in both swirling sprays and non-swirling sprays. Therefore, this vortex is not induced by the swirl motion imparted on the spray. It is more likely that the velocity profile relaxation of the liquid spray increases the thickness of the air entraining layer. This layer consists of a large coherent vortex structure with low velocity. The spray is also widened by the small droplets laden in this vortex. Accordingly, the spray angle is increased to 70°. This large scale vortex enhances the momentum transfer between the liquid droplets and surrounding gaseous phase, accelerating the spray dispersion process. As a result, a short penetration and a slightly increased spray angle are obtained.

In Figure 4 (a) and (b), larger droplets are shown at the spray leading edge than elsewhere. This result could be attributed to droplet coalescence, vanishing of the small droplets due to rapid vaporization, and large droplets with high initial momentum. Figure 5 shows the temporal variation of the SMD, spatially integrated at the plane 30 mm from the injector. At 0.6 ms after the SOI, the spray leading edge arrives at the sampling plane. The SMD of the spray leading edge is about 23 microns, becoming smaller with time as the main spray body passes the 30 mm plane. The average SMD of the main spray body is 12 microns. The SMD as a function of time and radial position in the same plane is shown in Figure 6. The large droplets reside on the spray outer edge as expected for a swirling spray. However, the number density of these large droplets is very low. Figure 7 shows the radially-integrated SMD as a function of the axial location and time. It is evident that the drop size of the spray leading edge is larger than the main spray. It increases with distance and time. DV90 is an appropriate term describing the large droplet characteristics of the spray.

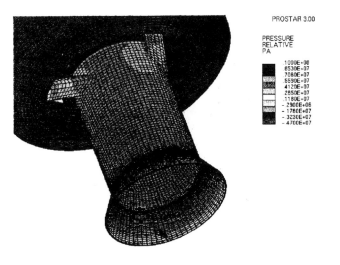

Figure 2: Star-CD modeled pressure distribution inside the Delphi DI injector.

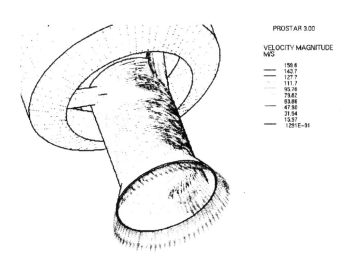

Figure 3: Velocity vectors of the injector internal flow predicted by the Star-CD injector flow model.

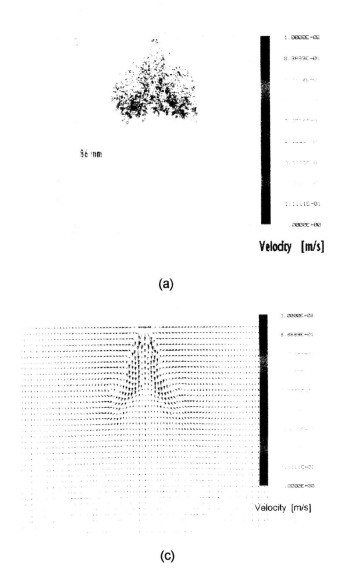

Velocity [m/s]

(a)

Velocity [m/s]

(c)

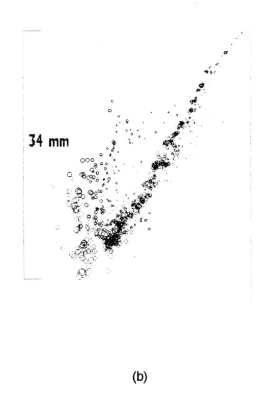

34 mm

(b)

Figure 4: KIVA-3 spray modeling results for the Delphi DI injector spray into 100 kPa absolute ambient pressure at 1 ms after the start of injection. (a) Spray droplet location, relative size and velocity; (b) Enlarged 2-D sectional view of the half spray; (c) Spray induced air flow velocities.

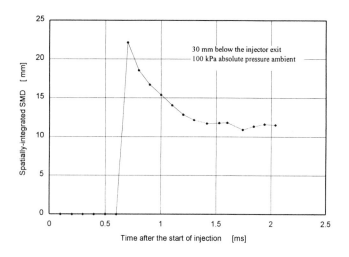

Figure 5: Spatially-integrated SMD as a function of time predicted by KIVA-3.

Table 2: The Delphi DI injector spray characteristics predicted by KIVA-3

Characteristic	P_a=100 kPa	P_a=1.5 MPa	Requirement/ Design target (100 kPa)
Penetration at 1 ms	34 mm	16 mm	30 mm
Spray angle at 1 ms	70°	76°	60°
SMD at 30 mm	13.4 μm	42 μm	15 μm
DV90 at 1 ms	28 μm	58 μm	40 μm
Rosin-Rammler N value	2.138	1.646	>1.9

The modeled spray characteristics are summarized in Table 2 with the DI spray requirements. Since the requirements are specified only for the spray injected into 100 kPa absolute pressure, the Delphi DI injector spray is verified to be sufficient. The spray injected into 1.5 MPa absolute pressure is very compact, even though the spray angle increased slightly. Both the axial penetration and the radial width of the spray are much smaller compared to the 100 kPa absolute pressure condition. The largely increased droplet size of the spray may be due to the increased probability of droplet collision and coalescence. However, in a real engine, high cylinder pressure is usually accompanied by high temperature during the compression stroke. The enhanced heat transfer and rapid evaporation rate could cancel some of the possible coalescence effects.

EXPERIMENTAL SPRAY ASSESSMENT

The Delphi DI fuel injector was tested per a standard protocol developed for evaluation of DI injectors. This protocol involves identifying the injector flow as a function of pulse width at the operating fuel pressure of 10 MPa and spray tests. Experimental

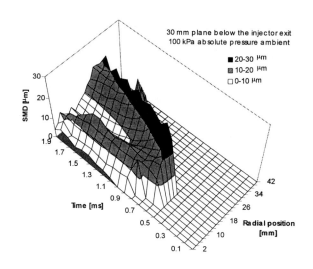

Figure 6: Spatially-resolved SMD as a function of time after the SOI predicted by KIVA-3.

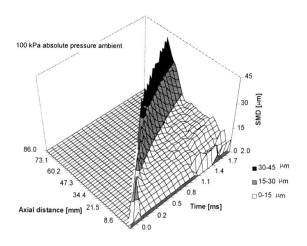

Figure 7: Spatially-integrated SMD as a function of axial distance and time after the SOI computed by KIVA-3.

assessments of sprays were conducted so as to identify the macroscopic spray parameters first, then to perform more detailed analyses by using the results obtained from the initial spray studies. An injector flow of 14 mg per pulse was selected for the spray testing. Spray axial symmetry was investigated to provide indication of mass distribution of the spray. The results of the spray symmetry evaluation guided the testing to characterize the spray structure by time-resolved still imaging techniques. The injector was oriented in a manner to expose the maximum maldistribution of the spray in the images. All subsequent spray testing was performed with the injector in this same orientation. Instantaneous images were sequentially recorded to visualize the spray structure and the development of the spray from the injector exit. Spray angle and penetration were determined from the images of the spray captured as a function of time after injector logic.

The spray droplet size distribution was evaluated in two ways: spatially-integrated and spatially-resolved. The spatially integrated droplet size measurement was performed along a line passing through the spray center and the spray's maximum flux region at an axial distance of 30 mm from the injector exit. The breakup of the liquid sheet occurred before the measuring plane. This was verified by the imaging spray structure evaluation. Using the results of the spatially-integrated droplet sizing, the appropriate optics and instrument settings were chosen for performing a spatially-resolved droplet size and axial velocity interrogation of the spray.

FLOW TESTING

Prior to spray evaluation, the flow performance of the injector was verified. Stoddard solvent at 10 MPa was supplied to the injector, and the injector steady-state flow rate of 11.5 g/s was verified with the pintle held open. Then, the dynamic response of the injector was determined. The injector logic pulse width corresponding to 14 mg of liquid per pulse was determined with the injector cycling at 100 Hz. An accelerometer was used to check the opening and closing response of the pulsing injector. All of the spray experiments were conducted at the 14 mg per injection pulse width, 1.30 ms, as determined by this procedure. The fluid used for all of the spray tests was California Phase II gasoline supplied to the injector at 10 MPa. The frequency of injections was varied for the spray tests to accommodate limitations in the measurement techniques and equipment, while allowing the spray data to be collected most efficiently. The operating characteristics of the Delphi DI injector are summarized in Table 3.

Table 3: Delphi DI injector operating characteristics

Fuel supply pressure	10 MPa
Operating voltage	13.5 volts
Steady-state flow rate	11.5 g/s
Opening and closing response	0.3 ms
Minimum flow	1.5 mg/pulse

SPRAY SYMMETRY ASSESSMENT

The injector spray symmetry was determined by use of a spray patternator. The patternator is approximately 90 mm in diameter and consists of 253, 4.5 mm diameter cells arranged in a hexagonal pattern to minimize the space between the cells. Each cell is shaped to allow the spray droplets which hit the patternator between the cells to be guided into the adjoining cells. The center-to-center distance between two adjacent cells is 5.5 mm. Testing was performed by centering the injector 30 mm above the patternator. The injector was cycled at a frequency of 25 Hz for 1000 pulses, so as to fill the cells as much as possible without overflowing any cell. Analysis of a typical Delphi DI injector spray is shown in Table 4. Some spray skew was observed. The orientation of the region of maximum flux was noted and used for subsequent experiments.

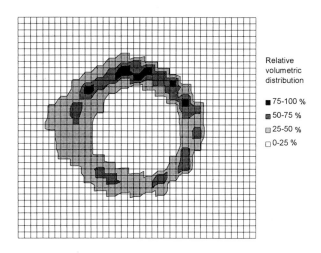

Figure 8: Patternator spray volume distribution for the Delphi DI injector.

Table 4: Spray patternator test results

Characteristic	Test result
Offset of spray centroid point	5.5 mm
10 % included volume radius	18.3 mm
50 % included volume radius	24.4 mm
90 % included volume radius	32.2 mm

The individual cell volume data was transformed to Cartesian coordinates and is shown in Figure 8.

SPRAY STRUCTURE AND DEVELOPMENT CHARACTERIZATION

The structure and development of the Delphi DI injector spray were assessed with imaging. The spray angle, penetration and breakup of the liquid sheet can be observed in the collected images. By capturing multiple images of the spray at various times after injector logic, it was possible to measure spray penetration versus time. The injector was housed in a nitrogen-purged windowed chamber capable of simulating engine cylinder pressures. Images of the spray at 100 kPa absolute and 1.5 MPa absolute chamber pressures were captured with a Kodak 8-bit, 1000 by 1000 pixel digital camera and a Bitflow frame-grabber board. These images were processed with Optimas image analysis software. The illumination source for the images was a 10 ns pulse, Nd:YAG laser. The laser beam was transformed optically to a vertical sheet of light approximately 0.5 mm thick. The light sheet was centered on the injector axis and the injector was rotated to allow the light sheet to intersect the spray's maximum flux region. The injection frequency was set at 4 Hz because of the required time for the laser and camera synchronization. Figure 9 shows an image of the Delphi DI injector spray at 1 ms after injector logic at 100 kPa absolute pressure. The large vortices predicted by the modeling (Figure 4) are apparent in the image. A comparison of Figure 4 and

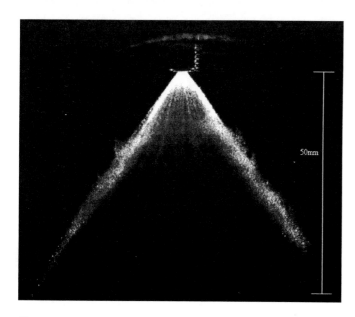

Figure 9: Laser sheet spray image at 1 ms after the SOI at 100 kPa absolute pressure ambient.

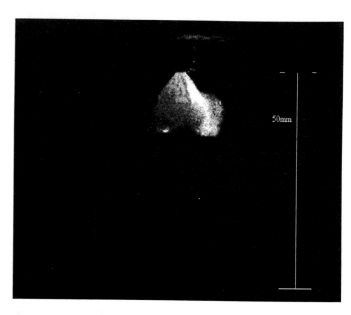

Figure 10: Laser sheet spray image at 1 ms after the SOI at 1.5 MPa absolute pressure ambient.

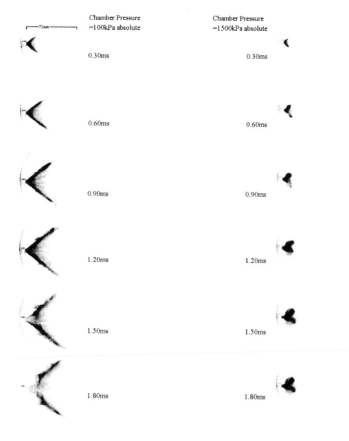

Figure 11: Sequential laser sheet spray images as a function of time at 100 kPa absolute and 1.5 MPa absolute ambient pressures.

Table 5: Spray characteristics as a function of ambient pressure as determined from imaging

Characteristic	P_a = 100 kPa	P_a = 1.5 MPa
Spray angle at 1 ms	74 °	75 °
Penetration at 1 ms	51.5 mm	14.4 mm

Figure 9 shows that the modeled spray is very similar to the spray observed in the physical testing. Figure 10 shows an image of the spray at 1 ms after injector logic at 1.5 MPa absolute chamber pressure. Table 5 shows the spray angle and penetration determined from these images.

Spray penetration as a function of time was determined from 30 sequential images of the spray captured beginning at 0.1 ms after injector logic and ending at 3 ms after injector logic. The penetration as a function of ambient pressure was obtained from images of the spray at 100 kPa absolute and 1.5 MPa absolute chamber pressures. Figure 11 shows two series of laser sheet illuminated digital images of the spray depicting the spray development as a function of time for the two chamber pressures. The axial penetration distance of the spray leading edge was measured from each image of the sequence. Figure 12 shows the axial spray penetration as a function of time. For comparison, the modeled spray penetration is included in this plot. There is good agreement between the modeled and measured penetrations for the two chamber pressures.

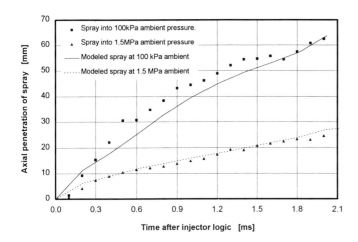

Figure 12: Measured and predicted spray penetration as a function of time at 100 kPa absolute and 1.5 MPa absolute ambient pressures.

SPATIALLY-INTEGRATED DROPLET SIZE DISTRIBUTION CHARACTERIZATION

The spatially-integrated droplet size distribution tests were conducted using a laser diffraction instrument. All of the droplet size measurements were performed with the injector spraying into air at 100 kPa absolute pressure, with the line of measurement centered 30 mm below the injector exit. The laser diffraction instrument used was a Sympatec model HELOS/Vario-KF fitted with a 5 mW, 632.8 nm Helium Neon laser. The receiver of this instrument has 31 semicircular detector rings and 3 centering elements. The receiver was fitted with a lens of 100 mm focal length, so that particle sizes from 0.9 to 175 microns could be measured. Testing was performed with laser beam diameters of 2.2 mm and 13 mm, with similar results. The injection frequency was 50 Hz. The laser diffraction instrument was set to sample the diffracted light signal for 10 seconds, thereby temporally averaging the spray droplet distributions of 500 injection cycles. The light intensities from the 31 detector rings were processed by the HELOS software to derive a spray droplet size distribution and characteristic sizes that are reported in Table 6.

Table 6: Laser diffraction spray sizing results

Characteristic	Test result
Arithmetic mean diameter	8 microns
SMD	15 microns
DV90	28 microns
DV50	12 microns
DV10	7 microns

SPATIALLY-RESOLVED DROPLET SIZE DISTRIBUTION CHARACTERIZATION

The spatially-resolved spray droplet size measurements were conducted using phase Doppler interferometry. Using the results of the spatially-integrated spray droplet size measurements, the appropriate optics and instrument settings were chosen for performing a spatially-resolved droplet size and axial velocity interrogation of the injector spray. The test conditions of the laser diffraction testing were reproduced for the phase Doppler testing. Pointwise measurements were performed along the same line of measurement as in the laser diffraction tests. The phase Doppler interferometry instrument used was an Aerometrics Phase Doppler Particle Analyzer (PDPA) operated so as to measure the size and axial velocity of the spray droplets. The instrument was used in forward scatter mode with the receiver offset 30 degrees from the axis of the transmitter. Focal lengths of the transmitting and receiving lenses were 250 mm and 500 mm, respectively. The droplet diameter range was set at 2 to 100 microns. The argon ion laser power was set so that the transmitted light intensities were 5 mW for the frequency shifted and unshifted beams. The PDPA was set to automatically adjust the photomultiplier tube high voltage. For the Delphi DI injector spray in the region of peak flux, the photomultiplier high voltage was 400 volts. Phase calibration of the instrument was performed at this setting. The PDPA's intensity validation function was enabled to remove false measurements due to light reflected from droplets. The injector was axially centered 30 mm above the PDPA probe volume. PDPA measurements were performed along a radius of the spray by moving the injector in 1 mm increments from the axial center point, through the region of peak flux, to the edge of the spray where the spray flux was less than 1% of the maximum measured spray flux. The number of droplet samples collected at each point was 10,000. Figure 13 shows the droplet size variation with radial

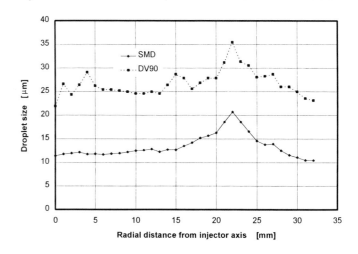

Figure 13: Radially-resolved SMD and DV90 for the Delphi DI injector spray measured by PDPA.

Table 7: Phase Doppler interferometry measured spray characteristics

Characteristic	Test result
Radially-integrated SMD	13.6 microns
Minimum SMD	10 microns
Maximum SMD	21 microns
Minimum DV90	22 microns
Maximum DV90	35 microns
Rosin-Rammler N value	2.2 - 3.3
Minimum axial velocity	0.2 m/s
Maximum axial velocity	23 m/s

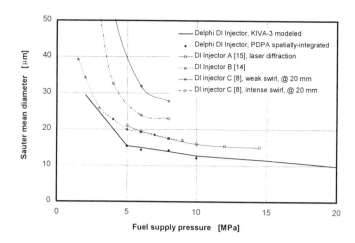

Figure 14: Effects of fuel pressure on SMD of DI sprays at 100 kPa absolute pressure ambient.

distance. Table 7 shows the PDPA measured spray droplet size and axial velocity. The averaged SMD value was radially-integrated from the PDPA measurements.

The physical assessment of the Delphi DI injector spray confirmed the CFD modeled results. The injector flow rate, spray angle, penetration, SMD and DV90 were accurately predicted by modeling. Table 8 shows a summary of the modeling results, the testing results and the DI spray requirements. The injector spray satisfies the stringent DI spray requirements.

Table 8: Comparison of modeling and testing results with DI requirements

Characteristic	Modeling result	Testing result	Requirement
Steady-state flow rate [g/s]	11.6	11.5	11.5
Spray maximum penetration [mm]	63	63	<70
Spray angle [°]	70	65	60~70
SMD [μm]	13.4	13.6	<15
DV90 [μm]	28	35	<40

DISCUSSION

Despite the many advantages of DI gasoline engines, one major obstacle is the increased cost of the fuel system compared to that of the PFI engines. This cost is assumed to be proportional to the system fuel pressure. If the flow and spray requirements can be met, the fuel pressure should be reduced as much as possible. Figure 14 shows the relation of SMD of the DI spray versus fuel supply pressure at 100 kPa absolute pressure and room temperature. The thick solid line represents the KIVA-3 modeling results. The discrete diamond symbols denote the values integrated from the PDPA measured radial SMD profile at 30 mm distance from the Delphi DI injector. For comparison, similar data from some inwardly opening DI injectors are also included [8, 14, 15]. In general, the SMD decreases with

increasing fuel pressure. In the low pressure range, the SMD is very sensitive to the fuel pressure. For the pressures greater than 5 MPa, further pressure increase does not lead to significant drop size reduction. The Delphi DI injector sprays show the same trend, however, the SMD values of these sprays are at least 4 microns smaller than from the inwardly opening swirl injector sprays. This spray drop size difference is considered to be a result of the smaller initial liquid sheet thickness of the Delphi DI injector sprays.

In Figure 14, the fuel pressure of 5 MPa seems to be high enough for producing an acceptable DI spray. Nonetheless, a higher fuel pressure is recommended for the following reasons. As the fuel pressure was increased from 5 MPa to 10 MPa, the SMD of the Delphi DI spray decreased from 15.4 microns to 13.6 microns. This difference does not seem significant, but the total surface area for 14 mg of fuel increased from 7411 mm^2 to 8392 mm^2. This surface area increase of 13% should lead to direct enhancements in heat transfer, vaporization and combustion. More importantly, high fuel pressure is required to reduce the maximum droplet size in the spray, expressed by DV90. Laser diffraction measurements have confirmed that the DV90 of the Delphi DI injector spray at 30 mm from the injector exit decreased from 40 microns to 28 microns, as the fuel pressure increased from 5 MPa to 10 MPa. Therefore, the DV90 is more sensitive to fuel pressure than the SMD.

The DI sprays, however, behave very differently in the high pressure, high temperature combustion chamber than they do in an atmospheric pressure, room temperature environment. The maximum cylinder pressure could be above 2 MPa at the end of the compression stroke. Since the pressure differential across the nozzle orifice dictates the spray quality, the spray may degrade significantly if the fuel pressure is not sufficiently high. According to KIVA-3 predictions, even though the Delphi DI injector is operated at 10 MPa

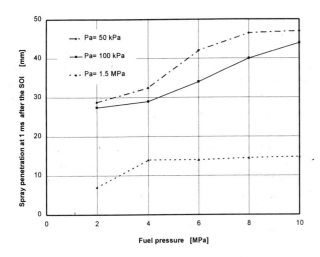

Figure 15: Effects of fuel pressure and ambient pressure on Delphi DI injector spray penetration measured at 1 ms after the SOI.

fuel pressure, the drop size will increase substantially with increased chamber pressure (Table 2). Essentially, the high fuel pressure is needed for fine sprays at the late injection stratified charge operation rather than the atmospheric air pressure or vacuum early injection conditions. In addition to its effects on the spray, the higher fuel pressure is also required for reducing the injector flow rate sensitivity to the injector valve stroke variations. Because of anticipated fuel pressure pulsation in the fuel system and unpredictable variations of the injector valve stroke, an adequate pressure drop must occur elsewhere in the fuel system besides the discharge orifice. In the Delphi DI injector, this pressure drop was used for swirl generation.

Fuel pressure and ambient pressure also have pronounced effects on spray penetration. Figure 15 shows both pressures' effects on the Delphi DI injector spray penetration at 1 ms after the start of injection. The spray penetration was measured from the laser sheet spray images. As fuel pressure was reduced and ambient pressure was increased, spray penetration was diminished. Penetrations of the sprays into a vacuum environment of 50 kPa absolute pressure were about 5 to 10 mm longer than those into 100 kPa absolute pressure. Sprays injected into 1.5 MPa absolute pressure had much shorter penetrations. At this ambient pressure, fuel pressures above 4 MPa did not increase the penetrations of the sprays at 1 ms.

In a DI gasoline engine, cylinder pressure varies with crank angle. The cylinder pressure's effect on the spray penetration depends on the injection timing and duration. In the early injection mode, the cylinder pressure is low, resulting in a long spray penetration. However, the distance from the injector exit to the piston is large. In the late injection mode, the increased cylinder pressure causes a shorter spray penetration. Short penetration is favorable because the distance from the injector exit to the piston is small. Some DI injection

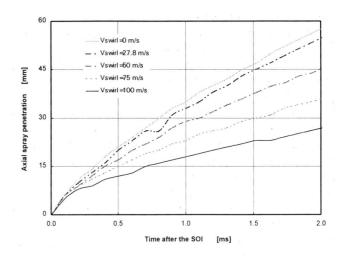

Figure 16: KIVA-3 predicted effects of initial swirl velocity at the injector exit on spray penetration as a function of time at 100 kPa absolute pressure ambient.

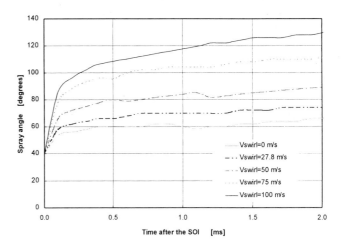

Figure 17: KIVA-3 predicted effects of injector initial swirl velocity on spray angle as a function of time at 100 kPa absolute pressure ambient.

systems require reduced fuel pressure to limit the in-cylinder spray penetration. It is important to note, however, that the use of reduced fuel pressure may compromise spray droplet size.

Swirl imparted on the spray droplets is known to be effective in reducing the spray penetration, with simultaneous enhancements to atomization. Figure 16 shows the relation between the spray penetration and the tangential velocity component at the Delphi DI injector exit, predicted by the KIVA-3 spray modeling. For the range of the swirl velocities less than 28 m/s, only a slight change in the penetration was noted. However, it was verified experimentally that the spray mass distribution of a swirl type Delphi DI injector was much more symmetric than one without swirl. The swirl motion in the azimuthal direction helped to evenly distribute the spray. As the tangential velocity was increased further, the spray penetration reduction became significant. For instance, when the tangential

209

velocity component was varied from 28 m/s to 50 m/s, the spray penetration at 1 ms after the SOI was reduced from 34 mm to 29 mm. However, the spray angle also was increased from 70° to 84°, as shown in Figure 17. If the increased spray angle is not desirable due to potential wall impingement in the spray radial direction, the injector pintle and seat angle could be reduced appropriately to compensate for this change. Since the penetration reduction by using swirl is always accompanied by increased spray angle, careful spray tailoring is required to avoid the wall impingement in radial and axial directions. Ideally, a promptly atomized spray with very small drop size and penetration, the "puff" spray, is required to achieve the maximum potential of the gasoline DI engine concept. The spray droplets are so small that their momentum are very quickly consumed by air drag. The KIVA-3 simulation also predicted a slight effect of the swirl intensity on the spray drop size. Comparing a spray with a tangential velocity component of 50 m/s and a non-swirling spray, only a 1 micron decrease in SMD was found in the swirling spray.

CONCLUSIONS

The outwardly opening Delphi DI injector has been developed to meet the stringent spray requirements of the direct injection gasoline engines. CFD modeling software has been effectively utilized to optimize the injector nozzle design towards specific performance and spray requirements. The experimental spray evaluation has verified the modeling accuracy. Confidence has been developed for the numerical modeling tools because of acceptable agreement between the modeling and test results. Regarding the nozzle concept selection, advantages of the outwardly opening nozzles and problems associated with the inwardly opening pressure-swirl nozzles were discussed. Strategies for separate control of the three major spray characteristics: droplet size, angle and penetration, have been described. The influences of the fuel pressure, ambient pressure and tangential swirl velocity on these characteristics have been investigated. In summary, the following main conclusions can be extracted from this work:

- The Delphi DI injector produces a fine spray which meets the stringent DI requirements. This design has flexibility for custom tailored sprays to satisfy various combustion systems.

- For the spray injected into 100 kPa absolute pressure, the radially-integrated SMD measured by PDPA, was 13.6 microns. The maximum DV90 along the radial scanning was 35 microns. The spatially-integrated DV90 measured by the laser diffraction technique was 28 microns. The spray angle was 74°. The spray penetration at 1 ms after the start of injection was 51.5 mm. The maximum penetration was about 70 mm.

- The spray became very compact with increased ambient pressures. The spray penetration did not exceed 16 mm at 1.5 MPa absolute ambient pressure at 1 ms after SOI. The hollow cone spray collapsed into a solid cone spray. KIVA-3 simulation showed the tendency of increased droplet size in the high ambient pressure conditions.

- Fuel pressure has significant influence on the spray droplet size. SMD decreases with increasing fuel pressure. As fuel pressure increased above 5 MPa, the SMD was reduced only slightly. However, the DV90 reduction due to the increased pressure was still notable in this range.

- Effects of the fuel pressure on the spray penetration are dependent on the ambient pressure. Generally, penetration increases with the fuel pressure, and decreases with the ambient pressure. The penetration in a vacuum of 50 kPa absolute was 5 to 10 mm longer than at 100 kPa absolute pressure. In 1.5 MPa absolute ambient pressure, fuel pressures greater than 4 MPa have negligible effects on the penetration at 1 ms.

- Increased swirl intensity reduces spray penetration, increases spray angle, and slightly reduces spray droplet size.

ACKNOWLEDGMENTS

The authors gratefully acknowledge the significant contributions and support from those who assisted in this work. Their help was invaluable in discussion, hardware preparation and testing. Particular thanks are extended to: R. Cooper, Y. Kazour, K. Keegan, E. Kobos, S. Maczynski, H. Mieney, D. Moran, E. Schneider, D. Varble and J. Zizelman.

REFERENCES

1. Zhao F.-Q., Lai M.-C., and Harrington D. L. "A Review of Mixture Preparation and Combustion Control Strategies for Spark-Ignited Direct-Injection Gasoline Engines," SAE Paper 970627.
2. Harada J., Tomita T., Mizuno H., Mashiki Z., and Ito Y. "Development of a Direct Injection Gasoline Engine," SAE Paper 970540.
3. Tomoda T., Sasaki S., Sawada D., Saito A., and Sami H. "Development of Direct Injection Gasoline Engine - Study of Stratified Mixture Formation," SAE Paper 970539.
4. Fraidl G. K., Piock W. F., and Wirth M. "Gasoline Direct Injection: Actual Trends and Future Strategies for Injection and Combustion Systems," SAE Paper 960465.
5. Han Z., Fan L., and Reitz R. D. " Multidimensional Modeling of Spray Atomization and Air/Fuel Mixing in a Direct-Injection Spark-Ignition Engine," SAE Paper 970884.
6. Solomon A. Panel Discussion on Direct Injection Gasoline Engines, ILASS-Americas '97.
7. Dodge L. G. "Fuel Preparation Requirements for Direct-Injected Spark Ignition Engines," SAE Paper 962015.
8. Kume T., Iwamoto Y., Lida K., Murakami M., Akishino K.,

and Ando H. "Combustion Control Technologies for Direct Injection SI Engine," SAE Paper 960600.

9. Parrish S., and Farrel P. V. "Transient Spray Characteristics of a Direct-Injection Spark-Ignited Fuel Injector," SAE Paper 970629.

10. Amsden A. A., O'Rourke P. J., and Butler T. D. "KIVA II - A Computer Program for Chemically Reactive Flows with Spray," Los Alamos Labs, LS 11560 MS, 1989.

11. Amsden A. A. "KIVA-3: A KIVA Program with Block-Structured Mesh for Complex Geometries," Los Alamos Labs, LS 12503 MS, 1993.

12. O'Rourke P. J. and Amsden A. A. "The Tab Method for Numerical Calculation of Spray Droplet Breakup," SAE Paper 872089.

13. Lai M.-C., Zhao F.-Q., Amer A. A., and Chue T.-H. "An Experimental and Analytical Investigation of the Spray Structure from Automotive Port Injectors," SAE Paper 941873.

14. Croissant K. and Kendlbacher C. "Requirements for the Engine Management System of Gasoline Direct Injection Engines," Presentation Technische Akademie Esslingen, December 1996.

15. Toyota Press Information '96, "Direct-Injection 4-Stroke Gasoline Engine," 1996.

Transient Flows in High Pressure Swirl Injectors

J. Cousin
CO.R.I.A. / UMR 6614 CNRS
Université et INSA de Rouen

W. M. Ren and S. Nally
Siemens Automotive

ABSTRACT

Gasoline direct injection requires that the injection time may be very short in duration, indicating that transient flow effects can have a strong influence on the flow behavior and on the spray properties. Consequently, a computational analysis of the dynamic flow in a high pressure swirl injector was conducted. In order to perform the flow simulation during a complete injection event, movement of the needle that controls the amount of fluid to be discharged has been considered and deduced from experimental data.

To validate the computational model, the predicted dynamic flow rate, temporal cone angle and instantaneous mass flow rate were compared to experimental data. The calculated results were found to be consistent with measurements.

The dynamic calculations allow a better understanding of the complex transient flow during one injection event and may be divided into four different stages where characteristics of the liquid emerging from the nozzle are completely different. In this paper, an attempt for determining the volume of fuel supplied during each stage is proposed.

INTRODUCTION

The Direct Injection SI engine is one of the most promising internal combustion engine designs for achieving lower fuel consumption and higher engine performance while maintaining low emissions and driving performance. Contrary to the port injected engine where the spray can evaporate on the warm inlet valve, the direct injection method injects and meters the fuel directly in the combustion chamber.

The control of the mixing process is a complex task to achieve. As mentioned by Zhao *et al* [1] in a complete review about direct gasoline engines, direct injection systems must have the capability of providing both late injection for stratified charge combustion at part load, as well as early injection for homogeneous charge combustion at full load. Due to this requirement, the fuel spray that is supplied from the injector must be characterized by a cloud of very small droplets. This is particularly critical for stratified charge where droplets have a very short time to evaporate.

High pressure injectors are recognized to be one of the best adapted for the direct injection spark ignited engines because they generally provide a fine and widely dispersed fuel spray with moderate injection pressure. Figure 1 shows the swirl generator inside a high pressure direct injector. It consists of a swirl chamber and a discharge orifice. During the injection process the pressurized fuel is forced to flow through tangential passages on the swirl disk into the swirl chamber, rotate in the chamber, then emerge from the discharge orifice in the form of a thin conical sheet.

Figure 1 : Swirl generator inside a high pressure injector

In the last 50 years, many studies on swirl injectors have been conducted. Effects of the design, operating conditions, fluid to be atomized and ambient conditions on the flow and the spray quality have been tested (see Lefebvre [2]). However, all these studies treat the injection during the steady state operation only. As a matter of fact, in many applications transient effects (beginning and end of injection) have negligible effects because the relative fuel volume supplied during these transient stages is small.

Unfortunately, this is not the case for this application where very short injections may be required. However, the steady state flow is still critical for GDI applications and a description of this portion of the flow will be useful for this study.

In steady state operations, swirl atomizers are characterized by the cone angle θ and the discharge coefficient Cd defined by :

$$Cd = \frac{q_m}{\pi r_0^2 \sqrt{2\rho\Delta P}} \qquad (1)$$

where r_0 and ΔP are the orifice radius and the pressure differential respectively.

In figure 2, classical evolution of Cd and θ as a function of the injection pressure are reported. This figure is very helpful in understanding this particular flow and is directly linked to the characteristics of the fluid emerging from the nozzle as described by Lefebvre [2] (fig. 3).

Figure 2 : Evolution of the discharge coefficient and the cone angle for a pressure swirl atomizer

Figure 3 : Evolution of the liquid system with the injection pressure (Lefebvre [2])

For very low injection pressure, the progression from the gutation regime (a) to the onion regime (c) passing through the jet regime (b) can be observed. These three regimes are characterized by liquid systems having a relative low velocity which results in a coarse atomization.

Between (a) and (b), Cd decreases because the injection pressure is not dissipated in friction only, but creates a "real" flow. At (c), Cd decreases slightly, this corresponds to the appearance of an air core around the axis of the chamber resulting from a high swirling motion. When ΔP continues to increase, Cd and θ reach an asymptotic value and become independent of the injection pressure : the injector is said to work in its stable zone. This zone corresponds to the most effective mode of atomization because a large cone angle and a fine spray are supplied. This result was experimentally observed by Doumas et al [3] and Dombrowski et al [4] who observed that Cd and θ depend only on the fluid properties and the injector design only. Typically, a good swirl injector is characterized by a small Cd (lower than .3) and a large cone angle in the stable zone. Moreover due to consideration of energetics, it is preferred to obtain the stable zone for small values of injection pressure.

In transient operating conditions, the behavior of the flow is similar to that in the description of figs. 2 and 3. At the start of injection, pressure below the sealing point equals atmospheric pressure. Then, injection pressure increases and the five regimes described previously are generally observed. The quality of the spray resulting from the first four regimes is known to be undesirable compared to the stable zone.

Therefore, as far as gasoline direct injection is concerned, it appears crucial to improve the knowledge of the start of injection. More precisely, the time scales of each regime and more importantly the amount of fuel supplied during each regime need to be known. Here, this understanding is achieved by computational fluid dynamics calculations. This allows possible improvements of the gasoline direct injectors, especially during operations at short pulse widths.

This work constitutes the second step of the work performed by Ren et al [5] concerning an extensive study of the flow behavior in steady state operations. Effects of the internal geometry as well as the injection pressure were considered and some of these results are used in this study.

COMPUTATIONAL MODEL

Flow inside high pressure swirl injectors is governed by the continuity and Navier-Stokes equations. For this application, the flow that takes place in the atomizer is assumed to be fully turbulent and therefore the simulation is performed using the standard k-epsilon model.

Figure 4 : Computational domain and mesh

Figure 4 shows the computational domain and the grid [5] when the needle that controls the amount of fuel to be discharged is at its upper position. Here, only a quarter of the injector is considered because of the symmetry of the injector.

The flow is assumed to be isothermal and incompressible. The working liquid is a solvent, called Stoddard, whose physical properties are close to the values of a standard gasoline (density ρ = 785 kgm⁻³, viscosity μ = 9.4 10⁻⁴ Nsm⁻²). The model contains about 68000 cells.

The pressure boundary conditions are applied at the inlet and outlet planes (see fig. 4). The inlet total pressure equals the fuel rail pressure (fixed at 70 bar in this study) and the relative outlet pressure equals zero. Cyclic boundary conditions are fixed at the two periodic planes.

The FIRE commercial code was employed to obtain the numerical solution. This code allows calculations with a moving, unstructured mesh. In this study, the movement of the needle is taken into account and the flow can be simulated during a complete injection event. In order to be as close as possible to the operation of a real injector, needle lift was deduced from experiments. Figure 5 illustrates the case of a 2 ms pulse width.

Figure 5 : Simulation of the needle lift with time (Ti = 2 ms, Ti is the fuel logic pulse width)

As the problem is posed, the mesh can not allow the simulation of the fully closed position. Therefore, injection is assumed to start and end with a small lift as indicated in fig. 5. This assumption is not too unrealistic for the start of the injection because the lift is so small that the flow is negligible. However, it has to be kept in mind that this assumption will sligthly overestimate the amount of fuel supplied during the start and the stop of injection.

The dynamic calculations performed in this study allow the prediction of the velocity field in the injector as a function of time. Values of the instantaneous mass flow rate are directly deduced from these calculations and the procedure to predict the cone angle and the sheet thickness will be presented in detail. In parallel, some comparisons with experiments will show the validity of this study.

RESULTS AND DISCUSSION

Figure 6 illustrates the temporal evolution of the total pressure in a periodic plane (see fig. 4). These pictures allow an observation of the flow during a complete injection. First, one can observe that, even if the air core is not taken into account, the fuel flow is mainly concentrated around the orifice wall. It is also evident that the steady flow is obtained very quickly : typically, before the fully open position of the needle is reached. For the case presented in fig. 6, the steady flow is obtained at approximately 0.13 ms after the start of calculation. At the end of the needle motion, even if velocities are smaller than during the steady state, the swirling motion remains present indicating an acceptable behavior of the flow and therefore a good spray quality.

Figure 6 : Evolution of the total pressure

MASS FLOW RATE

In figure 7, the temporal evolution of the relative mass flow rate is presented for a 1 ms pulse width. This relative value corresponds to the mass flow rate divided by the calculated value in steady state operation. The curve we obtained is similar to what was described about the evolution of the discharge coefficient as a function of the injection pressure (fig. 2). At the start of injection, a high value of q_m corresponding to the jet regime is observed. Even if velocities are not at their maximum values, q_m is quite high due to the fact that during this regime, the fuel is discharged from the complete diameter of the orifice. As shown in fig. 7, even if the time scale of the jet regime is rather small, some big droplets whose diameter is of the same order of magnitude of the orifice diameter are formed.

Figure 7 : Predicted instantaneous mass flow rate (Ti = 1.0 ms)

The decrease of the flow rate corresponds to the formation of an internal swirling flow. The sharp decrease that is observed is improved by the movement of the needle that seems to increase the effects of depression in the swirl chamber. Finally, as observed previously, the steady state is achieved before complete needle lift has occurred.

Due to the short time delay of one injection, no experimental device allows the measurements of the instantaneous mass flow rate. However, we can define an instantaneous mass flow rate from classical measurements. Dynamic flow rate measurements allow the determination of the total mass injected during one injection, mT, as a function of the pulse width Ti.

mT is mathematically expressed by :

$$mT = \int_{injection} q_m dt \quad (2)$$

and was measured by weighing over 3,000 injections at short injection times (Ti < 0.8 ms) and between 1,000 and 2,000 injections for the longer injection times.

Then, the entity defined by :

$$q_m' = \frac{dmT}{dTi} \quad (3)$$

has a dimension of a mass flow rate. q_m' is not a real instantaneous mass flow rate because for each value of the injection time, the fuel volume supplied during transient and continuous regimes are integrated into the total mass, mT. However, when the injection time is very short, the continuous regime is negligible and in this case, high values of relative q_m' are observed (fig. 8) as in the calculations. Finally, this constitutes a qualitative validation of our calculations.

Figure 8 : Experimental mass flow rate

Quantitative comparisons performed here are based on the dynamic flow rate. Figure 9 presents the experimental and calculated injected mass during one pulse as a function of Ti. The calculations are consistent with measurements even if values of mT are underestimated for the high values of the pulse width. Ren *et al* [5] observed also this discrepancy and they attributed it to the air core that was not taken into account in the calculations. In a recent paper, Chinn *et al* [6] propose a new way for the simulation of the flow in steady state operations. As a matter of fact, they simulate the air core as a free surface with a slip condition imposed. The results they found seemed to be in very good agreement with measurements. This approach has to be validated on high pressure swirl atomizers where dimensions are smaller and injection pressure is higher than in the case of the Chinn's study. Here, start of injection is mainly studied and short injection times are considered. For this reason, the steady state has not been considered for improvement in this study.

Figure 9 : Experimental and calculated dynamic flow rates

Instantaneous flow rate is of paramount importance because it allows the calculation of the mass of fuel supplied during the start of injection.

Figure 10 : Calculation of the transient masses

As indicated in fig. 10, we can calculate the mass m_1 supplied during the jet stage. The jet regime is assumed to stop when q_m is at its maximum, namely :

$$m_1 = \int_0^{t_{peak}} q_m dt \quad (4)$$

The mass m_2 is the mass supplied between the jet regime and the steady state, that is to say :

$$m_2 = \int_{t_{peak[}}^{t_{steady}} q_m dt \quad (5)$$

As seen in fig. 7, q_m fluctuates during the start of the steady state indicating that m_2 may be not accurately determined.

Figure 11 : Proportion of the poorly atomized volume

In fig. 11, the percentages of mass supplied during the start of injection are presented. This result indicates that the volume that comes from the jet regime increases dramatically when the pulse width is small. As the amount of fuel is poorly atomized during this stage, the development of the knowledge of this part of injection is needed in order to reduce this mass m_1. As far as the quality of the spray is concerned, the mass m_2 need less attention because this mass is better atomized with the help of the developed swirling motion.

AIR CORE

Although the air core is not taken into account in this study, an estimation is made possible with the help of the axial velocity at the orifice. As found in a previous study [5], a back flow is observed close to the axis of the injector and the air core is presumed to be located at the radial position where the axial velocity equals zero.

In fig. 12, the temporal evolution of the sheet thickness is presented. We can remark that, at the start of the injection, sheet thickness equals the orifice radius indicating that a cylindrical jet is supplied at the injector tip. A sharp decrease of the sheet thickness (or an increase of the air core radius) is observed, and finally its asymptotic value is obtained quite quickly.

Figure 12 : Temporal evolution of the sheet thickness

MEAN VELOCITIES

With the help of the value of the air core radius (r_{ac}), values of the mean axial and azimuthal velocity at the orifice may be found. The radial velocity was found negligible as in the previous Ren's calculations for the steady state operation [5].

Mean axial velocity W is estimated from the value of q_m and r_{ac} deduced from CFD calculations at each time step. Here, liquid is assumed to be discharged from the estimated liquid sheet only:

$$W = \frac{q_m}{\pi\rho\left(r_0^2 - r_{ac}^2\right)} \quad (6)$$

As indicated by Ren et al [5], r_{ac} is probably underestimated inducing an underestimation of W.

Mean azimuthal velocity is deduced from the conservation of the angular momentum with respect to the chamber axis :

$$V = \frac{2}{\left(r_0 + r_{ac}\right)} \frac{\sum_{i=1}^{n} r_i v_i}{n} \quad (7)$$

where r_i and v_i correspond to the values of the radius of the orifice and the azimuthal velocity on the n nodes along a radius of the orifice where the fuel sheet is assumed to be present.

Figure 13 : Mean velocities at the outlet

CONE ANGLE

As proposed in the literature (Lefebvre [2]), a theoretical cone angle θ_T can be calculated when mean velocities are known:

$$\theta_T = 2\mathrm{Arc}\tan\left(\frac{V}{W}\right) \qquad (8)$$

However, by assuming a constant profile of the flow at the orifice, this cone angle is highly overestimated (see Bayvel et al [7] for details).

In the case of a 2D axis-symmetric calculations, Dumouchel et al [8] proposed a corrective factor that was found experimentally :

$$\theta = 0.556\theta_T \qquad (9)$$

It was decided to apply this empirical relation here to predict the cone angle as a function of time. This calculated cone angle is compared with measurements in fig. 14. The visible spray angles were measured from photographs with a stroboscope as light source.

Figure 14 : Measured and calculated cone angle

In fig. 14, one can remark that the predicted cone angle is close to experimental data. The quick increase of the cone angle is well predicted as well as its value during the steady state. At the end of injection, the increase of the cone angle is also well predicted. As a matter of fact, at the end of injection, when the needle approaches the sealing point, the swirling motion is still present and the mean axial velocity decreases quicker than the mean azimuthal velocity inducing this increase of the cone angle.

The temporal discrepancy observed at the end of injection is mainly due to the simulation of the needle lift. As a matter of fact, in our calculations, the end of injection is slightly delayed compared to what was experimentally observed.

CONCLUSION

The flow field inside a high pressure swirl injector during a whole injection has been studied in details. The temporal evolution of the mass flow rate, cone angle and sheet thickness were examined. Results are summarized as follows :

1. The steady flow is obtained very quickly, typically before the fully open position of the needle. This means that the swirling motion, necessary for the production of a fine spray, is obtained very quickly.

2. At the start of injection, a cylindrical jet emerges from the nozzle. That volume of fuel supplied during this stage is not negligible when the injection time is small. The volume of fuel supplied during the start of injection must be precisely studied

because droplets issued from the breakup of this jet are likely to be detrimental to the combustion process.

3. At the end of injection, swirling motion remains present, indicating a well atomized fuel preparation.

REFERENCES

1. Zhao F.Q., Lai M.C., Harrington D.L., A Review of Mixture Preparation and Combustion Control Strategies for Spark-Ignited Direct Injection Gasoline Engines, SAE Technical paper, No 970627, 1997.
2: Lefebvre A.H., Atomization and Sprays, Hemisphere Publishing Corporation, 1989.
3. Doumas M., Laster R., Liquid-Film Properties for Centrifugal Spray Nozzles, Chem. Eng. Prog., pp 518-526, 1953.
4. Dombrowski N., Hasson D., AICHE Journ., Vol. 15, No 4, pp 604, 1969.
5. Ren W.M., Shen J., Nally J.F., Geometrical Effects on Flow Characteristics of Gasoline High Pressure Direct Injector, SAE Technical paper, No 97FL-95, 1997.
6. Chinn J.J., Yule A.J., De Keukelaere J.K., Swirl Atomizer Internal Flow : A Computational and Experimental Study, ILASS Europe, pp 41-46, Sweden 1996.
7. Bayvel L., Orzechowski, Liquid Atomization, Taylor and Francis, 1993.
8. Dumouchel C., Bloor M.I.G., Dombrowski N., Ingham D.B., Ledoux M., Viscous Flow in a Swirl Atomizer, Chemical Engineering Science, Vol. 48, No 1, pp 81-87, 1993.

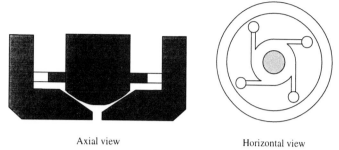

Axial view   Horizontal view

Figure 1

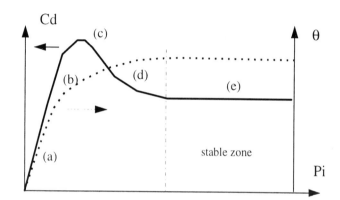

Figure 2

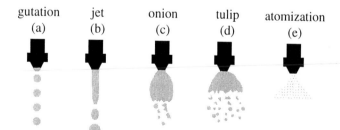

Figure 3

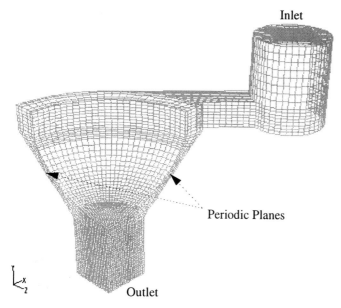

Figure 4

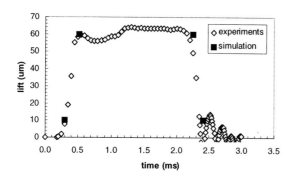

Figure 5

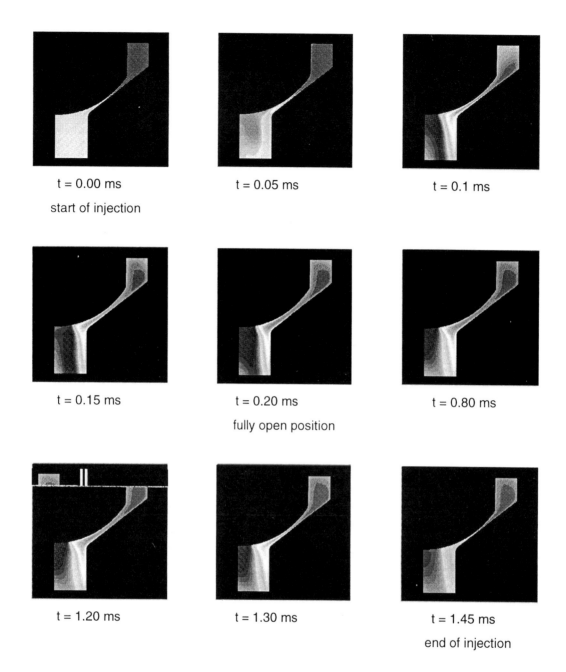

t = 0.00 ms

start of injection

t = 0.05 ms

t = 0.1 ms

t = 0.15 ms

t = 0.20 ms

fully open position

t = 0.80 ms

t = 1.20 ms

t = 1.30 ms

t = 1.45 ms

end of injection

Figure 6

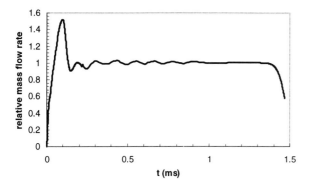

Figure 7

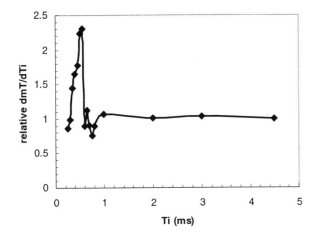

Figure 8

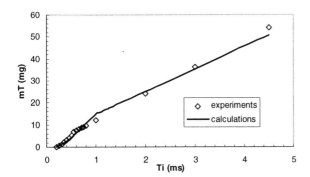

Figure 9

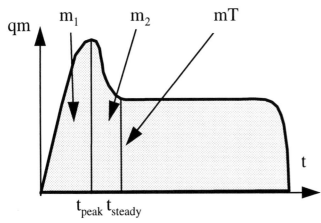

Figure 10

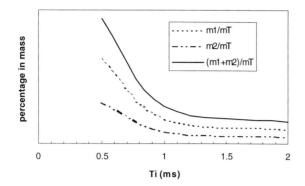

Figure 11

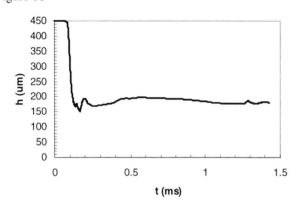

Figure 12

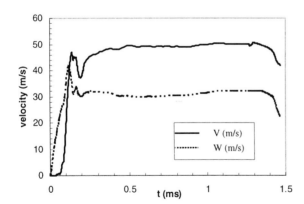

Figure 13

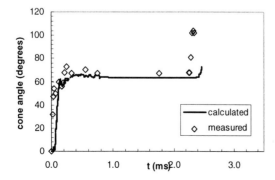

Figure 14

222

Multi-Dimensional Modeling of Direct-Injection Gasoline Engine Phenomena

Multi-Dimensional Modeling of Direct-Injection Gasoline Engine Phenomena

The physics of air-fuel mixture preparation and combustion in gasoline direct-injection engines is highly complex and nonlinear. Mixture preparation is affected by many parameters such as in-cylinder flow, chamber shape, fuel injection characteristics, chamber temperatures, and residual plus EGR content, and the interaction of these parameters makes the design of combustion systems for charge stratification a highly complex process. Gasoline direct-injection engines also have to operate with early injection for mid to high loads, which requires the charge to be well mixed. To understand these opposing requirements, engine designers have developed sophisticated numerical modeling tools.

Early multi-dimensional models were used to assist in the design of stratified charge rotary engines. More recently, OEMs are using multi-dimensional models to assist in engine combustion system design. Such models are mostly used for prediction of air-fuel mixture preparation, although they have also been used to predict combustion. There has been much effort in spray modeling for high-pressure injectors. This has resulted in higher quality predictions, although spray-wall impingement under engine-like conditions is still not fully understood.

Most multi-dimensional modeling codes use an averaged k-e turbulence model which results in the prediction of an average cycle. Some recent efforts involve the use of large eddy simulation (LES) codes which offer the promise of predicting the effects of cycle-by-cycle variation. Models are also being developed to address the effect of grid size on mixing. A typical grid size for a calculation is on the order of 2-3 mm, while mixing scales can be smaller. Some research is in process to develop sub-grid mixing models, which will allow a closer approximation of the actual physics. Many other improvements are also in process, which will further accelerate the use of multi-dimensional modeling for combustion system design.

The papers selected for this section cover early efforts on a stratified charge rotary engine as well as more recent research on gasoline direct-injection engine technology. These papers represent only a fraction of the many that have been published. There were many excellent papers to choose from, which made the task of selecting these few quite difficult. Judging from the increasing number of technical papers in this area in the last few years, it is safe to say that multi-dimensional modeling has become an integral part of IC engine combustion system design.

Fuel-Air Mixing and Distribution in a Direct-Injection Stratified-Charge Rotary Engine

J. Abraham
John Deere Technologies International
Wood-Ridge, NJ
F. V. Bracco
Dept. of Mechanical and Aerospace Engineering
Princeton Univ.
Princeton, NJ

A three-dimensional model for flows and combustion in reciprocating and rotary engines is applied to a direct-injection stratified-charge rotary engine to identify the main parameters that control its burning rate. It is concluded that the orientation of the six sprays of the main injector with respect to the air stream is important to enhance vaporization and the production of flammable mixture. In particular, no spray should be in the wake of any other spray. It was predicted that if such a condition is respected, the indicated efficiency would increase by some 6% at higher loads and 2% at lower loads. The computations led to the design of a new injector tip that has since yielded slightly better efficiency gains than predicted.

IN DIRECT-INJECTION ENGINES, such as Diesel engines and stratified-charge engines, atomization of the liquid, vaporization of the droplets, mixing of fuel and air, and the distribution of the fuel-air mixture within the combustion chamber ultimately determine both efficiency and emissions. Thus, it is necessary to understand those processes and to identify their controlling parameters in order to improve engine design. So far, engine improvements have primarily depended on experimental trial and error procedures. Such procedures may be efficient in the early stages of development of an engine, but are inefficient for the selection of engine strategies and for the refinement of well developed power plants. The direct-injection stratified-charge rotary engine is an example of an engine that has already undergone extensive development, both at Curtiss-Wright [1] and at John Deere Technologies International (JDTI) [2], and is the engine which is of interest in this work.

A three-dimensional model for flows, sprays and combustion in rotary engines was developed recently [3], and its accuracy has been assessed by comparisons with measured pressure in direct-injection stratified-charge engines [4], measured mean and turbulent velocities in a motored engine[5], and measured pressure in a premixed-charge engine[6]. In this work, the model is applied to study the processes that control vaporization, mixing and fuel-air distribution in a direct-injection stratified-charge engine. Sufficient progress is made to identify an injector configuration that should result in improved indicated efficiency.

The engine of interest is sketched in Fig. 1. It has a pilot injector with one hole and a main injector with six holes. The spark plug is located near the pilot injector. The engine parameters are given in Table 1.

The three-dimensional model is described in [3]. It is important to point out that the same set of model constants are used in all of our computations for both rotary [3-6] and reciprocating [7-10] engines. Thus, unless otherwise stated, no constants are adjusted in our computations. The numerical mesh is shown in Fig. 1. Most of the computations were made with 25 grid points along the rotor (I-direction), 8 points in the radial direction (K) and 13 points in the axial direction (J). Higher resolution gives more accurate results, but does not change the computed flowfield and our conclusions in any significant way.

RESULTS AND DISCUSSION

Five sets of computations are discussed. In the first set, combustion is computed for the JDTI engine with one of its current main-injector tips. For it, the orientation of the six holes was selected by the engine designer with the intent to distribute the fuel uniformly within the rotor cup. It is concluded that within the combustion chamber, there exist fuel-rich regions that slow down the burning rate significantly. It is also concluded that most of the combustion occurs after the end of fuel injection, so that analysis of fuel vaporization and mixing prior to combustion is appropriate. In the second set of computations, the fuel distribution prior to

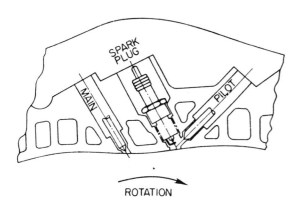

ROTATION

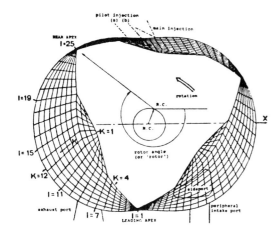

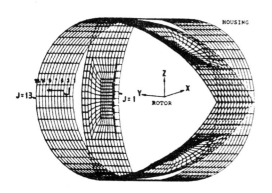

Fig. 1 - Sketch of the fueling and ignition parts of the engine and the three-dimensional grid of the computations.

TABLE 1

ENGINE PARAMETERS

Generating Radius = 10.64 cm; Eccentricity = 1.54 cm; Width = 7.71 cm; Displacement Volume = 663 cm^3; Compression Ratio = 7.5; Injector Orifice Diameter = 254 μm (main), 178 μm (pilot).

combustion is studied at two engine speeds, and two loads which are representative of the engine operating range. It is concluded that the fuel rich regions are present at all conditions, and should be reduced. In the third set, the objective of the computations, which are again without combustion, is to identify the processes that control fuel distribution in this particular engine. It is concluded that the orientation of the sprays from the main injector can be used to improve the fuel distribution. In a fourth set of computations, which are still without combustion, two configurations of the six holes of the main injector tip are analyzed, and one is selected for combustion computations. In the fifth set of calculations, the combustion results obtained with the new injector tip are compared with those of the standard injector tip, and it is concluded that with the new tip the indicated efficiency will increase throughout the computed range and by as much as 6%.

COMBUSTION WITH THE STANDARD INJECTOR TIP - The left column of Fig. 2 shows the computed flowfield in the symmetry plane (J=7) for case 1 of Table 2, which is used as the reference case. (The right column of Fig. 2 will be discussed later.) Gas velocity (a), turbulent diffusivity (b), liquid fuel parcels (c,f), gaseous fuel mass fraction (d), and gas temperature (e) are shown at five crankangles: 1035; 1065; 1080 (TDC); 15; 30. Fig. 2 (1035c) shows the location of the liquid fuel, i.e., the pattern of the sprays from the pilot and the main injectors. The pilot is located in the leading side past the major axis.

TABLE 2
COMPUTATION CONDITIONS

Case	Shaft rpm	Equiv. Ratio	Spark* Start		Spark* End		Pilot* Start	Pilot* End	Main* Start	Main* End
1	6000	0.60	1002	(1035)	1055	(1041)	1020	1048	1030	1080
2	5983	0.40	1001	(1027)	1054	(1033)	1012	1050	1025	1058
3	3004	0.65	1002	(1039)	1060	(1042)	1024	1052	1030	1055
4	3037	0.40	1016	(1045)	1062	(1048)	1030	1033	1035	1055
1'	6000	0.60	——		——		——	——	1030	1080
2'	6000	0.24	——		——		——	——	1030	1050
3'	3000	0.60	——		——		——	——	1025	1050
4'	3000	0.24	——		——		——	——	1025	1035

* Crankshaft Degrees (1080 = TDC)

228

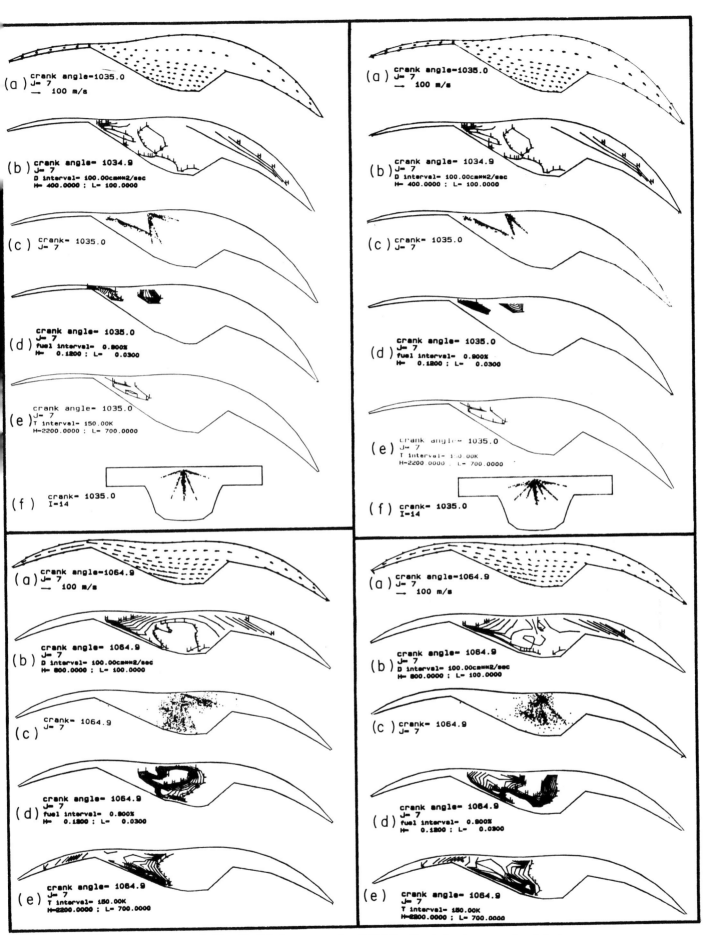

Fig. 2 - Computed flowfield for case 1 of Table 2 with the standard injector tip (left column) and with the staggered fan injector tip (right column) at the stated crank angles: (a) gas velocity; (b) turbulent diffusivity; (c,f) sprays; (d) gaseous fuel mass fraction; (e) gas temperature.

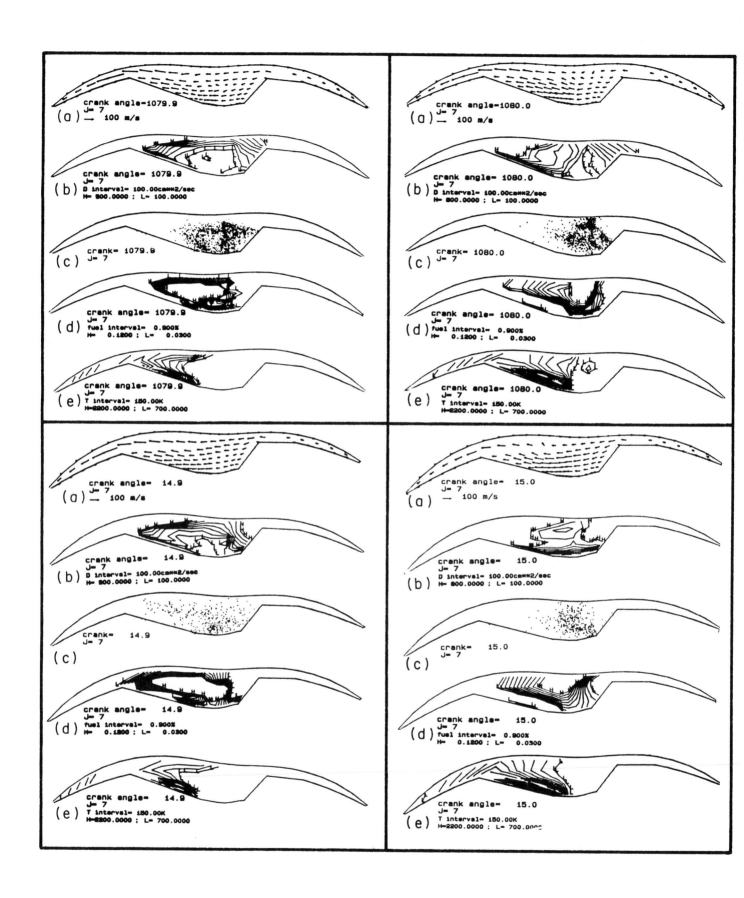

Fig. 2 (cont.) - Computed flowfield for case 1 of Table 2 with the standard injector tip (left column) and with the staggered fan injector tip (right column) at the stated crank angles: (a) gas velocity; (b) turbulent diffusivity; (c,f) sprays; (d) gaseous fuel mass fraction; (e) gas temperature.

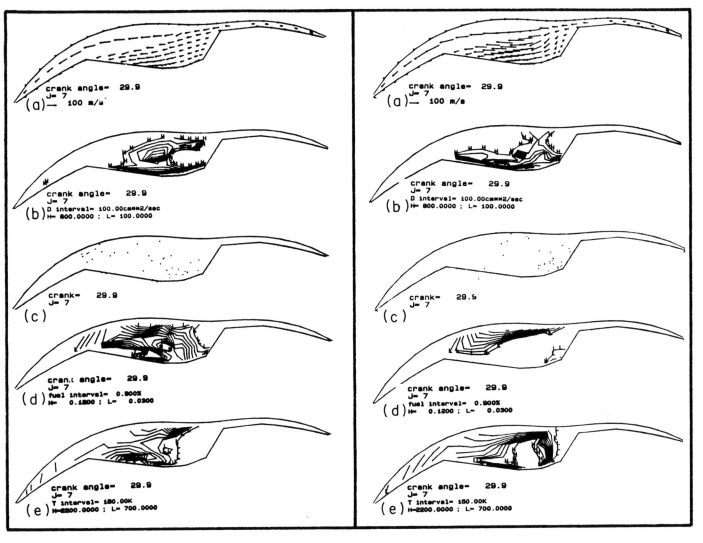

Fig. 2 (cont.) - Computed flowfield for case 1 of Table 2 with the standard injector tip (left column) and with the staggered fan injector tip (right column) at the stated crank angles: (a) gas velocity; (b) turbulent diffusivity; (c,f) sprays; (d) gaseous fuel mass fraction; (e) gas temperature.

The main is located in the trailing side. The orientation of its six holes can be visualized with the help of Fig. 2 (1035c,f).

Of particular interest are the gaseous fuel mass fraction contours at the five crankangles that are shown in Fig. 2 (1035d; 1065d; 1080d; 15d; 30d). The "H" contour indicates a gaseous fuel-air mixture which is close to the upper flammability limit, i.e., $\phi = 2$ for a stoichiometric gaseous fuel mass fraction of 0.06, and the "L" contour indicates a mixture which is close to the lower flammability limit, i.e., $\phi = 0.5$. In regions in which no contours are shown, the local and instantaneous equivalence ratio of the gaseous fuel and air mixture is either above 2.0 or below 0.5, both extremes being highly undesirable. Regions with $\phi > 2$ are seen in the middle of the combustion chamber at crankangles 1065, 1080 and 15 in the left column of Fig. 2 (1065d, 1080d, 15d). During combustion, the fuel-rich regions actually halt the propagation of the flame, as shown by the corresponding temperature contours of Fig. 2 (1065e, 1080e,15e). Indeed, by 1080 (TDC), nearly all of the fuel has been injected, but only 10% has burned, as shown in Fig. 3. The same figure also clearly evidences very slow burning for nearly 20 crankangle degrees. Obviously, there is a need for reducing the richness of the charge in the middle of the chamber, and the rest of the paper reports the search for a simple way of doing it. The fact that so little fuel is burned during most of the injection suggests that, initially, we may concentrate on the distribution of fuel prior to combustion, thus saving computer time.

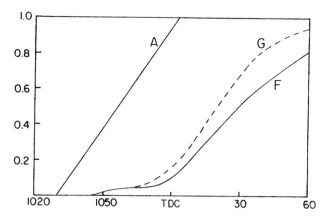

Fig. 3 - Injected fuel fraction (A) and burned fuel fraction versus crankangle for the standard (F) and the staggered fan (G) injector tip. Case 1 of Table 2.

It should also be pointed out that as much as 12% of the liquid fuel is predicted to impinge on the moving, hot wall of the rotor. The model assumes that gaseous fuel is generated at the wall at the same rate and location at which liquid fuel impinges on it. For the engine examined in this paper, the poor vaporization and mixing around the main injector is computed to be more influential on the heat release rate than the impingement of liquid fuel on the rotor wall. But no direct experimental evidence exists that the computed amount of fuel impinging on the wall and its computed evolution are either correct or incorrect.

FUEL DISTRIBUTION PRIOR TO COMBUSTION - It would help to know what specific distribution of ϕ one would like to establish within the combustion chamber, but there is no unique value of ϕ that simultaneously maximizes indicated efficiency and minimizes emissions of NOx, UH and soot, so that optimization of an IC engine is always a search for an acceptable compromise. Mixtures with $0.5 < \phi < 2.0$ may be considered acceptable in burning speed and completeness. To follow the progress of such mixtures, it is convenient to call the rich fraction that fraction of the vaporized fuel for which $\phi > 2$, the flammable fraction that for which $0.5 < \phi < 2$, and the lean fraction that for which $\phi < 0.5$. Also, the injected fuel fraction will be the fraction of the total fuel that has been injected up to a specific crankangle, and the vaporized fuel fraction will be the portion of the injected fuel fraction that has vaporized. Thus, in the absence of combustion, the rich, flammable, and lean fractions add up to unity at each crankangle.

Fig. 4(1) shows the injected fuel fraction (curve A), vaporized fuel fraction (curve B), rich fraction (curve C), flammable fraction (curve D), and lean fraction (curve E) versus crankangle (θ) for case 1' of Table 2. The conditions of case 1' are the same as those of case 1, except that there is no pilot injection and no combustion. The quantities of Fig. 4 are obtained by integrating the 3-dimensional results

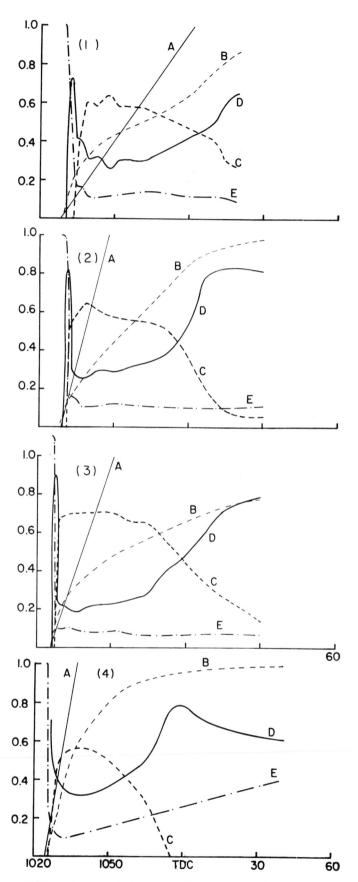

Fig. 4 - Injected fuel fraction (A), vaporized fuel fraction (B), rich fraction (C), flammable fraction (D), and lean fraction (E) versus crankangle for cases 1'-4' of Table 2 (without combustion).

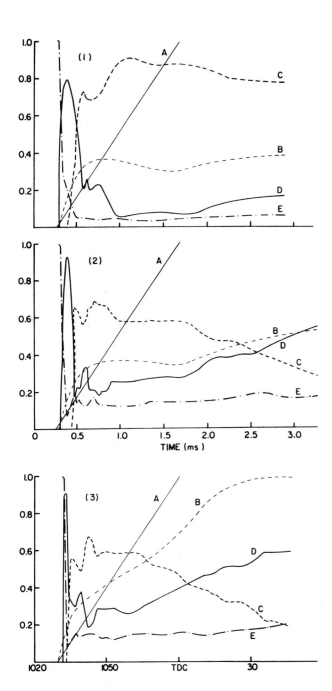

Fig. 5 - Injected fuel fraction (A), vaporized fuel fraction (B), rich fraction (C), flammable fraction (D), and lean fraction (E) versus time (1,2) or crankangle (3) for a single spray and different turbulent diffusivity and convection.

over the volume of the combustion chamber at selected crankangles. This type of plot will be called a f(fraction)-θ (crankangle) plot.

It may be seen from Fig. 4(1) that at the very beginning of injection, the lean fraction dominates, since the small amount of fuel that has vaporized is quickly convected away from the liquid source and mixed with a gas that is mostly air and residuals from the previous combustion

cycle. As more liquid fuel vaporizes, the lean fraction drops sharply and the flammable one increases rapidly, but only briefly, since fuel vaporization continues and a sharp rise of the rich fraction, and fall of the flammable fraction, follow. About 5 crankangle degrees after the beginning of injection, the rich fraction becomes greater than the flammable fraction. For the next 45 crankangle degrees (i.e., TDC), the rich fraction is greater than the flammable one by as much as a factor of 2. After that, injection ends, and diffusion and convection reduce the rich fraction, even while vaporization continues, since there is no combustion.

Fig. 4(2) shows the f-θ plot for case 2' of Table 3 in which the quantity of injected fuel is smaller (i.e., the load is lower) but the rpm is the same. The trends are similar, but the rich fraction is less than in case 1', and the flammable fraction starts to increase rapidly before TDC because the quantity of fuel is smaller and that of air is about the same. Figs. 4(3,4) show the f-θ plots for cases 3',4' that are at lower rpm, and the trends remain the same.

Thus, it is concluded that the rich fraction, that is established before much of the combustion takes place, ends up dominating combustion over a good part of the operating range of the standard engine.

PROCESSES CONTROLLING FUEL DISTRIBUTION - A trial and error procedure to find ways to reduce the dominance of the rich fraction would be too time consuming and expensive, even computationally. It is necessary first to isolate some of the parameters that control the mixing of the fuel and air; turbulent diffusivity and convection come to mind immediately.

The diffusivity plots and the gas velocity plots in the left column of Fig. 2 show that in the reference case, they can vary by a factor of 10 within the combustion chamber and at various crankangles. The sensitivity of the mixing to the two parameters is then of interest since both quantities can be controlled, if only moderately, by changing the shape of the rotor pocket, for example.

Fig. 5(1) shows a f-θ plot for a case where the rotor is fixed at crankangle 1035, i.e., the only gas velocity is that induced by the spray, and fuel is injected through a single hole. The initial chamber diffusivity is set at the low value of 2 cm²/s. During the injection and within the spray, the diffusivity increases to the order of 10 cm²/s. It may be seen that the rich fraction is very large and the flammable and lean fractions are very small. Fig. 5(2) shows the corresponding results for a case in which the diffusivity is kept everywhere at the high value of 200 cm²/s. As expected, the rich fraction decreases and the flammable and lean ones increase due to the greater mixing rate, but the flammable fraction increases by only a factor of 2 or 3 in response to a change by a factor of 100 in the gas diffusivity. Notice also that the vaporization rate is not much larger than that of Fig. 5(1).

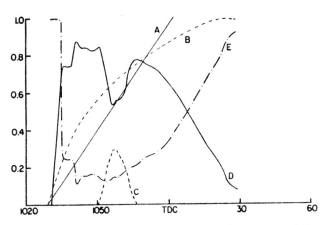

Fig. 6 - Injected fuel fraction (A), vaporized fuel fraction (B), rich fraction (C), flammable fraction (D), and lean fraction (E) versus crankangle when only one of the six holes of case 1' of Table 2 is active.

Thus, it is concluded that controlling the gas diffusivity is a way of increasing the flammable fraction, but it would not be a simple way, and the limited sensitivity of the flammable fraction to the diffusivity suggests that, for this engine, one should first try to identify alternative means.

In Fig. 5(3), the effect of convection is considered. The shaft speed is 6000 rpm, but the gas diffusivity is still fixed at 200 cm^2/s. It is seen that the vaporization rate is increased significantly, but the fraction of flammable mixture is not increased drastically.

Thus, it is concluded that tailoring the relative velocity between spray and gas is an effective way of changing the vaporization rate, but not a sensitive way of controlling the flammable fraction. Although increasing the vaporization rate can be important, and can be obtained easily by changing the angle of the spray, it was decided to pursue the control of the flammable fraction first.

The interactions among sprays are also likely to influence the formation of the rich, flammable and lean fractions. Fig. 6 shows the f-θ plot for the spray from a single hole of the multi-hole tip of case 1' injecting 1/6 of the fuel and without the pilot spray. By comparing Fig. 6 with Fig. 4(1), it is seen that the single hole gives a much smaller rich fraction and a much greater flammable fraction than the six holes. (Eventually, the lean fraction increases rapidly through convection and diffusion and on account of the large amount of excess air, but by this time, combustion would have altered the flowfield significantly.) When several sprays are present, the fuel which is convected and diffused away from one spray combines with that of neighboring sprays to increase the rich fraction and to reduce the lean one. The change of the flammable fraction probably is more sensitive to details. The significant difference between the single and the multiple-

hole configurations suggests that the number and orientation of the holes should be considered further; indeed, changing the holes in the injector tip is the simplest of modifications.

Fig. 7 shows two sprays, one behind the other with respect to the direction of the air flow and also side-by-side and separated by different angles. During injection, when one spray is behind the other, the rich fraction is significantly greater than for the single hole (Compare Fig. 7(1) with Fig. 6.). But when the two sprays are side by side, the difference is smaller, and if the angle between them is sufficiently large, the difference tends to disappear. (After the end of injection, two side-by-side sprays give a much higher flammable fraction than the single hole, but only on account of the smaller amount of excess air.)

It is concluded that it is practical to try and improve the fuel-air distribution by changing the number and orientation of the spray holes since the change is easy in practice and its effect is significant. In particular, injector tips in which the holes are side by side with respect to the air motion appear promising. We shall refer to them as fan tips.

FUEL DISTRIBUTION WITH FAN INJECTOR TIPS - Fig. 8 shows the f-θ plots for cases 1'- 4' of Table 2 without combustion and with an injector fan tip in which the six holes are side by side on the same plane and across the air stream. The angle between the sprays is 30°, and the angle of their plane to the local tangent to the trochoid surface of the housing is 65°. By comparing the results of Fig. 8 with those of Fig. 4, which pertain to the standard engine and for which some of the sprays are in the wakes of others, it is noted that, after initial transients of 5 to 10 crankangle degrees, the rich fraction is always smaller, and the flammable one larger, for the fan tip than for the standard tip. While injection continues, the lean fraction is about the same, constant and relatively small; thus, the flammable fraction has increased at the expense of the rich fraction, as desired. Also interesting is that the vaporized fuel fraction is initially always smaller with the fan tip than with the standard tip. (Later the trend reverses, but, again, by then combustion would have altered the flowfield.) The difference in vaporization rate is due to the relative velocity between the spray and the chamber gas which, in turn, is influenced by the angle between the spray and the local tangent to the trochoid surface. In the standard injector tip, three of the sprays are at 15°, two at 65°, and one at 73°; in the fan tip all six sprays are at 65°. For the sprays at 15° the relative velocity is greater and so is the vaporization rate.

For the particular engine which is considered in this study, combustion rate is generally low. Thus it is advantageous not only to increase the flammable fraction, but also the vaporization rate. Accordingly, at this point, we could have studied different angles for the plane of the spray flow, different hole sizes and number, different injection timing and rate,

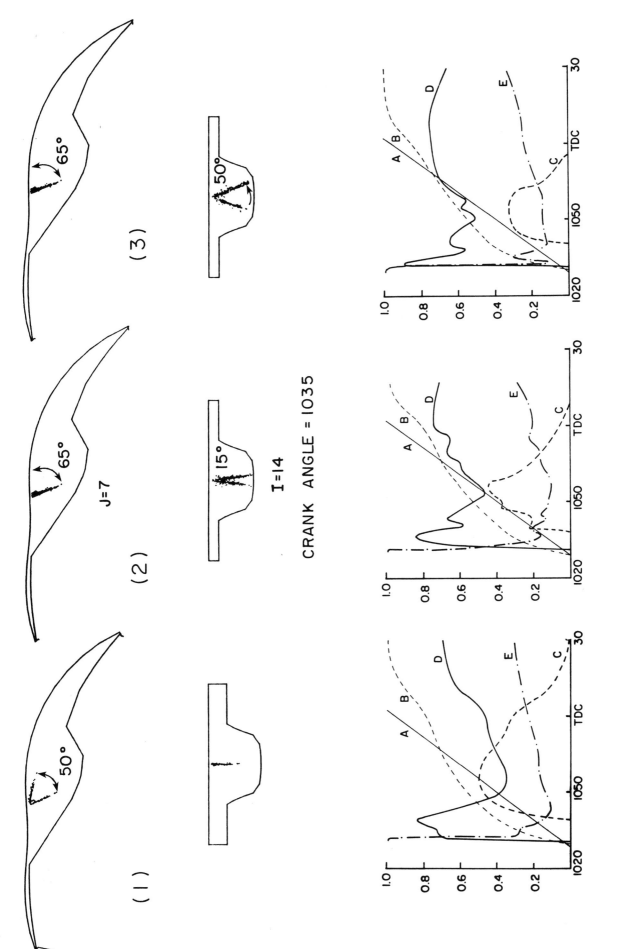

Fig. 7 - Injected fuel fraction (A), vaporized fuel fraction (B), rich fraction (C), flammable fraction (D), and lean fraction (E) versus crankangle for two sprays at different orientations with respect to the direction of the air flow.

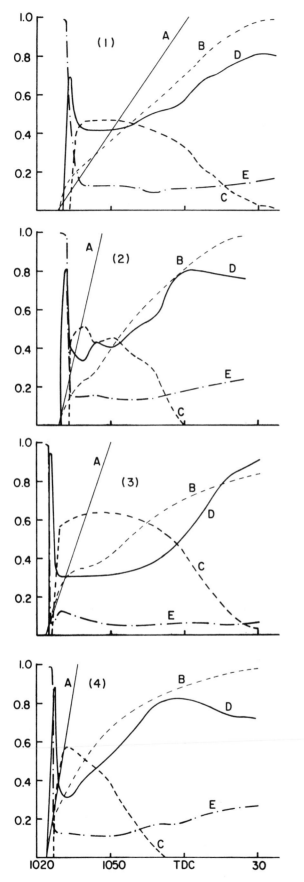

Fig. 8 - Injected fuel fraction (A), vaporized fuel fraction (B), rich fraction (C), flammable fraction (D), and lean fraction (E) versus crankangle for cases 1'- 4' of table 2, but with the staggered fan injector tip.

etc. Instead, we decided to investigate a tip configuration that is a compromise between the fan tip and the standard tip.

Since putting six holes on the same plane in the small tip of an injector can be tricky, an alternative design was explored in which two holes are on a plane at 45° to the trochoid surface, four are on a plane at 65°, and no spray is directly in the wake of any other. These angles were selected to minimize fuel impingement on the walls and at the same time ensure that the vaporized fuel from the pilot and main form a continuous mixture. However, they should be optimized through additional computations. This arrangement is called a staggered fan, and its rich fraction was still found to be smaller, and the flammable one larger, than those of the standard tip. The improvement was smaller than for the fan tip, but still noticeable, particularly at high load and speed. Thus combustion computations were made to see whether the staggered fan tip would yield higher indicated efficiency than the standard tip.

COMBUSTION WITH STANDARD AND FAN INJECTOR TIPS - Combustion computations were made for cases 1-4 of Table 2 with both the standard and the staggered fan injector tips. Jet-A fuel was used both in the tests and in the computations. The properties of the fuel were obtained from [11-12].

As stated in the introduction, we use the same three-dimensional model and set of model constants for both reciprocating and rotary engines. In general, we also use the actual spark time [6,9,10], but for the engine studied in this paper, the use of the actual spark time is not possible. It may be seen from Table 2 that in the engine, the spark starts several crankangle degrees before the pilot injection begins, i.e., when there is no fuel in the chamber. In the model, to simulate ignition, fuel is depleted in the spark cells over a few crankangle degrees. Thus, numerical (and actual) ignition can start only after the pilot has injected some fuel. The computational spark was started at the time when $\phi = 1$ in the spark cells. The crankangles at which numerical ignition starts and ends are given in parentheses in Table 2. The criterion that ignition starts when $\phi = 1$ in the vicinity of the spark is reasonable, but it is uncertain when combustion actually starts in this engine, and we must be ready to adjust numerical ignition if evidence suggests that we should. As it happened, no such adjustment was necessary.

Initially, a combustion computation was made for case 1 of Table 2 with the fan-tip injector. The gain in efficiency was found to be 7.8%, but as previously stated, the fan tip arrangement is not practical, and the rest of the computations were then made with the practical staggered fan-tip configuration.

For case 1 of Table 2, the combustion flowfield obtained with the staggered fan injection tip, and still using the injection and spark timings of the standard engine, is shown in the right column of Fig. 2. It can be compared

236

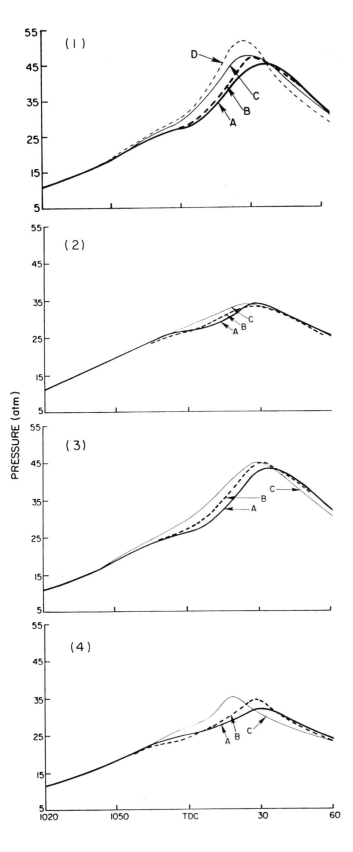

PRESSURE (atm)

(1)

(2)

(3)

(4)

Fig. 9 - Computed and measured pressure for cases 1-4 of Table 2: A, measured with standard injector tip; B, computed with standard injector tip; C, computed with staggered fan tip (and advanced timings, for cases 2 and 4); D, computed with fan tip.

directly with that obtained with the standard injector tip which is shown in the left column of the same figure. It is clear that at all crankangles, the rich fraction is smaller with the staggered fan tip than with the standard one, and that the propagation of the flame into the spray area is hampered less by the rich mixture; indeed the burning rate is faster, as shown in Fig. 3, particularly at the beginning of combustion. The computed pressures are compared in Fig. 9.

Combustion computations were also made with the staggered fan tip, but still using the injection and spark timing of the standard injector, for cases 2, 3, and 4 of Table 2. The corresponding predictions for the changes in indicated efficiency are: +6.1% (case 1); -1% (case 2); +4.3% (case 3); -2% (case 4). It is seen that a gain is predicted for the richer cases, and a loss for the leaner ones. The loss for the lean cases is due to the pressure peaking too late after TDC.

Since injection and spark timings are routinely optimized whenever changes are made to the injector tip, the combustion computations of cases 2 and 4 were repeated after advancing injection and spark timing by 10 crankangle degrees. The predicted changes in indicated efficiency became +2% and +1.5%, respectively.

In general, we do not expect our model to be accurate to within a few percent in its prediction of indicated efficiency. We use the model to interpret the flowfield of an engine and to identify strategies that should, on account of common sense, bring about desirable changes, such as an improvement in fuel economy. Thus our conclusion would be that by going from the standard injector tip to the staggered fan tip, we expect a gain in efficiency, particularly at higher loads. However, in this particular study, considering that we made no efforts to optimize injection and spark timing for the staggered fan tip, we may venture to expect a gain for the rich cases of at least 6%, and for the lean cases, of at least 2%. Considering that changing the injector tip is a minor modification, the expected improvement is quite significant and definitely worthy of a try.

SUMMARY AND CONCLUSIONS

A three-dimensional model for flows and combustion was applied to a direct injection stratified-charge rotary engine in order to try and improve its indicated efficiency. Five sets of computations were performed.

In the first set, combustion was computed for the standard engine at two speeds and two loads that are representative of the engine operating range. It was concluded that within the combustion chamber, there exist fuel-rich regions around the main injector that slow down the burning rate significantly.

In the second set of computations, the fuel distribution prior to combustion was studied. It was concluded that the fuel rich regions are formed quickly after the beginning of the main injection and disperse too slowly.

The third set of computations was to identify the processes that control fuel distribution in this particular engine. It was concluded that although turbulent diffusion and relative velocity between sprays and air are important, it is the orientation of the sprays from the main injector that can be modified most readily and to the greatest advantage.

In particular, it was found that rich regions are formed around the main injector when one spray is in the wake of another. Thus, the sprays from the main injector should be positioned side by side and across the air stream. We call such sprays "fan" sprays.

In a fourth set of computations, two fan spray configurations were studied and one was selected for combustion and efficiency studies.

In the final set of computations, the indicated efficiency of a staggered fan arrangement was compared with that of the standard engine, and it was concluded that the staggered fan would yield efficiency gains of at least 6% for richer mixtures and 2% for leaner mixtures.

Measurements made after this study was completed, and with an injector tip similar to the staggered fan tip, have shown indicated efficiency gains of 6% or better for richer mixtures, and of 3% or better for leaner mixtures.

ACKNOWLEDGEMENTS

Support for this work was provided by the Rotary Engine Division of John Deere Technologies International. John Abraham also wishes to acknowledge the support of NASA Lewis Research Center through the Rotary Engine Division.

REFERENCES

1. Jones, C., Lamping, H.D., Myers, D.M. and Loyd, R.W., "An Update of the Direct Injected Stratified Charge Rotary Combustion Engine Developments at Curtiss-Wright", SAE Paper 770044 (Also in SAE Trans., Vol. 86), 1977.

2. Jones, C., "A New Source of Lightweight, Compact Multi-fuel Power for Vehicular Light Aircraft and Auxiliary Applications", The John Deere SCORE^{TM} Engines, ASME Gas Turbine and Aeroengine Congress, Paper 88-GI-271, Amsterdam, Netherlands, June 1988.

3. Grasso, F., Wey, M.-J., Abraham, J. and Bracco, F.V., "Three-Dimensional Computations of Flows in a Stratified-Charge Rotary Engine," SAE Paper 870409, 1987.

4. Abraham, J., Wey, M.-J. and Bracco, F.V., "Pressure Non-Uniformity and Mixing Characteristics in Stratified-Charge Rotary Engine Combustion," SAE Paper 880624, 1988.

5. Abraham, J. and Bracco, F.V., "Comparisons of Computed and Measured Mean Velocity and Turbulent Intensity in a Motored Rotary Engine," SAE Paper 881602, 1988.

6. Abraham, J. and Bracco F.V., "Comparisons of Computed and Measured Pressure in a Premixed-Charge Natural-Gas-Fueled Rotary Engine," Paper to be presented at the 1989 SAE Congress and Exposition, February 1989.

7. Grasso, F. and Bracco, F.V., "Computed and Measured Turbulence in Axisymmetric Reciprocating Engines, " AIAA J., Vol. 21, No. 4, pp. 601-607, April 1983.

8. Hayder, M.E., Varma, A.K. and Bracco, F.V., "A Limit to TDC Turbulence Intensity in Internal Combustion Engines, " AIAA J. of Propulsion and Power, Vol. 1, No. 4, pp. 300-308, July 1985.

9. Abraham, J., Reitz, R.D. and Bracco, F.V., "Comparisons of Computed and Measured Premixed-Charge Engine Combustion, " Combustion and Flame, Vol. 60, No. 3, pp. 309-322, June 1985.

10. Abraham, J. and Bracco, F.V., "Comparisons of Computed and Measured Bulk Velocity and Turbulence in an I.C. Engine with Combustion," to appear in AIAA J. of Propulsion and Power.

11. Handbook of Aviation Fuel Properties. Coordinating Research Council Report No. 530, 1983. Distributed by SAE, Warrendale, Pennsylvania.

12. Selected Values of Physical and Thermodynamic Properties of Hydrocarbons and Related Compounds. Report of the American Petroleum Institute Research Project No. 44, Carnegie Press, 1953.

Three-Dimensional Computations of Combustion in Premixed-Charge and Direct-Injected Two-Stroke Engines

Tang-Wei Kuo and R.D. Reitz
General Motors Research Labs.

ABSTRACT

Combustion and flow were calculated in a spark-ignited two-stroke crankcase-scavenged engine using a laminar and turbulent characteristic-time combustion submodel in the three-dimensional KIVA code. Both premixed-charge and fuel-injected cases were examined. A multi-cylinder engine simulation program was used to specify initial and boundary conditions for the computation of the scavenging process.

A sensitivity study was conducted using the premixed-charge engine data. The influence of different port boundary conditions on the scavenging process was examined. At high delivery ratios, the results were insensitive to variations in the scavenging flow or residual fraction details. In this case, good agreement was obtained with the experimental data using an existing combustion submodel, previously validated in a four-stroke engine study. However, at low delivery ratios, both flow-field and combustion-model details were important, and the agreement with experiment was poor using the existing combustion submodel, which does not account for the effect of residual gas concentration.

To improve the agreement between modeling and experimental results, a modified combustion submodel was introduced that includes the effect of residual gas concentration on the laminar characteristic time. With the new submodel, agreement with the experiment has been improved considerably for all cases considered in this study. These levels of agreement between experiment and computations are similar to those found in previous applications of the laminar and turbulent characteristic-time combustion submodel to

four-stroke engine combustion. Further improvement of the combustion submodel was made difficult by the observed coupling between the in-cylinder flow-field and the combustion-model details at low delivery ratios.

Three-dimensional computer models have been applied to predict combustion in internal combustion engines. For example, computations of spark-ignited premixed-charge combustion in research and production-type four-stroke engines have been presented by Kuo and Reitz [1]*. Spark-ignited, premixed-charge and direct-injected rotary engine computations have been presented by Abraham and Bracco [2,3]. The latter computations suggested design changes that led to approximately 6 percent improvement in efficiency in a direct-injection stratified-charge rotary engine.

The above studies were conducted using a characteristic-time combustion submodel originally proposed by Abraham et al. [4]. In this model, chemical species approach their thermodynamic equilibrium with a rate that is a combination of the turbulent-mixing time and the laminar chemical-kinetics time. The combination is formed in such a way that the longer of the two times has more influence on the conversion rate. An additional element of the model is that the laminar-flame kinetics strongly influence the early flame development following ignition.

Comparisons with experimental engine pressure measurements indicate that the

---

*Numbers in brackets denote References at end of paper

model predictions agree reasonably well with measurements under normal engine operating conditions. The causes of discrepancies were discussed in detail by Kuo and Reitz [1,5] who observed that the level of agreement between the predictions and the experiments is consistent with the levels of uncertainty in the input parameters to the computations (e.g., there is some uncertainty about the gas temperature, turbulence intensity and length scale existing in the combustion chamber at the start of combustion, and also in the wall-heat-transfer and turbulence model constants).

There is current interest in engines that operate under dilute conditions either with air, internal residuals, or recirculated exhaust gas. Najt and Kuo [6] have successfully applied the laminar and turbulent characteristic-time combustion submodel to engines diluted with air. A major goal of the present study was to investigate the performance of the combustion submodel for engines diluted with residual gas. This condition is of interest in variable-valve-actuation engines with large valve overlap, and in premixed-charge and direct-injected two-stroke engines at low delivery ratios, which have high residual gas concentrations in the combustion chamber. For this study, the model was applied to crankcase-scavenged two-stroke engine combustion for which experimental data is available.

This paper is organized as follows. First, the experiments used to assess the performance of the combustion submodel are described. Then, details are given of the initial and boundary conditions, and the combustion and spray submodels. The results are divided into two parts. The first part explores the sensitivity of the results to the in-cylinder flow field and combustion model details in the premixed-charge engine. The second part presents a parametric study of the model performance in the premixed-charge and direct-injected engines.

EXPERIMENTS FOR COMPARISON

The data used in the model evaluation were obtained from tests of a three-cylinder crankcase-scavenged two-stroke engine operated with both premixed charge and direct fuel injection. The engine utilized a reed-valve induction system. The engine characteristics are summarized below for later reference.

ENGINE DESCRIPTION - The geometrical specifications for both the premixed-charge and direct-injected engines are listed in Table 1. The combustion chamber

Table 1 - Two-Stroke Engine Specifications

Nominal Compression Ratio	11.2
Effective Compression Ratio	6.5
Bore (mm)	84.0
Stroke (mm)	72.0
Cylinder Displacement (L)	0.4
Connecting Rod Length (mm)	135.9
Squish Height† (mm)	0.8
Squish Area¶ (%)	70.0
Exhaust-Port Opening	87.0° ATDC
Transfer-Port Opening	115.0° ATDC

† Squish height is defined as the distance between the piston-crown and bottom of the head-dome at TDC.

¶ % squish area is defined as 100 times the squish region shown in Fig. 1 divided by the piston-top area.

has an offset head-dome and a spherical-segment piston top, see Fig. 1. The modeled transfer and exhaust ports are shown schematically in Fig. 2. The model engine considered had three transfer ports (two side ports and one boost port) and one exhaust port. In the actual engine, the two side ports and the boost port are divided into two ports each, yielding a six-port system. The arrangement of the transfer ports around the cylinder liner is symmetric with respect to a diametrical plane through the center of the exhaust port. The flow angles of the ports are shown in Fig. 2. The angles for the side ports were obtained from considerations of LDV measurements in a similar engine [7], which were somewhat limited at the time this work was done. The angle for the boost port was estimated using the geometry of the port passage and was subsequently varied parametrically.

FUELING AND IGNITION - For the premixed-charge engine, the spark gap was located 3 mm below the top of the head dome as indicated in Fig. 1A. The fuel used was propane. For the direct-injected engine, the spark plug was located on the left-hand-side of the head dome as shown in Fig. 1B. The fuel was injected with a poppet nozzle that produced a pressure-atomized hollow-cone spray with a spray full angle of 78 degrees. Measured fuel flow rates and injection pressure histo-

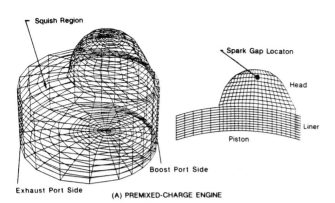

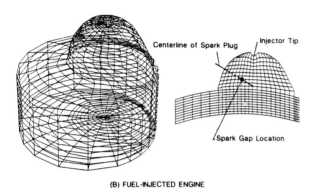

Fig. 1 - Details of combustion chamber geometry and computational grid used in the premixed-charge and fuel injected engines

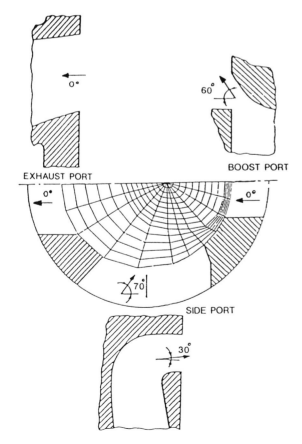

Fig. 2 - Details of the port flow angles and computational grid

ries were used to estimate the injection duration and the injection velocity. The mean injection pressure was about 4000 kPa. The injector tip location is also shown in Fig. 1B. The fuel used was Amoco-91 gasoline. This fuel was simulated as isooctane in the model.

TEST CONDITIONS - Table 2 summarizes the test conditions for both engines. For

Table 2 - Test Conditions

Case	Delivery Ratio	Overall Equivalence Ratio	Engine Speed (r/min)	Spark Timing (Deg BTDC)	Fuel Flow (g/s)	Air Flow (g/s)
Premixed-Charge Engine:						
1	1.0	1.0	1600	10.7	2.21	34.8
2	0.7	1.0	1600	13.7	1.54	24.2
3	0.5	1.0	1600	19.7	1.10	17.2
4	0.3	1.0	1600	29.7	0.66	10.4
5	1.0	1.0	800	8.9	1.18	18.5
Direct-Injected Engine:						
6	0.92	0.24	1600	20.2	0.61	34.2
7	0.66	0.27	1600	22.2	0.43	21.6

the premixed-charge engine, two effects were examined: variations in delivery ratio (DR = 0.3, 0.5, 0.7 and 1.0) and speed (800 and 1600 r/min). Delivery ratio is defined as the mass of air supplied to the cylinder divided by the theoretical mass of air that could be retained in the displacement volume at ambient conditions, taken to be standard pressure and temperature (1 atm and 298 K). The spark timing was MBT (Minimum spark advance for Best Torque), and the wall-temperature was assumed to be 400 K for all cases. The effect of delivery ratio was also examined for the direct-injected engine (DR = 0.92 and 0.66).

The predicted chamber-pressure histories were compared with available data from experiments. The measured air and fuel flow rates (see Table 2) were used to specify the operating conditions for a multi-cylinder engine-simulation program [8] that was used to determine initial and boundary conditions for the three-dimensional computations. The complexity of the flows into and out of the chamber during the scavenging event due to interactions between the three cylinders is evident in the mass flow rate histories shown in Fig. 3, calculated using the engine simulation program. Positive flow rates correspond to flow into the cylinder.

THE MODEL

The computations were made using the KIVA computer program, which solves

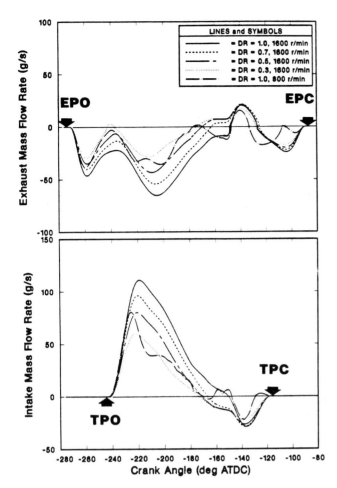

Fig. 3 – Calculated exhaust and intake
mass flow rates through the
ports using the multi-cylinder
engine-simulation program

EPO: exhaust port opening
EPC: exhaust port closing
TPO: transfer port opening
TPC: transfer port closing

three-dimensional equations of transient
chemically-reactive fluid dynamics. The
equations and the numerical-solution
method are discussed in detail by Amsden
et al. [9]. The finite-difference grid
used in the computations is given in
Fig. 1. Details of the combustion, ignition and spray submodels, and the initial
and boundary conditions used in the present study are described in the following
sections.

COMBUSTION SUBMODEL – The combustion
submodel of Abraham et al. [4] simulates
both the growth of the initial flame kernel into a fully developed turbulent flame
and the subsequent turbulent flame propagation. With this model, the time rate of
change of species mass fraction due to
conversion from one species to another is
given by

$$\frac{dY_i}{dt} = -(Y_i - Y_i^*)/\tau_{eq} \qquad (1)$$

where Y_i is the mass fraction of species
i, and Y_i^* is the local and instantaneous
thermodynamic equilibrium value of the
mass fraction. (The present computations
considered seven species: fuel, O_2, N_2,
H_2O, CO_2, CO and H_2.)

The characteristic time for
conversion, τ_{eq}, is a combination
of the local laminar-kinetics time, τ_L,
and the turbulent-mixing time, τ_T, (to
ensure that the longer of the two characteristic times has more influence), i.e.,

$$\tau_{eq} = \tau_L + f \tau_T \qquad (2)$$

where

$$\tau_L = A \, T \, P^{-(1+2\beta)} e^{(Eg(\phi)/T)}/ \, CF^2 \qquad (3)$$

and

$$\tau_T = C_{m2} \, \eta \, k/\epsilon \qquad (4)$$

The parameter η in the turbulent-mixing
time is equal to unity when h is ≥ 1 and η
= 1/h when h < 1, where

$$h = \frac{C_{m2}}{C_{m3}} \frac{(Y_p - Y_{ps})}{(Y_F - Y_F^* + Y_{O_2} - Y_{O_2}^*)}$$

The constants C_{m2} and C_{m3} are listed in
Table 3.

Table 3 - Combustion Model Constants

A	=	$1.54 \cdot 10^{-12}$ atm·s/K	(propane)
		$3.09 \cdot 10^{-12}$	(isooctane)
B	=	0.08	(propane)
		-0.08	(isooctane)
E	=	15098 K	
C_{m1}	=	1.4	
C_{m2}	=	0.058	
C_{m3}	=	0.097	
Pr	=	0.5	

Ignition was modeled by adding energy
to the charge in the computational spark
cell at a specified rate (504 W/mm³), as
described by Kuo and Reitz [1]. Laminar
kinetics effects are accounted for during
ignition with the delay coefficient, f in
EQ (2), specified as

$$f = 1 - e^{-(t - t_s)/\tau_d}$$

where $\tau_d = C_{m1} \ell/S_L$ and $(t - t_s)$ and ℓ are the time after spark and the turbulence length scale, respectively. The laminar flame speed, S_L, was determined using the correlation of Metghalchi and Keck [10], i.e.,

$$S_L = [B_m - B_2(\phi - \phi_m)^2] \, T_o^{\,\alpha} P_o^{\,\beta}(1 - 2.1 \, r) \qquad (5)$$

where $\alpha = 2.18 - 0.8(\phi - 1)$, $\beta = -0.16 + 0.22(\phi - 1)$ and T_o and P_o are the unburned gas temperature and pressure ahead of the flame (normalized by 298 K and 1 atm, respectively). The local residual mass fraction is r, and the constants B_m, ϕ_m and B_2 are 34.22, 1.08 and 138.65 for propane, and 26.32, 1.13 and 84.72 for isooctane, respectively. The model constant C_{m1} is also listed in Table 3. Its value, and those of C_{m2} and C_{m3}, were suggested previously by Kuo and Reitz [1,5] when a wall-boundary-layer Prandtl number (Pr) of 0.5 is used for the wall-heat-transfer calculations.

The function $g(\phi) = 1 + B|\phi-1.15|$ and the constants A, B, and E in the equation for the laminar-kinetics time, EQ (3), were obtained by calibrating computed and measured laminar flame speeds over a range of equivalence ratios (ϕ), pressures (P) and temperatures (T) [1]. These combustion-model constants are also listed in Table 3. The variable $\beta = -0.16 + 0.22(\phi - 1)$ in the laminar-kinetics time was introduced by Reitz and Kuo [5] to account for the effect of equivalence ratio on the pressure exponent that is consistent with the flame speed correlation of EQ (5) (a constant pressure exponent equals -0.75 was used previously in [1,4]). It was derived using the flame speed correlation of EQ (5) and the relationship [11]

$$\tau_L \propto D_L/S_L^2 \qquad (6)$$

using the fact that the laminar diffusivity, D_L, is inversely proportional to pressure.

The revised expression for the laminar-kinetics time that accounts for the effect of equivalence ratio on the pressure exponent is still inconsistent with the laminar-flame-speed correlation of EQ (5) since the effect of residual gas concentration is not considered, i.e., the equation considers air-fuel mixtures only. This situation corresponds to

$$CF = 1.0 \qquad (7)$$

in EQ (3). This does not affect conclusions reached in the previous studies [1,5] since the residual mass fraction was only about 10 percent for all cases examined.

In order to account for the residual gas effect on the laminar-kinetics time, one can derive the expression

$$CF = 1.27 \, (1 - 2.1 \, r) \qquad (8)$$

using Eqs. (5) and (6) the same way the effect of equivalence ratio was accounted for in the pressure exponent. The constant 1.27 in EQ (8) is introduced to ensure that the correction factor CF reduces to unity at a residual mass fraction of 10 percent. Thus, it is expected that the correction in EQ (8) will have minimal effect on the computed results in the previous studies [1,5] and in the present study for engine operation with high delivery ratio since these cases also have residual mass fraction levels close to 10 percent.

However, at low delivery ratios, the calculated residual mass fraction can reach values up to 45 percent. The flame speed correlation of Metghalchi and Keck [10] used to derive EQ (8) is based on measurements with residual mass fractions of less than 20 percent. In order to test the sensitivity of the results to the form of the residual fraction correction, a modified expression of CF was also considered in the present study as

$$CF = 1.39 \, (1 - 2.8 \, r) \quad \text{for } r < 0.3 \qquad (9a)$$

and

$$CF = 7.99 \, \exp(-11.92 \, r) \quad \text{for } r > 0.3 \qquad (9b)$$

The constant 1.39 in EQ (9a) is introduced to ensure that the correction factor CF also reduces to unity at a residual mass fraction of 10 percent. The form of EQ (9b) was chosen to give the same CF value as EQ (9a) at residual mass fraction of 0.3, and that CF remains positive for all residual mass fractions larger than

0.3. Equations (8) and (9) are plotted in Fig. 4, and they agree to within 15 percent for residual mass fractions less than 20 percent. This is well within the uncertainty of laminar-flame-speed data for hydrocarbon fuels [2,12].

SPRAY MODEL - The fuel-injection process was modeled using a computationally efficient stochastic parcel injection method [9]. In this technique, each computational parcel represents a group of drops with similar physical attributes. The drops exchange mass, momentum and energy with the gas through source terms in the gas equations. Turbulence dispersion, drop breakup and coalescence effects are included in the model as described by Reitz [13]. Further details about the spray model are given in Amsden et al. [9].

The spray drops were injected using the 'blob' injection procedure of Reitz and Diwakar [14]. The initial blob size was determined in the present study from the estimated instantaneous poppet lift. This initial drop size also agrees well with initial drop size values calculated from the empirical breakup formulae described by Reitz and Diwakar [15]. The injection velocity was specified and held constant at 60 m/s during the injection, based on line pressure measurements upstream of the injector. The injected liquid temperature was 375 K. Spray/wall impingement was also accounted for in the present study using the wall interaction submodel of Naber and Reitz [16].

INITIAL AND BOUNDARY CONDITIONS - Computations were started at the time of exhaust port opening. The initial thermodynamic conditions (mass, temperature and pressure) were assumed to be uniform in the cylinder and were specified using a multi-cylinder engine-simulation program [8]. The initial mean axial velocity component was assumed to vary linearly with distance from the piston velocity on the piston crown to zero on the head. The radial and tangential velocities were assumed to be zero. The initial value of the turbulence kinetic energy was assumed to be equal to 0.9 of the square of the mean piston speed [16]. The turbulence length scale at each point in the chamber was assumed to be proportional to the distance to the closest wall.

The mass flow rates through the transfer and exhaust ports, calculated from the engine-simulation program, were used to specify boundary conditions for

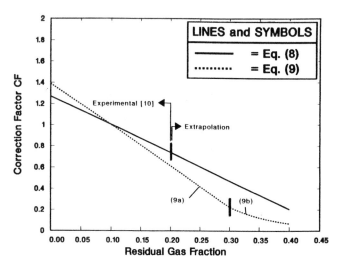

Fig. 4 - Comparison of the combustion submodel correction factor (CF) as a function of residual mass fraction. Solid line: EQ (8); dashed line: EQ (9)

the multi-dimensional computations (see Fig. 3). The velocity profiles at the ports were assumed to be uniform, and the intake flow angles are those shown in Fig. 2 unless otherwise stated. These angles were kept constant during the scavenging process as was assumed also by Ahmadi-Befrui et al. [17]. The intake flow turbulence intensity and length scale were determined using standard turbulent gas-jet equations [18].

No-slip boundary conditions were specified for the gas velocities on the chamber walls, and the boundary condition for the energy equation was a fixed wall temperature. The mass conservation equations used the boundary condition of zero mass flux through walls. Although piston-crevice-flow effects influence cylinder pressures somewhat [5], they were not considered in the present study for simplicity. The boundary conditions for the turbulence kinetic energy and dissipation rate (k and ϵ) equations were specified as described by Kuo and Reitz [1]. Wall functions were used to specify the fluxes of momentum and energy at the wall [1] and a wall-boundary-layer Prandtl number of 0.5 was used (see Table 3).

RESULTS AND DISCUSSION

This section is separated into two parts. The first part discusses the sensitivity of the computational results to variations in port inflow boundary conditions, delivery ratio and combustion model details. This study was performed using the premixed-charge engine, and the results were also used to suggest an improved combustion submodel. The second part presents parametric calculations of combustion using the improved combustion submodel for both premixed-charge and direct-injected engines.

SENSITIVITY STUDY - The calculations were performed using the three-dimensional computational mesh shown in Fig. 1A. At the start of the sensitivity study, the computations used the port inflow angles shown in Fig. 2 and the original combustion model with CF = 1.0 , EQ (7). The sensitivity study next explored the effect on combustion of changes in the scavenge-flow details. The scavenge-flow field was varied by using various boost-port tilt angles that ranged between 20 to 70 degrees from the horizontal plane (see Fig. 2). These cases are identified in Table 4 by the suffix "a", "b" and "c" for 60, 20 and 70 degrees, respectively. Cases with 40 degree boost-port tilt angle appear without a suffix.

Fig. 5 shows comparisons between measured and calculated cylinder pressures for Cases 1a and 4a (see Table 4), which correspond to delivery ratios of 1.0 and 0.3, respectively, with a boost-port tilt angle of 60 degrees. The results for Case 1a are in excellent agreement with the experiments, but there is considerable disagreement between the computations and the experiments for Case 4a (the peak cylinder pressure differs by as much as a factor of two).

Table 4 shows the gas temperature and residual mass fraction at the spark location at the time of the spark. The values for Case 1a are similar to those found in the four-stroke engine study of Kuo and Reitz [1] for which the combustion submodel constants were optimized. However, the present engine employs a very diffe-

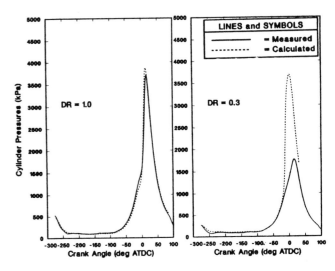

Fig. 5 - Measured and calculated cylinder pressures as a function of crank angle for Cases 1a and 4a with delivery ratio of 1.0 and 0.3, respectively (boost-port tilt angle of 60 degrees, combustion submodel correction factor CF = 1)

rent inflow process and has a much higher squish area than that of the four-stroke computations of Ref. [1], and the present good agreement between experiment and computation is encouraging.

The poor agreement between the computations and the experiments for Case 4a, DR = 0.3, of Fig. 5 may be due to its higher gas temperature since the laminar and turbulent characteristic-time combustion submodel is very sensitive to the value of the gas temperature as noted by Kuo and Reitz [1]. In particular, this case has slightly higher residual gas concentration, but much higher gas temperature (about 10 % higher) than a four-stroke engine case with similar spark timing considered by Kuo and Reitz [1] (Case 3, Table 7 of Ref [1] which has a

Table 4 - Effect of Boost Port Tilt Angle on Computed Results

Case	Delivery Ratio	Engine Speed (r/min)	Boost Port Tilt Angle (deg)	Residual Mass Fraction (%)	Gas Temperature (K)	Trapping Efficiency (%)
1	1.0	1600	40	10.6	776	44.5
1a	1.0	1600	60	6.1	742	46.8
4	0.3	1600	40	28.7	862	75.6
4a	0.3	1600	60	16.9	784	79.3
4b	0.3	1600	20	32.8	893	73.2
4c	0.3	1600	70	18.8	788	80.1

residual mass fraction of 14 percent and a temperature of 714 K). This could explain the very fast combustion noted for Case 4a.

It is also known that the predicted gas temperature and residual mass fraction at the spark location at the time of the spark depends critically on the port-inflow boundary conditions. This can be seen, for example, by comparing the Cases 1 and 1a, or 4 and 4a in Table 4, which have boost-port tilt angles of 40 and 60 degrees, respectively. The differences in gas temperature and residual mass fraction are most dramatic for the low-delivery-ratio Case 4 and are due to the qualitatively different flow fields that are produced during the scavenging process when the inflow angle is changed.

The effects of different boost port angles on the resulting flow field are shown in Fig. 6a, which gives velocity vectors, temperature and fresh-charge mass-fraction contours in the combustion chamber at 80 degrees BTDC (after the end of the scavenging process) for Cases 4 and 4a (40 and 60 degree boost-port angles, respectively). In particular, the high and low contours of temperature and fresh-charge mass fraction are seen to be situated in different parts of the combustion chamber for the two cases. These differences persist throughout the compression process as can be seen in Fig. 6b which shows the corresponding plots at the time of the spark (30 degrees BTDC). This indicates that charge stratification at the time of the spark is strongly influenced by the details of the scavenging process.

The influence on combustion of the qualitatively different flow field produced with the 40-degree boost-port angle is shown in Fig. 7. The agreement between experiment and computation for the high-delivery-ratio Case 1 is again excellent in spite of the slight differences in gas temperature and residual gas concentration obtained with the 40 degree port angle (see Table 4).

The discrepancy between the measured and calculated cylinder pressure is also large for Case 4, although the predicted peak pressure is somewhat lower for Case 4 than for Case 4a (see Fig. 5). This is

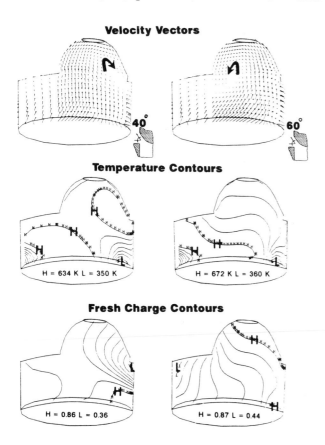

Fig. 6a - Computed velocity vectors, temperature and fresh charge contour plots at 80 degrees BTDC for Cases 4 and 4a (boost-port tilt angles of 40 and 60 degrees, respectively)

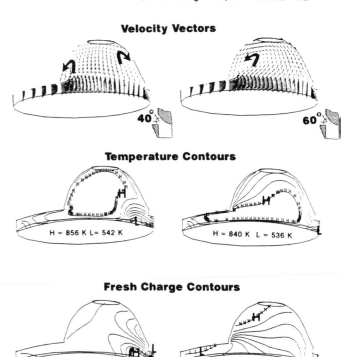

Fig. 6b - Computed velocity vectors, temperature and fresh charge contour plots at 30 degrees BTDC for Cases 4 and 4a (boost-port tilt angles of 40 and 60 degrees, respectively)

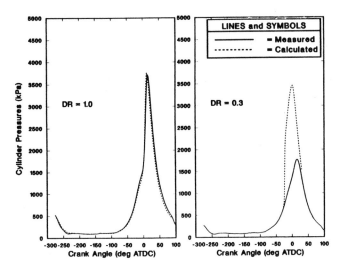

Fig. 7 - Measured and calculated cylinder pressures as a function of crank angle for Cases 1 and 4 with delivery ratio of 1.0 and 0.3, respectively (boost-port tilt angle of 40 degrees, combustion submodel correction factor CF = 1)

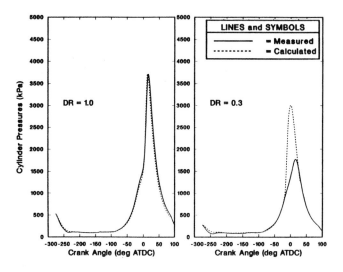

Fig. 8 - Measured and calculated cylinder pressures as a function of crank angle for Cases 1 and 4 with delivery ratio of 1.0 and 0.3, respectively (boost-port tilt angle of 40 degrees, combustion submodel correction factor CF using EQ (8))

mainly due to the fact that the 40-degree boost-port angle of Case 4 gives lower trapping efficiency than the 60-degree angle of Case 4a (see Table 4), which correspondingly lowers total energy available for combustion. However, close examination of Fig. 7 shows that the rate of pressure rise in the early stages of combustion is actually higher with the 40-degree boost-port angle due to the higher unburned gas temperature (see Table 4).

Evidently, differences in the flow field and charge stratification details generated by using the different port inflow angles are not sufficient to improve the agreement between the measured and the computed cylinder pressures with the present combustion model (CF = 1.0, see EQ (7)). Possible reasons are that the combustion model does not consider the influence of the residual gas concentration on the characteristic laminar time and the model is over-sensitive to temperature.

The effect of residual mass fraction is considered in the revised expression for the characteristic laminar time obtained using EQ (8) for CF. This correction accounts for the influence of residual gas concentration in a manner that is consistent with the laminar-flame-speed correlation of Metghalchi and Keck [10]. Computations made using this modi-

fied characteristic laminar time are shown in Fig. 8 for Cases 1 and 4 (DR = 1.0 and 0.3, respectively, 40-degree boost-port angle). As expected, the high-delivery-ratio results are not influenced because the correction is small near 10 percent residual gas concentrations. The low-delivery-ratio results are improved somewhat, but the agreement with the measurements is still poor (peak cylinder pressure differs by 70 percent).

As previously demonstrated, different port inflow boundary conditions give different in-cylinder gas temperature and residual gas concentrations. Can the agreement shown in Fig. 8 for the DR = 0.3 case be improved by only varying the flow-field details? Fig. 9 shows computational results obtained with boost-port angles of 20, 40, 60 and 70 degrees. The corresponding gas temperatures, and residual mass fractions at the time of the spark at the spark location are also given in Table 4. As can be seen in Fig. 9, the computed cylinder pressures are insensitive to the boost-port flow angle. (The small differences in cylinder pressure at exhaust port opening (87 degrees ATDC) are due to differences in trapping efficiency (see Table 4).) This indicates that modifications of flow-field details are not sufficient to explain the discrepancies, even when an attempt is made to account for the influence of residual gas concen-

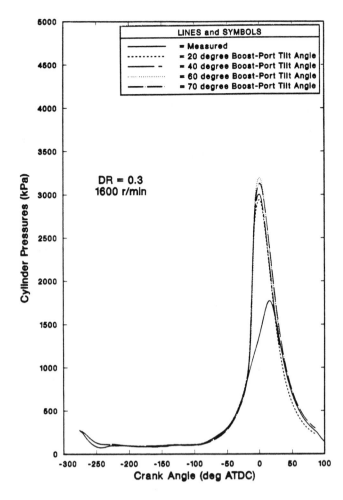

Fig. 9 - Effect of boost-port tilt angle
on calculated cylinder pressures
for Cases 4, 4a, 4b and 4c with
boost-port tilt angles of 40,
60, 20 and 70 degrees, respectively (combustion submodel correction factor CF using EQ (8))

tration in the combustion model by using
EQ (8).

However, the fact that the modified
combustion model which uses EQ (8) did
lead to some improvement in the results
warrants closer examination of the model.

It is through the laminar flame speed
that the correction to account for residual gas appears in the combustion model.
However, there is uncertainty about the
dependence of the laminar flame speed on
residual gas concentrations. For example,
the flame speed correlation of Metghalchi
and Keck [10] used to derive EQ (5) was
based on measurements made at residual
mass fractions only up to 20 percent. Use
of their correlation is thus questionable
in the present low-delivery-ratio computations whose residual mass fraction varies
from 15 to 45 percent in the chamber (see
Figs. 6a and 6b). Moreover, limited lami-

nar-flame-speed computations of Blint [19]
indicate that the slope of the flame speed
versus residual gas concentration curve
(see Fig. 4)) varies with unburned gas
pressure and temperature. This is not
accounted for in the Metghalchi-Keck correlation in which the slope is independent
of gas pressure and temperature.

Accordingly, several different
expressions for the correction factor CF
were considered by varying the slope of
the flame speed versus residual gas concentration systematically. The CF expression that works best is the one given in
EQ (9) and the results of Fig. 10 show
computations made with that CF correction
(see also Fig. 4, dashed line). It can be
seen that this modification of the combustion model improves the agreement with the
experiments considerably for low-delivery-
ratio Case 4 (boost-port tilt angle of 40
degrees). The difference between measured
and predicted peak pressure is now less
than 20 percent.

The low-delivery-ratio results of
Fig. 10 are reproduced in Fig. 11. This
allows comparison with results obtained
using a boost-port tilt angle of 60
degrees. Contrary to the findings of
Fig. 9, the comparison shows that different flow fields now have a stong effect
on the results when EQ (9) is used for CF
in the combustion model at low delivery
ratios. In particular, varying the

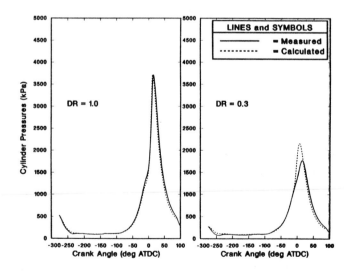

Fig. 10 - Measured and calculated cylinder
pressures as a function of crank
angle for Cases 1 and 4 with
delivery ratio of 1.0 and 0.3,
respectively (boost-port tilt
angle of 40 degrees, combustion
submodel correction factor CF
using EQ (9))

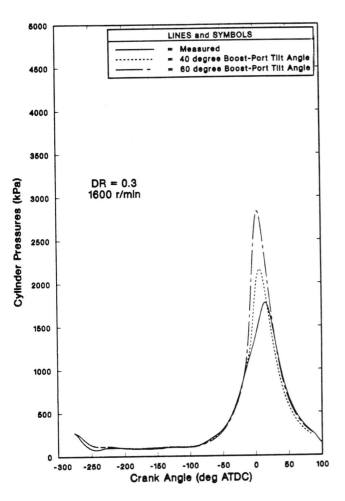

DR = 0.3
1600 r/min

Fig. 11 - Effect of boost-port tilt angle
on calculated cylinder pressures
for Cases 4 and 4a with boost-
port tilt angles of 40 and 60
degrees, respectively (combustion submodel correction factor
CF using EQ (9))

boost-port flow angle from 40 to 60
degrees leads to a considerable deterioration in the results. The high-delivery-
ratio results (not shown) were not influenced by the choice of boost-port angle
because the differences in gas temperature
and residual mass fraction were small (see
Table 4).

The results of Fig. 11 suggest that
it is now difficult to decouple the influence of changes in the combustion model
from changes in the flow field. Therefore, further improvement in the results,
by either optimizing port-inflow boundary
conditions or by using other laminar-
flame-speed/residual-fraction correlations, was not pursued.

In summary, this work indicates that
additional laminar-flame-speed data is
needed at high residual mass fractions to
further improve the modification of the

laminar-kinetics time in the combustion
submodel. Also, detailed measurements of
the scavenging flow fields are needed for
better characterization of the port-inflow
boundary conditions.

PARAMETRIC STUDY - Although the
adjustment to the laminar-kinetics time
with CF given by EQ (9) is not completely
satisfying, a parametric study was conducted to assess the overall performance
of the revised combustion submodel for the
premixed-charge and direct-injected two-
stroke engines. The port inflow angles
given in Fig. 2 were used, with the exception of the boost-port angle, which was
chosen to be 40 degrees. This selection
of angle was made since it gave the best
overall results in the sensitivity study.
The LDV measurements of Fansler [7] in a
motored engine also indicate that the
boost-port angle is about 40 degrees for
much of the scavenging period.

Premixed-Charge Engine - The cases
considered in the parametric study are
listed in Table 5. Cases 1 and 4 correspond to the cases already examined in the
sensitivity study. The effect of variations in delivery ratio is shown in
Fig. 12 which compares predicted and measured cylinder pressures for Cases 1, 2, 3
and 4 (DR = 1.0, 0.7, 0.5 and 0.3, respectively, and 1600 r/min). The agreement
between predicted and measured peak cylinder pressures is within 20 percent for all
cases and is comparable to the cyclic variation in the measured pressures (see
error bars, Fig. 12). This level of
agreement between the experiments and the
predictions is consistent with the levels
of uncertainty in the input parameters to
the computations [1,5] and in the combus-
tion-submodel and flow-field details
described previously. For example, use of
the same port-inflow boundary conditions
for the different delivery ratio and speed
cases is considered questionable. This
alone is enough to cause the observed 20
percent discrepancy between the experiment
and computation as shown in Fig. 11.

Table 5 - Parameters Calculated for Premixed-Charge Engine

Case	Delivery Ratio	Engine Speed (r/min)	Residual Mass Fraction (%)	Gas Temperature (K)	Trapping Efficiency (%)
1	1.0	1600	6.1	742	44.5
2	0.7	1600	15.7	842	52.9
3	0.5	1600	19.8	855	61.9
4	0.3	1600	28.7	862	75.6
5	1.0	800	9.6	773	45.8

249

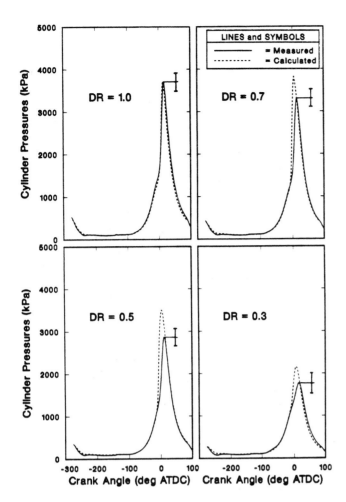

Fig. 12 - Measured and calculated cylinder pressures in the premixed-charge engine for Cases 1, 2, 3 and 4 with DR = 1.0, 0.7, 0.5 and 0.3, respectively (combustion submodel correction factor CF using EQ (9)). Error bar indicates plus/minus one standard deviation of the measured pressure at the location of peak pressure.

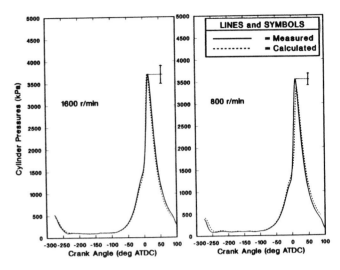

Fig. 13 - Measured and calculated cylinder pressures in the premixed-charge engine at two engine speeds for Cases 1 and 5 (1600 and 800 r/min, respectively, combustion submodel correction factor CF using EQ (9)). Error bar indicates plus/minus one standard deviation of the measured pressure at the location of peak pressure.

The effect of variations in engine speed is shown in Fig. 13 which compares predicted and measured cylinder pressures for Cases 1 and 5 (1600 and 800 r/min, respectively, and DR = 1.0). As expected, the agreement between the predicted and measured cylinder pressures is good since both have high delivery ratio of 1.0.

Details of the flame propagation in the combustion chamber are given in Fig. 14, which presents temperature contours in the diametrical plane through the center of the exhaust port at -5, 0, 5 and 10 degrees ATDC for Case 1. The flame is seen to be convected counter-clockwise during the combustion by the squish-generated mean flow. The retardation effect of the residual gas on flame propagation is

seen in Fig. 14 at 10 degrees ATDC. The regions of slow burn correspond to the regions of high residual gas concentration seen in Fig. 6b. Flame propagation results obtained for the other cases in Table 5 were similar to Fig. 14 except that the flame thickness (the region between the 2000 and 1200 K contour lines) increases with decreasing delivery ratio.

Direct-injected Engine - The operating conditions for the two direct-injected engine computations, Cases 6 and 7 (DR = 0.92 and 0.66, respectively, and 1600 r/min) are given in Table 6. It should be emphasized that these computations were made using the same model constants and boundary condition treatments as those of the premixed-charge engine calculations. Details of input parameters to the spray computations are described in the model section. These parameters were taken from the spray modeling study of Reitz and Diwakar [15] which used an injection system similar to that employed in the current study.

Fig. 15 shows liquid and vapor distributions existing in the combustion chamber at the end of fuel injection (30 degrees BTDC, see Table 6). The fuel droplet distributions are a line-of-sight plot of all computational parcels existing

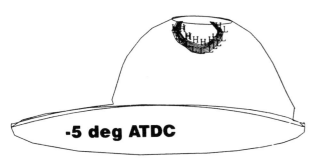

-5 deg ATDC

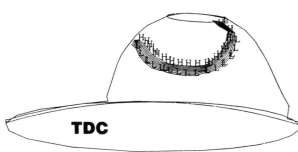

TDC

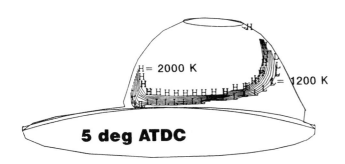

H = 2000 K
L = 1200 K

5 deg ATDC

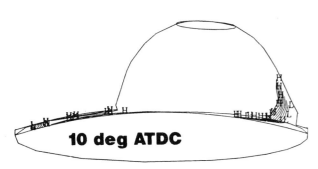

10 deg ATDC

Fig. 14 – Details of flame propagation as represented by computed gas temperature contour plots at -5, 0, 5 and 10 degrees ATDC for Case 1 (boost-port tilt angles of 40 degrees, combustion submodel correction factor CF using EQ (9)). H = 2000 K and L = 1200 K.

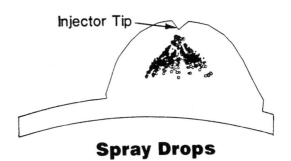

Injector Tip

Spray Drops

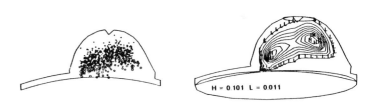

H = 0.16 L = 0.018

Fuel Vapor Contours

Fig. 15 – Computed fuel droplet and vapor distributions at the end of fuel injection (30 degrees BTDC) for Case 6 (DR = 0.92)

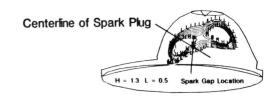

H = 0.101 L = 0.011

Fuel Droplets   **Fuel Vapor Contours**

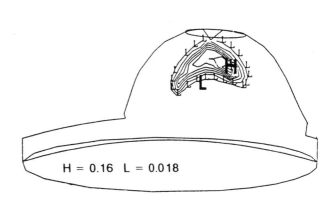

Centerline of Spark Plug

H = 1.3 L = 0.5   Spark Gap Location

Equivalence Ratio Contours

Fig. 16 – Computed fuel droplet, vapor and equivalence ratio distributions at the start of combustion (20 degrees BTDC) for Case 6 (DR = 0.92)

Table 6 - Parameters Used and Calculated in Direct-Injected Engine

Case	Delivery Ratio	Residual Mass Fraction (%)	Gas Temperature (K)	Trapping Efficiency (%)	Injection Timing (deg BTDC)	Duration (deg)
6	0.92	14.6	710	50.9	35.0	5.0
7	0.66	33.5	796	61.2	34.0	4.0

in the chamber. Each dot or open circle represents the location of the parcel. The circle diameter, however, represents the size of the droplet contained in the parcel. The squish flow causes the fuel vapor contour to rotate counter-clockwise slightly. The effect of the squish flow on both the spray and vapor distributions is more pronounced by the time of the spark as can be seen in Fig. 16 (spark timing is 20 degrees BTDC, see Table 2). At this time the spray also begins to impinge on the piston crown.

The corresponding equivalence ratio plot is also shown in Fig. 16, together with the location of the spark gap used in the experiments. The spark is located in a region where the equivalence ratio is between 1.3 and 0.5, which is within the range needed for successful ignition of isooctane.

As can be seen in Fig. 17, by 10 degrees BTDC the center of the developing flame has been convected toward the center of the chamber by the squish flow. The flame propagates radially outward toward the chamber walls as can be seen in the contour plots at TDC. At this point, about half of the injected fuel has burned and the liquid drops have nearly all vaporized, as can be seen in Fig. 18. Fig. 18 also gives details of the liquid and vapor histories (normalized by the total amount of fuel injected) in the chamber as a function of crank angle. Only 10 percent of the injected liquid has been vaporized by the end of injection (30 degrees BTDC). The amount of fuel vaporized increases to 55 percent of the total injected fuel at the time of the spark (20 degrees BTDC).

Predicted and measured cylinder pressures are given in Fig. 19 for the high- and low-delivery-ratio Cases 6 and 7 (see Table 6). The agreement between measurement and computation is better than 10 percent. It is possible that the choice of fuel injection parameters such as initial drop size and temperature, injection velocity, spray angle and injection rate shape will influence the level of agreement shown in Fig. 19. However, as previously concluded, there are uncertainties in the combustion-model and flow-field details, and thus attempts to further improve the results were not pursued.

In summary, although the comparison between experiments and computations is not completely satisfactory, it should not

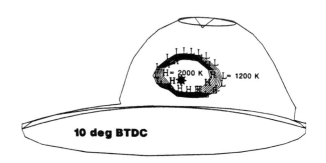

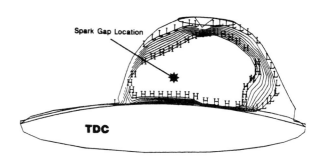

Fig. 17 - Details of flame propagation as represented by computed gas temperature contour plots at 10 and 0 degrees BTDC for Case 6 (DR = 0.92). H = 2000 K and L = 1200 K.

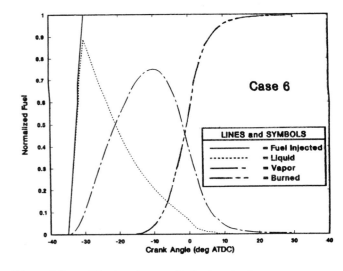

Fig. 18 - Histories of injected, vaporized, non-vaporized, and burned fuel amounts in the combustion chamber as a function of crank angle for Case 6 (DR = 0.92, normalized by total amount of injected fuel)

discourage the potential user of the model. Rather than assuming a totally accurate predictive model, the user should consider the model an evolving tool which can be used to numerically perform certain experiments not readily performed in the

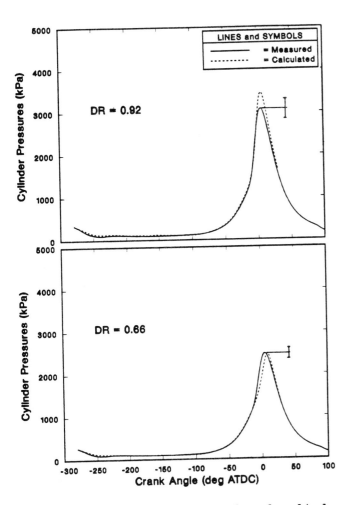

Fig. 19 - Measured and calculated cylinder
pressures in the direct-injected
engine at two delivery ratios
for Cases 6 and 7, DR = 0.92 and
0.66, respectively (boost-port
tilt angle of 40 degrees, combustion submodel correction factor CF using EQ (9)). Error bar
indicates plus/minus one standard deviation of the measured
pressure at the location of peak
pressure.

laboratory.

SUMMARY AND CONCLUSIONS

Combustion and flow were calculated
in a spark-ignited two-stroke crankcase-
scavenged engine using a laminar and turbulent characteristic-time combustion submodel in the three-dimensional KIVA code.
Both premixed-charge and direct-injected
cases were examined. The premixed-charge
engine used propane as fuel. The direct-
injected engine used a pressure-atomized
spray and isooctane as fuel. A multi-cylinder engine simulation program was used
to specify initial and boundary conditions
for the computation of the scavenging pro-

cess. The test conditions corresponded to
both light- and heavy-load engine operation with delivery ratios between 0.3 and
1.0, and engine speeds of 800 and 1600
r/min.

A sensitivity study was conducted
using the premixed-charge engine data.
The influence of different port boundary
conditions on the scavenging process was
examined. At high delivery ratios, the
results were insensitive to variations in
the scavenging flow or residual fraction
details. In this case, good agreement was
obtained with the experimental data using
an existing combustion submodel, previously validated in a four-stroke engine
study (measured and computed peak cylinder
pressures agreed to within 12 percent).
However, at low delivery ratios, both
flow-field and combustion-model details
were important, and the agreement with
experiment was poor using the existing
combustion submodel, which does not
account for the effect of residual gas
concentration (measured and computed peak
cylinder pressures differed by as much as
a factor of two).

To improve the agreement between
modeling and experimental results, a modified combustion submodel was introduced
that includes the effect of residual gas
concentration on the laminar characteristic time. With the new submodel, agreement with the experiment has been improved
considerably for all cases considered in
this study (e.g., measured and computed
peak cylinder pressures agreed to within
20 percent). These levels of agreement
between experiment and computations are
similar to those found in previous applications of the laminar and turbulent characteristic-time combustion submodel to
four-stroke engine combustion. Further
improvement of the combustion submodel was
made difficult by the observed coupling
between the in-cylinder flow-field and the
combustion-model details at low delivery
ratios.

The improved combustion submodel was
also applied to the direct-injected engine
using the same set of model constants as
those used for the premixed-charge engine.
Similar levels of agreement between measured and computed cylinder pressures were
obtained as in the premixed-charge engine
study.

The results of this work indicate
that additional laminar flame speed data
is needed at high residual mass fractions
to improve the specification of the lami-

253

nar characteristic time in the combustion submodel. The present study also highlights the importance of the flow field and residual gas concentrations on combustion in two-stroke engines. Further investigation of these effects is needed in order to further understand two-stroke engine performance.

ACKNOWLEDGMENT

The authors wish to acknowledge the following colleagues from the General Motors Research Laboratories for their contributions to this work:

G. Szekely, R. Otto, A. Solomon and S. De Nagel of the Thermosciences Department for providing the two-stroke engine data used in this study.

R. Krieger of the Thermosciences Department for his help in running the multi-cylinder engine-simulation program.

E. Groff of the Thermosciences Department and R. Rask of the Engine Research Department for helpful comments.

REFERENCES

1. Kuo, Tang-Wei and Reitz, R. D., "Computation of Premixed-Charge Combustion in Pancake and Pent-Roof Engines," SAE Paper No. 890670, 1989.

2. Abraham, J. and Bracco, F. V., "Comparisons of Computed and Measured Pressure in a Premixed-Charge Natural-Gas-Fueled Rotary Engine," SAE Paper No. 890671, 1989.

3. Abraham, J. and Bracco, F. V., "Fuel-Air Mixing and Distribution in a Direct-Injection Stratified-Charge Rotary Engine," SAE Paper No. 890329, 1989.

4. Abraham, J., Bracco, F. V., and Reitz, R. D., "Comparisons of Computed and Measured Premixed Charge Engine Combustion," Combustion and Flame, Vol. 60, pp. 309-322, 1985.

5. Reitz, R. D. and Kuo, Tang-Wei, "Modeling of HC Emissions due to Crevice Flows in Premixed-Charge Engines," SAE Paper No. 892085, 1989.

6. Najt, P. M. and Kuo, Tang-Wei, "An Experimental and Computational Evaluation of Two Dual-Intake-Valve Combustion Chambers," SAE Paper No. 902140, 1990.

7. Fansler, T. D. and French, D. T., "The Scavenging Flow Field in a Crankcase-Compression Two-Stroke Engine -- A Three-Dimensional Laser-Velocimetry Survey," to be presented at SAE International Congress and Exposition, Detroit, MI, 1992.

8. Krieger, R. B., and Rask, R. B, GMR, private communication, May 1989. (The two-stroke engine-simulation program was based on the model given in SAE Paper No. 690135.)

9. Amsden, A. A., Ramshaw, J. D., O'Rourke, P. J., and Dukowicz, J. K., "KIVA: A Computer Program for Two- and Three-Dimensional Fluid Flows with Chemical Reactions and Fuel Sprays," Los Alamos Scientific Laboratory Report LA-10245-MS, February 1985.

10. Metghalchi, M. and Keck, J., "Burning Velocities of Mixtures of Air with Methanol, Isooctane, and Indolene at High Pressure and Temperature," Combustion and Flame, Vol. 48, pp. 191-210, 1982.

11. Williams, F. A., "Combustion Theory," Benjamin-Cummings, Menlo Park, CA, 1985.

12. James, E. H., "Laminar Burning Velocities of Iso-Octane-Air Mixtures - A Literature Review," SAE Paper No. 870170, 1987.

13. Reitz, R. D., "Modeling Atomization Processes in High-Pressure Vaporizing Sprays," Atomisation and Spray Technology, Vol. 3, pp. 309-337, 1988.

14. Reitz, R. D. and Diwakar, R., "Structure of High-Pressure Fuel Sprays," SAE Trans. Vol. 96, Sect. 5, pp. 492-509, 1987.

15. Reitz, R. D. and Diwakar, R., "The Effect of Drop Breakup on Fuel Sprays," SAE Trans. Vol. 95, Sect. 3, pp. 218-227, 1986.

16. Naber, J.D. and Reitz, R.D., "Modeling Engine Spray/Wall Impingement," SAE Paper No. 880107, 1988.

17. Ahmadi-Befrui, B., Brandstatter, W. and Kratochwill, H., "Multidimensional Calculation of the Flow Processes in a Loop-Scavenged Two-Stroke Cycle Engine," SAE Paper No. 890841, 1989.

18. Kuo, Tang-Wei and Bracco, F. V., "On the Scaling of Impulsively Started Incompressible Turbulent Round Jets," J. Fluids Engineering, Vol. 104, No. 2, pp. 191-197, 1982.

19. Blint, R. J., "Flammability Limits for Exhaust Gas Diluted Flames," Proceedings of the 22nd Symposium (International) on Combustion/The Combustion Institute, pp. 1547-1554, 1988.

NOMENCLATURE

A	Combustion-submodel constant EQ (3), Table 3 (atm·s/K)
ATDC	After top dead center
B	Combustion-submodel constant EQ (3), Table 3
B_2	Flame-speed constant, EQ (5)
B_m	Flame-speed constant, EQ (5)
BTDC	Before top dead center
CF	Correction factor, Eqs. (3), (7), (8) and (9)
C_{m1}	Combustion-submodel constant, Table 3
C_{m2}	Combustion-submodel constant, Table 3
C_{m3}	Combustion-submodel constant, Table 3
D	Diffusivity
DR	Delivery ratio
E	Activation energy divided by the gas constant EQ (3), Table 3
f	Delay coefficient EQ (2)
g	Function of ϕ in EQ (3)
h	See description of EQ (4)
k	Turbulence kinetic energy
ℓ	Integral length scale
P	Pressure
Pr	Prandtl number
r	Residual mass fraction
S	Flame speed
t	Time
T	Temperature
TDC	Top dead center
Y	Mass fraction
ϵ	Turbulence kinetic energy dissipation rate
η	$\eta = 1$ when $h > 1$, $\eta = 1/h$ when $h < 1$
ϕ	Fuel-air equivalence ratio
τ	Characteristic time
α	Function of ϕ in EQ (5)
β	Function of ϕ in EQ (3) and (5)
ϕ_m	Flame-speed constant, EQ (5)

Superscript

*	Local thermodynamic equilibrium value

Subscript

d	Delay
eq	Thermodynamic equilibrium
F	Fuel
i	Species
L	Laminar
o	Reference value ahead of the flame
O_2	Oxygen
p	Product
ps	Residual or product at spark time
s	Spark
T	Turbulent

*R. D. Reitz is now at the University of Wisconsin-Madison.

Numerical Prediction of Fuel Secondary Atomization Behavior in SI Engine based on the Oval-Parabola Trajectories(OPT) Model

Ken   Naitoh
Yasuo Takagi
Hiroko Kokita
NISSAN Motor Co., Ltd.
Kunio Kuwahara
Institute of Space and Astronautical Science

ABSTRACT

A theoretical model based on a nonlinear ordinary differential equation was developed, which can estimate the atomization process of fuel droplets after the wall impingement. The phase-space trajectory of the equation for droplet deformation and oscillation varies from oval to parabola with increasing impact velocity. Four different regimes for droplet diameter distribution are derived from this complex feature of the equation. The amount of liquid film remaining on the wall and the number of droplets are estimated from the related mass and energy conservation laws. The model is called the Oval-Parabola Trajectories (OPT) model in the present report.

Comparisons made with some fundamental experimetal data confirm that this mathematical model is effective in a velocity range from 2m/s to 40m/s and in a diameter range below 300 micrometers.

A previously reported numerical code based on the multi-level formulation and the renormalization group theory is combined with the OPT model and the TAB model. The visualizations reemerged by computations indicate that secondary atomization behavior on valve surfaces plays a significant role in the fuel mixture formation in the cylinder of spark-ignition engine.

INTRODUCTION

Mathematical models of the following four aspects are required to perform the numerical predictions of the liquid fuel distribution in gasoline engines;
(1) breakup process caused by air-droplet interaction,
(2) breakup process caused by droplet-wall and droplet-droplet interactions,
(3) film flow on walls, and
(4) phase transitions between vapor and liquid.

Recent studies for the air-droplet interaction have produced some effective mathematical models based on linear analysis [1][2].

The droplet-wall interaction mechanism, similar to the "milk crown" formation, plays a decisive role in the liquid fuel distribution in spark-ignition engines. The development of the theoretical model for droplet-wall interaction becomes a bottleneck for realizing an entire model of engine performance predictions. A factor disturbing the development is "complexity" due to the strong nonlinearity lying on the phenomenon.

In the present research, a theoretical model for the droplet-wall interaction was developed by

---

* Numbers in brackets designate references at end of this paper.

deriving a nonlinear ordinary differential equation governing the distortion and oscillation of droplets. This nonlinear ordinary differential equation is a non-conservative type. The trajectory of this equation system varies from an oval to a parabola in the phase space of droplet distortion and its speed, due to the increase in the initial velocity and distortion. The following four regimes for the diameter distribution of child droplets can be predicted from the complex feature of the equation.

(Regime A) Permanent coalescence regime
(Regime B) Weak breakup regime with large child
droplets of a single size
(Regime C) Medium breakup regime with two
different size of droplets
(Regime D) Strong breakup regime with small
droplets

The amount of liquid film remaining on the wall and the number of the broken droplets are estimated by the mass and energy conservation laws.

The resulting theoretical model is called the Oval-Parabora Trajectories (OPT) model from the trajectory characteristics.

Comparisons with some fundamental experimetal data indicate that this mathematical model can predict correctly the mass remained on walls, and the child droplet diameters for the regime of initial velocities of 2 - 40 m/s and for the region of the initial diameters below 300 micrometers.

The OPT model and the TAB model [2] are combined with the flow code based on the multi-level formulation [3,4] and the renormalization group theory [5]. The computational results obtained with the code indicate that the secondary atomization behavior on valve-surfaces plays a significant role in the fuel mixture formation in the cylinder of spark-ignition engine.

PHYSICAL PROCESS OF WALL IMPINGEMENT OF DROPLETS

For a wide range of initial parameters of the

impaction velocity and diameter of a parent droplet, the child droplet size, the number and velocity of child droplets, amount of liquid remaining on the wall, and the interval to droplet breakup must be obtained by solving the related mass, momentum, and energy conservation laws.

The physical process to be modeled is shown in Fig.1. The process from the wall collision to breakup consists of three phases.

At the first stage (Phase A), the initial parent droplet with mass m_{d1} and radius r_{d1} collides with the initial droplet on the wall with mass m_{d2}, initial thickness Xo, and equivalent radius r_{d2}. The velocity of the initial parent droplet and the component normal to the wall are Up and Uo. Each droplet is deformed and oscillated after collision.

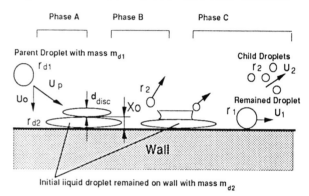

Fig.1 Breakup process and film flow after wet-wall impingement of a parent droplet

The second stage (Phase B) shows the physical situation near the breakup occurrence. As the parent-droplet thickness approaches the minimum value d_{disc}, the oscilating droplets with a spheroidal shape produce the child droplets with a smaller radius. The physical process is similar to the " milk crown " formation.

At the last stage (Phase C), the child droplets with the mass of m_2, the velocity of U_2, and the radius of r_2 fly after breakup. Then a new droplet remaining on the wall with the mass m_1 and the radius r_1 moves, until it will meet the next collision.

In the following section, a nonlinear ordinary differential equation is presented, which governs the oscillating motions after droplets collide and after the impingement of droplet to the liquid film on the wall. The characteristics of the equation is examined.

QUASI-CHAOTIC ORDINARY DIFFERENTIAL EQUATION (QCODE) GOVERNING DROPLETS AND DROPLET-WET WALL INTERACTIONS

Equation system for the physical proces - The following nonlinear ordinary differential equation system describes the deforming and oscilating motions of two interacting droplets.

$$m_{d1}\ddot{X}_{d1} = -C_c\rho_L r_{d1}^2 (\dot{X}_{d1})^2 - C_c\rho_L r_{d2}^2 (\dot{X}_{d2})^2$$
$$- m_1 C_k \frac{\sigma}{\rho_L r_{d1}^3} X_{d1}$$

$$m_{d2}\ddot{X}_{d2} = -C_c\rho_L r_{d1}^2 (\dot{X}_{d1})^2 - C_c\rho_L r_{d2}^2 (\dot{X}_{d2})^2$$
$$- m_2 C_k \frac{\sigma}{\rho_L r_{d2}^3} X_{d2}$$

$$(1)$$

The first and second terms on the right-hand side in Eq.(1) represent the convection contribution, while the third term for surface tension.

σ, ρ_L, C_c, and C_k are the surface tension, the droplet density, and the artificial constants to adjust two contributions, respectively.

X_{d1}, $\dot{X}_{d1}$, $\ddot{X}_{d1}$, X_{d2}, $\dot{X}_{d2}$, and $\ddot{X}_{d2}$ demonstrate the distortion, the distortion speed, and the distortion acceleration of two droplets.

The surface tension contribution term in Eq. (1) is depicted in Ref. [2]. The form of first and second term in the right hand side of Eq.(1) is considered from the fact that convection contribution is the square of velocity. Details of the derivation of Eq. (1) are demonstrated in Appendix A.1.

Equation (1) is transferred to the following first- order system by a simple variable transformation.

$$\dot{X}_{d1} = \frac{1}{r_{d1}}Y_{d1}$$
$$\dot{Y}_{d1} = -\frac{A}{r_{d1}^2}(Y_{d1})^2 - \frac{A}{r_{d2}^2}(Y_{d2})^2 - \frac{B}{r_{d1}^2}X_{d1}$$

$$\dot{X}_{d2} = \frac{1}{r_{d2}}Y_{d2}$$
$$\dot{Y}_{d2} = -\frac{A}{r_{d1}^2}(Y_{d1})^2 - \frac{A}{r_{d2}^2}(Y_{d2})^2 - \frac{B}{r_{d2}^2}X_{d2}$$

$$(2)$$

$$A = \frac{3}{4\pi}C_c, \quad B = C_k\frac{\sigma}{\rho_L}$$

$$(3)$$

The following modification of Eq. (2) expresses the oscilating motion after the collision of a droplet with the liquid film remaining on the wall.

$$\dot{X}_{d1} = \frac{1}{r_{d1}}Y_{d1}$$

$$\dot{Y}_{d1} = -\frac{A}{r_{d1}^2}(Y_{d1})^2 - \frac{A}{r_{d2}^2}(Y_{d2})^2 - \frac{B}{r_{d1}^2}X_{d1}$$

$$\dot{X}_{d2} = \frac{1}{r_{d2}}Y_{d2}$$

$$\dot{Y}_{d2} = -\frac{A}{r_{d1}^2}(Y_{d1})^2 - \frac{A}{r_{d2}^2}(Y_{d2})^2 - \frac{B}{r_{d2}^2}(\dot{X}_{d2}-X_0)$$

$$(4)$$

Trajectories of the equation system in the phase space - Figure 2 shows the trajectories for four initial conditions. Iso-Octane is used here as the fuel. The set of Case A1 and Case A2 is for the collision of two equal-sized droplets. The trajectory for Case A1 with a low impact velocity is a closed oval, while that for Case A2 with a high impact velocity is an open parabola. The set of Case A3 and Case A4 is for the collision of two droplets of different size. It is seen that Case A3 and Case A4 show complex trajectories in the phase space. Figure 3 shows a Poincare section of Y1=Y2=0 for Case A3. When Y1=Y2=0, X1 and X2 show considerable large distortion values near the edges of the trajectory. It is thought that the breakup process occurs around these points, since the pressure difference within a droplet becomes large.

As a result of performing the numerical

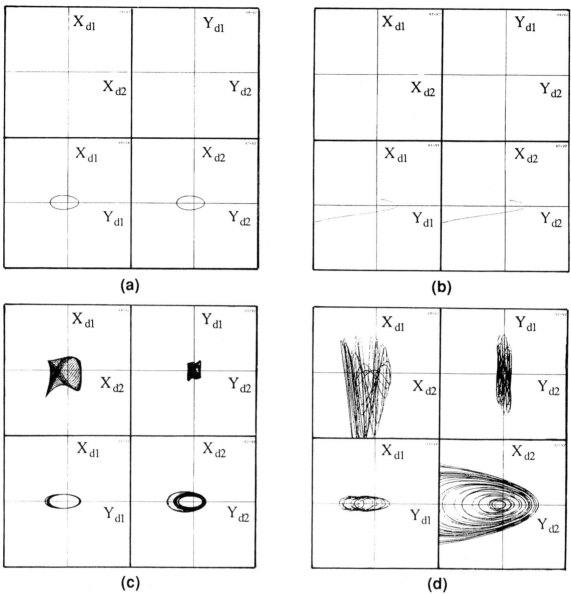

Fig. 2 Trajectory of QCODE

 (a) Case A1 : Droplet diameter= 200micrometers, Uo=3.0m/s

 (b) Case A2 : Droplet diameter= 200micrometers, Uo=9.0m/s

 (c) Case A3 : Droplet diameter= 200 and 260 micrometers, Uo=3.0m/s

 (d) Case A4 : Droplet diameter= 200 and 340 micrometers, Uo=3.0m/s

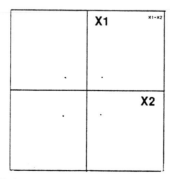

Fig. 3 Poincare section of Eq. (2) (Case A3)

calculations until t = 3.0 seconds with 10,000,000 time steps, it was confirmed that the Lyapounov exponents are close to (+, 0, 0, -) for Case A3, while the exponents are close to (0, 0, 0, 0) for Case A1. The total value of the exponents is close to be zero. This feature suggests that the system is chaotic. However, the trajectory of this equation system goes to infinity for large impact velocities. Further analysis is needed to decide whether this equation system really shows a chaotic behavior for small initial values of velocity and distortion. This system is called a quasi-chaotic ordinary differential equation (QCODE) from the trajectory behavior.

A detailed equation system can be obtained by including the dissipation term in the O'Rourke-Amsdem model [2]. However, the trajectory even goes to infinity for cases of large initial velocities. This characteristics of openness may correspond to the transition from oscillation to breakup. Or it may be due to the fact that Eqs. (2) and (4) do not include turbulent dissipation for cases of large initial values of diameter and velocity. In a later section, it is shown that energy dissipation due to droplet deformation during collision gives the maximum distortion limit and the smallest size of droplets. Then the trajectories of Eqs. (2) and (4) never go to infinity.

Equations (2) and (4) are important in estimating the diameters of child droplets after droplets collide and after droplet impinges wall, repsectively.

Since the surface tension of gasoline fuel is very small, the thickness of the fuel film is very thin. Then it can be confirmed easily by Eq. (4) that the maximum distortion is very close to the dry wall case. In the next section, a concrete model is derived for determining the diameter and number of child droplets after impingement on a dry wall.

AN OVAL-PARABOLA TRAJECTORIES (OPT) MODEL OF BREAKUP AFTER DRY-WALL IMPINGEMENT BASED ON SIMPLIFIED QCODE

The breakup process after the impingement of a droplet on a dry wall is modeled by simplifying the QCODE. The physical situation to be modeled is shown in Fig.4. For this situation, Eq. (2) is contracted to a two variable system.

In Phase A, the maximum distortion is estimated from the momentum equation. The relation between the child droplet size and the maximum distortion is then derived from mass and energy conservation laws around the edge of the distorted droplet in Phase B. Finally, the number of child droplets and the amount of liquid film remaining on the wall are estimated from mass and energy conservation laws between the condition before the impingement and the situation after the breakup in Phase C.

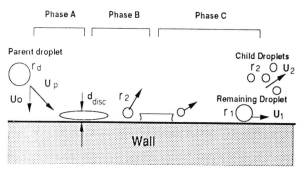

Fig.4 Breakup process after dry-wall impingement

Determination of child droplet size -
Maximum distortion of parent droplet (Phase A)
The following equation system is obtained by simplifying Eq. (1).

$$m \ddot{X} = - C_c \rho_L r_d^2 (\dot{X})^2 - m C_k \frac{\sigma}{\rho_L r_d^3} X \quad (5)$$

$$\dot{X}_{t=0} = U_0 , \quad X_{t=0} = 0 \quad (6)$$

$$m = \frac{4}{3} \pi \rho_L r_d^3 \quad (7)$$

Equation (5) describes the deforming and oscilating motion of a parent droplet from wall-impingement to breakup. Equation (6) expresses the initial conditions for the distortion and distortion speed. Details of the derivation of Eq. (5) are shown in Appendix A.1.

These equations are simplified to the following first-order ordinary differential equation system.

$$\dot{X} = Y \tag{8}$$

$$\dot{Y} = -A Y^2 - B X \tag{9}$$

$$A = C_c \frac{\rho_L r_d^2}{m}, \quad B = C_k \frac{\sigma}{\rho_L r_d^3} \tag{10}$$

In the present study, C_c and C_k are set to be 0.15 and 2.0, respectively. These two values are determined from the comparisons with the experimental data, which are shown in the latter section.

Fortunately, the present equation system with Eqs. (8) and (9) has an analytical solution in the phase space. The trajectory of the system shows a discontinuous transition due to increasing initial velocity. For small initial values of X and Y, the trajectory is an oval in the phase space. For large initial values, the trajectory is a parabola [6].

$$U_o^2 < \frac{B}{2A^2}: \quad \text{Oval}$$

$$U_o^2 > \frac{B}{2A^2}: \quad \text{Parabola} \tag{11}$$

When Y equals zero in Fig.5, X gives the value of maximum distortion.

$$X - \frac{1}{2A} \left[\left(\frac{2A^2}{B} U_o^2 - 1 \right) \exp(-2AX) + 1 \right] = 0 \tag{12}$$

Equation (12) gives the maximum distortion of the parent droplet. Two finite modes of maximum distortion appear for oval oscillation, while one finite mode and another infinite one appear for parabola. In case of the parabolic trajectory, the maximum distortion is limited by the energy conservation law with dissipation. The details on this limitation are described in the latter section.

The deformed droplet shape is assumed to be a disc in the present study. Then the minimum thickness of the disc is determined by Eq.(13).

$$d_{disc} \pi (r_d + 0.5 |\overline{X}|)^2 = \frac{4}{3} \pi r_d^3 \tag{13}$$

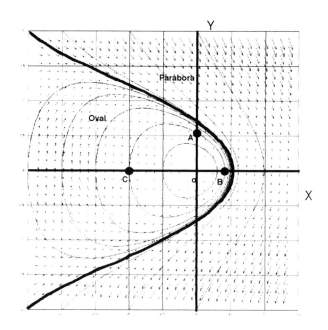

Fig.5 Trajectories in the phase space of the simplified QCODE

Relation between maximum distortion and child droplet size (Phase B) - It is roughly assumed that, when the representative distortion X is over one-half of the initial droplet diameter, the parent droplet breaks up into smaller child droplets and film flow. The bouncing phenomenon in the low Weber number regime is not modelled in the present report, since the regime is very narrow for fuel droplets motions in engines. The usefulness of the present assumption is examined in the comparisons with experimental data in the later section.

The relation between the distorted disc size and the broken droplet size is given by mass and energy conservation laws around the localized edge of the disc. The situation is shown in Fig. 6.

$$V = d_{disc} (\pi r_b^2) = \frac{4}{3} \pi r_2^3 \tag{14}$$

$$E = \pi r_b^2 \sigma = 4 \pi r_2^2 \sigma \tag{15}$$

In Eq. (15), the surface energy of the side ring is eliminated. Then

$$r_2 = 3 \, d_{disc} \tag{16}$$

Equation (16) gives the resulting relation between

the size of a child droplet and the thickness of the distorted droplet.

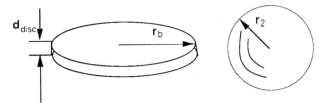

Fig.6 Relationship between disc thickness and child droplet size

It is assumed only in Eq. (16) that breakup occurs when the droplet velocity becomes zero, since the child droplet velocity is very small in comparison with the parent droplet.

It is understood from Fig. 5 that two possibilities exist for the radius of the child droplet. Then the following four regimes can be proposed as characteristic breakup patterns.

(1) Permanent coalescence regime: the maximum distortion is smaller than one-half of the parent droplet size,

(2) Weak breakup regime with large child droplets of a single size: produced by weak oval oscillation,

(3) Medium breakup regime with two different droplet size droplets: caused by strong oval oscillation,

and

(4) Strong breakup regime with small droplets : produced by parabolic oscillation.

Determination of the number of child droplets and amount of liquid film remaining on the wall (Phase C) - It is indicated from the QCODE that the breakup is possible to occur in two different modes. It is assumed in the present modeling that either one mode among two is selected randomly for each impingement.

Then mass and energy conservation laws between the initial condition before impingement and the situation after breakup are described as follows.

$$\frac{4}{3} \pi \rho_L r_d^3 = n(\frac{4}{3}\pi \rho_L r_2^3) + \frac{4}{3}\pi \rho_L r_1^3 \qquad (17)$$

$$4\pi r_d^2 \sigma + (1-C_{dis}) \frac{1}{2}(\frac{4}{3}\pi \rho_L r_d^3) U_o^2$$
$$= n (4\pi r_2^2 \sigma) + 4\pi r_1^2 \sigma \qquad (18)$$

In Eq. (18), C_{dis} and n represent the dissipation rate of the kinetic energy of parent droplet and the number of child droplets, respectively. C_{dis} includes the kinetic energy of child droplets. It is reported in Ref. 8 that C_{dis} is constant for the low Weber number regime.

Equations (17) and (18) are simplified to the following third-order equation.

$$XX^3 - \frac{r_2}{r_d} XX^2 + \frac{r_2}{r_d} [1 + \frac{1}{6\sigma} \rho_{lr} r_d U_0^2 (1-C_{dis})] - 1 = 0 \qquad (19)$$

$$XX = \frac{r_1}{r_d} \qquad (20)$$

$r_2 = 0$ and $r_1 = r_d$ satisfy the equation under a limited condition of Uo = 0 m/s.
Under another limited condtion of infinite impingement velocity, Eq.(21) gives the critical droplet size.

$$(r_2)_{crit} = \frac{6\sigma}{1-C_{dis}} \frac{1}{\rho_L U_0^2} \qquad (21)$$

Equation 21 expresses the smallest child droplet size within the limit allowed by energy dissipation.

The amount of mass remaining on the wall is obtained from Eqs. (19) and (20), after the child droplet size has been determined.

Determination of velocity of child droplets - It is assumed that the child droplet moves with the one-tenth kinetic energy of the parent droplet. The kinetic enegy of child droplets is roughly decided from the PDPA data. The child droplets move in a random forward direction. The flight direction of the child droplets is shown in Fig. 7. Equation (22) gives the components of the velocity vector of the child droplets.

$$U_n' = \frac{1}{\sqrt{10}} * \sqrt{\text{rans}1} * U_n$$

$$U_{h1}' = \frac{1}{\sqrt{10}} * U_{h1} + \frac{1}{\sqrt{10}} * \sqrt{\text{rans}2} * U_n * \cos(2\pi*\text{rans}3)$$

$$U_{h2}' = \frac{1}{\sqrt{10}} * \sqrt{\text{rans}2} * U_n * \sin(2\pi*\text{rans}3)$$

(22)

Where rans1, rans2, and rans3 represent the random numbers between zero and one.

Determination of the interval to breakup- The time from impingement to breakup is determined by solving Eq. (5) numerically. The intervals involved are the ones from point A to point B and from point A to point C in Fig.5.

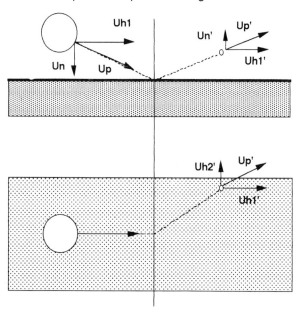

Fig. 7 Flight direction of child droplets

FORMATION OF NUMERICAL CODE FOR FUEL-MIXING PROCESS IN ENGINE

The numerical model for predicting the fuel motions in engines consists of four parts: the subroutine group I for the air flow calculation, subroutine group II for spray calculation as that in KIVA II [2] (subroutines for breakup by air-droplet interaction, evaporation-condensation, and drag force), subroutine group III for droplet-wall interaction (the OPT model), and subroutine groups IV for the vapor fuel flow calculation. The Yakhot-Orszag model [5] is employed for the subgrid turbulence estimation. Details of subroutines I are reported in Refs. 3 and 4. The effectiveness of subroutine groups II is tested in Appendics. Subroutine group III is checked in the next section. The CIP method [12] is used to discretize the convective terms in the species equations. The usefulness of the scheme for discontinuity is shown in the appendics.

FUNDAMENTAL EVALUATIONS OF THE OVAL-PARABOLA TRAJECTORIES (OPT) MODEL

Some computations for the wall-impingement of equal-sized droplets are compared here with the corresponding experimental data in order to confirm the effectiveness of the OPT model.

Wall impingement (Up=12.0m/s, initial diameter =160 μm, Impimgement angle = 20, 30, 60 degree) - Droplet impingemenst at three inclined angles were examined first. Figure 8 shows the histograms of the predicted and measured child droplet sizes, which indicate a bi-modal distribution. Table 1 gives the diameters at the peak points for 30degree. This characteristics of bi-modal distribution is explained by the trajectory characteristics of the QCODE. Another experimental data [15] has also shown the existance of two peaks.

Figure 9 shows the visualization of the breakup processes. Very small droplets are produced in the impinging angles of 30 and 60 degrees. The parent droplets do not break into small ones for the impinging angle of 20 degree, both in computation and experiment. The flying directions of child droplets are also predicted well.

Table 1   Child droplet size

	Predicted	Experimental
Peak D1	22 μm	24 μm
Peak D2	38	34

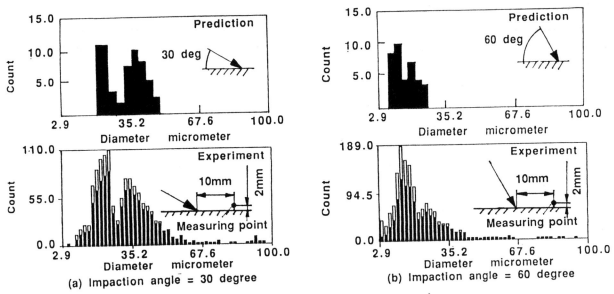

Fig. 8 Distribution of child droplet size

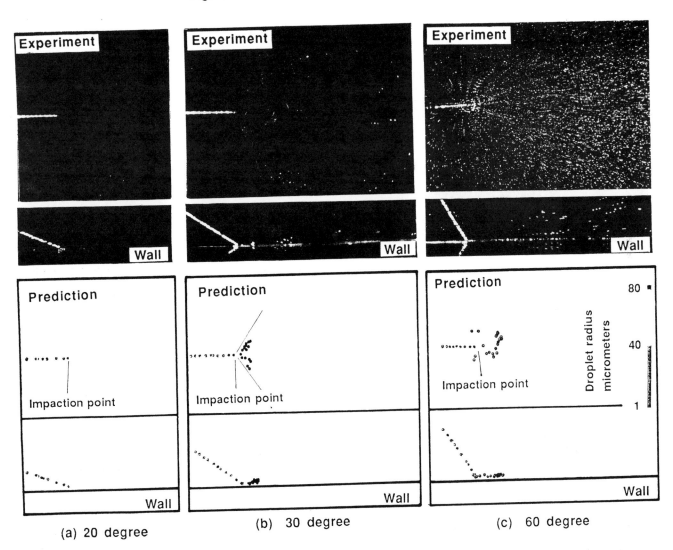

Fig. 9 Visualization of the breakup process after wall impingement

Wall impingement (Uo=12, 18, 28, 40 m/s, initial diameter =100 μm) - Next, the mass rate remaining on wall is examined by the comparison with the experimental data [7]. Predictions were obtained only with using the OPT model without flow field calculations. Table 2 shows the mass rates of child droplets for some values of Uo. Both the calculated and the experimental results indicate an increase in percent mass of child droplets produced with increasing impact velocity. When the impaction velocity is smaller than 3.5 m/s, no child droplets are generated in the prediction. The discrepancy between the prediction and the experiment in high-impaction speed regime may be explained by the fact that predictions in Table 2 doesnot include influence of gravity, evaporation, and turbulene calculations. Further studies on mass remained on wall are needed for confirming the effectiveness of the OPT model.

Table 3 shows the size of the child droplet for the high initial velocity of 40 m/s. The value of Peak D2 is the limitation from Eq. (21). Both the predicted and experimentl data show that very small droplets are produced with the size below 10 micrometers.

Table 2   Mass rate of child droplets

	Predicted		Experimental
	Peak D1	Peak D2	
Uo= 3.50 (m/s)	0%	0%	-
12.0 (m/s)	64.5%	26.0%	22%
18.0 (m/s)	81.0%	11.0%	38%
28.0 (m/s)	100.0%	100.0%	48%
40.0 (m/s)	100.0%	100.0%	63%

Table 3 Droplet size (Uo=40m/s I. D. =100 μm)

	Predicted	Experimental
Peak D1	2.73 μm	0.0 - 10.0 μm
Peak D2	1.97	

Collision of two droplets (Uo=2.3m/s, initial diameter=300 μm) - Finally, a case in which two droplets of same size and velocity collide head on, was examined. The experimental data taken by Umemura is used [8]. Table 4 shows the child droplet size. The number of the child droplets is given in Table 5.

Table 4   Child droplet size

	Predicted	Experimental
Peak D1	144-186μm	110 μm
Peak D2	110-150	

Table 5   Number of child droplets

Predicted	Experimental
1 - 14	1

It is clear from Table 5 that the nature of this case is around the critical point from permanent coalescence to breakup.

In the region of Uo of 2 - 40 m/s and the region of initial diameter below 300 , the OPT model predicts several features observed in the experiments including a bimodal distribution of child droplet sizes and an increase in percent mass of child droplets produced with increasing impact velocity.

APPLICATION TO THE FUEL SECONDARY ATOMIZATION PROCESS AND MIXTURE FORMATION IN A SPARK-IGNITION ENGINE

In the present section, the fuel motion in a spark-ignition engine is examined by using the entire numerical code outlined in the former section.

Figure 10 shows the engine grid system used in the calculations. The number of grid points is 65 x 35 x 37. The specifications for the injector and for the engine are listed in Table 6.

Figures 11(a) and (b) demonstrate fuel droplets distributions, calculated with the OPT model and without including the model, respectively. It is understood from Figs. 11(a) and (b) that the air-droplet interaction at the throat of valve is only an assistance for fuel atomization process. The droplet-wall interaction has an decisive role for the formation of fuel vapor cloud in the cylinder. It is also understood that a part of fuel droplets entering into th cylinder impinges to the cylinder wall.

Table 7 shows the mass rate of fuel entering in cylinder. The corresponding experimental data is obtained at an accelerated condition of the engine. The fuel mass taken into the cylinder directly will be 30-40% of the total quantity injected.

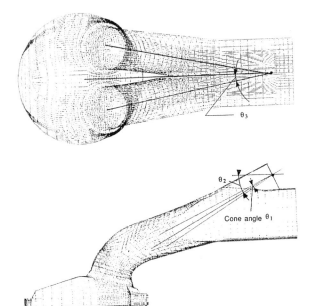

Fig. 10 The grid system for spray calculation
Grid number : 65 x 35 x 37 points

Table 6 Specifications of injector and engine

Injector (with two spray cones)		
D10		73.0 μm
D32		122.0 μm
Cone angle	θ_1	12.0 degree
Injection velocity		20.0 m/s
Injection timing		0-30 degATDC
Injection angle	θ_2 θ_3	29 deg, 13 deg
Engine		
Bore x Stroke		83.0 x 86.0 mm
Engine speed		1200 rpm
A/F ratio		14.8 : 1
Compression ratio		9.0 : 1
Wall temperature		400K

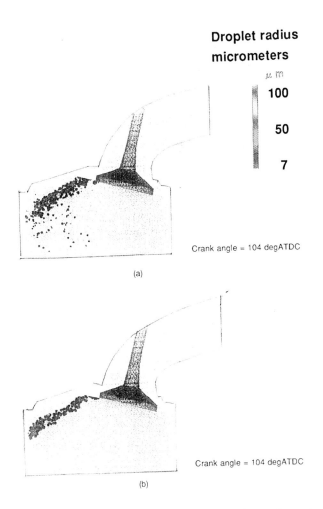

Droplet radius
micrometers
μ m
100
50
7

Crank angle = 104 degATDC
(a)

Crank angle = 104 degATDC
(b)

Fig. 11 Fuel droplets motions in the intake ports and cylinder (a) With OPT (b) Without OPT

Table 7 Amount of fuelentering into cylinder directly

	Predicted		Experimental
Without OPT	With OPT		
15%	34.0%		39.0%

Figure 12 shows the details of the droplets distribution and the vapor cloud of air-fuel ratio under 14.8. A part of the injected fuel droplets breaks up into sub-droplets smaller than ten micrometers, due to the wall impingements. Thus the broken droplets in the cylinder move riding on the air motion. It is also understood that the lean cloud exists around the cylinder-wall of the intake valve side near the end of the compression process.

Figure 13 demonstrates the fuel droplets remained on the port-wall and the valve surface.

267

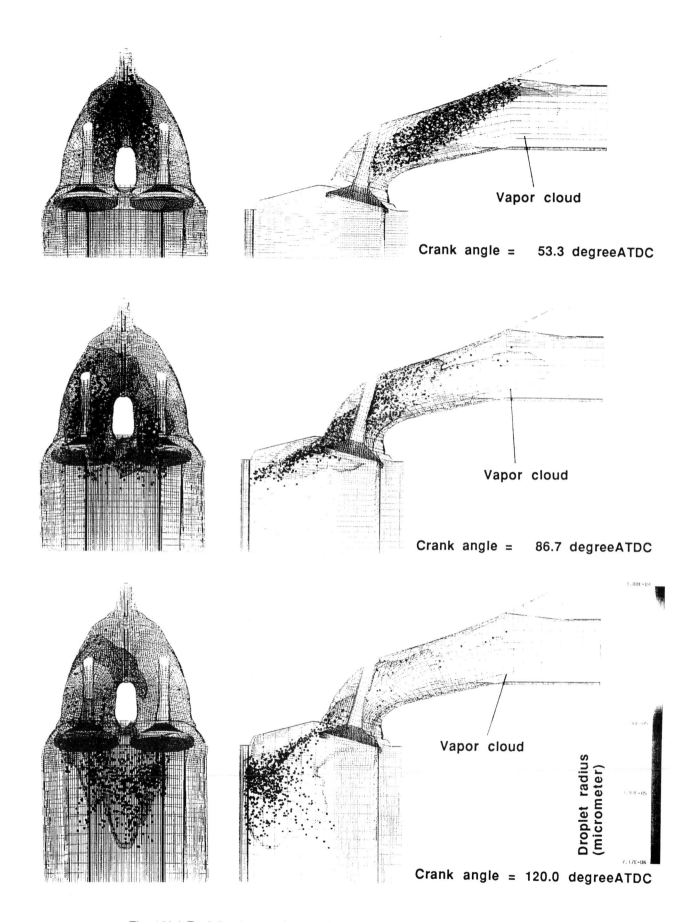

Crank angle = 53.3 degreeATDC

Crank angle = 86.7 degreeATDC

Crank angle = 120.0 degreeATDC

Fig. 12(a) Fuel droplets motions and vapor cloud in the intake process

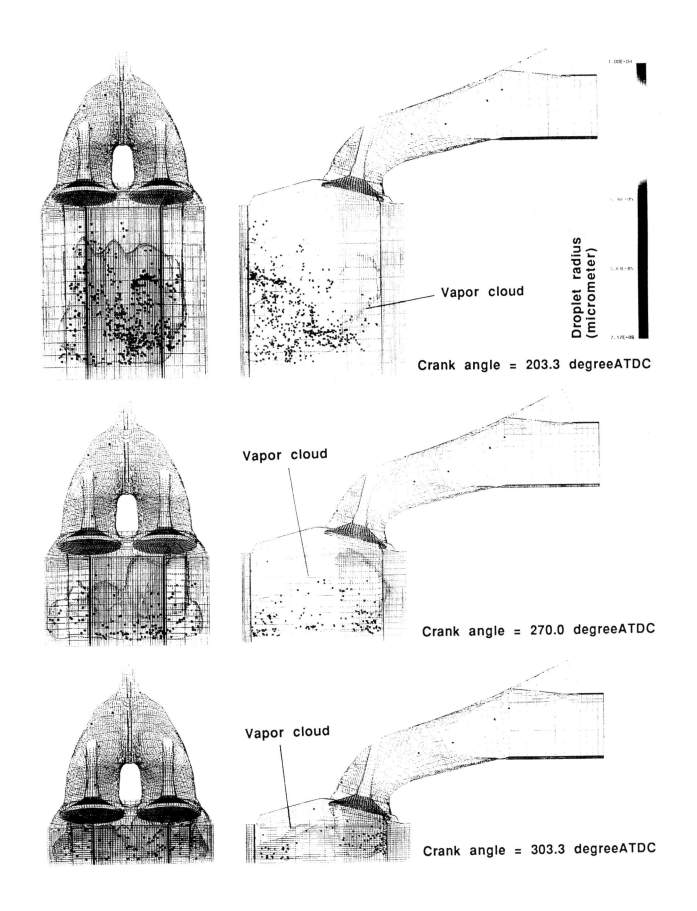

Fig. 12(b) Fuel droplets motions and vapor cloud in the compression process

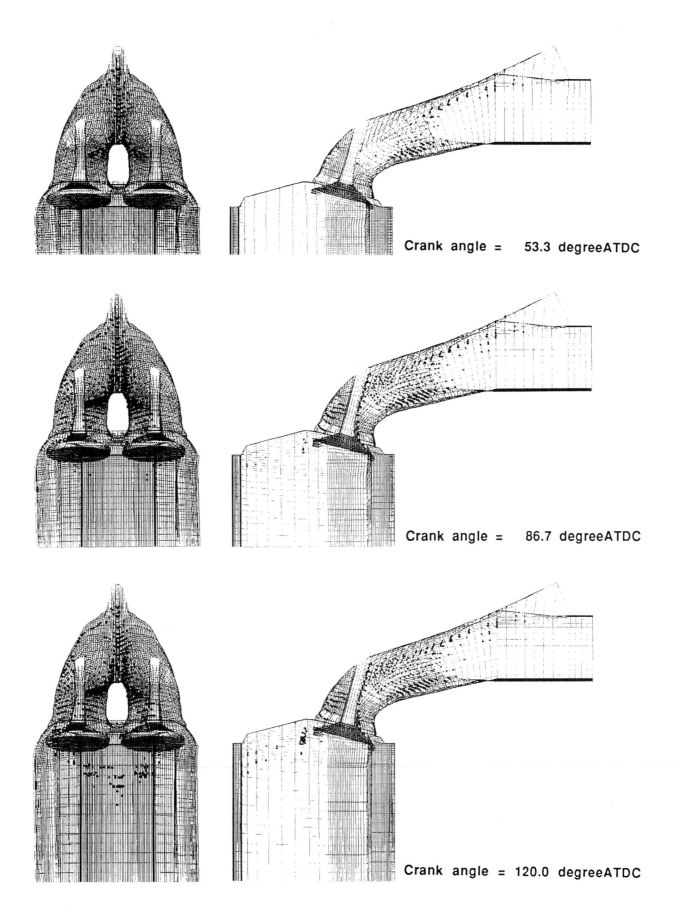

Crank angle = 53.3 degreeATDC

Crank angle = 86.7 degreeATDC

Crank angle = 120.0 degreeATDC

Fig. 13 Fuel droplets remained on the walls

These droplets remained on walls are supplied as a homogeneous mixture in the later engine-cycle after evaporation.

By using the present code, the fuel motion in the cylinder and the mass remained on valves and on walls can be examined for several variations of engine-configuration and operating condition.

CONCLUSIONS AND DISCUSSIONS

Conclusions - A nonlinear ordinary equation system was discovered, which approximately expresses the deforming and oscilating motion by droplet-droplet and droplet-wet wall interactions. The equation is called the Quasi-Chaotic Ordinary Differential Equation (QCODE) from the characteristics of the trajectories.

The equation system for oscilating motion is open in the phase space. However, the trajectory of the equation does not go to infinity because of the presence of maximum distortion due to the energy dissipation.

Based on a simplified version of the QCODE, a concrete model for the breakup process after the dry wall impingement of a droplet is developed, which is called the Oval-Parabora Trajectories (OPT) model. The OPT model demonstrates four different regimes for the diameter distribution of the broken droplets and the mass remained on walls.

It is noteworthy to mention that the model can predict the size of child droplets after wall impingement and the amount of film flow remaining on the wall, for a Uo range of 2 - 40 m/s and a region of initial diameter below 300 micrometers.

The present theoretical model can be regarded as a subscale model for phenomena below the droplet size. The present type of subscale modeling based on nonlinear ordinary differential equations hold out the possibility for two-phase flow numerical simulations in the future. So far algebraic models were insufficient for such complicated phenomena as two phase flows. Models based on ensemble-averaged partial differential equations [16] can not easily express "spatial and temporal discontinuities" such as

droplet breakup and the vapor-liquid interface. Direct numerical simulation (DNS) of two phase flows is definitely impossible to be performed because the deterministic physical scale is very small in comparison with the largest scale of the system. The situation seems to be worse than in the case of DNS of combustion calculations.

The model described here can be used to predict the nature of the fuel in engines, by combining with the TAB model and the previous-reported flow code.

The computational results presented in a visualized image indicate that the droplet-wall interaction plays an important role for the secondary atomization process in spark-ignition engines. A part of the droplets entering into the cylinder becomes below ten micrometers after the impingement to the valve surfaces. Thus the child droplets can move riding on the air motion. The fuel amount, which comes into the cylinder during the injected cycle, is about 30-40% in the injected total mass.

While recent efforts have been devoted to the development of measuring methods of the spatial distribution of fuel, it takes even a long time to quantify the characteristics involved. Simultaneous studies should be performed by using both the present numerical code and the experimentation.

Discussions -

The decision of the child droplet size - It is approximated in the present report that breakup occurs when the droplet velocity becomes zero. The child droplet size and velocity should be decided from Eqs.(9), not from Eq. (12), in order to estimate more accurately.

Prediction of bounding regime - Reference [8] shows that there is another regime between Regime A and Regime B. Further study of Eq. 19 may give us the modeling of bouncing regime, as is shown in Table 2.

Empirical coefficients Cc and Ck - The three-dimensional phenonena are approximated by the zero-dimensional mode herel. Thus some artificial constants are needed.   These are determined in the present report so that predictions by using the OPT model agrees with experimental data.

Characteristics of and further possibilities of the QCODE - First, the difference between the Tayler-analogy equation [2] and the QCODE is discussed. Of course, each model treats different physical phenonenon.   However,   the only essential difference between two is that the kinetic energy term in the QCODE is time-dependent, while a constant in the Taylor analogy equation. The difference causes a thick border wall between linear phenomenon and nonlinear one.

The trajectory of the equation is varied from a closed oval to an open parabola depending on increasing impact velocity.   A jump in the trajectory may imply the transition to turbulence in the liquid motion within a droplet.   To shed some light on this point,   further mathematical analysis should be performed, which is based on the derivation of the QCODE from the Navier-Stokes equations with the proper boundary conditions, so that the Lorenz model was derived [9].

In the near future, a more advanced model should be created with the QCODE in order to estimate the droplets and droplet-film interactions. Such a model may also be useful for the predictions of the spray-spray impact and dense sprays, not only for the droplet-film interaction.   In order to realize the model, detailed experimental analysis is demanded for droplets interaction mechanism. Further experimental analysis for two droplets interaction produces us useful information.   Spray conditions involving several droplet diameters are too complex to understand the essential dynamics.

Bi-modal distribution of droplet size - It was found from the present model and from the experimental data that the droplet size histgram has a bi-modal distribution.   However, further study should be performed to confirm this characteristic, since the PDPA data shows some unrealistic peaks due to errors eventually. The diameter distribution may be concluded from direct experimental visualizations.

Here, the physical meaning of the bi-modal distribution is discussed by considering the droplet shapes just before breakup occurs. A thin spheroid was assumed in the present report. However, the direct photographs taken by Umenura [8] showed that the sattelite droplets is possible to emerge from "pole", which is stirred toward impaction direction. Thus, it should be tested whether Eq. (13) can be changed to an equation for the pole situation.

Motion of droplets remaining on the wall - Since computations were performed only for short interval, the film flow is not calculated. Several types of approaches for the process will be useful, such as the smooth particle hydrodynamics (SPH) [10] and the particle in cell method (PIC) [11].

During next ten years, the cycle-resolved computation [4] for the turbulent flow, combustion, and fuel preparation will be accelerated to be developed. One reason is that calculations over ten cycle should be performed in order to know the fuel mixture distribution in the cylinder at steady condition of spark-ignition engines. The other is an essential demand that the cyclic variations of combustion should be predicted.   The computer suppliers say " Teraflops machine before the next century ".   When the machine is realized,   100-1000 cycle computation may be possible to be performed in ten CPU hours.

Influence of the wall temperature to the breakup process - The temperature on valve surfaces may be below the critical point of boiling occurrence for the present engine.   The produced model must be modified when the surface temperature is above the critical point.

Modeling of the flying direction of child droplets - Equation (22) gives the fyling direction of the child droplets and the velocity magnitude in the present report.   Further studies are neccessary to model accurately the flying pattern of child droplets.

ACKNOWLEDGEMENTS

The authors would like to thank Prof. K. Yamamoto of Waseda University for his much advice on the mathematical aspects. The authors are indebted to Dr. T. Itoh, Mr. T. Nakada, Mr. Y.

Amenomori and Mr.S. Takayanagi for their assistance of experiments. Thanks are also done to Mr. K. Yoshida for making a figure.

REFERENCES

[1] R.D. Reitz and F.V. Bracco: Mechanisms of Atomization of a Liquid Jet, Phys. Fluids 25(10), 1982.

[2] P.J. O'Rourke and A.A. Amsden: The TAB Method for Numerical Calculation of Spray Droplet Breakup, SAEpaper 872089, 1987.

[3] K. Naitoh and K. Kuwahara: Large Eddy Simulation and Direct Simulation of Compressible Turbulence and Combusting Flows in Engines based on the BI-SCALES method, J. Fluid Dynamics Research, 1992.

[4] K. Naitoh, Y. Takagi, and K. Kuwahara: Cycle-resolved computation of compressible turbulence and premixed-flame in an engine, J. Comp. and Fluids., 1993.

[5] V. Yakhot and S. Orszag: Renormalization Group Analysis of Turbulence. I. Basic Theory, J. Sci. Comp., 1986.

[6] A.A. Andronow and C.E. Chaikin: Theory of Oscilations : Princeton University Press, 1949.

[7] J. Suzuki, H. Shimoda, and H. Kodama: Experimental Study on the Atomization of the Array of Droplets Impinging upon the Solid Surface, Proceedings of the 10th Conference on Liquid Atomization and Spray Systems in Japan, 1982.

[8] A. Umemura: Collision Behavior of Hydrocarbon Droplets, Proceedings of Tsukuba International Workshop on Mechanics of Reactive Flows, 1990.

[9] P. Berge, Y. Pomeau, and Ch. Vidal: L' ORDRE DANS LE CHAOS, Paris, Hermann, 1984.

[10] J.J. Monagam: Particle Methods for Hydrodynamics, Computer Physics Report, no.3, 1985.

[11] A.A. Amsden: The Particle-In-Cell method for the Calculation of the Dynamics of Compressible Fluids, LA report- 3466, 1966.

[12] H.Takewaki, A. Nishiguchi, and T. Yabe: Cubic Interpolated Pseudo-particle Method (CIP) for Solving Hyperboric-Type Equations. J. Comp. Phys. 61, 1985.

[13] T. Yabe: A Universal Cubic Interpolation Solver for Compressible and Incompressible Fluids, Shock Waves, Vol.1, No.3, 1992.

[14] K. Naitoh and K. Kuwahara: Numerical Simulation of the Compressible Turbulence around a Circular Cylinder, presented in the 4th Conference on Fluid Dynamics, Sendai, 1993.

[15] K. Takeuchi, J. Senda, and Y. Sato: Proceedings of the 9th Japan Conf. on Liquid Atmisation and Spray System, 1981.

[16] M. Ishii: Thermo-fluid dynamic theory of two-phase flow, Eyrolles, 1975.

[17] Chandra et al: Proc. R. Soc. Lond. A, Vol.432, pp. 13-41, 1991.

APPENDICS

The derivation of Eqs. (1) and (4)- The derivations of the Eq. (4) are described here. It is since Eq. (1) is a simple extension of Eq. (4).

The second term in Eq. (4), which is due to surface tension, is derived in the previous report[2]. The first term, which depicts the nonlinear contribution by the convection, is discussed.

The momentum equation obtained by assuming the irrotational flow is expressed as follows.

$$\frac{\partial}{\partial t} u_i = -|p + \frac{1}{2}(u_i)^2|_{,i}$$

(A.1)

Next, it is assumed that the droplet shape is oval. And Eq. (A.1) is integrated in the whole region within the droplet. Then the convection contribution is expressed by $C_c(\dot{X})^2$. It is understood that Eq. (4) expresses approximately the distorted motion of the droplet impinging the wall. Cc and Ck are artificial constants connecting with the distortion pattern.

Further formulation on the derivation of Eqs. (1) and (4) should be performed in order to determine artificial constants theoretically.

Evaluation of the TAB model at the initial stage of fuel injection - Figure A.1 shows the visualizations of the initial development process of the fuel spray. The predictions of diameters and the the corresponding PDPA data are depicted in Table A.1. Emphasis is placed on the reasonable values of D10 and D32. The penetration length and the droplet diameters agree with the experimental data by choosing some parameters for cone angles, initial velocity, and initial diameter in the subroutines for the spray calculations in KIVA II.

Table A.1   Droplet diameter (D10, D32)

	Predicted	Experimental
D10	58.0 μm	73.0 μm
D32	120.0	122.0

Estimation of the drag force due to the relative velocity between air and droplets - Figure A.2 shows the spray shape in a side wind. Comparisons were made for two wind speed conditions. It was confirmed that subroutine groups II can calculate the drag force for a side-wind velocity below 120 m/s with fairly good accuracy.

CIP method to capture the discontinuity of scalor quantities - The distribution of the vapor fuel eventually becomes discontinuous spatially. The usefulness of the CIP method[12] employed in this study is shown here for the discontinuous phenomena. It is demonstrated in Fig. A.3 that a supersonic turbulence around a cylinder is calculated

by the CUP method [13], which is an upgrade version of the CIP method. The oscilating shock waves and the unsteady vortices are reemerged [14].

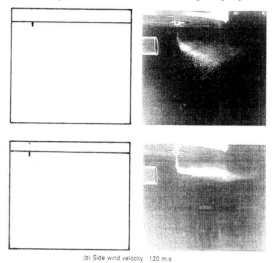

(b) Side wind velocity : 120 m/s

Fig. A2   The spray shape in the side wind

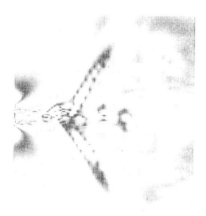

(a) Calclated density gradient
(Re = 127,000, M = 0.95)

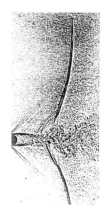

(b) Shadowgraphs (M = 0.95)
[A. Dyment et al. : Euromech Colloq. no. 135, Marseilles, France, 1980]

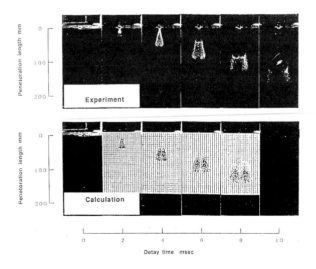

Fig. A1 The initial development of the spray

Fig. A3   Oscilating shock-waves and vortices

Effects of Injection Timing on Air-Fuel Mixing in a Direct-Injection Spark-Ignition Engine

Zhiyu Han and Rolf D. Reitz
University of Wisconsin-Madison

Jialin Yang and Richard W. Anderson
Ford Motor Co.

ABSTRACT

Multidimensional modeling is used to study air-fuel mixing in a direct-injection spark-ignition engine. Emphasis is placed on the effects of the start of fuel injection on gas/spray interactions, wall wetting, fuel vaporization rate and air-fuel ratio distributions in this paper. It was found that the in-cylinder gas/spray interactions vary with fuel injection timing which directly impacts spray characteristics such as tip penetration and spray/wall impingement and air-fuel mixing. It was also found that, compared with a non-spray case, the mixture temperature at the end of the compression stroke decreases substantially in spray cases due to in-cylinder fuel vaporization. The computed trapped-mass and total heat-gain from the cylinder walls during the induction and compression processes were also shown to be increased in spray cases. These thermodynamic features are shown to change with injection timing and they indicate the potentials of increasing engine thermal efficiency and power density in a direct-injection spark-ignition gasoline engine.

INTRODUCTION

Recently there have been intensive efforts to develop novel four-stroke direct-injection (DI) gasoline engines due to potential benefits of improved engine fuel economy and reduced pollutant emissions. The benefits have been demonstrated to be substantial in recent prototype engines such as a Ford Direct Injection Spark Ignition (DISI) engine [1] and a Mitsubishi direct injection SI engine [2].

Comparing with a port-injection engine, DI gasoline engines can avoid rich mixture operation during cold-start, which helps to reduce HC emissions [1]. At partial load, unthrottled operation offers DI gasoline engines a diesel-like fuel economy [2] because of the avoidance of pumping losses. At the wide-open-throttle (WOT; high load) condition, thermodynamic analyses have indicated a substantial reduction in charge temperature that leads to an improvement in the engine volumetric efficiency by 2-3% and offers an opportunity to increase the engine compression ratio [1]. As a result, an improvement in WOT output of 5-10% over a port fuel injection engine is achieved [1].

While most recently reported DI gasoline engines use a stratified-charge lean-burn strategy at partial load which favors high fuel economy, stratified-charge combustion presents technical challenge in controlling HC emissions. In the Ford DISI engine design, instead, stoichiometric combustion is adopted so that a conventional 3-way catalyst can still be used [1]. On the other hand, at high load, almost all reported designs have adopted early fuel injection (fuel injection occurs during the induction stroke) with a stoichiometric or slightly rich air-fuel ratio so that the air can be fully utilized and maximum engine power output can be obtained.

Understanding air-fuel mixing phenomena in a DISI engine is important for better combustion control. For this purpose multidimensional models were developed and computations were carried out of the in-cylinder spray atomization and air-fuel mixing processes. In this paper, the formation of air-fuel mixtures under early injection conditions and the effects of injection timing are discussed. In a separate paper [3], the physical submodels and numerical details are discussed and the gas flow and air-fuel mixing in a particular engine geometry is characterized.

The computational results of Han et al. [3] have revealed that the intake-generated gas flow interacts with the injected spray drops when an early fuel injection scheme is adopted. Under the impact of the intake-generated central gas jet between the two intake valves, the centrally injected spray is deflected and redirected. On the other hand, the spray induced flow also affects

---

* Numbers in brackets designate References at the end of the paper.

the gas flow such as to suppress the intake-created tumble motion [3]. All these phenomena are found to have a substantial influence on distributions of the liquid and vaporized fuel in the combustion chamber, and therefore on air-fuel mixing.

Since the momentum of the intake flow changes during the induction process, it is expected that the dynamic interactions between the spray and the gas flows vary with the start of fuel injection (SOI). The effects of injection timing on the evolution of the fuel distribution and air-fuel mixing are addressed here. The changes of thermodynamic characteristics of the in-cylinder charge (including the trapped fresh mass, wall heat transfer and temperature), due to in-cylinder fuel vaporization, are also discussed.

In the present work the KIVA-3 code with added moving-valve capability in the realistic engine geometry [4,5] was used. Improved submodels of gas turbulence, wall heat transfer, hollow-cone spray atomization and vaporization, and geometry representation were implemented. The details of these models can be found in the work of Han et al. [3] and will be not repeated here.

ENGINE AND COMPUTATIONAL CONDITIONS

The engine studied was a 0.62 L 4-valve single-cylinder engine. It has a 93.6 mm bore and a 90.6 mm stroke. The engine features a pent-roof combustion chamber with two intake valves and a flat piston crown, as depicted in Fig. 1 in which a cut-away view is given to show the intake valves. The exhaust valves/ports are not shown since the exhaust process was not modeled. The modeled engine has a compression ratio of 10.5:1.

The engine was operated at a high load and the detailed operation conditions are listed in Table 1. 45.7 mg of fuel (modeled as iso-octane) is injected within 44 crank angle degrees. The overall equivalence ratio is 1.18 (or A/F is 12.75).

Details of the computational mesh are given in Ref. [3] and in Fig. 2. The figure shows a cut-away perspective view (the cutting plane is located at the front valve center line) so that the interior grids can be seen. There are some 85,000 cells and the typical cell size is 3x3x3 mm in the cylinder. Pressure-inflow boundary conditions were imposed at the open ends of the runners. Computations were started at the intake TDC (0 degree) and ended at the compression TDC (360 degrees). The initial thermodynamic and turbulence parameters were specified to be uniform in the cylinder and in the runners separately. The parameters were set by referring to cycle simulation results of a similar single-cylinder DISI engine.

The considered engine uses a pressure-swirl injector that is located near the center of the chamber and fuel is injected near vertically. The injection pressure is about 5 MPa. The spray injected under the mentioned injection pressure and at atmospheric back pressure is illustrated in Fig. 3. In this figure, measured spray image

using a CID camera imaging system [6] and computed spray outline using the present spray model are given. Both experiment and computation show that the spray has a hollow-cone structure with an average cone angle of 55° and drop Sauter mean diameter of 30 μm when fuel is injected under the mentioned injection pressure and at atmospheric back pressure. Detailed discussion of characteristics of the spray and comparisons between the present model and experimental spray penetration and drop size data for a wide range of operating conditions are given in the work of Han et al. [7].

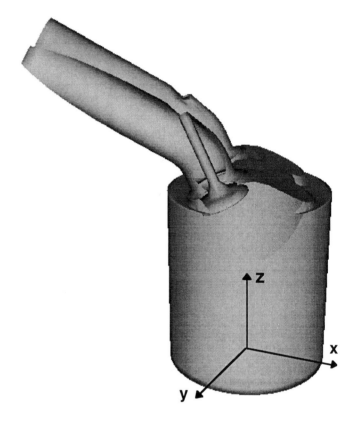

Fig. 1 Schematic of the chamber and intake valves/ports geometry of a DISI engine.

Table 1 Computational conditions for a DISI engine.

Bore (mm)	93.6
Stroke (mm)	90.6
Cylinder displacement (L)	0.62
Compression ratio	10.5
Engine speed (rev/min)	1500
Equivalence ratio	1.18
Intake pressure (MPa)	0.1
Intake gas temperature (K)	310
Residual gas (%)	5.8
Residual gas temperature (K)	900
Piston surface temperature (K)	516
Head surface temperature (K)	473
Cylinder liner temperature (K)	416
Fuel	iso-octane

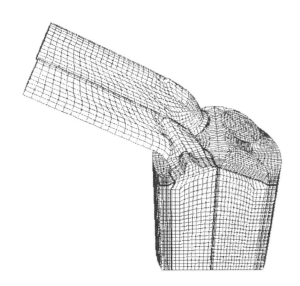

Fig. 2 Perspective view showing outline of the computational mesh at BDC. Slice in the foreground shows detail of the valve stem and port.

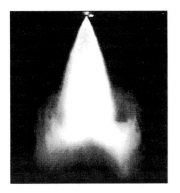

Fig. 3 Comparison of measured (Left) and computed (Right) spray (3.7 msec after the start of injection) resulting from an automotive pressure-swirl injector.

RESULTS AND DISCUSSION

As discussed previously, a pressure-swirl injector was used in the considered engine and a hollow-cone spray is formed to promote fast mixing. The hollow-cone structure of the injected spray can be also seen in the engine when it is injected during the induction stroke, as demonstrated in Fig. 4 in which the spray droplets are represented by the particles. In this figure, two 'lighting' sheets cut through the spray. One is on the axial plane of the intake valves and the other one is on a horizontal plane downstream of the injector. Liquid density contours in the two planes are shown. The shapes of these contours clearly indicate that the spray in the cylinder generally maintains its hollow-cone structure, as it also exhibits in a quiescent environment [7].

However, it was found that unlike in a quiescent environment, the spray in the engine cylinder is deformed and redirected under the interactions of the intake-generated gas flows and the spray. During the intake process, a gas jet-like flow is formed from the valve curtain flows in the central region between the two intake valves [3]. This gas jet has relatively high momentum and directly impacts the spray which is injected in this central region. As a result, fuel distribution is affected. In the following discussion, the influence of fuel injection on intake-generated gas flow fields is addressed first, then the impact of the gas flow on the spray is discussed.

The computed total mean kinetic energy of the gas in the combustion chamber of a non-spray case (i.e., keeping other conditions the same but no fuel is injected) as well as of cases with various SOI conditions are plotted in Fig. 5. The effects of fuel injection on the total mean kinetic energy are clearly seen. The momentum carried by the injected spray is partially transferred to the surrounding gases and that increases the kinetic energy of the charge soon after the fuel is injected. The spray-introduced enhancement of gas mean motion helps the air-fuel mixing. However, the increased kinetic energy quickly decays when the piston moves up in the compression stroke and the changes in the kinetic energy over the non-spray case at TDC become relatively insignificant.

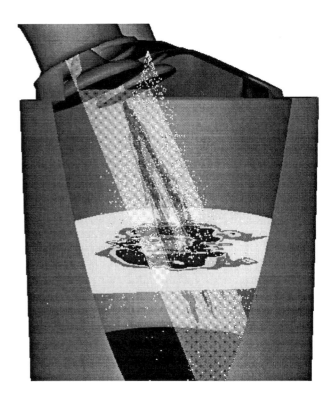

Fig. 4 Spray structures of a direct-injection spark-ignition engine.

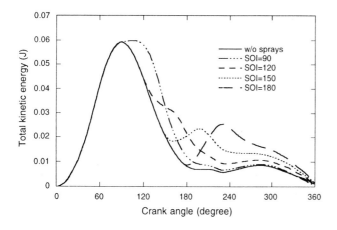

Fig. 5 Effects of fuel injection on the total gas mean kinetic energy.

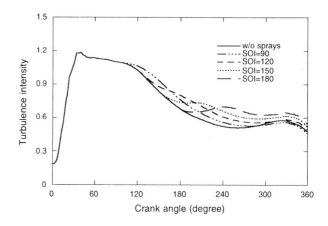

Fig. 7 Effects of fuel injection on gas turbulence intensity. The intensity is normalized by the mean piston speed.

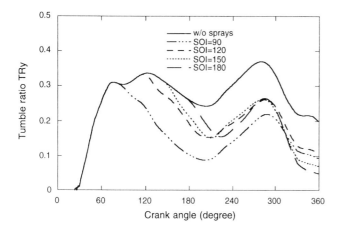

Fig. 6 Effects of fuel injection on the tumble ratio.

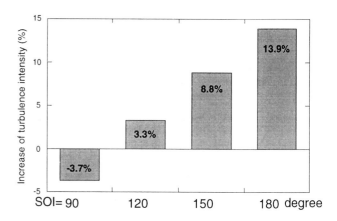

Fig. 8 Relative turbulence levels at the compression TDC showing the increases over the non-spray case at various injection timings.

Fuel injection is also found to influence the gas rotational-motion behavior. This phenomenon is examined in Fig. 6 in which the tumble ratio TRy (tumble rotation around an axis parallel to the y-axis, see Fig. 1) is plotted. The other two rotation components TRx (tumble rotation around an axis parallel to the x-axis and SR (swirl rotation around z-axis) are not shown since they are insignificant due to the symmetric design of the valves and chamber of the considered engine configuration [3]. The tumble ratios are evaluated at the instantaneous mass center [8]. It is seen from Fig. 6 that the intake-created tumble motions are significantly suppressed by the spray-induced flows in all the injection cases considered. This tumble-suppression may have negative effects on the air-fuel mixing and the mixture distributions at the end of compression. It was found that the rotational motions were helpful to enhance the air-fuel mixing [8].

The effects of fuel injection on gas turbulence is shown in Fig. 7. The intensity given is mass-averaged over the entire chamber. As can be seen, fuel injection enhances the turbulence of the in-cylinder charge. It is

interesting to notice that, although the spray-generated turbulence also decays during the compression stoke, the remaining level of turbulence intensity is substantially higher in late injection cases than that in the non-spray case. This is summarized in Fig. 8 which shows the increased turbulence levels over the non-spray case at the compression TDC. The trend is that the later the start of the injection, the higher the turbulence intensity at TDC. The implication of this is that a late injection scheme would have faster combustion in a DISI engine. A very late injection scheme, however, could result in worse combustion due to poor air-fuel mixing.

It has been experimentally shown that the unburned gas turbulence level is very important to the combustion in spark-ignition engines. The flame speed has been found to be strongly related to the turbulence intensity of the mixture [9], and hence the burning rate can be increased by the enhanced turbulence. The enhanced turbulence of the mixture has been also shown to increase engine cyclic stability [9]. In a port fuel injection (PFI) engine, enhancing the turbulence level of the charge is very commonly achieved by promoting

Fig. 9 Perspective views of the computed sprays at a delay time equal to 30 degrees (3.33 msec) after the start of each injection. Form (a) to (d): SOI=90, 120, 150 and 180 degrees.

rotational (swirl or tumble) flows through various intake-port and chamber geometry designs [10]. However, much of the turbulence generated during the induction stroke decays towards the end of the compression stroke when combustion is ready to take place. Also, a dedicated port-design for swirl-generation purposes usually sacrifices engine volumetric efficiency. In the present computed DISI engine, about 10 percent extra turbulence intensity is generated by the spray when the SOI is retarded to be later than 150 degrees in the same engine configuration. This spray-generated turbulence would not be present in a PFI engine since the fuel is injected against the intake valve surfaces, and even if some extra turbulence is generated, it would not persist to the end of the compression stroke, as indicated by the present 90-degree SOI case in which the spray-generated turbulence is diffused out (in fact, a slightly lower lever of turbulence intensity is seen compared with the non-spray case due to the turbulence energy dissipated in dispersing the spray drops).

The degree to which the intake-generated flows impact the injected sprays also varies with the start of fuel injection. Figure 9 shows the spray in the cut-away engine geometry at 30 degrees after the start of each respective injection. It is seen that the spray is deflected the most (away from the intake-valve side of the liner) when the SOI is 90 degrees since it encounters the gas jets during their highest-momentum time period. The spray deflection is reduced when the injection is retarded since the effects of the intake-generated flows on the spray drop trajectories are reduced.

The computed spray tip penetrations of the main spray (following the initial slug, see Fig. 9) in the cylinder axial direction versus the delay time after the start of injection are given in Fig. 10. It is seen that the spray-tip penetration is also affected by the intake flow details. Compared with that of the 180-degree SOI case, the penetration is increased as the injection is advanced due to the increased effects of the intake gas jets. For the same reason, the spray-tip axial velocity is increased when the injection is advanced beyond that of the 180-degree case. This is shown in Fig. 11 in which the velocity is deduced from the spray tip penetration data in Fig. 10.

Figure 12 illustrates the histories of the liquid fuel quantities on the wall surfaces for the various injection timing cases giving the temporal details of the liquid accumulation on the walls. In the figure, the amount of liquid fuel on a solid wall surface (piston or liner) is defined as the amount of fuel within a 0.1 mm layer immediately adjacent to the wall. The computations reveal that the injection timing is important to the spray/wall impingement details which are not only dependent upon the relation of the spray and piston trajectories, but also is influenced by the details of the gas flows that the spray encounters. It is seen that wall-wetting is reduced to a minimum when the injection is started at 180 degrees. In this case, although the piston is moving toward to the injector nozzle, the reduced axial penetrating velocity of the spray due to the increased

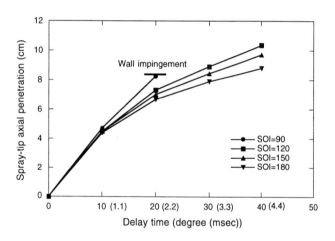

Fig. 10 Computed spray-tip axial penetration distance versus crank angle.

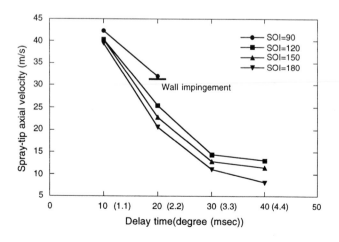

Fig. 11 Computed spray axial penetration velocity versus crank angle.

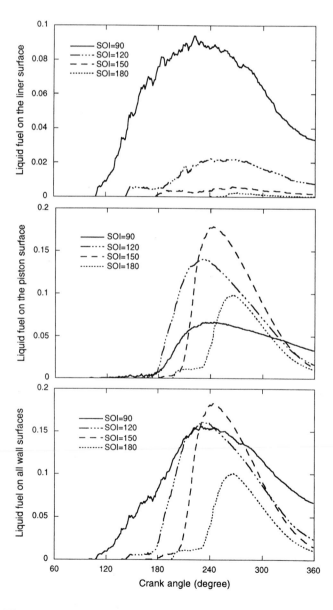

Fig. 12 Computed histories of the liquid fuel (normalized by the total injected fuel) on the liner surface (Top), piston surface (Middle) and the sum of the two (Bottom).

gas density helps to reduce the wall-wetting. For this particular engine, substantial wall impingement occurs about 50-60 degrees (5.6-6.7 msec) after the start of the injection, as indicated by the sharp rise in the curves in Fig. 12, except for the 90-degree case. In that 90-degree case, a substantial amount of the fuel impinges on the liner starting at about 15 degrees after the start of injection, as seen in top part of Fig. 12. The liquid fuel stays near the liner surface in this case and is seen to reach about 9% of the total injected fuel at its maximum, due to the effects of the intake gas jet as discussed above. It was found that the majority of the fuel is located in the corner region between the liner and the piston which slows down fuel vaporization significantly for this case (see Fig. 10, SOI=90). When the injection is retarded, the liner-wetting is significantly reduced. Hence, the computational results suggest that to avoid liner-wetting in this engine and at this high load condition, fuel should not be injected before 120 degrees.

Fuel vaporization histories at the various injection timings are illustrated in Fig. 13. The vapor is normalized by the total injected fuel. It is seen that the vaporization rate in the 90-degree SOI case is slowed down after about 140 degrees due to the reasons mentioned above, and about ten percent of the total injected fuel remains in the liquid phase at the end of the compression stroke. When the injection is retarded to after 120 degrees, fuel vaporization is essentially finished by TDC for all the three cases (SOI=120, 150 and 180). It should be noticed that the time available for evaporation is only one factor for reaching this completion, the charge temperature which the spray encounters is also important. For the 180-degree SOI case, the charge temperature is increased due to the compression which enhances the fuel vaporization rate.

The evolution of the in-cylinder mixture in different equivalence ratio ranges versus crank angle are shown in Fig. 14 for the cases with different SOI. It is seen that the mixture formation histories vary with the injection timings, however, the distribution of the mixture

in the three equivalence ratio ranges at TDC is not altered very much for the cases having a SOI later than 90 degrees. This can be traced to the fact that the remaining liquid in these cases (about 2-3%) is located near the piston surface and the rich mixture in these regions is less affected by the flow motions.

The computed spatial A/F ratio distributions at TDC for these cases are illustrated in Fig. 15 in which A/F ratio contours on three cutting planes are shown. The computations indicate that there exists A/F ratio stratification regardless of the injection timings. For the present particular engine configuration, generally, the mixture is leaner in the chamber region with the A/F ratios ranging from 12 to 20 (and up to a maximum of 24, not shown), while it is richer in the squish regions with the A/F ratios less than 12 (and down to a minimum of 8, not shown). Although the varied injection timings

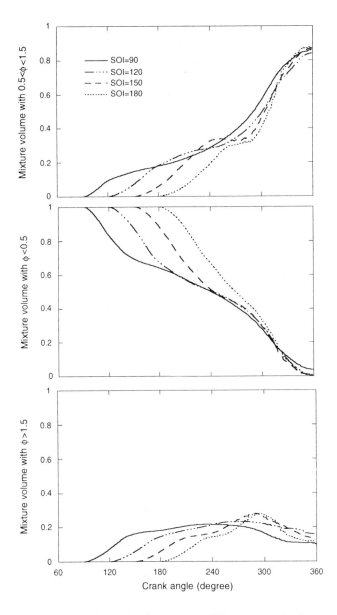

Fig. 14 Evolution of mixtures in different equivalence ratio ranges for various injection timings.

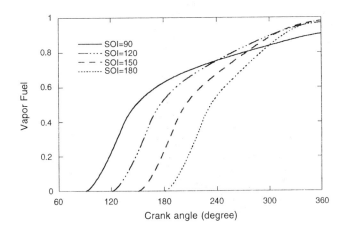

Fig. 13 Computed vaporization history versus crank angle for various injection timings.

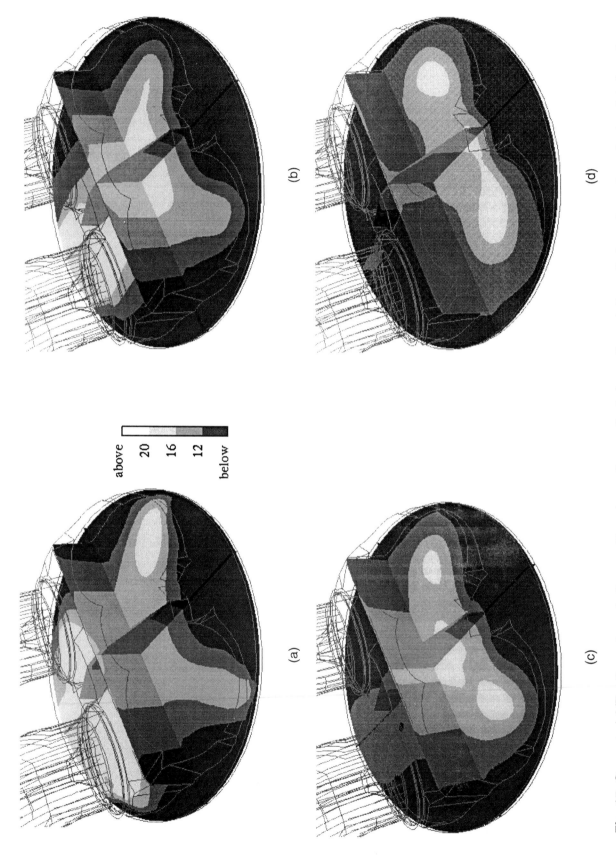

Fig. 15 Computed A/F ratio contours is three cutting planes (two vertical planes through the cylinder axis, and one horizontal plane 1 mm above the piston surface) at compression TDC for various injection timings. From (a) to (d): SOI=90, 120, 150 and 180 degrees.

above
20
16
12
below

(a)

(b)

(c)

(d)

result in different charge stratifications, the general trend with respect to the locations of rich and lean regions is not altered. The gross features of the charge distribution are set up by the injection and intake flow orientations.

The predicted A/F ratio stratification needs additional discussion. In traditional stratified-charge spark-ignition engines, HC emissions are generally high and one of the reasons cited for this is the flame-quenching phenomenon which occurs where the mixture is leaner than the lean-burn limits. However, the mixture stratification computed in the present DISI engine does not necessarily imply that high levels of HC emissions are to be expected since there are no mixtures with equivalence ratio less than 0.5 existing in the chamber at the end of the compression stroke when fuel is completely evaporated (See Fig. 14).

In addition, engine experiments in a single cylinder DISI engine have found that engine-out HC emissions are comparable when the same engine is operated with an in-cylinder-injection or a port-injection scheme at high loads [1]. However, these results do not necessarily demonstrate that the charge is homogeneous in a DISI engine since there is no direct experimental evidence to show the mixture in a PFI engine is itself uniform. In fact, recently, Berckmuller and co-workers [11] measured in-cylinder fuel concentrations in a modern 4-valve pent-roof PFI engine under firing engine conditions using a planar laser-induced fluorescence imaging method. Their results clearly show that the A/F ratio is non-uniformly distributed within the cylinder at the time of ignition. In this experiment, the A/F ratio varies locally up to 25 although the average A/F ratio is 14.4. Hence, additional experiments are recommended to directly measure A/F ratio distributions in a DISI engine and to relate them to engine-out HC emissions. These studies would lead to further understandings of the HC mechanisms for DISI engines.

Bonneau and co-workers [12] have also recently compared HC emissions from an engine using two different mixture preparation methods. For the overall A/F ratio ranging from 11.5 to 16, the HC emissions have been found to be the same using a carburetor fuel system and using a homogeneous mixture provided by an air-fuel mixing tank. This experiment implies that the charge inhomogeneity, if it exists in the carburetted engine, is not a noticeable source of HC emissions.

Next, the effects of in-cylinder fuel injection on the thermodynamic characteristics of the in-cylinder charge are addressed. Again, the following discussion is based on comparisons of the previously defined non-spray case (i.e., pure air) and various spray cases. Figure 16 compares the computed average gas temperature of the non-spray case and of a spray case (SOI=120 degrees). In the computed DISI engine, the evaporating fuel drops absorb heat from the in-cylinder gases, and as a result, the in-cylinder charge temperature decreases as indicated in Fig. 16. Detailed gas temperatures at 230 degrees (the intake valve closure time) for various cases are listed in Table 2. Normalized vapor fuels are also listed for reference. It is

Table 2 Comparison of gas temperature at IVC.

Case	Gas temperature (K)	Temperature decrease over the non-spray case (^{0}C)	Vapor fuel (normalized by the total fuel) (%)
Non-spray	374.2		
SOI=90	358.6	15.6	73.0
SOI=120	357.8	16.4	72.4
SOI=150	358.9	15.3	63.5
SOI=180	362.0	12.2	47.4

seen that although the charge cooling results in the spray cases are dependent on detailed vaporization and other factors (e.g., wall heat transfer), the charge temperature drop is roughly about 15 ^{0}C by the end of the induction process.

Some benefits can be gained from the charge temperature-decrease phenomenon. Specifically, the gas temperature at the end of the compression stroke is seen to be reduced significantly (see Fig. 16). For example, it is decreased from 812.5 K in the non-spray case to 696.5 K when the fuel is injected at 120 degrees - a decrease of 116 K. The decreased gas temperature makes an increase of the engine compression ratio possible in a DI gasoline engine and it will help to improve the engine fuel economy. It is known that one of the obstacles that limits an increase of the compression ratio in a conventional PFI engine is knocking combustion. It is the result of the end-gas auto-ignition due to high charge temperature and is harmful to engine performance [8]. A further increase of the compression ratio in a PFI engine would result in a higher gas temperature (since significant amount of the injected fuel absorbs heat for evaporation from the runner/port/valve walls) and hence an increase in the knocking tendency. Since the gas temperature is reduced in a DISI engine, the compression ratio of the engine can be possibly increased from the current levels of conventional PFI engines. Results in a single cylinder DISI engine have shown this potential [1].

It should be pointed out that the mentioned 116 ^{0}C temperature decrease is based on the comparison of a pure-air (non-spray) compression and a fuel-air mixture compression. It could be larger than what would be expected in comparison of a PFI and a DISI engine. One of the reasons is that in a PFI engine the compressed charge is air-gasoline mixture which has a smaller specific heat ratio due to its larger molecular weight in comparison with the pure air. The smaller specific ratio of the compressed charge results in a lower temperature at the end of compression. Also, in a PFI engine some of the fuel absorbs heat from the surrounding gases when they are evaporating, hence, the gas temperature should be lower than that of the non-spray case. On the other hand, in reality the

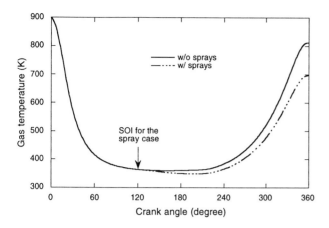

Fig. 16 Effect of fuel vaporization on the in-cylinder gas temperature.

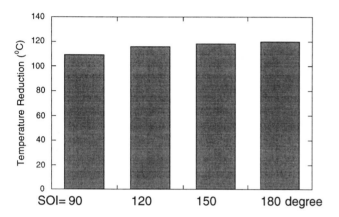

Fig. 17 Predicted gas temperature reduction at TDC for various injection timings in comparison with the non-spray case.

impinging spray drops also absorb some heat from the walls; since the present computation does not include the conductive heat transfer from the wall to these drops, instead, they absorb heat completely from the gas that can result in a lower gas temperature in the computation.

The temperature reductions over the non-spray case for the various injection timings are shown in Fig. 17. It is seen that there is only a minor difference in the temperature-decrease among these injection-timing cases, except possibly for the 90-degree SOI case. In that case, it was shown in Fig. 13 that about ten percent of the injected fuel had not yet evaporated by TDC and this part of the fuel made no contribution to the temperature-reduction effect. While in the other three cases, almost all of the injected fuel was vaporized by TDC and that leads to about the same gas temperature at TDC for these cases.

Another benefit of the temperature-decrease effect is that the decrease of the charge temperature can result in a cylinder pressure drop during the intake process. Figure 18 illustrates the computed average cylinder pressure in which the bottom plot is a blow-up to show the pressures during the intake process. As can be

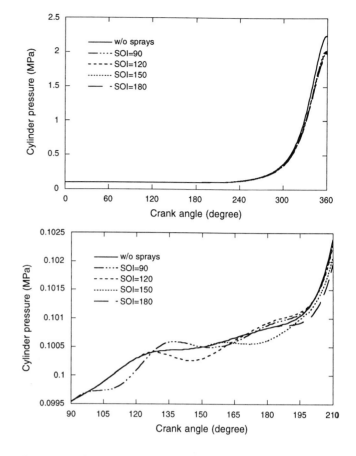

Fig. 18 Effect of fuel vaporization on the cylinder pressure. The bottom figure shows the pressures during the induction process using an enlarged scale.

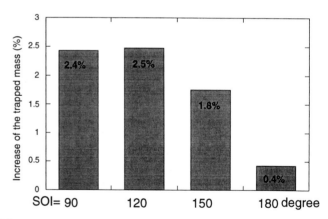

Fig. 19 Predicted increases in the trapped mass over the non-spray case at various injection timings.

seen, the cylinder pressure is reduced after the fuel is injected relative to that of the non-injection case. Some interesting dynamic oscillations in the gas pressure of the injection cases are also seen.

The decrease of the in-cylinder pressure can result in an increased in-flow mass during the induction process, and a reduced out-flow mass (due to the back-flow) near the end of the intake stroke. As a result, the in-cylinder trapped mass can be increased and the engine volumetric efficiency can be improved. The

284

computed increases of the amount of trapped mass is shown in Fig. 19 in comparison with the non-spray case. It is seen that the trapped mass can be increased by as much as 2.5 percent for the 120-degree injection case. However, the extra gain of the trapped mass generally decreases at a retarded SOI. This is due to the fact that a retarded injection experiences less pressure-reduction time during the induction process, as can be seen in Fig. 18. In particular, the injection-introduced volumetric-efficiency benefit will vanish when the fuel is injected during the compression stroke. On the other hand, when the fuel is injected at a very early time, the increased wall-wetting also slows down the fuel vaporization process, resulting in a smaller temperature-decrease and hence a smaller pressure-reduction, and the trapped mass is therefore reduced. This sequence of events is demonstrated by the 90-degree injection case. The present computational results are consistent with the experimental observations made in a single cylinder DISI engine in which a 2-3% increase of the volumetric efficiency was found [1].

Figure 20 shows the predicted heat transfer integrated over the entire chamber wall-surface versus crank angle. A negative value represents heat transferred from the wall to the in- cylinder gas (heat gain) and a positive one represents heat transferred from the gas to the wall (heat loss). Figure 20 indicates that the heat transfer characteristics of a non-spray case is altered by in-cylinder fuel injections. Generally, since the modeled high-load engine has relatively high wall temperatures (see Table 1), the in-cylinder gas temperature (see Fig. 16) becomes lower than the wall temperatures soon after the intake valves are opened and heat is transferred from the wall to the charge. In the late compression stroke, the charge is compressed and its temperature is raised to values higher than the wall temperatures, and then heat is transferred to the walls. In the fuel injection cases, under the assumed constant wall temperature conditions, the decreased charge temperature leads to increased temperature differences between the gas and the wall during the induction and early compression processes which increases the heat gain. It also results in reduced temperature differences in the late compression stroke which, on the other hand, decreases the heat loss. Hence, benefits can be obtained from the increase in the overall heat gain of the engine in the induction and compression processes, by as much as 34.5 percent in the 120-degree SOI case over the non-spray case, as indicated in Fig. 21.

Variations of the extra heat gain over the non-spray case at various injection timings are also given in Fig. 21. In all the cases the wall temperatures were assumed the same. Figure 21 suggests that the heat-transfer advantage in a fuel-injection case diminishes as the injection is retarded. This is, again, related to the charge temperature history, and the later the injection, the shorter the time-period with the reduced gas temperature and the less the total heat gain.

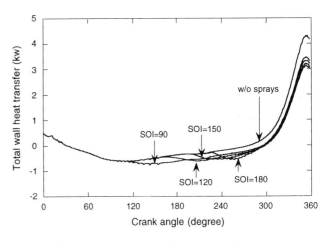

Fig. 20 Effect of fuel vaporization on wall heat transfer.

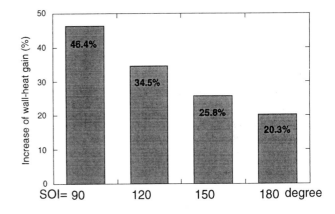

Fig. 21 Predicted increases in wall heat gain during the induction and compression strokes over the non-spray case at various injection timings.

The computed results shown in Fig. 21 are consistent with experimental results from a single cylinder DISI engine. It was found in the experiments [1] that the knock limited spark advance (KLSA) of the DISI engine can be increased as fuel injection is retarded. This experimental result suggests that retarding the injection timing reduces the charge heating by the walls and results in a lower mixture temperature at TDC, as indicated by the computational results shown in Fig. 21 and 17, respectively.

It should be point out that the above results are based upon computations of the induction and compression strokes only. While they do suggest the potentials that the total heat loss could be reduced in a DISI engine compared with a PFI engine (which has similar conditions to those in the non-spray case) in the light of the increased heat gain during the induction and compression strokes, investigation on the heat-transfer behavior of a DISI engine in the combustion and expansion processes should be carried out in order to reach definitive conclusions.

SUMMARY AND CONCLUSIONS

Multidimensional modeling was made of air-fuel mixing processes in a DISI engine. The considered engine features a 4-valve and pent-roof combustion chamber geometry and uses a pressure-swirl injector that forms hollow-cone sprays. Emphasis is placed on the effects of the start time of fuel injection on spray behavior and mixing.

A particular engine configuration with a high-load operating mode was considered. Fuel injection occurred in the induction stroke and was varied from 90 to 180 degrees after the intake TDC. It was found that the intake-generated gas flow influences the injected spray so that the spray is deflected and redirected. The spray-tip axial penetration can also be increased by the intake gas jet. As a combined result of the deflection and increased-penetration, the spray can impinge on the cylinder liner when fuel is injected between 90 to 120 degrees, even though the spray is injected vertically. The intake-flow has greatest influence on the spray when the spray is injected at 90 degrees and the influence diminishes when the injection is retarded.

The spray-induced gas motion affects the large-scale gas flow structures as well. In particular, it increases the mean velocities of the gases in the spray region and suppresses significantly the intake-generated tumble flow under all of the injection timings considered. On the other hand, fuel-injection generates enhanced turbulence of the charge, and the remaining levels of the turbulence intensity are substantially higher in the late injection cases (later than 150 degrees) than that of the non-spray case. The trend is that the later the start of the injection, the higher the turbulence intensity at TDC. The spray-enhanced turbulence intensity can be beneficial for a DISI engine since previous studies have shown that enhanced gas turbulence can increase the fuel burning rate and decrease engine cyclic variations.

Spray/wall impingement was seen to occur in the computed engine. Cylinder liner-wetting was also observed when fuel was injected earlier than 120 degrees due to the redirection of the spray by the intake gas jet. However, liner-wetting no longer occurred when the fuel injection was such that it took place in the late intake stroke. The instantaneous liquid-fuel amount on the wall surfaces reached as high as 10-18% of the total injected fuel, depending upon the injection timings, during the injection period and then decreased due to vaporization. The wall-impinged fuel droplets had lower vaporization rates and this led to the formation of relatively rich vapor regions near the piston surface during the compression stroke.

Air-fuel ratio stratification of the mixture was seen in the combustion chamber by the end of the compression process regardless of the injection timings computed. For the considered cases, generally, the mixture was relatively leaner in the combustion chamber region and richer in the squish regions, with the A/F ratio ranging from 8 to 24.

Regarding the thermodynamic behavior of the mixture due to in-cylinder fuel injection, the charge temperature of a spray-injection case was found to decrease 15 ^{0}C by the time of intake valve closure because of fuel vaporization effects. The charge cooling result suggests an opportunity to increase the compression ratio in a DISI engine over that of a conventional PFI engine. The decreased charge temperature during the intake process also results in an increased trapped cylinder mass by about 2% over the non-spray case (i.e., a volumetric efficiency increase). However, this advantage vanishes when the injection is retarded to occur at the end of the induction stroke. The computed total heat gain from the chamber walls in the spray-injection cases was also seen to increase substantially over the non-spray case during the induction and compression stroke. This suggests that the total heat loss could be reduced in a DISI engine compared with that in a PFI engine.

ACKNOWLEDGMENTS

This work was supported by Ford Motor Company. Additional support was provided by the US Army Research Office. Helpful comments of Drs. D. Brehob (Ford), J. Martin and P. Farrell (ERC) are appreciated.

REFERENCES

1. Anderson, R. W., Yang, J., Brehob, D. D., Vallance, J. K., and Whiteaker, R. M. (1996). Understanding the Thermodynamics of Direct Injection Spark Ignition (DISI) Combustion Systems: An Analytical and Experimental Investigation. SAE Paper 962018.

2. Kume, T., Iwamoto, Y., Iido, K., Murakami, M., Akishino, K., and Ando, H. (1996). Combustion Control Technologies for Direct Injection SI Engine. SAE Paper 960600.

3. Han, Z., Fan, L., and Reitz, R. D. (1997). Multi-dimensional Modeling of Spray Atomization and Air-Fuel Mixing in a Direct-Injection Spark-Ignition Engine. SAE Paper 970884.

4. Amsden, A. A. (1993). KIVA-3: A KIVA Program with Block-Structured Mesh for Complex Geometries. Los Alamos National Labs., LA-12503-MS.

5. Amsden, A. A. (1996). Private communication.

6. Parrish, S. E. (1996). Spray Characteristics of a Direct-Injection Spark-Ignited Engine. Ph.D. Thesis, University of Wisconsin-Madison, in preparation.

7. Han, Z., Parrish, S. E., Farrell, P. V., and Reitz, R. D. (1996). Modeling Atomization Processes of Pressure-Swirl Hollow-Cone Fuel Sprays. Submitted to *Atomization and Sprays*.

8. Han, Z., Reitz, R. D., Claybaker, P. J., Rutland, C. J., Yang, J., and Anderson, R. W. (1996c). Modeling the Effects of Intake Flow Structures on Fuel/Air Mixing in a Direct-Injected Spark-Ignition Engine. SAE Paper 961192.

9. Heywood, J. B. (1988). *Internal Combustion Engine Fundamentals*, McGraw-Hill Company, New York.

10. Hill, P. G., and Zhang, D. (1994). The Effects of Swirl and Tumble on Combustion in Spark-Ignition Engines. *Prog. Energy Combust. Sci.* **20**, 373.

11. Berckmuller, M., Tait, N. P., Lockett, R. D., Greenhalgh, D. A., Ishii, K., Urata, Y., Umiyama, H., and Yoshida, K. (1994). In-Cylinder Crank-Angle-Resolved Imaging of Fuel Concentration in a Firing Spark-Ignition Engine Using Planar Laser-Induced Fluorescence. *25th Symposium (International) on Combustion*, the Combustion Institute, 151.

12. Bonneau, R. J., Cunningham, M. J., and Martin, J. K. (1995). Emissions and Combustion Characteristics from Two Fuel Mixture Preparation Schemes in a Utility Engine. SAE Paper 952081.

Multidimensional Modeling of Spray Atomization and Air-Fuel Mixing in a Direct-Injection Spark-Ignition Engine

Zhiyu Han, Li Fan, and Rolf D. Reitz
University of Wisconsin-Madison

ABSTRACT

A numerical study of air-fuel mixing in a direct-injection spark-ignition engine was carried out. In this paper, the numerical models are described and grid generation methods to represent a realistic port-valve-chamber geometry is discussed. To model a vaporizing hollow-cone spray resulting from an automotive pressure-swirl injector, a newly developed sheet spray atomization model was used to compute the processes of disintegration of the liquid sheet and breakup of the subsequent drops. Computations were performed of a particular 4-valve pent-roof engine configuration in which the intake process and an early fuel injection scheme were considered. After an analysis of the intake-generated flow structures in this engine configuration, the spray behavior and the spatial and temporal evolution of fuel liquid and vapor phases are characterized. It was found that the intake-generated flow interacts with the injected spray drops which strongly influences fuel (both liquid and vapor phases) distributions and air-fuel mixing.

INTRODUCTION

Internal combustion engines have played a dominant role in the fields of power, propulsion and energy. They have been widely used in commercial vehicles and other domestic and industrial applications. However, recently transportation engines, and the industries that develop and manufacture them have been greatly challenged as the issues of shrinking resources, preserving the environment and market competition become increasingly important.

Hence, engine makers, driven by strict pollutant emissions regulations and fuel economy demands, are considering novel combustion systems to reduce engine-out emissions, i.e., unburned hydrocarbons (HC), oxides of nitrogen (NOx), carbon monoxide (CO) and particulates (especially for diesel engines) and to improve fuel economy. In the automotive industry research on modern four-stroke direct-injection (DI) gasoline engines has become active again due to their high-efficiency and low-emissions potentials. In addition, recent advances in fuel injection technology coupled with improved tools for understanding engine combustion have made the DI gasoline engine a very promising new combustion system option [1-3]*.

The direct-injection concept has been demonstrated again to improve engine fuel economy and power density in recent designs that feature increased compression ratios and improved volumetric efficiency, e.g., in a Ford Direct Injection Spark Ignition (DISI) engine [1] and in a Mitsubishi direct injection SI engine [2]. At partial load, unthrottled operation offers DI gasoline engines a diesel-like fuel economy [2], mainly due to the avoidance of pumping losses. At the wide-open-throttle (WOT; high load) condition, thermodynamic analyses have indicated a 50-60 ^{0}C reduction in charge temperature that leads to an improvement in the engine volumetric efficiency by 2-3% and offers an opportunity to increase the engine compression ratio [1]. As a result, an improvement in WOT output of 5-10% over a port fuel injection (PFI) engine has been reported [1].

Comparing with PFI engines, DI gasoline engines can avoid rich mixture operation during cold-start, which helps to reduce HC emissions [1]. The reduction of HC emissions during cold-start will significantly improve the total engine emissions level from an emission test cycle due to the fact that gasoline engines produce much more HC emissions during cold-start than during any other modes of operation. Direct injection also avoids the wall-wetting phenomena prevalent in PFI engines, which helps with fuel control during the cold-start, warm up, and other transient operations. This has been also demonstrated in the Ford DISI engine [1]. In addition, a DI gasoline engine has been shown to reduce NOx emissions due to the

---

* Numbers in brackets designate References at the end of the paper.

decreased charge temperature and possible operation with the use of large amounts of EGR [1,2].

While most recently reported DI gasoline engines use a stratified-charge lean-burn strategy at partial load which favors high fuel economy, stratified-charge combustion presents technical challenge in controlling HC emissions. In the Ford DISI engine design, instead, stoichiometric combustion is adopted so that the conventional 3-way catalyst can still be used [1]. On the other hand, at high load, almost all reported designs have adopted early fuel injection (fuel injection occurs during the induction stroke) to form a nearly homogeneous mixture with a stoichiometric or slightly rich air-fuel ratio so that the air can be fully utilized and maximum engine power output can be obtained.

Precise control over the fuel injection and air-fuel mixing process to form the desired in-cylinder mixture is one of the key issues for a successful DI gasoline engine design. Previous studies indicate that one of the most challenging problems is to create and stabilize a stratified mixture in the vicinity of the spark plug in stratified-charge combustion systems by matching the flow and sprays. This was achieved by using a hollow-cone spray and strong swirl motion in the Ford PROCO system [4,5], or by injecting a jet spray tangentially into the piston bowl with high swirl motion to stabilize the mixture in the Texaco TCCS system [6]. More recently, Mitsubishi reports the use of a hollow-cone spray and strong reverse tumble motion combined with a complex-shaped piston crown to realize this goal [2]. This design utilizes the piston crown shape, with the help of the tumble flow motion, to redirect the fuel spray that is injected onto the piston surface toward the vicinity of the spark plug where a relatively rich mixture is achieved. In the Ford DISI engine, it has been also shown that a careful selection of fuel injection scheme and intake system design is needed to obtain a nearly homogeneous mixture and to avoid wall-wetting so that optimized combustion and improved engine-out emissions can be achieved [1,7,8]. All these examples not only show the importance of air-fuel mixing for stable combustion, but also indicate the complexity of obtaining the desired air-fuel ratio distribution. In general, the in-cylinder mixing process is affected in a very complicated manner by large-scale gas motions, the fuel injection scheme used and the spray characteristics. It is also influenced by small-scale turbulent diffusion, chamber geometry and engine thermal conditions. Therefore, an in-depth study of mixing phenomena is necessary.

While gas flow structures have been shown to play an important role in air-fuel mixing, efforts to look for more suitable injectors and improved sprays for DI gasoline engines have been made [3,10]. Results to date have indicated that hollow-cone sprays with wide spray cone angles (~55^0) are more favorable [1,2]. These sprays are typically created by a pressure-swirl injector operating under moderately high injection pressures (~5 MPa).

As novel injectors and flow controlling technologies are introduced in future DI gasoline engine designs, the goal of improved mixture preparation and better control of combustion will be made more difficult by the fact that new design variables are added. Hence, it is helpful to simulate the engine processes with the use of computational models in order to aid engine development efforts. Multidimensional modeling can provide detailed temporal and spatial information which helps to visualize and analyze the complicated and highly interacting engine in-cylinder processes on a three-dimensional and dynamic basis. It also helps to identify important influential factors through parametric studies and offers an additional tool for engine designers to be able to quickly compare different engine designs at low cost.

A numerical study of air-fuel mixing phenomena in a DISI engine was carried out using multidimensional models [10]. In this paper mathematical models and other numerical details used in the study are reported. After a discussion of the numerical models, results of a particular engine configuration with an early fuel injection scheme are presented. Detailed flow structures, spray behavior and their effects on air-fuel mixing are characterized.

NUMERICAL MODEL

Computations were performed using an improved version of KIVA-3 [11], a CFD code for transient reactive flows with sprays. KIVA-3 uses a block-structured mesh and permits flexibility for modeling complex geometries with various boundary conditions.

In the present study, the original KIVA submodels for gas turbulence, wall heat transfer and hollow-cone spray atomization and vaporization are replaced by improved models. The KIVA grid generator, K3PREP, is also modified to add more capabilities for the typical modern gasoline engine geometry with a pent-roof chamber and an elbow-shaped port/runner assembly. The details are described next.

TURBULENCE AND HEAT TRANSFER - The standard k-ε model in KIVA-3 was replaced by a modified RNG k-ε model [12], specifically adapted for engine flows. KIVA-3 uses a law-of-wall model for wall shear and a wall function based on the Reynolds Analogy for gas/wall heat transfer [11]. These models were also replaced by the models of Han and Reitz [13] that account for variable-density effects. Han and Reitz used a new method to calculate the wall shear and considered the variations of gas density and turbulent Prandtl number in the gas/wall heat transfer computations. Due to the gas density variation, the wall heat flux is found to be proportional to the logarithm of the ratio of the flow temperature to the wall temperature instead of to the arithmetic difference between the two temperatures, which is the case for incompressible flows. With the use of the new model, the predictions of wall heat flux in a gasoline engine were shown to be improved significantly over the original KIVA model [13]. Further details of the RNG k-ε and heat transfer models can be found in References [12] and [13)]

SPRAY ATOMIZATION - Hollow-cone sprays resulting from an automotive pressure-swirl atomizer were considered in this study. In a pressure-swirl atomizer angular momentum is imposed on the liquid to form a swirling motion. Under the action of centrifugal forces the liquid spreads out in the form of a conical sheet as soon as it leaves the orifice and a hollow cone spray is formed due to breakup of the sheet.

The hydraulic processes inside a pressure-swirl injector are complicated and vary, in particular, with different design details for the generation of the angular momentum. To simplify the problem, processes inside the injector are not modeled in the present study. Spray atomization is considered to be only due to the liquid sheet breakup.

A sheet atomization model was developed to compute the liquid sheet disintegration processes [14]. Detailed comparisons of computed and experimentally determined spray characteristics of sprays resulting from a prototype pressure-swirl injector have showed good agreements [14].

The basic idea of the model is summarized here. It is assumed that a conical liquid sheet with a length L and a thickness h is formed at the exit of the nozzle, as illustrated in Fig. 1. The average cone angle 2θ is defined by the nozzle internal geometry. Instead of assuming an intact sheet, discrete 'blobs' are injected that have a characteristic size equal to the thickness of the sheet. One of the advantages of this treatment is that the computationally efficient stochastic spray model [15, 11] can be utilized. Since blobs or parcels are used to represent the liquid sheet, they are assumed to experience negligible dynamic drag force and are not affected by turbulence dispersion. However, these parcels are subject to breakup once certain breakup criteria are satisfied according to the breakup model. The children drops formed due to a parent blob breakup

will be treated as ordinary drops that experience drag forces and are affected by gas turbulence. A blob will be also treated as an ordinary drop once its traveling distance is greater than the sheet breakup length L.

The sheet velocity V is defined as

$$V = K_v (2(p_1 - p_2)/\rho_l)^{0.5} \qquad (1)$$

where p_1 is the fuel pressure within the injector, p_2 is the environmental pressure, ρ_l is the liquid density and K_v is the velocity coefficient. K_v can be derived based on inviscid analysis as [16]

$$K_v = C \left(\frac{1-X}{1+X} \right)^{0.5} \frac{1}{\cos\theta} \qquad (2)$$

where C=1.1 is used to account for discrepancy between the theory and experiments [16], and

$$X = \left(1 - \frac{2h}{d_0} \right)^2 \qquad (3)$$

where d_0 is the nozzle orifice diameter.

The thickness of the sheet is estimated by [17]

$$h = \left[A \frac{12}{\pi} \frac{\mu_l \dot{m}_l}{d_0 \rho_l (p_1 - p_2)} \frac{1+X}{(1-X)^2} \right]^{0.5} \qquad (4)$$

where $\dot{m}_l$ is the liquid mass-flow rate and μ_l is the dynamic viscosity of the liquid. The constant A is related to the nozzle geometry and is set to be 400 in this study. It was found that the downstream drop sizes were not very sensitive to the sheet thickness since the 'blobs' soon breakup into small drops [14].

The sheet breakup length is calculated according to Clark and Dombrowski [18] who performed non-linear stability analyses of a planar liquid sheet. For a conical sheet it is recast as

$$L = B \left[\frac{\rho_l \sigma \ln(\eta/\eta_0) h \cos\theta}{\rho_g^2 U^2} \right]^{0.5} \qquad (5)$$

where ρ_g is the environmental gas density, σ is the liquid surface tension coefficient, U is the sheet-gas relative velocity, η is the wave amplitude when the sheet breaks up, and the parameter $\ln(\eta/\eta_0)$ is determined experimentally to be equal to 12 [18]. B is a constant and is set to be 3 [14].

The above formulations provide the initial blob information. A model is still needed to compute the breakup processes of the blobs and the resulting drops. In the present study, the TAB model of O'Rourke and Amsden [19] was modified and used for this purpose.

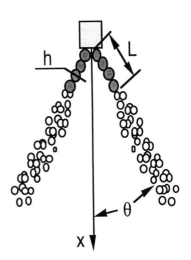

Fig. 1 Schematic diagram showing sheet disintegration processes.

The TAB model is based on the Taylor's analogy between an oscillating and distorting drop and a spring-mass system. The external forces acting on the mass, the restoring force of the spring, and the damping force are analogous to the gas aerodynamic force, the liquid surface tension force, and the liquid viscosity force, respectively. The parameters and constants in the TAB model equations have been determined from theoretical and experimental results.

In the TAB model, the force balance gives

$$\frac{d^2y}{dt^2} + \frac{5\mu_l}{\rho_l r^2}\frac{dy}{dt} + \frac{8\sigma}{\rho_l r^3}y - \frac{2}{3}\frac{\rho_g U^2}{\rho_l r^2} = 0 \qquad (6)$$

where t is time and y is the normalized (by the drop radius) drop distortion parameter. It is assumed that breakup occurs if and only if $y>1$. When this condition is reached, the droplet breaks up into smaller droplets with sizes determined by an energy balance taken before and after the breakup as [19]

$$\frac{r_1}{r_2} = \frac{7}{3} + \frac{1}{8}\frac{\rho_l r_1^3}{\sigma}\left(\frac{dy_1}{dt}\right)^2 \qquad (7)$$

where r is the drop radius and subscripts 1 and 2 represent conditions before and after the breakup process, respectively.

In the original TAB model, it is assumed that after a parent drop breaks up, the sizes of the children drops are distributed according to the χ^2 function with r_2 assumed to be the Sauter mean radius. This distribution was chosen based on diesel jet spray measurements. However, computations of pressure-swirl atomized hollow-cone sprays indicate that the χ^2 distribution results in an over-estimated population of large-size drops and this directly influences the spray structure and tip penetration. Instead, a Rosin-Rammler distribution was used which improves spray predictions [14].

The Rosin-Rammler cumulative distribution has the general form

$$V = 1 - \exp\left(-(D/\overline{D})^q\right) \qquad (8)$$

and the corresponding volume distribution is

$$\frac{dV}{dD} = \frac{qD^{q-1}}{\overline{D}^q}\exp\left(-(D/\overline{D})^q\right) \qquad (9)$$

where V is the fraction of the total volume contained in drops of diameter less than D, q is the distribution parameter which was set to be 3.5 in this study, and $\overline{D}$ is a characteristic mean drop size that is related to the Sauter mean diameter by

$$\overline{D} = D_{32}\,\Gamma\left(1 - q^{-1}\right) \qquad (10)$$

where Γ is the gamma function and D_{32} (SMD) is given by Eq. (7) (where $D_{32}=2r_2$).

Detailed comparisons between the present model and experimental spray penetration and drop size data are given in the work of Han et al. [14]. An example of the good agreement is given in Fig. 2 which shows a comparison between measured and computed spray outlines for a 4.76 MPa injection at 3.7 msec after the beginning of the injection.

The original KIVA evaporation model was also replaced by an improved version. In the improved model, the effects of fuel vapor on the surrounding gases are introduced and more accurate transport and thermo-physical properties of the liquid, vapor and environmental gas and air-vapor mixture rules are used. The model details are given by Han et al. [7]. A modified drop dynamic drag force model reported by Han et al. [7] was also used. When spray impinges on the engine walls, spray/wall interactions were modeled according to Naber and Reitz (20). Like other spray studies, spray drop collisions and coalescences, turbulence dispersion and modulation were also considered using the KIVA models [11].

INTAKE FLOW - Intake flow modeling is complicated by grid generation and the need to be able to model moving valves. There are several strategies to model intake flows for different research needs. One simplified method is to specify a prescribed velocity profile at the valve exit annuli in the head that represent the intake valve openings [7]. This approach eliminates the need for computations of flow in the intake manifolds and valve motions. Thus, the computational time can be reduced significantly and the grid generation difficulties caused by resolving the moving valves can be avoided.

The most comprehensive intake flow model reproduces the valve events during the computations. In KIVA-3, a snapper procedure is used to move the piston. In a recent version of KIVA-3 [21] (which was used in this work), this snapper algorithm is extended to model moving intake valves based on the work of Hessel and Rutland [22].

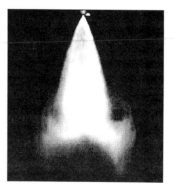 

Fig. 2 Comparison of measured (Left) and computed (Right) spray resulting from an automotive pressure-swirl injector.

292

The snapper algorithm deactivates grid planes when the piston moves up, and activates planes when the piston moves down. For a valve, two moving surfaces are assigned to the top and the bottom of the valve face, respectively. The valve stems and sides are considered to be solid walls and do not move. These valve surfaces move according to the valve lift profiles. Grid points at the moving planes coincide with the moving valve face surfaces and they are moved while the grid deforms. After a certain amount of deformation occurs, these grid points are remapped. The valves are closed by building a zero thickness wall about the periphery of the valve top surface when its assigned minimum valve lift value (usually about 0.5 mm) is reached and the cell boundary conditions are changed.

The advantages of using the snapper technique is that no mesh regeneration is needed during the run. Remapping of grid and cell quantities is only performed for cells near the moving surfaces when the snapping occurs. This approach requires a complete mesh from BDC (bottom dead center) to TDC that is provided by a grid generator. Disadvantages include the fact that particular grid line settings are required in order to perform the snapping. For example, grid lines must be normal to the valve centerline in the valve moving region, and horizontal grid lines are required around the periphery of the cylinder to move the piston. Although this grid limitation is relaxed in a vertical valve setting, it poses difficulties in grid generation for pent-roof chamber and inclined-valve geometries. The quality of the grid deteriorates unavoidably in the chamber region.

GRID GENERATION - A grid generator, K3PREP, which is provided in the KIVA-3 package, was used to create the computational mesh. K3PREP can define a variety of block shapes and patch them together to generate a moderately complex mesh that KIVA-3 is capable of running.

Some limitations were found in the original K3PREP that restrict its grid capability. Hence, K3PREP was modified to add more capabilities for the typical gasoline engine geometry with a pent-roof chamber and an elbow-shaped port/runner assembly.

Major modifications in the present work include enhancement of the chamber shape capability, implementation of a new method for the port/runner and some refinements of local grid quality. K3PREP flags and tilts the chamber region to form the pent-roof shape and the interior grids are interpreted accordingly along the z-axis. The flag method was tailored so that the complete chamber region could be tilted. Using the modified K3PREP, realistic squish and mask regions (to block parts of the intake flow path) that are located between the valve seats in the considered engine can be represented. The interior grid interpretation was modified so that the non-convex grids resulting from the chamber tilting at the chamber corners were removed.

The original K3PREP only allows rectangular runners to patch on cylindrical ports, as illustrated in Fig. 3 (a). This not only limits the capability of the grid representations but also causes a geometrical

deformation at the intersections of the two parts. A new surface mapping method [23] was then implemented in K3PREP to generate grids for the port/runner blocks.

The idea is that the port/runner logical mesh blocks are first generated using K3PREP. Then, these blocks are replaced by separately generated new mesh blocks that have the physical dimensions using either user subroutines or a commercial code. The corresponding flag indices of these blocks are kept the same.

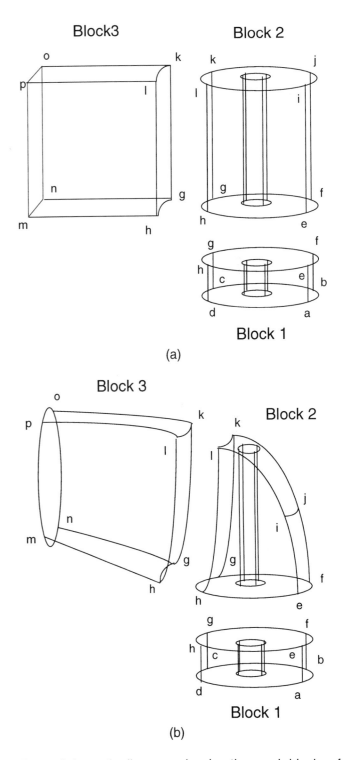

Fig. 3 Schematic diagrams showing the mesh blocks of the port/runner grid system.

In this study, the physical port/runner blocks were generated by user subroutines. In the mapping process, Block 1 is kept to be identical to its logical block created by K3PREP. The bottom face (e-f-g-h) of Block 2 (see Fig. 3 (b)) is also kept the same as its logical block face. The left face (g-k-i-h) of Block 2 was created by mapping the bottom one and the indices are resigned. Faces e-f-j-i and i-j-k-l were reshaped according to the physical geometry. The interior grids were generated by interpolating the dimensions within these four faces. Block 3 was made such that its right face h-g-k-i are corresponding to the left face (g-k-i-h) of Block 2, so that these two blocks can be patched together.

The valve stems were realized using the reshaping function in K3PREP originated by Hessel and Rutland [22]. Valve surface shapes were defined using the design curves. Both valves and stems are defined as ghost blocks.

ENGINE AND COMPUTATIONAL CONDITIONS

The engine studied was a single-cylinder Ricardo optical engine with a DISI engine head. It has a 93.6 mm bore and a 90.6 mm stroke. The engine features a pent-roof combustion chamber with two intake valves and a flat piston crown, as depicted in Fig. 4 in which a cut-away view is given to show the intake valves. The exhaust valves/ports are not shown since the exhaust process was not modeled. A pressure-swirl injector is located near the center of the chamber and fuel is injected with the injector oriented near vertically. The compression ratio of the actual optical engine is about 5.6:1. However, the compression ratio was increased to 10.5:1 by reducing the squish height in some computational cases to simulate a fired engine.

The outline of the created mesh at BDC is illustrated in Fig. 5. The figure shows a cut-away perspective view (the cutting plane is located at the front valve center line) so that the interior grids can be seen. The mesh system consists of fifty-one logical blocks. There are some 85,000 cells and the typical cell size is 3x3x3 mm in the cylinder. The realistic valve shapes are represented. However, some unavoidably skewed cells are seen to exist in the chamber region due to the limitations associated with the grid snapper procedure.

Pressure-inflow boundary conditions were imposed at the open ends of the runners. Computations were started at the intake TDC (0 degree) and ended at the compression TDC (360 degrees). The initial thermodynamic and turbulence parameters were specified to be uniform in the cylinder and in the runners separately. Other operating conditions are listed in Table 1 for both the low and the high compression ratio cases. In the low compression ratio case, the parameters were set based on available engine-out measurements [24] under motoring engine conditions. The parameters were set in the high compression ratio case by referring to cycle simulation results of a similar single-cylinder DISI engine under WOT conditions.

Fig. 4 Schematic diagram of the chamber and intake valves/ports geometry of a DISI engine.

Table 1 Computational conditions for a DISI engine.

	Low CR case	High CR case
Engine condition	Motored	High Load
Compression ratio	5.6	10.5
Squish height (mm)	12.55	2.35
Engine speed (rev/min)	1500	1500
Equivalence ratio	N/A	1.18
Intake pressure (MPa)	0.1	0.1
Intake gas temperature (K)	300	310
Residual gas (%)	18.3	5.8
Residual gas temperature (K)	340	900
Piston surface temperature (K)	380	516
Head surface temperature (K)	380	473
Cylinder liner temperature (K)	380	416
Fuel	N/A	iso-octane

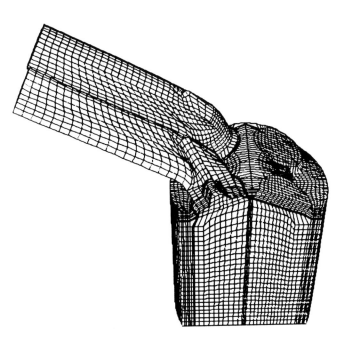

Fig. 5 Perspective view showing outline of the computational mesh at BDC. Slice in the foreground shows detail of the valve stem and port.

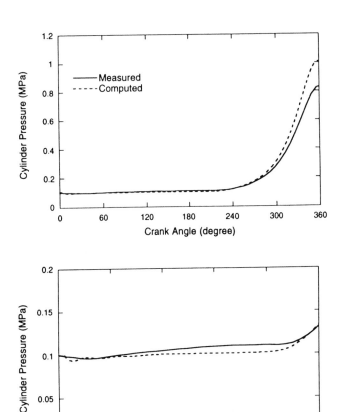

Fig. 6 Comparison of computed and measured cylinder pressure. The bottom figure shows the pressures during the induction stroke using an enlarged scale.

MODEL VERIFICATION

The optical engine with compression ratio of 5.6:1 was first considered to verify the intake flow modeling using available experiments. Figure 6 shows cylinder pressures versus crank angle for the experiment [24] and computation during the induction and compression processes. The bottom of Fig. 6 shows a blow-up of the pressure traces during the induction process. As indicated in this figure, the maximum difference is about 6% during the induction stroke that could be due to the omission of intake-exhaust valve overlaps. However, omission of a crevice flow and blow-by calculation results in an increased difference between the computation and measurement toward the end of compression process, as is seen in the top figure of Fig. 6. Although it is believed that fuel evaporation and air-fuel mixing are not significantly affected by the observed pressure difference in the late compression stroke (since most fuel is evaporated by then for an early fuel injection which is studied in this work), implementation of a crevice model [25] would be preferred for late (near compression TDC) injections and combustion simulations. It should be noted that the experimental optical engine is expected to have high crevice flow and blow-by rates due to the artificially large piston ring top land in the optical engine.

More evidence of the high blow-by phenomena that exist in the optical engine is shown in Fig. 7 in which the same pressure traces shown in the compression stroke in Fig. 6 are plotted versus the computed cylinder volume. Linear curve fits of both the measured and computed pressure traces are superposed in this figure and the indicated polytropic exponent n is obtained from the curve fit. It is seen that the measured pressure trace gives a polytropic exponent of 1.24. This value is low compared with the range of 1.3~1.38 found in conventional SI engines [26]. This indicates that there exists substantial blow-by and crevice flow in this optical engine. In the computations the resulting polytropic exponent is 1.38. This number is reasonable since wall heat transfer is minimal due to the small temperature difference between the gases and cylinder walls under motoring engine conditions, and this also agrees with the literature data.

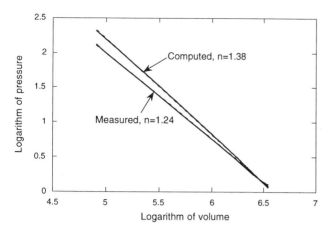

Fig. 7 Pressure versus cylinder volume during the compression stroke showing the polytropic process behavior. Linear curve fits are represented by the dashed lines.

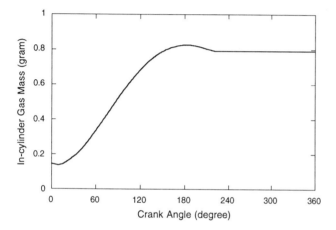

Fig. 8 Computed in-cylinder mass history. 0 degree is the intake top dead center.

The computed in-cylinder mass history is shown in Fig. 8. The in-cylinder mass continuously increases during the induction process by BDC (180 degrees). It is also seen that there is evidence of backflow into the intake port prior to valve closure which results in a decrease of the in-cylinder mass beyond 180 degrees.

The above discussions give some confidence in the computations. However, more comparisons with experiments are planned once measured data become available.

CHARACTERIZATION OF GAS FLOW

It has been reported that DISI engines have the potential to use higher compression ratios over conventional PFI engines [1]. Hence, it was interesting to study the in-cylinder air-fuel mixing processes at a higher and more realistic compression ratio and under firing engine conditions. A compression ratio of 10.5:1 was chosen for the present mixing simulations and the other considered conditions are, again, given in Table 1.

The in-cylinder mean flow field is first analyzed since the gas flow impacts spray development and air-fuel mixing. In the first case, no spray was injected in the computation. The computed in-cylinder flow field at three different crank angles in the front-valve axial cutting plane is illustrated in Fig. 9. The velocity fields are represented by the vectors (arrows) whose lengths are proportional to the local velocity magnitudes and the velocity directions are specified by the direction of the arrows.

These velocity fields indicate the presence of a structured flow comprised of one pair of counter-rotating vortices under the intake valve during the early induction process (90 degrees in Fig. 9). These flow structures remain visible for most of the intake process. As the piston moves towards BDC, the near-center vortex develops into a larger tumbling motion that dominates the flow structure in this cutting plane. It is also seen in Fig. 9 that some gases start to flow out of the cylinder and into the intake port at BDC due to the higher gas pressure in the cylinder over the intake manifold pressure by this time. This result explains the reduction of the in-cylinder mass during the early compression (before the intake closure) shown in Fig. 8. During the compression stroke, the dominant feature of the mean flow field continues to be the main tumbling flow on this cutting plane, as evident in the figure (270 degrees in Fig. 9), which is compressed and weakened when the piston moves up.

Figure 10 illustrates the velocity fields at the same times as in Fig. 9 but on an axial cutting plane which is located between the two valves. The flow structures in this plane are more interesting since the central gas flow directly interacts with the fuel sprays in the current injector location design. During the induction stroke, instead of the one-pair-vortex structure under the valves in Fig. 9, a jet-like flow in the center region dominates the flow structures on this cutting plane. This center gas jet flow moves from the top-left to the bottom-right on this cutting plane, as shown in Fig. 10 (90 degrees) and has a relatively high velocity magnitude. As will be discussed later, it interacts with the fuel sprays and modifies the trajectories of the drops. By the end of the induction process (180 degrees in Fig. 10), the central gas flow becomes weakened and a large vortex flow structure is formed. This vortex structure, affected by the flows under the valves, develops into a coherent tumbling motion during the compression stroke as indicated by Fig. 10 (270 degrees).

The formation of the tumbling motion under the valves and the center jet motion during the induction process can be visualized in Fig. 11 in which velocity magnitude contours at crank angle of 90 degrees on two cutting perpendicular planes are shown in the 3-D engine geometry. The front transparent plane cuts through the two valves and coincides with the valve axes, and is perpendicular to the back valve-axial plane. The contour values in Fig. 11 are selected so that the high velocity regions represented by the dark areas are clearly shown. As can be observed from this figure, the

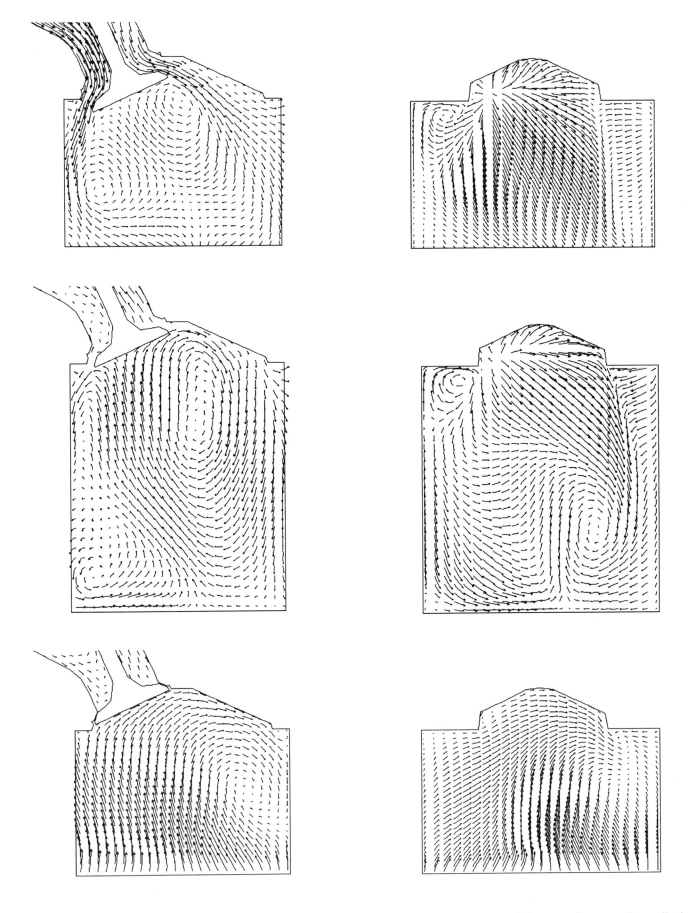

Fig. 9 Development of the gas flow in the front-valve-axial cutting plane. From left: crank angle=90, 180, 270 degrees and the corresponding velocity magnitudes, Vmax=63.4, 9.6 and 9.0 m/s, respectively.

Fig. 10 Development of the gas flow on the cylinder-axial cutting plane. From left: crank angle=90, 180, 270 degrees and the corresponding velocity magnitudes, Vmax=43.0, 9.1 and 8.7 m/s, respectively.

in-cylinder flow is characterized by the presence of two distinct jet-like flows emerging from the valve curtain area seen on the back plane and of one central jet flow shown on the front plane (indicated by the arrows). It should be noticed that these two distinct jet flows shown on the back plane are parts of conical jets produced by the intake valve openings. However, it is visualized as two separated jets on a 2-D cutting plane and these two jet flows are named as wall jets here. It is also worthwhile to point out that the same flow structures seen at the back valve are also present at the front valve. It is interesting to see that air flows into the cylinder and is confined by the cylinder liner and the chamber roof, forming the wall jets, during the intake process. The competing flow from the two jets leads to the dominant flow structure seen at the end of the intake process. In the present engine case, the jet along the chamber roof wall has higher momentum than that along the left-hand-side of the liner so that a dominant organized tumbling flow structure results by BDC (see Fig. 9). On the other hand, gas flows from the two valve openings impinge on each other in the region between them as indicated in the front cutting plane in Fig. 11. As a result, these flows merge together to form the central jet flow that is redirected toward the right side of the cylinder. It is noticed that this gas jet has higher velocity (and therefore momentum) at 90 degrees and then becomes weaker as the piston moves downward. The gas jet structure disappears during the compression stroke since there is no inflow to sustain it after the intake valves are closed. This behavior can also be seen in Fig. 10.

Another feature of the in-cylinder flow is the highly symmetrical structure of the flow that is demonstrated in Fig. 12. In this figure instantaneous streamlines at 60 degrees, represented by the ribbons, are shown. There are three streamlines that start at the open side of each runner. The same starting location arrangement was used for both runners. A cut-away side-view of the cylinder is given in the top part of Fig. 12 in which the wall jet flow structure is seen again. The top-view of the 'transparent' engine geometry is given in the bottom part of Fig. 12. As indicated by the symmetrical projections of the streamlines, the flows are seen to be very symmetrical about the center plane between the two valves due to the symmetrical runner/port/valve setting. This symmetrical flow structure implies that the bulk flows from the two valve openings do not mix together significantly. It was interesting to notice in the computation that this symmetry feature was preserved throughout the entire induction and compression processes.

The rotational flow behavior can be also characterized by global tumble and swirl ratios, as shown in Fig. 13. The tumble and swirl ratios are calculated based on the instantaneous mass center [7]. It is seen that the dominant tumble component, TRy, about the y-axis (see Fig. 4) increases during the early induction period, reaches a peak, and then decays around BDC. It is interesting to notice that TRy then

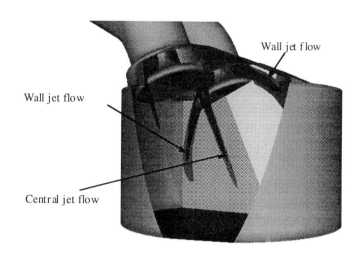

Fig. 11 Velocity contours on two cutting planes at 90 degrees. The darker color represents the higher magnitude.

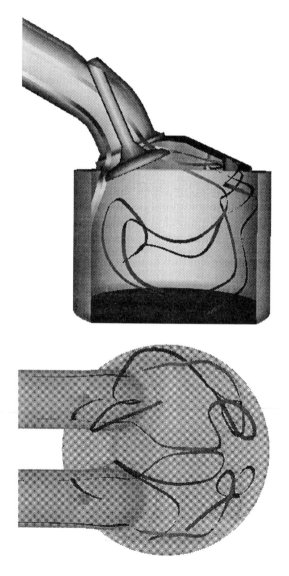

Fig. 12 Computed streamlines (ribbons) at 60 degrees showing the flow structure.

increases again during the middle part of the compression stroke. This is due to the fact that the moment of inertia of the in-cylinder fluid decreases as the piston moves up, while the fluid tends to maintain its angular momentum. The final decay in the tumble motion (which starts at about 270 degrees) is due to the breakdown of the tumble and its dissipation into turbulence as the chamber volume is decreased.

The evolution of the mass-averaged in-cylinder turbulence intensity for this engine is illustrated in Fig. 14. The intensity is normalized by the mean piston speed. It is indicated that high turbulence is generated due to the gas jet flows during the middle of the induction stroke and the generated turbulence then decays significantly during the late induction and early compression stroke. The turbulence intensity is then amplified in the late compression process. As alluded to before, this increase of turbulence intensity is related to the tumble flow breakdown. The tumble motion persists into the compression stroke and is destroyed by the compression process (see Fig. 13). This leads to the increase of the turbulence intensity toward to the end of the compression process.

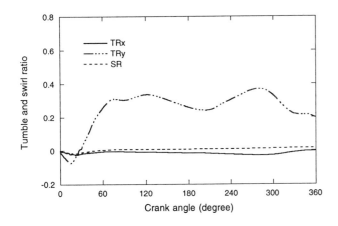

Fig. 13 Tumble and swirl ratios during the induction and compression strokes.

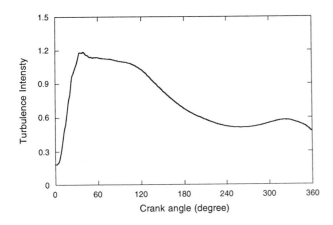

Fig. 14 Averaged in-cylinder turbulence intensity (normalized by the mean piston speed).

CHARACTERIZATION OF AIR-FUEL MIXING

Computations were performed to simulate the in-cylinder air-fuel mixing process of the high-compression-ratio engine. The emphasis here was to characterize the sprays/gas interactions and to show the general dynamic behavior of the liquid sprays and fuel vapors.

In the case discussed here, the start of fuel injection (SOI) was 120 crank angle degrees and the other engine conditions were kept the same as those used in the last section. The fuel injection pressure is 4.76 MPa and the injection duration is 44 degrees. Both experimental measurement and computation show that the average drop Sauter mean diameter of the spray is about 30 µm when it is injected into an open vessel under this injection pressure [14]. The amount of fuel injected (modeled as iso-octane) is 45.7 mg that gives an overall equivalence ratio of 1.18 or A/F ratio of 12.75.

The hollow-cone structure of the injected spray is illustrated in Fig. 15 in which the spray droplets are represented by the particles. Two 'lighting' sheets cut through the spray. One is on the axial plane of the intake valves and the other one is on a horizontal plane downstream of the injector. Liquid density contours in the two planes are shown. The shapes of these contours clearly indicate that the spray in the cylinder generally maintains its hollow-cone structure, as it also exhibits in a quiescent environment [14]. However, the spray structure is deformed by the gas flows, which is discussed as follows.

Figure 16 shows the computed spray drop locations versus crank angles in perspective views. The engine cut-away geometry is also shown to indicate the relative locations of the sprays. The spray droplets are again represented by enlarged particles (so that some particles appear to be outside the cylinder walls). Interesting spray behavior can be observed from Fig. 16. First, an initial fuel slug is modeled which leads the main spray (see Fig. 16 (a)). This initial part of the spray has been experimentally observed when fuel is injected into a bomb [27] and into an engine [8]. It is believed that the initial spray is formed during the initial (startup) stages of the injection at which time the angular momentum of the liquid within the injector passages has not yet been fully built up.

In the present work, the startup process of the sprays was modeled in an *ad hoc* way. Drops with sizes comparable to the nozzle diameter (400 µm) were introduced into the computation. The modified TAB model was then used to compute the breakup process of these (large) drops. The injection duration time of this initial part of the spray was set to be 0.1 ms, based on experimental observations. This treatment was shown to work well, as is evidenced by a good agreement between computed and measured sprays [14].

It is shown that the initial fuel slug penetrates more quickly than the main body of the spray. One of the reasons is that this part of the spray contains larger

Fig. 15 Spray structures of a direct-injection spark-ignition engine.

droplets (represented by dark colors in Fig. 16). The initial spray soon impinges on the piston surface within 15 crank angle degrees (1.7 msec) after the start of injection. Secondly, the main spray is seen to be deformed by the gas flows so that its tip is deflected toward the exhaust valve side. Although the fuel is injected vertically, the central intake-created gas jet (as shown in Fig. 11) imposes its momentum on the spray and hence the spray is deflected. The deformation is seen to spread out the fuel drops which also influences the vapor distribution and may help fuel vaporization. However, it redirects some of the fuel drops toward the cylinder liner, as indicated in Fig. 16 (b). Thirdly, by the time of BDC, noticeable spray/piston impingement is seen to have occurred.

In order to further study the observed gas/spray interactions, the detailed flow structures were analyzed. Figure 17 compares the flow velocity fields of two cases with and without spray injection at 150 crank angle degrees. The velocity vector fields on two orthogonal cylinder axial cutting planes are shown. The coordinate system can be referred to Fig. 4. In the non-spray case, the central gas-jet dominated flow structure is, again, seen in the top of Fig. 17. On the y-z cutting plane, the gas jet induces three pairs of vortices with about the same strength and opposite rotating direction that have zero net contribution to the tumble ratio TRx, as was indicated in Fig. 13. However, when fuel was injected at 120 degrees, the intake-created gas jet flows and the spray-induced flows interact with each other which

results in the enhanced central bulk flow that can be seen in the bottom of Fig. 17. In addition, the spray injection causes air entrainment which is clearly shown in the bottom-right plot. Another important phenomenon due to the spray injection is that the intake-created tumble motion of the gas flows is significantly suppressed by the spray-induced strong vertical flows. This can be observed by comparing the top-left and bottom-left plots in Fig. 17.

The velocity magnitudes along the line 30 mm below the head surface (as indicated by the solid horizontal line in the bottom plots in Fig. 17) are plotted in Fig. 18 in which zero-x represents the cylinder axis and the negative values of x-coordinate represent the distances away from the cylinder axis on the intake-valve side. These figures indicate that the gas velocity is amplified by as much as a factor of 3.5 times by the injected spray in the core region of the cylinder for this particular engine configuration and injection scheme.

Fuel injection also affects the gas turbulence fields which is demonstrated in Fig. 19. Turbulence kinetic energy contours on the same cutting planes as in Fig. 17 are shown in Fig. 19. It is seen that the gas turbulence field is altered by the spray. Generally, the turbulence is enhanced by the spray-induced velocity gradients in the spray region. The increase of gas turbulence can be also seen in Fig. 20 in which turbulence intensity is plotted on the same horizontal lines as those in Fig. 18. Although turbulence intensity is seen to be decreased along the cylinder axis in Fig. 20, the overall turbulence kinetic energy or averaged intensity is increased due to the fact that turbulence is enhanced in the vicinity of the spray periphery. The spray-enhanced turbulence persists to the end of the compression process which would be helpful to speed up flame propagation and to reduce the combustion duration.

Since the momentum of the central intake-created gas jet changes during the induction process (it reaches a maximum at 90 degrees), it is expected that the interactions between the gas flow and the sprays are altered when the injection timing is varied. The influence of various injection timings is addressed in a separate paper [28]. The evolution of the liquid and vapor fuel distributions is discussed next.

Figures 21 and 22 illustrate the spatial liquid and vapor density distributions versus crank angle, respectively. Liquid and vapor densities are shown on three cutting planes whose locations can be seen by referring the superposed engine geometry outline. Some important features can be observed in these figures for this particular case considered. First, liquid fuel is distributed in a hollow-cone shape in the cylinder some tens of crank angle degrees after the start of the injection (Fig. 21, Top-left). However, the vapor phase tends to distribute itself into a solid-cone shape due to the effect of gas entrainment motion, as discussed in Fig. 17. The same phenomena were found in the computation of sprays in a quiescent environment [14]. Secondly, the bulk of the liquid spray is noticed to

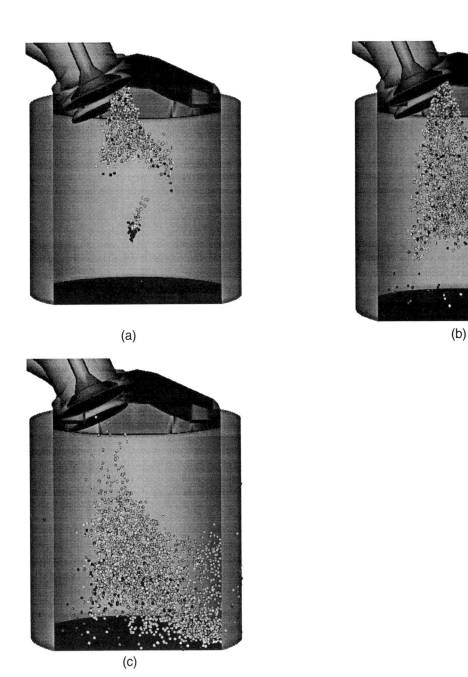

(a)

(b)

(c)

Fig. 16 Computed spray development versus crank angle. SOI=120 degrees. From (a) to (c): 130, 150 and 180 degrees.

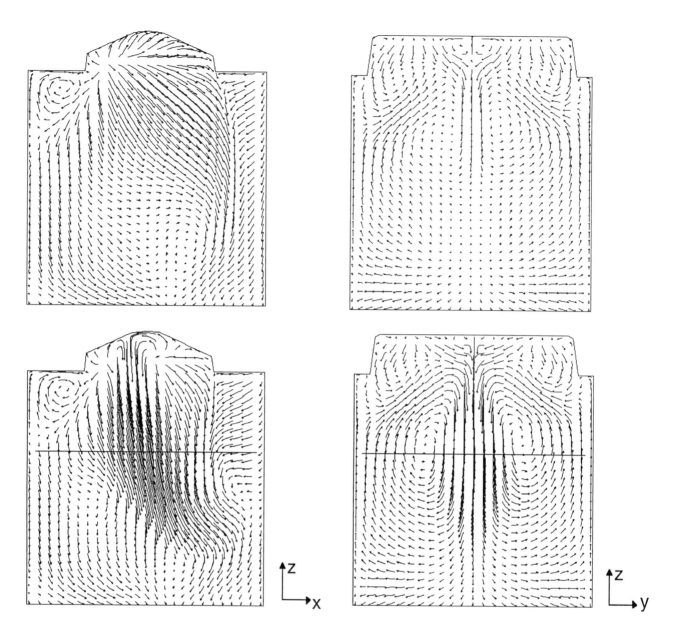

Fig. 17 Computed velocity vector fields in the cylinder-axial cutting planes. Top: non-spray case. Bottom: spray case (SOI=120 degrees).

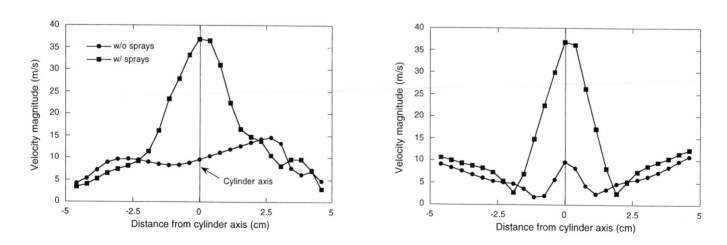

Fig. 18 Effects of spray injection on gas mean velocity magnitude. Monitoring line is located on the cylinder x-z axial plane (Left) and on the y-z plane (Right), 30 mm below the head surface.

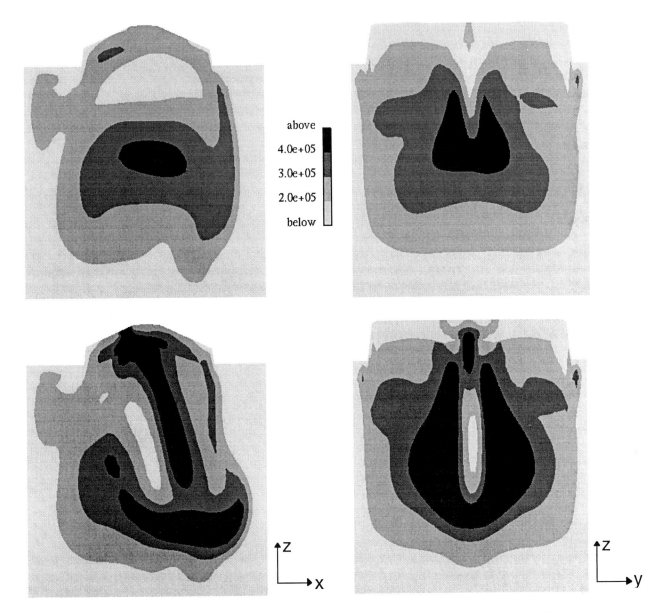

Fig. 19 Computed turbulence kinetic energy contours on the cylinder-axial cutting planes. Top: non-spray case. Bottom: spray case (SOI=120 degrees).

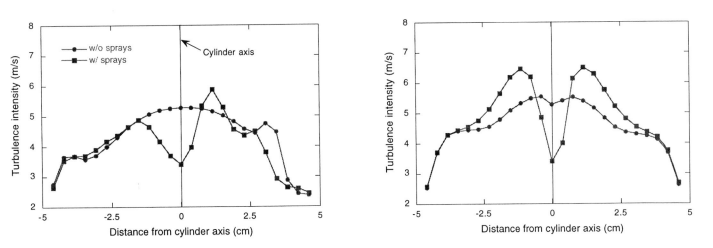

Fig. 20 Effects of spray injection on gas turbulence intensity. Monitoring line is located on the cylinder x-z axial plane (Left) and on the y-z plane (Right), 30 mm below the head surface.

303

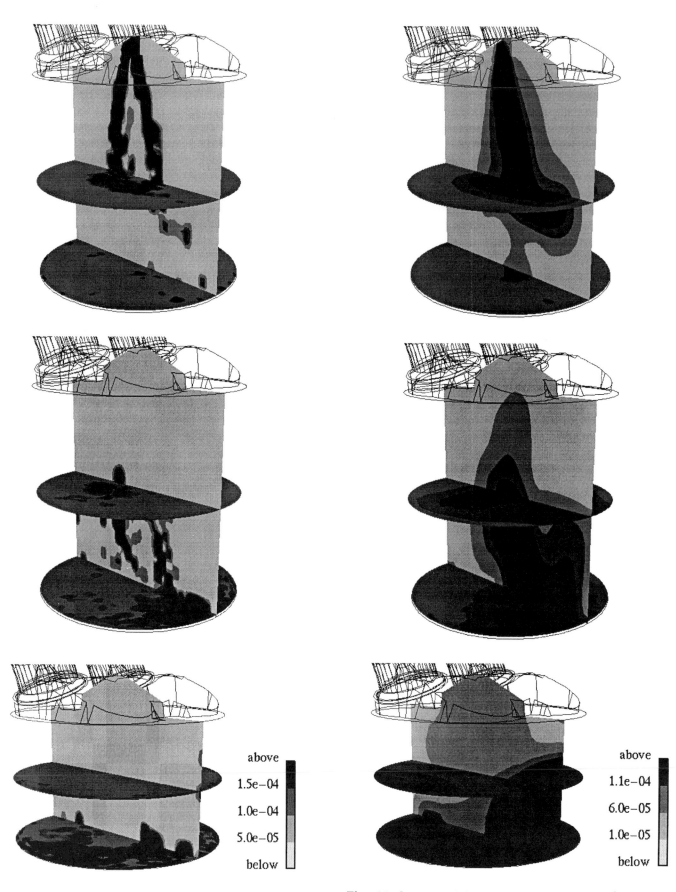

Fig. 21 Computed fuel liquid density (g/cm³) contours versus crank angle. SOI=120 degrees. From top: 150, 180 and 270 degrees. The upper horizontal cutting plane is located in the middle distance between the head and piston surfaces.

Fig. 22 Computed fuel vapor density (g/cm³) contour versus crank angle. SOI=120 degrees. From top: 150, 180, 270 degrees. The upper horizontal cutting plane is located in the middle distance between the head and piston surfaces.

impinge on the piston surface by about 170 crank angle degrees (50 degrees after SOI, the initial injected slug impinged earlier, at about 15 degrees after the SOI). The spray tip impinges first on the piston surface on the right-hand side of the cylinder (the exhaust-valve side) under the influence of the intake flows, as discussed before. The high vapor density region is seen to follow the liquid distribution which confirms the importance of the liquid fuel distribution on the vapor distribution. Thirdly, in the late compression stroke, the majority of the remaining liquid fuel is located near the piston surface as indicated by the bottom plot and by Fig. 23. Some liner-wetting is seen at the same time which is due, in part, to the redirection effects of the gas flows on the spray drops. However, the amount of fuel impinged on the cylinder liner is reduced substantially in the retarded injection cases since the effects of the intake flows on the spray become less and less when fuel injection is retarded [28].

The temporal variation of the instantaneous amount of liquid fuel located on (or near) the cylinder walls (as defined in Section 5.1) is shown in Fig. 23. The amount of liquid fuel on a solid wall surface (piston or liner) is defined as the amount of fuel within a 0.1 mm layer immediately adjacent to the wall. It is seen that significant wall impingement does not occur until by about 180 degrees for the considered case. The liquid fuel on the wall surfaces increases at first due to the continuous wall impingement of the droplets, and then it decreases due, mainly, to vaporization. In this particular case, up to 16 and 2 percent of the total injected fuel were found on the piston and liner surfaces, respectively, by 240 degrees during the early compression period. However, these amounts were reduced to less than 2 and 1 percent, respectively, by the end of the compression stroke.

The above discussion suggests that the vapor distribution and therefore the air-fuel mixing is mainly influenced by both the liquid distribution and the large-scale gas flow structures. Intake-created gas motion also impacts the liquid distribution. In this particular case, the liquid distribution is influenced significantly by the gas motions during the induction stroke, while in the late compression stroke, the rich-vapor region is generally seen to remain in the exhaust-valve side corner near the piston surface. This is because the tumble motion is suppressed by the spray-induced flows starting from the late induction process, as seen in Fig. 17. In a preliminary study [7], it was also found that intake flow structures have a strong influence on air-fuel mixing when injection starts at 90 degrees. A strong tumble or swirl motion helps to spread out both the liquid and vapor fuel distributions and this enhances air-fuel mixing.

The global fuel evaporation history is shown in Fig. 24 in which the fuel is normalized by the total injected fuel. Fuel continues to vaporize and over 97 percent of the total injected fuel is vaporized by the end of the compression stroke in the computation. It is seen that the evaporation rate decreases after about 180 degrees by comparing the slope of the vapor phase

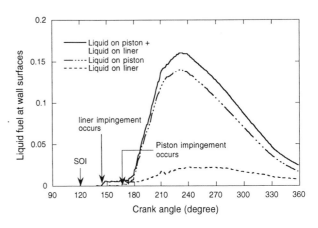

Fig. 23 Computed temporal variation of wall-wetting on the piston and liner surfaces (SOI=120 degrees). Liquid fuel is normalized by the total injected fuel.

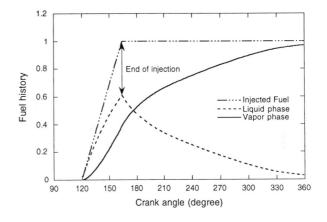

Fig. 24 Computed vaporization history (SOI=120 degrees). Fuel is normalized by the total injected fuel.

curve at different times during the cycle. This is partially because the amount of liquid that exists in the form of impinged droplets is increased after this time. These impinged droplets undergo decreased heat and mass transfer due to the reduced motion that slows down the vaporization process. Another reason is that the vapor density near the piston surfaces is high during the compression stroke, and a high vapor concentration in the environment gases also reduces the local heat and mass transfer rates of the evaporating drops.

In the computations, drop impingement hydrodynamics (rebounding and sliding) was considered. Convective heat transfer between the gas and the wall and that between the impinging drops and the gases were also computed in the near-wall region. However, the present model does not consider the conductive heat transfer between the liquid and a solid wall and other liquid/wall effects (such as film flow and drop breakup) which may affect the near-wall fuel evaporation. Further study of the effects of liquid film and liquid/wall heat transfer [29,30] on the near-wall fuel vaporization is recommended for future work.

The fuel vapor mixes with air to form the air-fuel mixture. The local mixture equivalence or A/F ratio varies spatially and temporally due to the inhomogeneously distributed liquid fuel and the anisotropic gas motion. To characterize the mixture quality, the evolution with crank angle of the local mixture volumes in different equivalence ratio regimes is given in Fig. 25. It is seen that the amount of the flammable mixture ($0.5 \leq \phi \leq 1.5$) increases and the extra-lean mixture ($\phi < 0.5$) decreases during the induction and compression processes since the fuel vapors are continuously replenished through fuel vaporization. Notice that after about 270 degrees, the formation rate of the flammable mixture is increased considerably, accompanied by a rapid decrease of the amount of the extra-lean mixture. This indicates that the elevated compression temperatures due to the change of the cylinder volume, and diffusion effects become important to air-fuel mixing during the late compression stroke. By the end of the compression stroke, the majority of the mixture, i.e., about 84 percent of the total chamber volume, is in the flammable regime. While no appreciable extra-lean mixture remains by the end of the compression stroke, about 16 percent of the mixture is in the extra-rich regime ($\phi > 1.5$). This part of the mixture accumulated by about 240 degrees and then kept almost constant throughout the compression process. The major contribution to this part of the mixture is from the near piston surface region where more vapor is located during the late compression stroke.

It may be speculated that in the actual engine part of the rich mixture would enter the ring crevice region in the compression stroke and would subsequently flow back into the cylinder, contributing to unburned hydrocarbon emissions from the engine late in the expansion and exhaust stroke [25]. An improved engine design would attempt to reduce the amount of rich mixture in the piston-liner regions, possibly by altering the details of the intake and port flows.

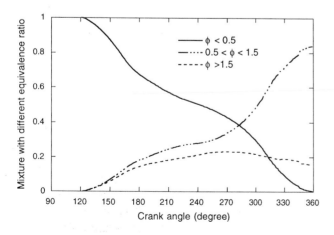

Fig. 25 Evolution of the total in-cylinder gas mixture volume (normalized by the instantaneous cylinder volume) in different equivalence ratio ranges.

SUMMARY AND CONCLUSIONS

Multidimensional modeling was used to study spray atomization and air-fuel mixing in a direct-injection spark-ignition engine using the KIVA-3 code, supplemented with improved turbulence, heat transfer, spray atomization and vaporization models. The considered engine features a 4-valve and pent-roof combustion chamber geometry and uses a pressure-swirl injector that forms a hollow-cone spray.

Intake flow was modeled in the realistic valve-port-runner geometry with moving intake valves. To improve computational accuracy, a new surface mapping method was implemented in the KIVA-3 grid generator so that the typical gasoline engine geometry (slant valves, pent-roof chamber and elbow-shaped ports/runners) can be represented.

A particular DISI engine configuration was computed. Complicated intake-generated flow structures were shown. The in-cylinder flow fields without fuel injection were found to be dominated by intake-created large-scale tumble gas motions up to the end of the compression that are the result of two competing wall jets. There was little or no organized large-scale swirl flow motion in this engine configuration. It was also found that there exists a gas jet in the center region of the cylinder during the induction process. This jet results from the impingement of the gases flowing through the two valve openings and influences the trajectories of spray drops injected into this region.

The modeled engine was operated at a high-load operating condition and fuel was injected at 120 degrees after the intake TDC. The computations reveal how the intake-generated flows interact with the injected spray drops. It is shown, on one hand, that the sprays are deflected and redirected; and on the another hand, that the intake-generated tumble flow is suppressed and the gas turbulence is enhanced.

The liquid-fuel distribution was found to be set up by the injected sprays and to be also influenced by the intake-generated flows. The rich vapor-fuel regions generally follow the liquid-fuel distribution in the present engine configuration that has an inherent fairly weak tumble. Fuel vaporization is essentially completed by the end of the compression under the considered injection conditions. Spray/wall impingement was also seen to occur in the computed engine. Slight liner-wetting was observed when fuel was injected at 120 degrees due to the redirection of the spray drops by the intake gas jet. However, the degree of wall-wetting and the evolution of air-fuel mixing vary with the start of fuel injection, which is further discussed in a separate paper [28].

ACKNOWLEDGMENTS

This work was supported by Ford Motor Company. Additional support was provided by the US Army Research Office. Help from A. Amsden and N. Johnson (LANL), R. Anderson and J. Yang (Ford), and C. Rutland (UW) is gratefully acknowledged.

REFERENCES

1. Anderson, R. W., Yang, J., Brehob, D. D., Vallance, J. K., and Whiteaker, R. M. (1996). Understanding the Thermodynamics of Direct Injection Spark Ignition (DISI) Combustion Systems: An Analytical and Experimental Investigation. SAE Paper 962018.

2. Kume, T., Iwamoto, Y., Iido, K., Murakami, M., Akishino, K., and Ando, H. (1996). Combustion Control Technologies for Direct Injection SI Engine. SAE Paper 960600.

3. Fraidl, G. K., Piock, W. F., and Wirth, M. (1996). Gasoline Direct Injection: Actual Trends and Future Strategies for Injection and Combustion Systems. SAE Paper 960465.

4. Simko, A. O., Choma, M. A., and Repko, L. L. (1972). Exhaust Emission Control by the Ford Programmed Combustion Process - PROCO. SAE Paper 720052.

5. Scussel, A. J., Simko, A. O., and Wade, W. R. (1978). The Ford PROCO Engine Update. SAE Paper 780699.

6. Alperstein, M., Schafer, G., and Villforth, F. (1974). Texaco's Stratified Charge Engine - Multifuel, Efficient, Clean and Practical. SAE Paper 740563.

7. Han, Z., Reitz, R. D., Claybaker, P. J., Rutland, C. J., Yang, J., and Anderson, R. W. (1996). Modeling the Effects of Intake Flow Structures on Fuel/Air Mixing in a Direct-Injected Spark-Ignition Engine. SAE Paper 961192.

8. Salters, D., Williams, P., Greig, A., and Brehob, D. (1996). Fuel spray characterization within an optically accessed gasoline direct injection engine using a CCD image system. SAE Paper 961149.

9. Spiegel, L., and Spicher, U. (1992). Mixture Formation and Combustion in a Spark Ignition Engine with Direct Fuel Injection. SAE Paper 920521.

10. Han, Z. (1996). Numerical Study of Air-Fuel Mixing in Direct-Injection Sprak-Ignition and Diesel Engines. Ph. D. Thesis, University of Wisconsin-Madison.

11. Amsden, A. A. (1993). KIVA-3: A KIVA Program with Block-Structured Mesh for Complex Geometries. Los Alamos National Labs., LA-12503-MS.

12. Han, Z., and Reitz, R. D. (1995). Turbulence Modeling of Internal Combustion Engines Using RNG k-ε Models. *Comb. Sci. Tech.* **106**, 207.

13. Han, Z., and Reitz, R. D. (1996). A Temperature Wall Function Formulation for Variable-Density Turbulent Flows with Application to Engine Convective Heat Transfer Modeling. *Int. J. Heat Mass Transfer*, in press.

14. Han, Z., Parrish, S. E., Farrell, P. V., and Reitz, R. D. (1996). Modeling Atomization Processes of Pressure-Swirl Hollow-Cone Fuel Sprays. Submitted to *Atomization and Sprays*.

15. Dukowicz, J. K. (1980). A Particle-Fluid Numerical Model for Liquid Sprays. *Journal of Computational Physics.* **35**, 229.

16. Lefebvre, A. H. (1989). *Atomization and Sprays*, Hemisphere Publishing Corp., New York.

17. Rizk, N. K., and Lefebvre, A. H. (1985). Internal Flow Characteristics of Simplex Swirl Atomizers. *AIAA. J. Propul. Power*, **1**, 193.

18. Clark, C. J., and Dombrowski, N. (1972). Aerodynamic Instability and Disintegration of Inviscid Liquid Sheets. *Proc. Roy. Soc. Lond. A.* **329**, 467.

19. O'Rourke, P. J., and Amsden, A. A. (1987). The TAB Method for Numerical Calculation of Spray Droplet Breakup. SAE Paper 872089.

20. Naber, J., and Reitz, R. D. (1988). Modeling Engine Spray/Wall Impingement. SAE Paper 880107.

21. Amsden, A. A. (1996). Private communication.

22. Hessel, R. P., and Rutland, C. J. (1993). Intake Flow Modeling in a Four Stroke Diesel Using KIVA3. AIAA Paper 93-2952.

23. Meintjes, K. (1996). Private communication.

24. Backer, W. B. (1996). Planar Laser Techniques for Determining In-Cylinder Mixing Rates. MS thesis, University of Wisconsin-Madison, in preparation.

25. Reitz, R. D. and Kuo, T. W. (1989). Modeling of HC Emissions due to Crevice Flows in Premixed-Charge Engines. SAE Paper 892085.

26. Obert, E. F. (1973). *Internal Combustion Engines and Air Pollution*. Harper & Row, New York.

27. Parrish, S. E. (1996). Spray Characteristics of a Direct-Injection Spark-Ignited Engine. Ph.D. Thesis, University of Wisconsin-Madison, in preparation.

28. Han, Z., Reitz, R. D., Yang, J. and Anderson, R. W. (1997). Effects of Injection Timing on Air-Fuel Mixing in a Direct-Injection Spark-Ignition Engine. SAE Paper 970625.

29. Eckhause, J. E., and Reitz, R. D. (1995). Modeling Heat Transfer to Impinging Fuel Sprays in Direct Injection Engines. *Atomization and Spray Technology.* **5**, 1.

30. Liu, Z., and Reitz, R. D. (1995). Modeling Fuel Spray Impingement and Heat Transfer Between Spray and Wall in D.I. Diesel Engines. *Numerical Heat Transfer, Part A*, **28**.

981440

Numerical Optimization of the Fuel Mixing Process in a Direct-Injection Gasoline Engine

Ken Naitoh, Yasuo Takagi, Nissan Motor Co., Ltd.

Kunio Kuwahara, Institute of Space and Astronautical Science

ABSTRACT

The spray formation and mixing processes in a direct-injection gasoline engine are examined by using a sophisticated air flow calculation model and an original spray model. The spray model for a spiral injector can evaluate the droplet size and spatial distribution under a wide range of parameters such as the initial cone angle, back pressure and injection pressure. This model also includes the droplet breakup process due to wall impingement. The arbitrary constants used in the spray model are derived theoretically without using any experimental data. Fuel vapor distributions just before ignition and combustion processes are analyzed for both homogeneous and stratified charge conditions.

INTRODUCTION

Development work has been proceeding in recent years on a direct-injection gasoline engine, fitted with spiral fuel injectors that form a hollow cone spray, with the aim of reducing carbon dioxide emissions and improving power output. Numerical analysis is expected to play an important role in optimizing the performance of this engine.

A fundamental design parameter of the initial droplet breakup process is the initial cone angle of the hollow cone spray. Varying that angle makes it difficult to predict the droplet size or spatial distribution. For that reason, experimental data are ordinarily used as the initial data in a spray analysis.

This paper first describes a theoretical spray model which includes the initial breakup process of the hollow cone spray and which can be derived without the use of experimental data. The fuel distribution predictions obtained with the model are presented along with the experimental validation of the model. The spray model is then combined with a model of droplet breakup due to wall impingement and a model of multi-component fuel evaporation to analyze the fuel mixing process in a direct-injection gasoline engine.

MODELING AND ANALYSIS OF HOLLOW CONE SPRAY

When the injection pressure is in a range of approximately 5-20 MPa and a fuel injector that produces a spiral flow inside the nozzle is used, a hollow cone liquid film is formed following discharge which then proceeds to break up into liquid droplets. Fundamental fuel spray parameters include the initial cone angle (i.e., initial divergence angle of the spray following discharge), fuel injection pressure, back pressure and injection rate. It is necessary to be able to predict how changes in these parameters will affect (1) the breakup distance of the liquid film, (2) the degree of reduction in the cone angle of the spray following discharge, (3) droplet size and (4) spatial distribution. We have developed an integrated model of the liquid droplet/liquid film breakup process which takes into account the effects of changes in these parameters.

The proposed model is based on the following fundamental assumptions.

(1)   A liquid film is considered as an array of connected droplets.

(2)   A droplet retains its spheroidal shape while oscillating about its equilibrium configuration.

(3)   The flow in a droplet is incompressible and irrotational.

The constants used in the model can be derived theoretically without using experimental data. A detailed explanation of the model is given in reference (1), but it should be noted that the liquid fuel jet at the onset of injection has not been considered. The model employed to calculate the flow of air and gaseous fuel is explained in reference (2).

The present model including air and liquid motions was used to calculate the fuel droplet distribution relative to the back pressure and initial cone angle. The calculated results based on this model are compared with visualized representations of the experimental data in Figure 1. The motions of the initial hollow cone shape of the liquid film and the breakup process of the film are shown. The driving pressure and the fuel amount per cycle were set to 5 MPa and 10.0 mm³, respectively. SHELL-LAWS was used as the working fluid and vaporization was not included in the computations. It is seen that the initial cone angle and the spray penetration are reduced when the back pressure is high and that greater spray penetration is obtained with a spray pattern having a small initial cone angle.

In a previous report,[1] we showed that the droplet size (Sauter mean diameter) at a point 50 mm downstream of the point of injection increases at this point when the back pressure is high. This increased size is attributed to the collision and coalescence of fuel droplets. However, it should be noted that the mean diameter of all the droplets in the suspension decreases with increasing back pressure.[1]

MODEL FOR PREDICTING IN-CYLINDER MIXTURE FORMATION PROCESS

A model for predicting the in-cylinder mixture formation process in a direct-injection gasoline engine was created by adding the following two models to the fuel spray and air flow models.

FUEL SPRAY WALL IMPINGEMENT MODEL—With a fuel injection pressure in a range of 5-20 MPa, a portion of the fuel spray impinges on the piston and forms a film and the rest atomizes as a result of breaking up further. These phenomena were examined with the OPT model presented in a previous report[3] which is capable of calculating the droplet size following further breakup and the amount of fuel adhering to the wall. The computational results obtained without the OPT model were far from the experimental data with respect to the amount of fuel entering the cylinder during the injection cycle.[3] It should be noted that this model assumes the absence of film boiling.

MULTI-COMPONENT FUEL EVAPORATION MODEL—The use of a single-component fuel results in a large discrepancy for the vapor amount present near the ignition timing. A four-component model was developed, consisting of 25% hexane, 38% heptane, 19% toluene and 18% 3M-benzene, under the assumption of an ideal solution[4] and incorporated in the overall model. The mass ratios used for the components were the same values as those of the zero-dimension evaporation model employed for A/F ratio control under transient engine operating conditions.[5] This evaporation model is applicable to both the fuel droplets in the air and the fuel film on the wall.

ANALYSIS UNDER A CONDITION OF STRATIFIED CHARGE COMBUSTION

Figures 2, 3, 4, 5, and 6 show the spatial distributions of fuel droplets and vapor and instantaneous stream lines calculated with and without horizontal swirl when fuel was injected in the latter half of the compression stroke. The conditions used in the calculation are given in Table 1. (The word "TDCF" implies "Top Dead Center of Firing".)

It is seen in Figs. 2, 3, 4, 5, and 6 that swirl induces upward air flow from the vicinity of the piston surface toward the spark plug. Then this air flow serves to transport the fuel to the vicinity of the spark plug. Figure 7 shows the velocity vectors in the two cases. Stronger upward flow is confirmed in the case with horizontal swirl flow. The differences observed in the air flow in Figures 2, 3, 4, 5, and 6 can be explained by the general concept illustrated in Figure 8. Firstly, the center axes of the bowl and cylinder (head) are offset so that the center of their respective swirl differs. Secondly, the pressure at the center of the swirl in the cylinder is reduced, while the pressure directly below it near the bowl wall is relatively high. This pressure differential produces upward air flow from near the bottom of the bowl toward the spark plug. The fuel vapor distribution that was visualized experimentally by using the LIF technique for this condition is shown in Figure 9. It is clearly seen that the fuel vapor is transported upward when swirling air motion is present.

The relation between the calculated A/F ratio near the spark plug at the time of ignition and the experimental results for the cycle-to-cycle variation in combustion pressure is shown in Figure 10. The value Cpi indicated on the y axis in the figure is defined as

$$Cpi = \Delta Pi / Pi \times 100$$

where Pi and ΔPi denote the cylinder pressure averaged during n cycles and the standard deviation of the cylinder pressures averaged during one cycle, respectively.

These results were obtained for different air flow conditions and piston geometries. Although the calculated values tend to differ somewhat from the experimental data on the lean mixture side, the two sets of results correlate fairly well.

ANALYSIS FOR FUEL INJECTION IN THE INTAKE STROKE

Figure 11 presents computational results showing the effect of the injector mounting angle on the amount of liquid fuel adhering to the cylinder wall, piston surface, and head wall, in cases where fuel is injected during the intake stroke. Table 2 shows the conditions used in the computations. The results indicate that the amount of liquid fuel adhering to the walls is greatly reduced by increasing the injector mounting angle to 36 degree.

Figure 12 shows the fuel vapor distribution in the latter half of the compression stroke for two different piston shapes, when fuel was injected in the intake stroke under wide open throttle (WOT) operation. The conditions used in the calculation are given in Table 3. The results indicate that the piston shape influences the mixture formation.

CONCLUSION

It has been shown that analyses conducted with the

proposed model are effective in accomplishing the conflicting objectives of improving mixture formation when fuel is injected in the intake stroke and optimizing swirl motion for stratified charge combustion.

One function of horizontal swirling flow with respect to engine performance has been made clear, which is the generation of upward air flow from the piston surface to the ignition plug.

Tasks that must be addressed in future work include calculation of the flow inside the injector and the modeling of liquid fuel film boiling.

ACKNOWLEDGMENTS

The authors would like to thank Dr. Egon Krause of the Aerodynamisches Institut, RWTH Aachen, Germany for his encouragement and advice during this research. Thanks are also due Keiichi Yamada and Yuji Furutani of Zexel Corporation for their invaluable cooperation in connection with the preparation of the visualized fuel spray photographs and the droplet size data. The authors would like to thank Mr. Atsushi Teraji of Nissan Motor Co., Ltd. for his help of the visualizations of computational results.

REFERENCES

(1)   K. Naitoh, et al., Synthesized spheroid particle method for calculating spray phenomena in direct-injection SI engine, Trans of SAE Paper 962017, 1996.

(2)   K. Naitoh and K. Kuwahara, Large eddy simulation and direct simulation of compressible turbulence and combusting flows in engines based on the BI-SCALES method, Journal of Fluid Dynamics Research, 10, 1992.

(3)   K. Naitoh, et al., Numerical prediction of fuel secondary atomization behavior in SI engine based on the Oval-Parabola Trajectories (OPT) model, Trans of SAE Paper 940526, 1994

(4)   A. Harajima, Thermodynamics and Statistical Mechanics, Baifukan, Tokyo, 1984 (in Japanese).

(5)   H. Iwano, et al., SAE Paper 912348, 1991.

Table 2 Calculation conditions
Engine: 4valves engine
Bore x stroke: 86 mm x 86 mm
Piston geometry: bowl in piston
Intake port: without swirl control valve
Engine speed: 1400 rpm
Charging efficiency: 40 %
A/F ratio: 14.5:1.0
Injector: Hollow cone type
Injection pressure: 7 MPa
Initial cone angle (ICA): 72 degrees at 1 atm
Injection timing: 90 degATDC

Table 3 Calculation conditions
Engine: 4valves engine
Bore x stroke: 86 mm x 86 mm
Piston geometry: bowl in piston / flat piston
Intake port: without swirl control valve
Engine speed: 6000 rpm
Charging efficiency: 85 %
A/F ratio: 14.5:1.0
Injector: Hollow cone type
Injection pressure: 7 MPa
Initial cone angle (ICA): 60 degrees at 1 atm
Injection timing: 90 degATDC

Table 1 Calculation conditions
Engine: 4 valves engine
Bore x stroke: 86 mm x 86 mm
Piston geometry: bowl in piston
Intake port: with/without swirl control valve
Engine speed: 1400 rpm
Charging efficiency: 75 %
A/F ratio: 40.0:1.0
Injector: Hollow cone type
Injection pressure: 7 MPa
Initial cone angle (ICA): 60 degrees at 1 atm
Injection timing: 60 degBTDCF

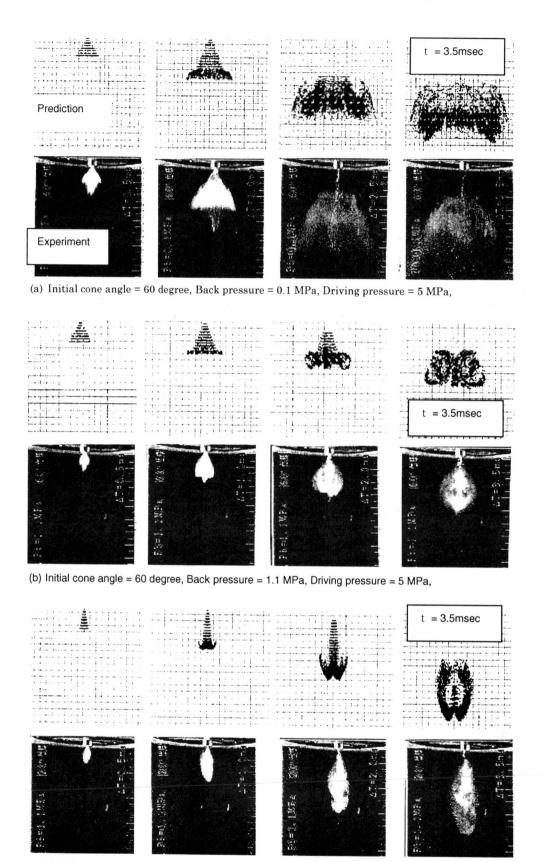

(a) Initial cone angle = 60 degree, Back pressure = 0.1 MPa, Driving pressure = 5 MPa,

(b) Initial cone angle = 60 degree, Back pressure = 1.1 MPa, Driving pressure = 5 MPa,

(c) Initial cone angle = 20 degree, Back pressure = 1.1 MPa, Driving pressure = 5 MPa

Fig. 1   Predicted spatial distribution of fuel spray droplets injected at a constant volume and corresponding experimental visualizations [1].

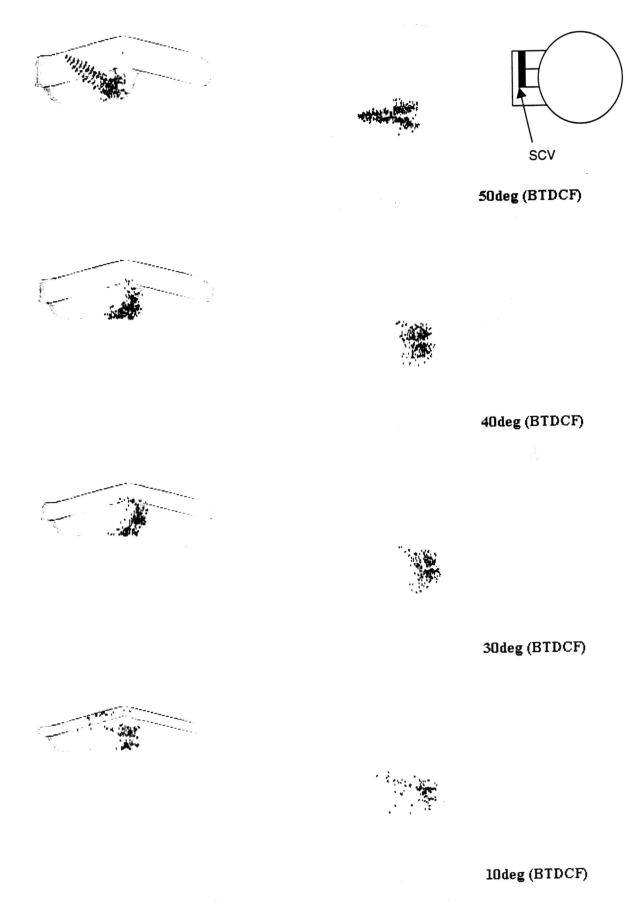

SCV

50deg (BTDCF)

40deg (BTDCF)

30deg (BTDCF)

10deg (BTDCF)

Fig. 2 Spatial distribution of fuel droplets with swirl control valve
(Engine speed = 1400 rpm, Charging efficiency = 75%, Injection timing = 60 degBTDCF,
Port shape: with swirl control valve)

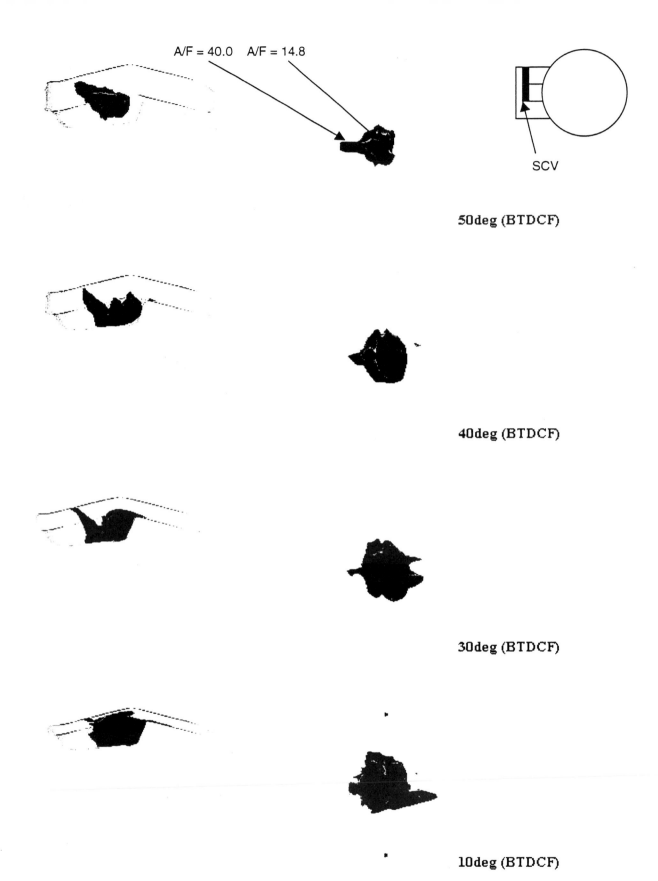

A/F = 40.0 A/F = 14.8

SCV

50deg (BTDCF)

40deg (BTDCF)

30deg (BTDCF)

10deg (BTDCF)

Fig. 3 Spatial distribution of vapor fuel with swirl control valve
(Engine speed = 1400 rpm, Charging efficiency = 75%, Injection timing = 60 degBTDCF,
Port shape: with swirl control valve)

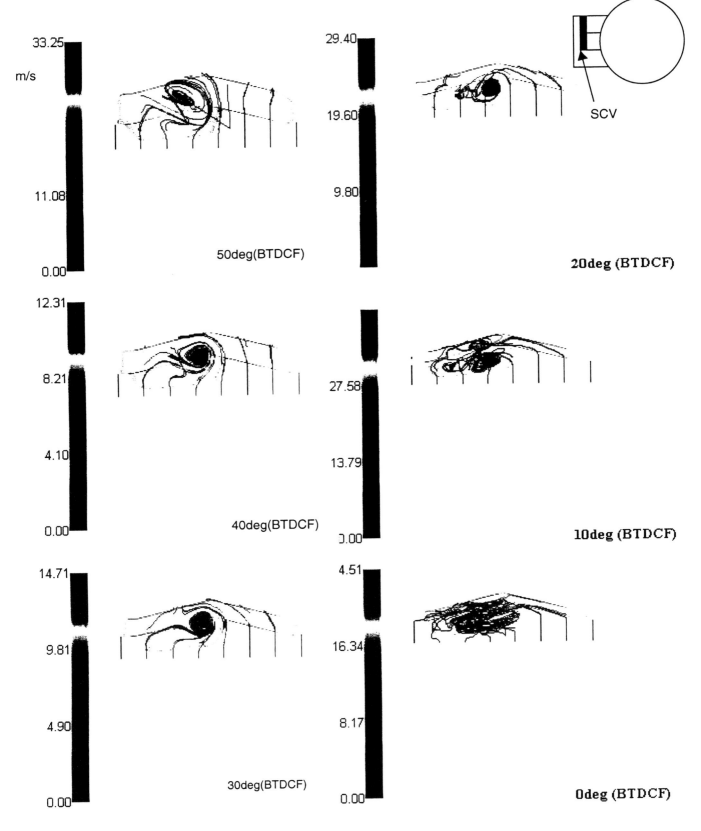

33.25

m/s

11.08

0.00

50deg(BTDCF)

29.40

19.60

9.80

20deg (BTDCF)

SCV

12.31

8.21

4.10

0.00

40deg(BTDCF)

27.58

13.79

0.00

10deg (BTDCF)

14.71

9.81

4.90

30deg(BTDCF)

4.51

16.34

8.17

0.00

0deg (BTDCF)

Fig. 4   Instantaneous stream lines with   swirl control valve
(Engine speed = 1400 rpm, Charging efficiency = 75%, Injection timing = 60 degBTDCF,
Port shape: with swirl control valve)

315

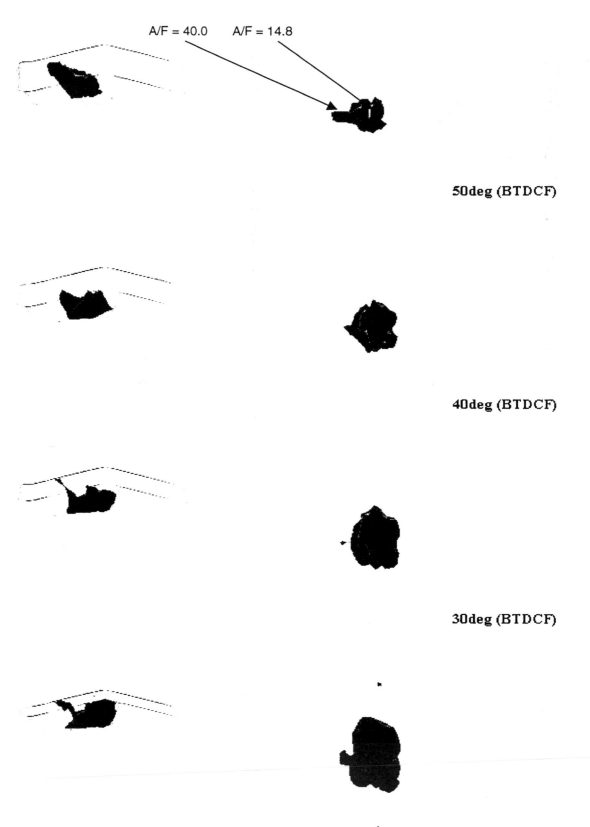

A/F = 40.0   A/F = 14.8

50deg (BTDCF)

40deg (BTDCF)

30deg (BTDCF)

10deg (BTDCF)

Fig. 5   Spatial distribution of vapor fuel without swirl   control valve
(Engine speed = 1400 rpm, Charging efficiency = 75%, Injection timing = 60 degBTDCF,
Port shape: without swirl control valve)

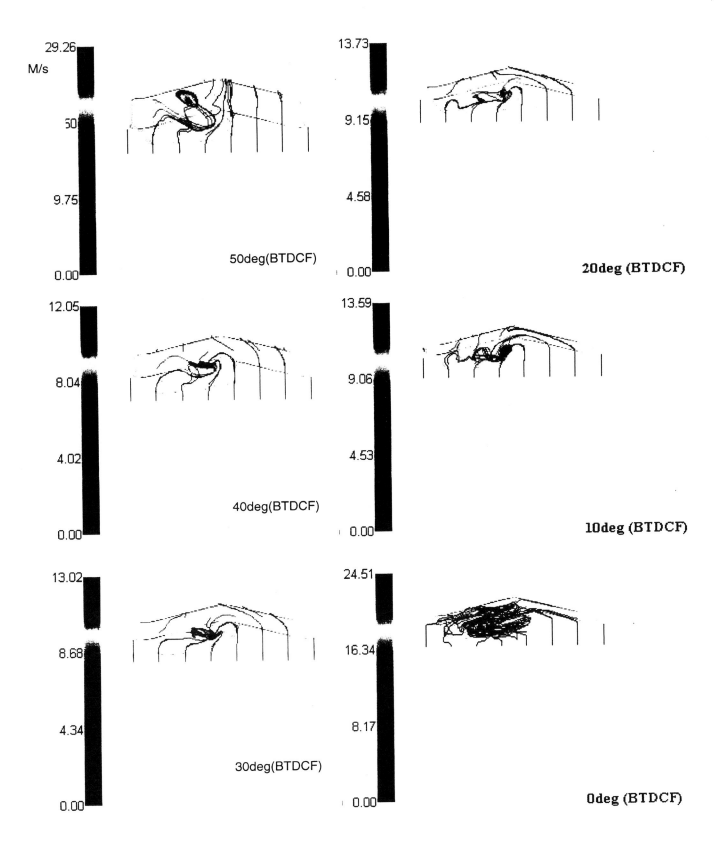

Fig. 6 Instantaneous stream lines without swirl control valve
(Engine speed = 1400 rpm, Charging efficiency = 75%, Injection timing = 60 degBTDCF,
Port shape: without swirl control valve)

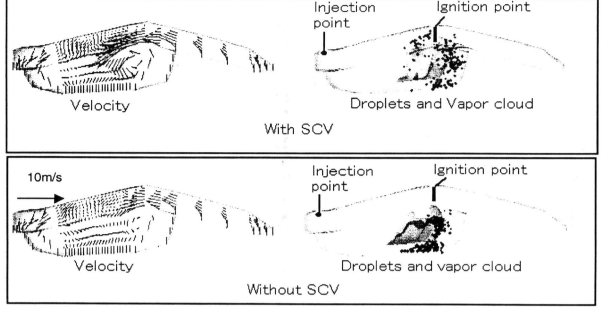

Fig. 7   Influence of horizontal swirling flow on vertical motion and fuel distribution at ignition
(30 degBTDCF). Engine speed = 1400 rpm, Charging efficiency = 75%, Injection timing =
60 degBTDCF

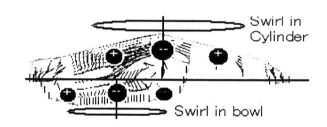

Fig.8 Representation of pressure distribution generated by the horizontal swirling
flows

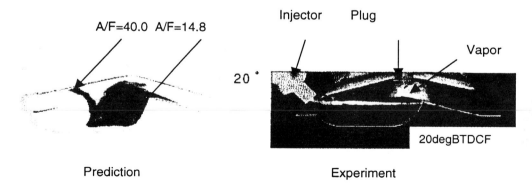

Fig. 9   Predicted vapor fuel distribution for a stratified-charge condition and corresponding LIF
experimental visualization (Engine speed = 1400 rpm, Charging efficiency = 75%, Injection
timing = 60 degBTDCF, Port shape: with swirl control valve)

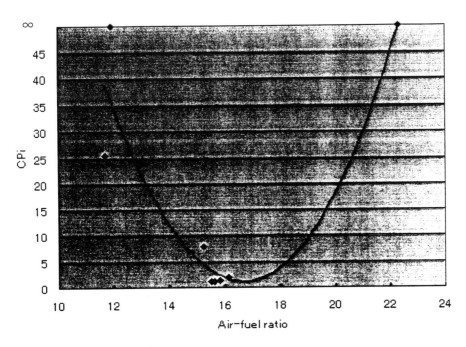

Fig.10 Relation between predicted A/F ratio and experimental combustion stability

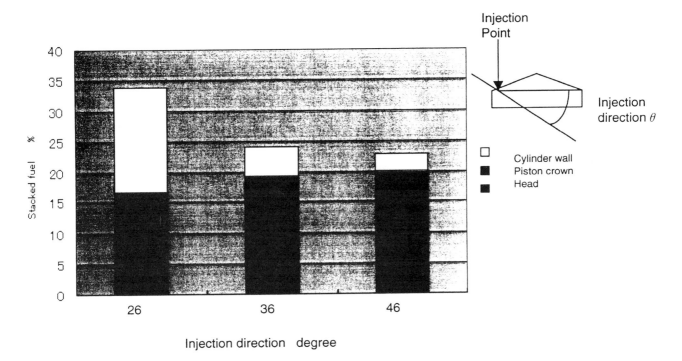

Fig.11 Influence of injection direction on the amount of fuel remaining on the cylinder head, piston crown, and cylinder walls

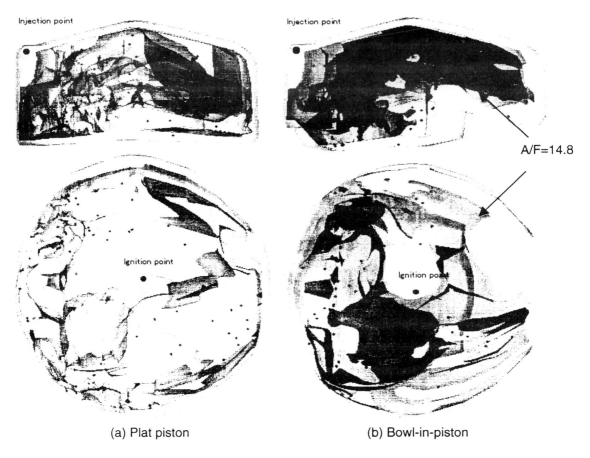

(a) Plat piston   (b) Bowl-in-piston

Fig.12 Iso-density surface of vapor fuel injected in the intake stroke (at 270 degBTDCF)

1999-01-0172

NSDI-3: A Small Bore GDI Engine

S. Henriot, A. Chaouche, E. Chevé, J. M. Duclos, P. Leduc,
P. Ménégazzi, G. Monnier and A. Ranini
Institut Français du Pétrole

ABSTRACT

Gasoline Direct Injection (GDI) is today more regarded as a suitable technology for relatively high displacement engines. The literature shows that the R&D effort on GDI engines is generally made for bores larger than 80 mm. But because GDI appears to be the most relevant way to improve fuel efficiency of S.I. engines, it should also be considered for small bore engines (bore below 75 mm). Nevertheless, locating an injector in already congested cylinder heads, with ultra lean stratified combustion capability while maintaining high engine specific power and proper cylinder head cooling is a real challenge. For such an engine, IFP "narrow spacing" proposal is a 3-valve per cylinder layout or NSDI-3 concept, with a spark-plug-close-to-the-injector design and a suitable piston to confine the fuel spray within the vicinity of the ignition location. This paper describes stage by stage the prototype engine realization using this novel concept. In the early stage, computer aided design and three-dimensional computational fluid dynamics tools were used in a step-by-step iteration cycle to characterize and validate the concept at low cost. One-dimensional calculations were carried out to design the entire intake system and valve timing as well as three-dimensional calculations were conducted to optimize the cylinder head water cooling. In-cylinder flow, injection timing and combustion characteristics were examined with the help of three-dimensional calculations for the stratified charge operation in order to evaluate the criteria required to optimize this crucial process. Special emphasis was given to the mixture formation, to spray and in-cylinder flow interaction as well as to combustion and pollutant formation process. Finally, corresponding engine tests confirmed the potential of the concept for stratified combustion.

INTRODUCTION

The development of an efficient GDI engine is undoubtedly a great challenge for the automotive industry. Recent work on GDI engines has shown their great potential in terms of fuel consumption economy and performance in comparison to conventional engines. Direct injection enables realization of stratified charges and seems the most promising way [1, 2] of improving fuel economy of spark-ignition (SI) engines in comparison to other solutions like homogeneous lean-burn conditions, variable valve timing, supercharging with displacement reduction or with a "start-stop device". Extensive research made by manufacturers showed numerous advantages of GDI in comparison to Port Fuel Injected (PFI) engines [3, 4, 5, 6, 7, 8]. Most are linked to the direct injection itself and the lean-burn combustion. However several technical difficulties still remain. The most challenging aspect of GDI is probably the specific stratified charge mixture operation at part-load which means that an adequate combustion chamber design and a suitable injection strategy are required to avoid excessively rich or lean areas around the spark-plug. In order to meet these targets and also to satisfy production constraints, various concepts are now proposed in the automotive market.

Side injection from an injector located between two intake valves with a curved-top piston to control the shape of the air-fuel mixture defines the "wide spacing" concept and has been adopted by several manufacturers. This approach is strongly influenced by a large number of parameters and the role of the in-cylinder charge motion (swirl or tumble) as well as the injection strategy were found to be essential in the mixing process. However, the suitable mixture strategy over a wide range of operating conditions requires a careful optimization of the piston shape and the in-cylinder flow. Both aspects create additional design costs and optimizing the fluid motion can reduce the GDI potential for increased volumetric efficiency. This is especially true in the case of swirl which may also require an auxiliary throttle leading to a more costly and complex engine.

Since the objective is to obtain a sufficient rich mixture in the compression stroke around the spark-plug, the "narrow spacing" concept, characterized by an injector close to the spark-plug position while the charge motion is used to constrain the mixture around this position can be another way to stratify the charge and is now matter of extensive research.

Though it represents a promising solution, the "narrow spacing" suffers from heavy requirements on injector spray (fine atomization and perfect axial-symmetry of the plume). Because of possible liquid fuel deposits on the spark plug, the risk of spark plug fouling is also high which may lead to misfiring. Nevertheless, experience showed that the ultra lean-burn potential of this concept is higher than the wide spacing concept, with already proven stratified operation with labmda value greater than 8. For "wide spacing", the removal of the spark plug from the injector often causes fuel dispersion.

Undoubtedly GDI will involve the whole range of engine sizes, although today most of the efforts of automotive manufacturers focuses on relatively large scale engines with bores greater than 80 mm. The IFP proposal for a small bore engine applies to a GDI combustion chamber with a bore below 75 mm. It is an in-line four cylinder engine and a "narrow spacing" concept was chosen in order to reach ultra lean-burn capabilities.

The basic concept, the combustion chamber layout and the different optimization stages are reported here after.

ENGINE BASIC SPECIFICATIONS

A three valves per cylinder arrangement is selected to deal with the small bore. No further auxiliary valve to control swirl, no variable valve timing nor variable length of the induction system are needed. To take full advantage of the direct injection fuel economy potential, the concept engine designate as NSDI-3 has to be able to run in stratified charge mode.

Intake and exhaust valves are sized to reach the 50 kw/l power target. To reach this value, two intake valves are located at a nearly central position on the cylinder head in order to allow for a large valve diameter. Intake valves stems are vertical and as a result valve seats are located in a plane perpendicular to the cylinder axis. The single exhaust valve located in a pentroof of the combustion chamber should give quicker catalyst light-off than a two exhaust valve layout.

To reduce heat transfer from the flame to the wall and to improve knock resistance the spark plug is conventionally centrally placed but because of the small bore an eccentric location is less constraining. As a result, and to deal with a congested cylinder head, the "narrow spacing" is realized by placing the spark plug on the intake side and the injector underneath, both far away from high temperature zones and easily accessible in a vehicle with the exhaust on the back side.

For ultra lean burn operation, the confinement of the fuel spray around the spark plug is ensured by a lateral piston bowl facing the injector (Figure 1). The bowl has an inclined axis to restrict the fuel-air mixture around the ignition position and to guide it to the spark-plug.

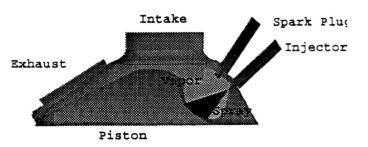

Figure 1. Basic concept

PRE-SCREENING WITH THREE-DIMENSIONAL MODELING

The development of favorable internal fluids dynamics and the formation of a suitable stratified mixture before ignition are essential stages of the NSDI-3 combustion chamber design. During the compression phase, the flow characteristics in the chamber are essentially governed by the interaction between the previous Bottom Dead Center (BDC) flow pattern and the piston bowl shape. This interaction determines the air-fuel distribution around the spark plug near TDC.

Basically three aspects must be taken into account: the in-cylinder air flow and the piston bowl profile characteristics, both should contribute to the mixture confinement; the fuel vaporization process during injection, which should be adequately fast to avoid presence of liquid on the chamber's wall near TDC; and finally the stratification obtained at the spark-plug location, which should not include as much as possible excessively rich or lean areas.

Taking theses aspects into consideration, three-dimensional calculations were performed in a step-by-step iteration cycle with the IFP Design Office to accurately define the adequate combustion chamber design for ultra lean burn operation. Special emphasis was given to the bowl in piston shape and the in-cylinder flow. The CFD

code used is KMB and it is a multi-block version of KIVA-2 [9, 10]. It solves the full 3D averaged compressible Navier-Stokes equations coupled with spray and combustion equations in a finite volume formalism. The multi-block approach developed at IFP allows computation of complex meshes with a low number of inactive cells. The WAVE-FIPA break-up model [12] is used to describe the injection process and the ECFM combustion model [13] is used to computer stratified combustion..

In order to reduce definition costs and time development only the compression phase is computed in this validation stage. Simulations started at BDC and ended at TDC. The in-cylinder flow at BDC is initialized by imposing a centered uniform solid body air flow in a half chamber. The half chamber is described by a mesh with 100000 vertices corresponding to circa 4 CPU hours on NEC SX4. Swirl, conventional and reverse (clock-size motion) tumble with various intensities (weak or strong) were imposed at BDC in order to characterize the future design of suitable intake process. Based on experimental visualizations, injection is simulated at the injector nozzle by a 70° hollow cone and a Sauter mean radius set to 10 μm. Injection is started at about -60 crank angle degree (cad) before TDC. Engine operating conditions used in this stage are summarized in Table 1.

Speed (rpm)	2000	Chamber Mean Turbulence at BDC	30 m²/s²
Volumetric efficiency	0.90	Chamber pressure at BDC	0.9 bar
Chamber temperature at BDC	350 Kelvin	Fuel/Air equivalence ratio	0.34
Injection duration	12 cad	Fuel mass rate (kg/h)	1.75

Table 1

Several combustion chambers were designed and for each in-cylinder air flow imposed at BDC, injection timing and injector inclination were adjusted in order to achieve the most suitable distribution of fuel in the vicinity of the spark plug.

The latest iteration of the combustion chamber design provides a good air-fuel equivalence ratio at the spark-plug as shown on Figure 2 and is obtained with an initial reverse tumble at BDC. Indeed, during the compression phase in-cylinder air flow is characterized by the progressive displacement of the tumble vortex toward the piston bowl. Near TDC, the vortex remains over the bowl and this structure contribute to the confining of the air-fuel mixture around the spark-plug (Figure 3).

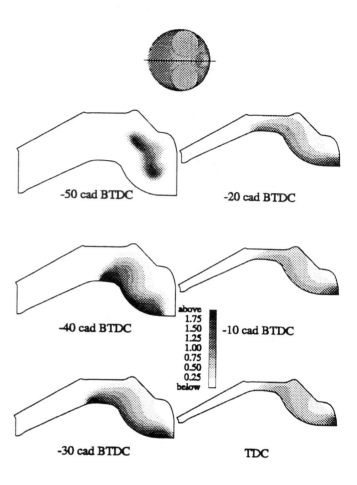

Figure 3. NSDI-3 prototype chamber design: Fuel-Air equivalence ratio fields after injection (-60 cad BTDC).

At the opposite, a conventional tumble vortex at BDC is characterized by the displacement of the main vortex towards the exhaust side and as a result an air-fuel distribution around the spark plug near TDC somewhat unfavorable to ignition.

According to theses conclusions upright intake ports well suited to develop a reverse tumble were designed and the basic combustion chamber design is shown in Figure 4.

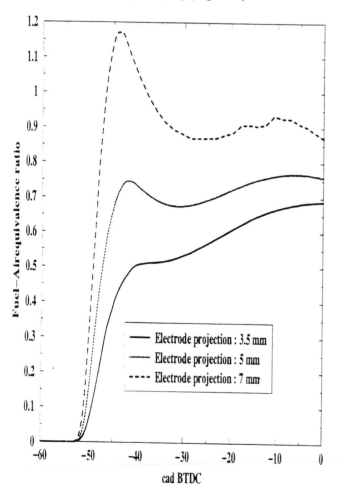

Figure 2. NSDI-3 prototype chamber design: equivalence ratio at the spark plug for various electrode projection.

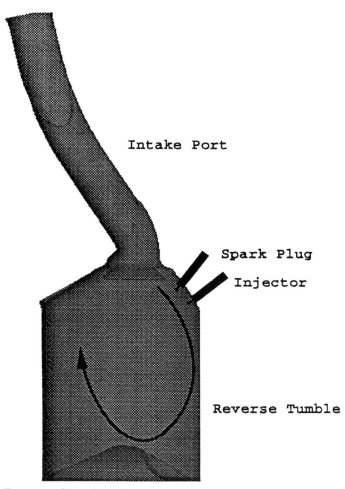

Intake Port

Spark Plug

Injector

Reverse Tumble

Figure 4. Chamber side view.

INTAKE SYSTEM AND VALVE TIMING OPTIMIZATION

A scale model of the NSDI-3 cylinder head was realized and the air-permeability of the ports equipped with valves was measured. Combining these results with a complete modeling of heat transfers fitted on the IFP existing engine of same bore, a 1D simulation of the whole engine was realized.

The aim was to determine the length of the manifolds and the valve timing in order to reach the power target.

Intake manifolds optimization

A combination of a 470 mm manifold and an intake valve closure (IVC) of 35 cad performs the power target. An IVC of 35 also gives a relatively high volumetric efficiency for low engine speed (Figure 5).

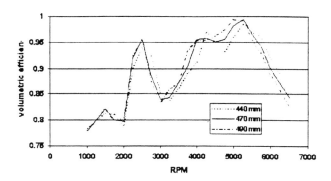

Figure 5. Optimization of the manifolds length (calculated).

Valve timing definition

Calculations show that a restricted duration of exhaust valve opening (200 cad corresponding for example to a valve timing of 25/-5 or 30/-10) gives a slight increase in torque at low speed with no penalty on maximum power as compared to a more extended valve opening (210 cad, for example 35/-5). A restricted duration will have a positive effect on knock at low speed because of reduced exhaust cylinder to cylinder interaction. But the EVC should not be too early to avoid increased pumping work (Figure 6).

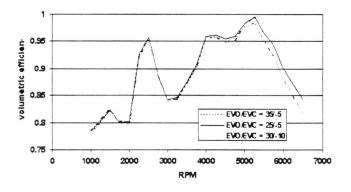

Figure 6. Optimization of the exhaust valve timing (calculated).

A valve timing of IVO/IVC=-5/35 and EVO/EVC=25/-5 appears to be a suitable configuration. The fitting of an optimized resonator has smoothing effect on torque over the entire engine speed range (Figure 7).

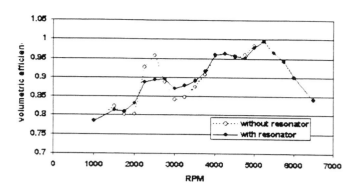

Figure 7. Resonator smoothing effect (calculated).

Power target

The calculations predict that the power target should be achieved from 5250 to 6000 rpm (Figure 8).

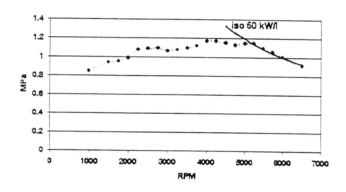

Figure 8. Final BMEP (calculated).

INTERNAL COOLING CIRCUIT OPTIMIZATION

Due to the high power density of the NSDI-3 engine, the internal cooling circuit optimizing is a quite important task, particularly for the engine head where the heat flux is maximum. The objective is here to ensure, with a minimal circuit volume, an effective cooling of highly thermally loaded areas. Moreover, we want to achieve an homogeneous temperature distribution for all cylinders.

The optimization step is based on 3D flow calculations and the use of Fluent/UNS, a general-purpose CFD code well suited for turbulent incompressible flow. Fluent/UNS uses a pressure-based segregated finite-volume method solver that can work with unstructured meshes for complex geometry modeling. In the framework of the study,

the tedious problem of mesh creation is solved in two steps: first, a surface triangle mesh is generated on the CAD surfaces of the geometry, directly within CATIA CAD software. From this surface triangle mesh, an unstructured mesh of tetrahedra is then generated with the Fluent meshing tool. Figure 9 shows the computational mesh obtained with this process. The cell number is about 600000.

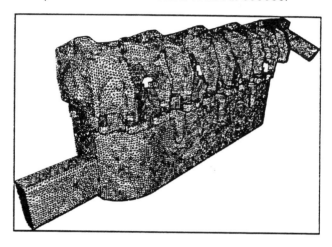

Figure 9. Computational grid for the internal cooling circuit.

Using this methodology, several geometries are analyzed in terms of velocity distribution and heat transfer. For each configuration, the critical maximum engine power case corresponding to the maximum coolant flow rate, is simulated: a 7000 l/h flow rate is imposed at the inlet and a reference pressure is given at the outlet. For the thermal problem, the fluid, supposed to be water, is taken at 363 K at the inlet. Wall temperatures are supposed at 383 K everywhere. For each case, the steady state flow is reached within 60 CPU hours, corresponding to 6 elapsed time hours, on 12 processors of a SGI Origin 2000 parallel computer.

As a result of the computational approach and based on experience of similar problems [11], an optimized circuit of the engine head is obtained. Figure 10 presents the velocity magnitude contours at the circuit wall of the block and the engine head. One can see that an adapted design of the coolant passages makes it possible to meet the main objectives: low speed areas (< 0.16 m/s) are avoided and high speeds are ensured around exhaust seats where the heat flux is maximum. The classical problem of unbalanced flow, due to the end location of the outlet, is here addressed through an adapted gasket passage design. The velocity and transfer coefficient fields are then quite similar from one cylinder to another, making the heat transfer more homogeneous in the cylinder head (Figure 11).

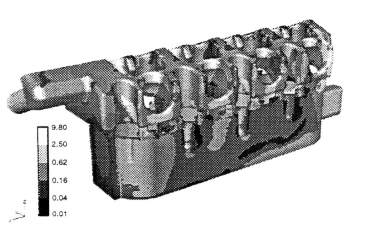

Figure 10. Exhaust side view of velocity magnitude contours (m/s)

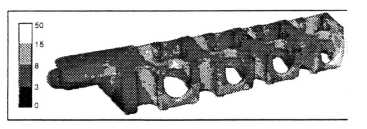

Figure 11. Exhaust view of heat transfer coefficient contours (KW/m2-K)

3D MODELING OF INTAKE, INJECTION AND COMBUSTION

In order to reach the fuel economy goal, the engine has to be able to work at a very low Fuel-Air equivalence ratio. Thus, we have to verify as a preliminary phase to the engine prototype building that the complete geometry studied allows stratified operating conditions by taking into account the whole engine cycle.

Three-dimensional computation of injection and combustion is an effective way to validate the choices made as it enables to understand how the combustion occurs in the combustion chamber. Of course, it requires a computational code able to describe precisely both fuel spray and combustion of the stratified load. KMB includes various physical sub-models. Such as the WAVE-FIPA breakup model and the ECFM combustion model [13] that allow stratified combustion in SI engines description.

Spray modeling

The gasoline stratification in the cylinder is, to a large extent, a function of the characteristics of the spray. Since the combustion process widely depends on the shape of the fuel cloud, the modeling of fuel injection has to be done carefully.

The spray model parameters were fitted on experimental results. Visualizations of the spray in a pressurized vessel were performed for various injection pressures, various injection duration and various vessel pressures. Then, the key parameters of the model were adjusted to reproduce the measurements.

The simulated injector is a Siemens Deka DI used with an injection pressure of 10 MPa. The resulting injection speed is around 120 m/s. For the operating conditions and according to spray visualizations, we consider that the injected mean Sauter diameter is about 10 μm and the initial cone angle is 90 degrees. We used a Rosin Rammler distribution for droplet mass [13].

Since the piston is rather close to the injection point, droplets impinge upon it. We describe this impingement according to Naber and Reitz [15] with heat flux between wall and droplets modeled as in Eckhause and Reitz [16].

Combustion and pollutant modeling

As underlined previously [13], the ECFM model does not require any specific tuning for this engine. It has just been used with the same parameters as in [13]. The ECFM model includes a conditioned burned / unburned gases description and pollutant computation.

In this phase, NOx is the only kinetically computed pollutant. CO is calculated through an equilibrium with CO2 (the cut off temperature is 1200K). Unburned hydrocarbons originate only from a reduction in the rate of combustion caused by excess dilution or temperature decrease during the expansion stroke. There is no activated crevices model. Anyway, fuel trapping by crevices is less likely to happen under stratified operating conditions, and so the computed HC should not be very far from what is measured.

Operating conditions

Two operating conditions were computed for a complete process (intake, compression, injection, ignition, combustion). As some experimental results were available for these points, we took the test rig engine parameters for our computations. Their characteristics are summarized in Table 2. Both correspond to ultra lean charge operations.

Case	1	2
Speed (rpm)	1200	2000
Volumetric efficiency	0.84	0.88
Fuel/Air equivalent ratio	0.38	0.35
Injection timing (cad BTDC)	-40	-45
Ignition timing (cad BTDC)	-5	-15

Table 2

One can notice that, as the piston bowl is much smaller than the Mitsubishi GDI piston bowl [13], the NSDI-3 engine requires a later injection timing than the Mitsegine. This concept allows also leaner combustion.

Results and discussion

For both cases, during the intake phase, the average flow motion is rather weak and the resulting flow is composed of two vortices: the first one over the bowl and the second on the exhaust side (Figure 12 and Figure 13). The configuration presented has been optimized for volumetric efficiency and the natural drawback is this lack of intensity in the mean flow motion. Consequently, turbulence intensity is not very high during the whole process.

In both cases computed in-cylinder flow pattern, injection characteristics and fuel distribution are very similar. Near the combustion top dead center, a small reverse tumble motion appears over the bowl (for case 2, see Figure 14). This vortex is mainly due to the shape of the piston bowl and to the effect of air motion on the spray. It has been observed experimentally that the engine works better at higher injection pressure. This might be directly linked to the flow motion and the turbulence generated by the spray.

For load stratification, as expected, because of the late injection timing, the fuel is rather concentrated in the piston bowl for both cases. The reverse tumble motion also helps to confine the fuel. Despite the important initial cone angle of the spray there is no fuel that directly impacts the bottom of the piston or the cylinder head. Indeed, as the density in the cylinder is rather high when injection starts, the spray tends to narrow. As a remark, the narrowing of the spray has not been observed in the Mitsubishi GDI engine to this extent [13]. The first reason is that the mean density of the medium when the injection occurs is three times greater in the NSDI-3 than in the Mitsubishi engine. The second reason is that the injection speed of this DekaDI with injection pressure of 10MPa is twice more elevated. As a result, the droplets size is much smaller and then the coupling between droplet and medium gas is more intense. This intense coupling creates a strong toroidal vortex which confines the spray and helps penetration. In the Mitsubishi GDI case, a toroidal vortex was observed, but it was much weaker. Note that this strong narrowing of the spray has been observed experimentally for the Mitsubishi GDI but only for late injection timing, when medium density at least twice the medium density the spray faces after injection in the engine.

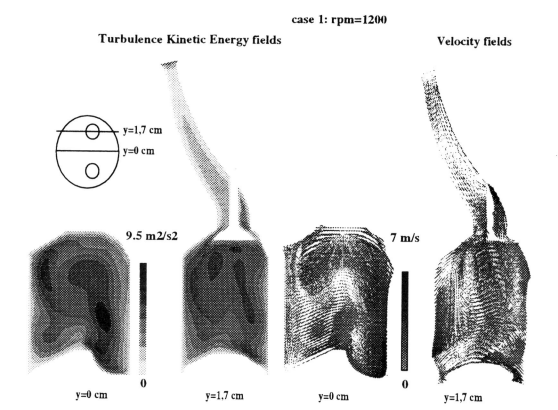

Figure 12. Turbulence kinetic energy and velocity fields at BDC for case 1.

case 2: rpm=2000

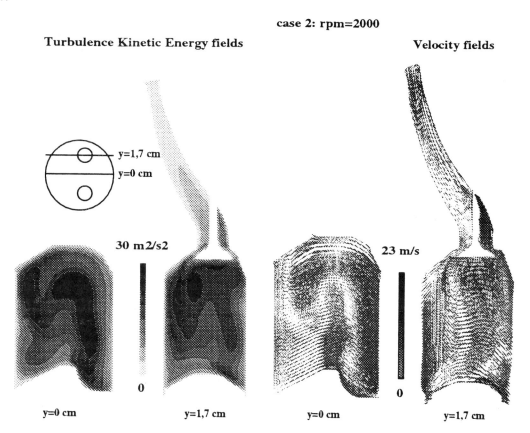

Figure 13. Turbulence kinetic energy and velocity fields at BDC for case 2.

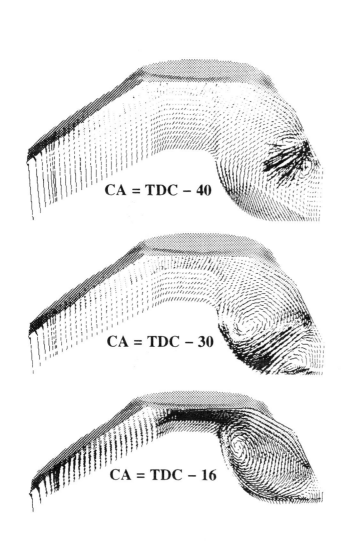

Figure 14. Case 2: velocity fields near TDC.

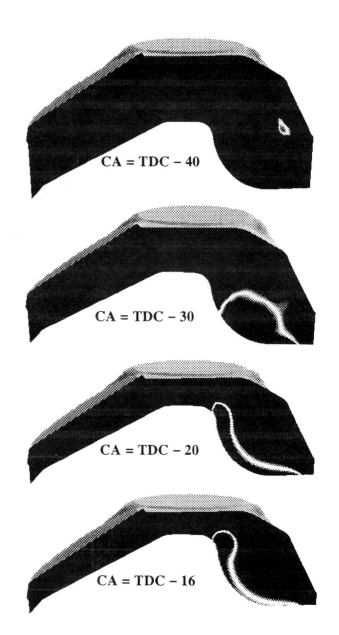

Figure 15. Fuel mass fraction during the compression phase (min=0 and max=0.2).

The shape of the fuel distribution is very similar in both cases 1 and 2, then we will only presents results about case 2 (Figure 15). The maximum of Fuel-Air equivalence ratio is located at the top of the bowl. Even if the stratification observed is satisfying, one would wish for a smaller peak equivalence ratio, with a wider area in the bowl occupied by the fuel, and a smaller amount of fuel located in the immediate vicinity of the piston. Compared with preliminary computations, the load is more confined. This is partly due to the injection timing needed by the engine on the test rig. The late timing prevents the fuel to flow out of the bowl.

Thus limiting fuel dispersion also limits fuel dilution and consequently HC emissions. The increase of the confinement might also be caused by the narrowing of the spray. The preliminary computations (Figure 3) did not exhibit the strong narrowing of the spray observed here: the injection speed and the density of the medium at the moment of injection were smaller. The ignition delay is small owing to the location of the spark plug in a rather rich area and to the high cylinder pressure. As the flame grows up (Figure 16), it pushes the gases, including fuel, out of the bowl and a great amount of CO is created just at the top of the piston. The spot of CO is the direct image of the spot of high fuel concentration that existed at the top of the piston bowl before ignition.

CA = TDC + 5

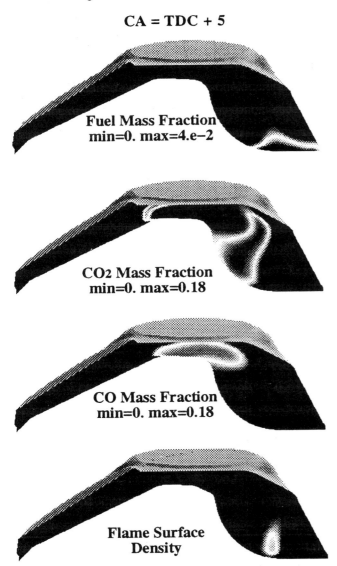

Figure 16. Combustion characteristics computed at TDC+5.

For the combustion phase, as measurements of cylinder pressure were already available, combustion and pollutant computations are used here to understand what happen in the cylinder when the flame appears and propagates. The pressure laws on Figure 17 shows some differences between computation and measurements, but the IMEP are rather close (less than 5%). For case 2, we have also plotted a computation result for a different spark plug location (2 mm closer of the head) and as a result the

quicker increase of the cylinder pressure is due to a better ignition.

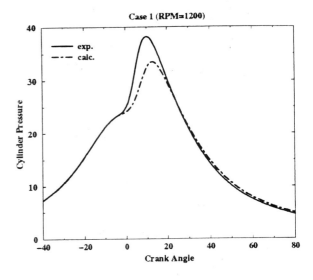

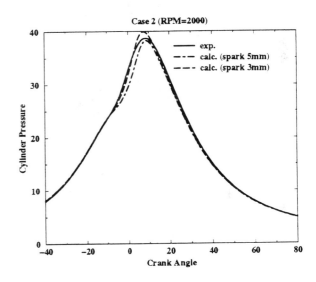

Figure 17. Comparison between experimental and computations.

Figure 18 presents the concentration of NOx, CO, CO2 and Fuel 100 crank angle degrees after TDC in the symmetry plane. The CO concentration traces the former high Fuel-Air equivalence ratio areas.

331

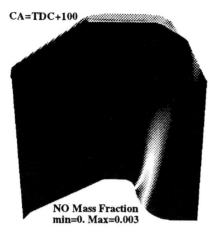

NO Mass Fraction
min=0. Max=0.003

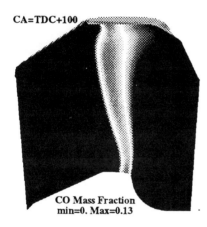

CO2 Mass Fraction
min=0. Max=0.2

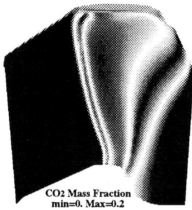

CO Mass Fraction
min=0. Max=0.13

Fuel Mass Fraction
min=0. Max=0.015

Figure 18. Fields 100 cad ATDC for case 2.

The following tables sum up the pollutant emissions computed and measured.

	NO (ppm)	HC (ppmc)
Computed (electrode projection 3 mm)	470	4100
Experiments	750	5600

Table 3. Case 1

	NO (ppm)	HC (ppmc)
Computed-1 (electrode projection 3 mm)	310	2950
Computed-2 (electrode projection 5 mm)	340	2750
Experiments	480	3500

Table 4. Case 2

For the cases studied, computed and experimental results follow the same trend. However, the computations under-predict NOx and HC levels. NOx under-prediction might be due to the stratification which is not very accurately described (NOx formation is extremely sensitive to the local fuel/air ratio). HC levels might be underestimated by the computations because of the actual chemical model and by the fact that crevices are not taken into account.

CONCLUSIONS

A complete set of computational tools has been used to define the global characteristics of a small bore GDI engine with a well adapted three valve per cylinder design. The IFP proposal is a combination of "narrow-spacing" layout and piston bowl confinement. Indeed, the spark-plug close to the injector arrangement seems to be the most promising way of achieving ultra lean-burn conditions.

A step-by-step process permitted to demonstrate

the ability of computational tools to the design of the whole engine in order to obtain a good stratification capability and potentially high performances. The engine was designed to reach 50 kW per liter displacement with relatively high BMEP over the entire engine speed range. The internal cooling circuit was also improved via numerical tools. Finally, ultra lean-burn operation computations of the complete engine cycle showed interesting results in same trend with preliminary engine tests.

A future small car fitted with an improved engine is now under intense development at IFP and its objectives are to meet emission standards with only a conventional three-way catalyst for year 2000 and for 2005 regulations a DeNox catalyst with 70% efficiency will be required.

REFERENCES

[1]. Ingénieurs de l'Automobile, no716, October 1997.

[2]. Croissant, K., and Kendlbacher, C., "Requirements for the engine management system of gasoline direct injection engines, "Direkteinspritzung im Ottomotor" Congress, Essen, Germany, 1997.

[3]. Harada, J. Tomita, T., Mizumo, H., Mashiki, Z., and Ito, Y., "Development of direct injection gasoline engine", SAE 970540, 1997.

[4]. Kume, T., Iwamoto, Y., Murakami, M., Akishino, K. and Ando, H.: "Combustion control technologies for direct injection SI engine", SAE 960600, 1996.

[5]. Karl, G., Kemmler, R., Bargende, M. and Abstoff, J.,"Analysis of a direct injected gasoline engine", SAE 970624, 1997.

[6]. Tomoda, T., Sasaki, S., Sawada, D., Saito, A., and Sami, H.: "Development of direct injection gasoline engine - study of stratified mixture formation", SAE paper no 970539, 1997.

[7]. Automotive engineering, December 1997.

[8]. Ando, H., Noma, K., Lida, K., Nakayama, O. and Yamauchi, T.: "Mitsubishi GDI engine strategies to meet the European requirements", Engine and Environment Conference 97, Graz, Austria, 1997.

[9]. Amsden, A., O'Rourke, P.J. and Butler T.D.: , "KIVA-2 a computer program for chemically reactive flows withh sprays", Report LA-11560-MS, Los Alamos National Laboratory, 1989.

[10]. Habchi, C. and Torres, A., "A 3D multi-block structured version of the KIVA-2 code", First European CFD Conference proceedings, 1992.

[11]. Porot, P., Ménégazzi, P. and Ap, N.S., "Understanding and improving evaporative engine at high load, high speed by engine tests and 3D calculations", 3rd Int. Conf. on Vehicle Thermal Management System (VTMS), SAE paper no971792, Indianapolis, USA, 1997.

[12]. Habchi, C., Verhoeven, D., Huynh Huu C., Lambert, L., " modeling atomization and break-up in high-pressure diesel sprays", SAE paper no 970881, 1997.

[13]. Duclos, J.M. and Zolver, M.,"3D modeling of intake, injection and combustion in DI-SI engine under homogeneous ans sratified operating conditions", 4th Int. Symposium on Diagnostics and modeling of Combustion in Internal Combustion Engines, COMODIA 98, Kyoto, Japan, 1998.

[14]. Han, Z. Parrish, S., Farrel, P.V. and Reitz, R.D.: ,"Modeling atomization processes of pressure swirl hollow cone fuel sprays", Atomization and sprays, vol. 7, pp. 663-684, 1997.

[15]. Naber, J.D. and Reitz, R.D.,"modeling engine spray/ wall impingement", SAE paper no880107, 1988.

[16]. Eckhause, J.E. and Reitz, R.D.,"modeling heat transfer to impingement fuel sprays in dirct-injection engines", Atomization and Sprays, vol. 5, pp. 213-242, 1995.

Diagnostic Methods for Direct-Injection Gasoline Engines

Diagnostic Methods for Direct-Injection Gasoline Engines

This section offers a selection of papers that focus primarily on in-cylinder diagnostic measurements in direct-injection engines. The development and application of various diagnostic techniques is considered key to the successful development of GDI combustion systems. Sometimes diagnostic measurements are used to directly observe and interpret engine performance, while other times the data from the measurements are used indirectly to tune computational fluid dynamic analytic models of the combustion process. In both cases, the end goal is to gain a more complete understanding of the combustion process.

The papers in this section were selected not only for their high quality and the valuable insight they provide into the combustion process, but also because of their uniqueness. High-speed shadow, Schlieren and visible-flame photography, measurements of laser-induced fluorescence and flame spectra, laser-Doppler velocimetry, particle image velocimetry, and the use of fiber-optic spark-plug probes and fast-response flame-ionization detectors are some of the techniques represented by the selected papers.

A combustion system can be improved only after it has been properly understood. Thus, diagnostic tools that help us in understanding the combustion process play an invaluable role in the development of direct-injection engines. It is noted that a single diagnostic tool usually cannot reveal all of the complex details of the physical and chemical processes involved in the combustion process. This is because no single tool has enough temporal and spatial resolution with regards to all relevant physical and chemical parameters of all the species (air, fuel-liquid, fuel-vapor, and recirculated exhaust gas) involved in the combustion process. A number of tools must be used, and a complete understanding is gained when all of the data from all of the tools are brought together and integrated with other engine performance and emissions data.

Mixing Control Strategy for Engine Performance Improvement in a Gasoline Direct Injection Engine

Kazunari Kuwahara, Katsunori Ueda and Hiromitsu Ando
Mitsubishi Motors Corp.
Japan

ABSTRACT

Spray motion visualization, mixture strength measurement, flame spectral analyses and flame behavior observation were performed in order to elucidate the mixture preparation and the combustion processes in Mitsubishi GDI engine. The effects of in-cylinder flow called reverse tumble on the charge stratification were clarified. It preserves the mixture inside the spherical piston cavity, and extends the optimum injection timing range. Mixture strength at the spark plug and at the spark timing can be controlled by changing the injection timing. It was concluded that reverse tumble plays a significant role for extending the freedom of mixing.

The characteristics of the stratified charge combustion were clarified through the flame radiation analyses. A first flame front with UV luminescence propagates rapidly and covers all over the combustion chamber at the early stage of combustion. Then, the combustion of rich mixture proceeds in the reaction zone behind a second flame front with thermal radiation. The second flame front is propagated into the post flame zone of the first flame front filled with the products of first flame such as radicals and CO. Soot generated in the rich mixture zone is burned-up in this radical rich zone .

Based on this finding, a new mixing control strategy for knock suppression named "two-stage mixing" was proposed. A first injection is performed during the early stage of the intake stroke to prepare the very lean premixed mixture and a second injection is performed during the later stage of the compression stroke to prepare the distinctively stratified mixture. The premixed mixture is too lean to induce knock and the stratified mixture does not have the enough time to proceed the incubation reaction for knock. What is interesting is that soot generated in the rich mixture zone is not emitted at all. In the case of the conventional stratified charge rich combustion, soot behind the flame front propagates to the air zone, and cooled to generate cold soot that will not be reburned. In case of the two-stage mixing, soot in the rich mixture zone propagates to the very lean premixed mixture zone, where soot plays a role of the ignition site and ignites the lean mixture. Soot in the lean mixture zone is burned-up efficiently utilizing the air, radicals and heat generated by the combustion of the lean mixture.

INTRODUCTION

Mitsubishi Motors Corporation had proposed a new mixing control strategy to realize a stable and distinctive stratification for its GDI (Gasoline Direct Injection) engine [1-5]. This concept adopts a wide spacing layout shown in **Figure 1**. First, the fuel spray is directed toward a spherical piston cavity, then, redirected towards the spark plug after the impingement on the cavity surface. As compared with the close spacing layouts the interval between the fuel injection and the spark ignition can be extended. Fuel vaporization and fuel air mixing take place during this interval. Thereby, fuel air mixture with the controlled mixture strength can be prepared around the spark plug at the spark timing.

At low engine speeds, the primary factor promoting mixing is the reflection on the piston surface and the air

---

* Numbers in parentheses designate references at end of paper.

motion induced by the spray. These factors do not depend on engine speed.

Upright straight intake ports are adopted to generate the reverse tumble with the rotational direction opposite to that of the conventional tumble. The reverse tumble flowing along the cavity surface effectively transports the fuel in the cavity to the spark plug. Upward piston motion also assists the fuel transport to the spark plug. Effects of the reverse tumble and the piston motion to mixture formation are enhanced at higher engine speed with the increase of the velocities of intake air and the piston. Thereby, the kinetic energy prepared by the reverse tumble and the piston motion becomes the primary factor for fuel air mixing at higher engine speeds. Many factors, that is, spray motion, reverse tumble and the piston motion cooperate to prepare the adequately dispersed mixture around the spark plug in wide engine speed range.

In this study, the characteristics of mixing and combustion of the GDI engine were investigated. Fuel spray motion was visualized by laser shadowgraphy. Mixture strength was measured by LIF and a new method using the spectrum of the spark discharge emission. Instantaneous spectra of the light emitted from the combustion zone and the sets of simultaneous flame photographs of UV and visible ranges were used for the combustion analyses.

Based on the findings of the combustion analysis, a mixing control strategy named "two-stage mixing" to suppress the knock will be proposed at the end of this paper.

EXPERIMENTAL PROCEDURE

OPTICAL ENGINE - A single-cylinder optical engine shown in **Figure 2** with bore x stroke of 85 mm x 88 mm was used. The GDI engine originally has the spherical piston cavity. However, because the phenomena in the spherical cavity is not optically accessible, the cylindrical piston surface configuration with the same design of cross section as the central cross section of the spherical cavity was adopted in place of the spherical cavity in some experiments. The mixing and the combustion process in the cylindrical chamber can be observed through the quartz windows on the side walls of a pentroof combustion chamber.

The compression ratio was reduced to 8.0:1 from 12.0:1 of the original GDI engine by changing the piston surface configuration. Intake air was heated to compensate the compression temperature decrease caused by the com-

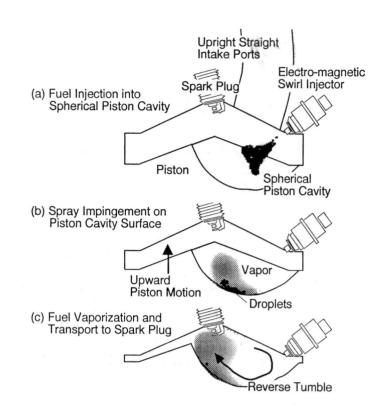

Fig. 1 Mixing Control Concept for the GDI Engine

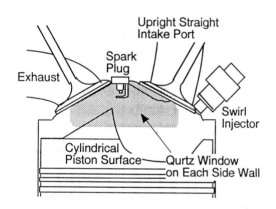

Fig. 2 Optical Access to Cylindrical Chamber

pression ratio reduction.

The intensity of the reverse tumble was controlled by intake valve shrouds. In the case of reverse tumble, the rotational speed of the reverse tumble was about six times as high as engine speed during the late stage of the compression stroke. In the case without reverse tumble, any large scale rotational air motion could not be observed.

The engine was operated by late injection. Operating conditions are noted below each figure. Before the measurements, it was operated by early injection for several minutes to heat the piston surface.

DIAGNOSTICS OF MIXING - The fuel spray reflecting on the piston surface was visualized by shadowgraphy with Ar ion laser through a pair of the quartz windows on both side walls. A high-speed video was used to image the motion of the fuel spray.

For the purpose of LIF measurement, a small cylindrical lens was mounted at the center of the pentroof combustion chamber in place of the spark plug. A substitute spark plug with long projection was installed on one of the side wall. A KrF excimer laser sheet was introduced into the cylindrical chamber through the cylindrical lens. Gasoline LIF distribution was observed through the side window by a high sensitivity image intensifier and a framing camera.

A new technique using the spark discharge spectra was adopted for the mixture strength measurement. The mixture strength at the spark plug at spark timing was measured quantitatively.

DIAGNOSTICS OF COMBUSTION - In order to analyze the combustion in engines, a high-speed OMA (Optical Multichannel Analyzer) illustrated in **Figure 3** was developed [6]. A pair of side-by-side cells in a 512-cell photodiode array was used to shorten the scanning time to 40 μs while maintaining the sensitivity. The flame spectra of every 1 °CA was acquired by this OMA.

UV flame luminescence was visualized by an image intensifier with short decay phosphor and an high speed video. Visible flame emission was observed by the high speed video.

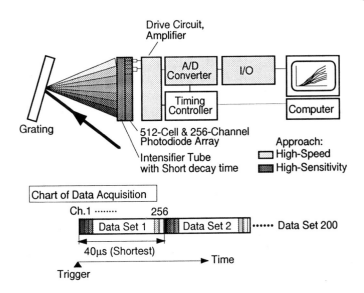

Fig. 3 High-Speed Optical Multichannel Analyzer

Soot was visualized by Ar ion laser light extinction measurement through a pair of the side wall windows.

CHARACTERISTICS OF MIXING ON GDI CONCEPT

FUEL SPRAY BEHAVIOR - **Figure 4** shows the contours of shadowgraphs of fuel spray. The influences of injection start and the reverse tumble on spray behavior were analyzed.

By comparing spray motions in the cases with and without reverse tumble, the following role of the reverse tumble on spray motion were confirmed:

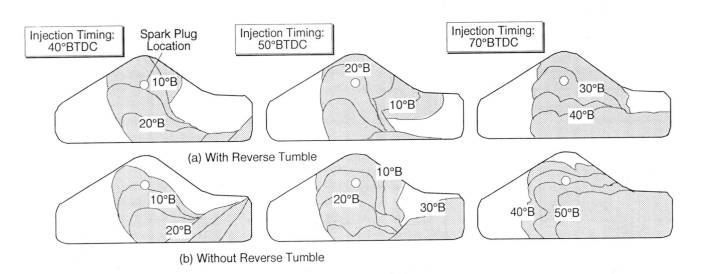

Fig. 4 Influences of Injection Start Timing and Reverse Tumble on Fuel Spray Motion
(Engine Speed: 1000min⁻¹, WOT, Injection Pressure: 5MPa, Injection Period: 1.67ms, A/F: 35)

(1) Spray motion is accelerated by the air flowing toward the spark plug on the piston surface.

(2) Fuel air mixing is promoted by the rotational flow in the combustion chamber.

(3) The dispersion of the fuel outside of the combustion chamber near the injector is suppressed by the flow descending in the intake side.

(4) Two-dimensional structure of the reverse tumble with the smaller velocity component in the axis direction attenuates the fuel dispersion in this direction.

When the reverse tumble is applied, 40 °BTDC is the retard side injection timing limit for the reflected fuel spray to arrive at the spark plug by the spark timing of 15 °BTDC, and 70 °BTDC is the advance side injection timing limit for the fuel spray to impinge on the piston surface. When the reverse tumble is not applied, fuel spray injected at 40 °BTDC can not arrive at the spark plug by the same spark timing, and the fuel spray injected at 70 °BTDC does not impinge on the piston surface. In short, when the reverse tumble is applied, available injection timing will range from 40 to 70 °BTDC. When the reverse tumble is not applied, the behavior of the fuel spray is controlled only by the reflection, and the available injection timing will be narrow. When the injection timing is too early, fuel will be over pen-etrated beyond the combustion chamber, when injected too late, fuel will not arrive at the spark plug.

In conclusion, it was clarified that the reverse tumble accelerating the spray motion extends the available injection timing range. What is important is that the mixture strength at the spark plug is affected by the injection timing. When the fuel is injected early, it is dispersed and the leaner mixture will be prepared, when injected late, richer mixture will be prepared. Extension of the available injection timing range by the reverse tumble means an increase of the freedom of the mixing control.

FUEL AIR MIXING - In order to clarify the influence of the injection timing on mixing control, cross-sectional structures of fuel spray were visualized by gasoline LIF distribution.

Results are shown in **Figure 5**. Because the laser is considerably absorbed by rich mixture, and temperature decrease caused by the fuel vaporization influences the quantum yield of LIF, it was impossible to convert LIF intensity into quantitative mixture strength. However, it had been confirmed that the LIF intensity correlates with the mixture strength by the preliminary experiments using the premixed mixtures.

When the fuel is injected at 40 °BTDC, the front contour of the reflected fuel spray arrives at the spark plug at the spark timing of 15 °BTDC. When fuel is injected at 50 °BTDC, the roll-up vortex region inside the reflected spray locates at the spark plug at the spark timing, in which the

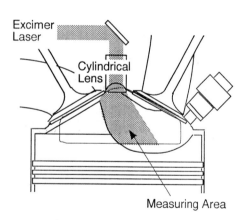

(a) Gasoline LIF Measuring Area

Fig. 5 Measurement of Gasoline LIF Distribution in Cylindrical Chamber
(A KrF excimer laser sheet was introduced into the cylindrical chamber through a cylindrical lens at the center of the pentroof combustion chamber.)

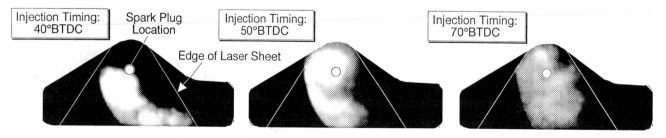

(b) Comparison of Gasoline LIF Distributions in the Cases of Different Injection Start Timings
(with Reverse Tumble, Imaging Timing: 15°BTDC, Other Conditions: See Fig. 4)

roll-up vortex promotes the air entrainment into the fuel spray. Consequently, the mixture strength around the spark plug is lower than that in the case of the injection timing of 40 °BTDC.

When the fuel is injected at 70 °BTDC, the fuel spray is widely dispersed in the combustion chamber by the spark timing. The mixture strength around the spark plug is the lowest.

QUANTITATIVE MIXTURE STRENGTH MEASUREMENTS - The spectra of spark discharge were detected with the temporal resolution of 0.5 °CA by the high-speed OMA. Discharge light in mixture is composed of the emissions mainly from OH band, CN band and the broad band NO-O recombination emission. **Figure 6 -(a)** shows the discharge light spectra in uniform mixture at the spark timing of 15 °BTDC. The spectra were analyzed at the moment that the intensity of the CN band emission is the maximum before the combustion. The dependence of spectral profile on mixture strength can be found.

The correlation between the ratio of the intensity in CN band emission with that in OH band emission and the mixture strength in uniform mixture is shown in **Figure 6 -(b)**. A distinct correlation can be observed. Based on this correlation, the mixture strength at the spark plug can be derived from discharge light spectra.

This technique was applied to the mixture strength measurement in GDI engine. In order to verify the result from the LIF measurement, the influence of injection timing on the mixture strength at the plug at the spark timing was examined. The results are shown in **Figure 6 -(c)**. By changing the injection timing from 50 to 70 °BTDC, the average ratio of the luminescence intensities decreases from 3.6 to 1.9. The equivalence ratio of the mixture at the spark plug decreases from 1.9 to 1.3 . The cycle by cycle variance in the equivalence ratio is in the relatively narrow range.

This technique could not be applied in the case of the injection timing of 40 °BTDC because the rich mixture did not arrive at the spark plug at the spark timing and the capacitive discharge is performed in the mixture too lean to emit the discharge light including the CN band emission. Even in this condition, however, stable combustion can be realized since the flammable mixture arrives at the spark plug before the inductive discharge period.

CHARACTERISTICS OF COMBUSTION

STABLE COMBUSTION ZONE - The influence of the reverse tumble on the stable combustion conditions was investigated by cylinder pressure analysis. **Figure 7** shows the stable combustion conditions on the map of the injection and the spark timing. The stable combustion zone was defined as the condition in which the COV of imep is less than 10 % and no misfire takes place.

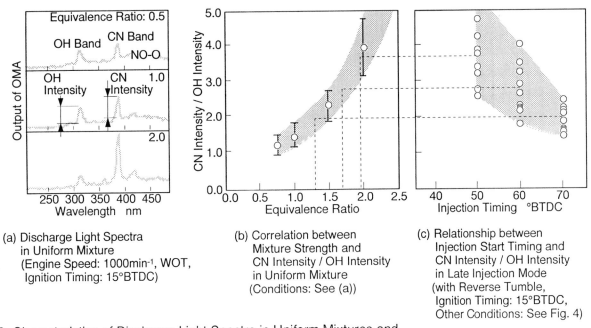

(a) Discharge Light Spectra
in Uniform Mixture
(Engine Speed: 1000min⁻¹, WOT,
Ignition Timing: 15°BTDC)

(b) Correlation between
Mixture Strength and
CN Intensity / OH Intensity
in Uniform Mixture
(Conditions: See (a))

(c) Relationship between
Injection Start Timing and
CN Intensity / OH Intensity
in Late Injection Mode
(with Reverse Tumble,
Ignition Timing: 15°BTDC,
Other Conditions: See Fig. 4)

Fig. 6 Characteristics of Discharge Light Spectra in Uniform Mixtures and Influence of Injection Start Timing on Mixture Strength on Spark Plug at Spark Timing in Late Injection Mode

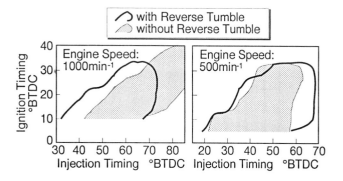

Fig. 7 Combustion Limits Defined by
COVimep of 10% or Miss Fire Rate of 1%
(WOT, A/F: 35)

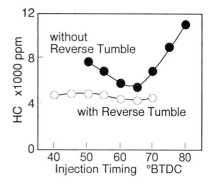

Fig. 8 Variation of HC Emission Level with
Change of Injection Start Timing
(Ignition Timing: 15°BTDC,
Other Conditions: See Fig. 4)

Results shown in the figure seems to show that the reverse tumble is not influential with the combustion stability in such a low engine speed conditions of 500 or 1000 min⁻¹. This results shows that the major factor controlling the mixing is the kinetic energy carried by the fuel spray itself, since the kinetic energy of reverse tumble in such a low engine speed is not significant.

However, as shown in **Figure 8**, the reverse tumble gives a large influence on HC emission. When the reverse tumble is not applied, a large amount of HC is emitted except in the very narrow injection timing range. When the reverse tumble is used, however, the HC emission is maintained low over the wide range of injection timing available for the stable combustion. This result confirms that the reverse tumble is effective to suppress the fuel spray dispersion to outside of the combustion chamber. In other words, when tumble is not used, air fuel mixture overpenetrate in the direction of the tumble axis to escape from the combustion chamber. The fuel dispersed out of the combustion chamber becomes too lean to be burned by the flame propagation.

Results of the cylinder pressure analysis for the cases with reverse tumble are shown in **Figure 9**. When the injection timing is later than 45 °BTDC, the induction period of combustion becomes longer. This may be the result of poor spark phenomena caused by the absence of the flammable mixture at the timing of the capacitive discharge. When the injection timing was earlier than 50 °BTDC, the influence of the injection timing is small. When the fuel is injected at 70

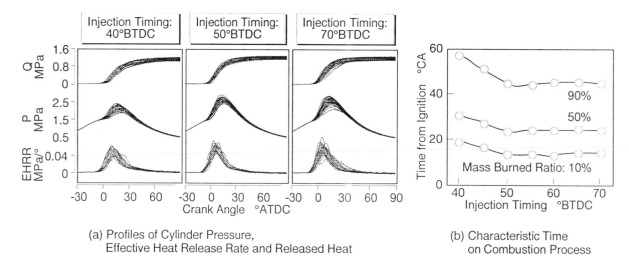

(a) Profiles of Cylinder Pressure,
Effective Heat Release Rate and Released Heat

(b) Characteristic Time
on Combustion Process

Fig. 9 Influence of Injection Start Timing on Combustion Process
(with Reverse Tumble, Ignition Timing: 15°BTDC, Other Conditions: See Fig. 4)

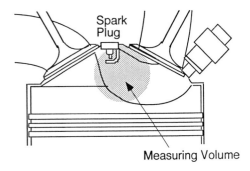

(a) Flame Emission Measuring Volume

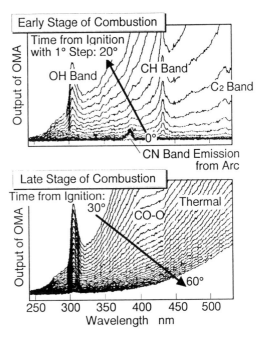

(b) Temporal Variation in Flame Spectra
(with Reverse Tumble,
Injection Start Timing: 50°BTDC,
Ignition Timing: 15°BTDC,
Other Conditions: See Fig. 4)

Fig. 10 Flame Spectral Analysis by High-Speed OMA

°BTDC, however, the cycle by cycle variance of the combustion rate is somewhat larger, probably because of the over mixing as had been shown in Figure 6 -(c).

In the case of the injection timing of 50 °BTDC, the combustion is the most stable. It can be considered that the cycle by cycle variance in equivalence ratio shown in Figure 6 -(c) is within the range of mixture strengths adequate to the stable combustion.

FLAME SPECTRA - The characteristics of the stratified charge combustion were clarified through the flame radiation analyses by the high speed OMA. **Figure 10** shows the flame spectra of every 1 °CA. Flame emission was integrated over the large space in the cylindrical chamber.

The major component of flame spectra is the thermal radiation from the soot generated in rich mixture zone all over the combustion process.

In the early combustion process, the chemiluminescences mainly from OH, CH and C_2 radicals are observed. After the disappearances of CH and C_2 chemiluminescences, the continuous emission from the broad band on CO-O recombination appears. The thermal radiation attenuates in relatively short period, showing the rapid burn-up of the soot.

FLAME BEHAVIOR - Simultaneous observations of the UV flame luminescence and the thermal radiation were performed. UV luminescence was observed in the wavelength from 270 to 370 nm in which the intensity of chemiluminescence is comparable to that of the thermal radiation . Thermal radiation was observed in the visible range from 510 to 590 nm. Results are shown in **Figure 11**.

At the early stage of combustion, a first flame front with UV luminescence propagates rapidly and covers all over the combustion chamber. Then, the combustion of the rich mixture proceeds in the reaction zone behind a second flame front.

Figure 12 shows the first flame contour extracted from UV luminescence. Referring the fuel spray behavior which had been shown in Figure 4, it can be found that the flame propagation is controlled by the mixture formation. For example, in the case of the injection timing of 40 °BTDC, the flame propagation toward the piston surface is faster because a large amount of fuel exists between the piston surface and the spark plug at the spark timing. In the case of the injection timing of 50 °BTDC, the flame propagation follows the fuel transport in the direction to intake side. In any injection timing, it seems that the UV flame propagates to the end of combustion chamber in a short time.

In the visible range, luminous flame with the wrinkled structure can be observed. It can be considered that the air entrainment into fuel spray and combustion zone is supported by eddies with the scale characterizing this structure. With the advance of the injection timing, that is, with the promotion of the mixing to prepare the leaner mixture around the spark plug, the local brightness and the area of the luminous flame are reduced.

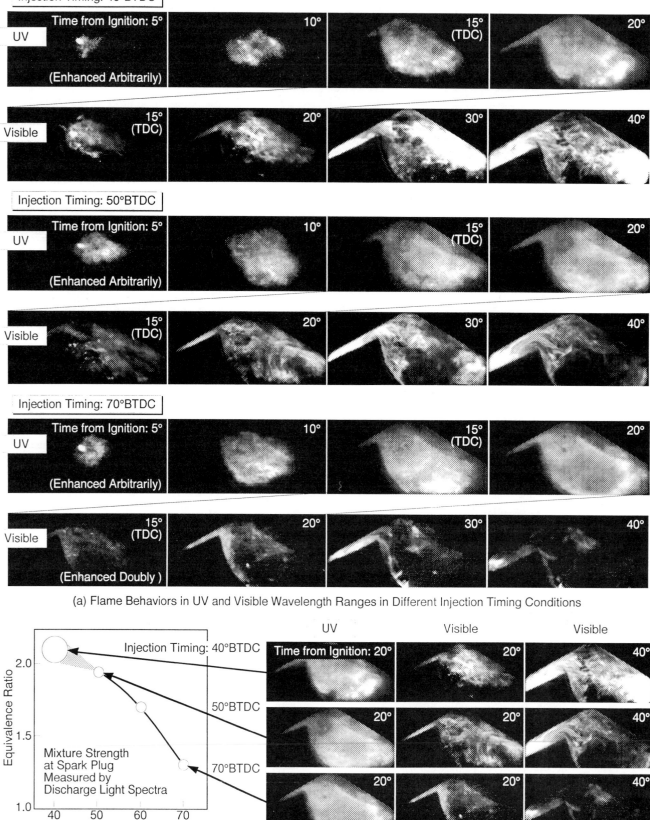

Injection Timing: 40°BTDC

UV | Time from Ignition: 5° | 10° | 15° (TDC) | 20°
(Enhanced Arbitrarily)

Visible | 15° (TDC) | 20° | 30° | 40°

Injection Timing: 50°BTDC

UV | Time from Ignition: 5° | 10° | 15° (TDC) | 20°
(Enhanced Arbitrarily)

Visible | 15° (TDC) | 20° | 30° | 40°

Injection Timing: 70°BTDC

UV | Time from Ignition: 5° | 10° | 15° (TDC) | 20°
(Enhanced Arbitrarily)

Visible | 15° (TDC) | 20° | 30° | 40°
(Enhanced Doubly)

(a) Flame Behaviors in UV and Visible Wavelength Ranges in Different Injection Timing Conditions

Injection Timing: 40°BTDC
50°BTDC
70°BTDC

Mixture Strength at Spark Plug Measured by Discharge Light Spectra

Equivalence Ratio
2.0
1.5
1.0
40 50 60 70
Injection Timing °BTDC

UV | Visible | Visible
Time from Ignition: 20° | 20° | 40°
20° | 20° | 40°
20° | 20° | 40°

(b) Influence of Injection Start Timing on Flame Behavior

Fig. 11 Influence of Injection Start Timing on Flame Behaviors in UV and Visible Wavelength Ranges
(with Reverse Tumble, Ignition Timing: 15°BTDC, Other Conditions: See Fig. 4)

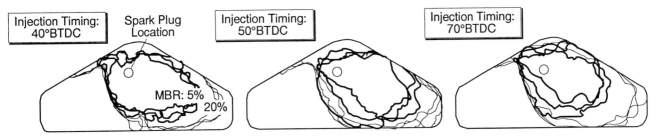

Fig. 12 Comparison between Flame Front Defined by UV Luminescence in Different Injection Timing Conditions (with Reverse Tumble, Ignition Timing: 15°BTDC, Other Conditions: See Fig. 4, Cycle-by-Cycle Variance of Three Cycles)

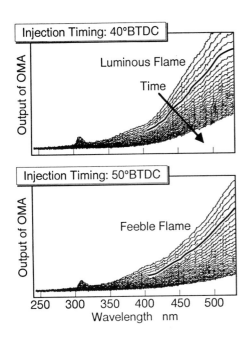

Fig. 13 Influence of Injection Start Timing on Flame Spectra of Late Combustion (with Reverse Tumble, Ignition Timing: 15°BTDC, Other Conditions: See Fig. 4)

The intensity of the thermal radiation is determined by the soot concentration and its temperature. The spectra of the thermal radiation from luminous and feeble flames are compared in **Figure 12**. In the case of the luminous flame observed with the injection at 40 °BTDC, the inflection point of the spectra seems to locate at the shorter wave length than that of the feeble flame observed with the injection at 50 °BTDC, suggesting that the radiation temperature of the soot in the richer mixture has higher temperature. This shows that the heat generation in the local area close to the soot is active in the case of the richer mixture. This will be result of the promoted burning up of the soot.

Comparing the flame behaviors in the UV and the visible ranges, it can be observed that the UV flame propagation with chemiluminescence precedes the appearance of the thermal flame. In the UV flame propagation, because the combustion zone is under rich condition and the flame propagation speed is very high, a large amount of fuel is left behind the UV flame front. After the UV flame propagation, the thermal flame appears in rich mixture zone and propagates to the area behind the UV flame front. The combustion of the rich mixture proceeds in the reaction zone behind the thermal flame front. The burning up of the soot generated in the rich mixture is completed principally by the air entrainment into the combustion zone promoted by the reverse tumble and the squish flow near the ceiling of the pentroof combustion chamber. The thermal flame propagates to the mixture zone in a conventional premixed engine. It propagates to the air zone in a diesel engine. In the case of direct injection gasoline engine, however, the ther-

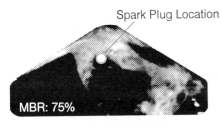

(a) with Reverse Tumble

(b) wthout Reverse Tumble

Fig. 14 Influence of Reverse Tumble on Thermal Radiation Distribution (Engine Speed: 500min^{-1}, Injection Start Timing: 30°BTDC, Ignition Timing: 10°BTDC, Other Conditions: See Fig. 4)

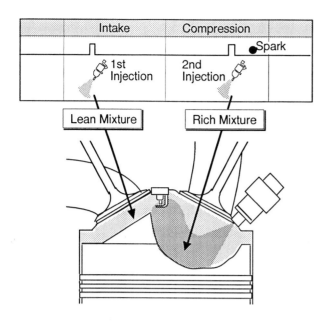

Fig. 15 A Mixing Control Strategy of
Two-Stage Mixing

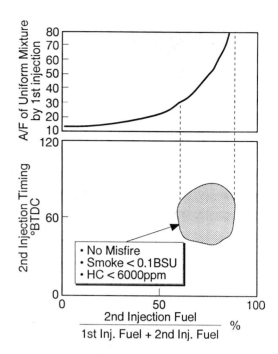

Fig. 16 Effective Conditions of Two-Stage Mixing
(Engine Speed: 600min⁻¹, WOT,
1st Injection Start Timing: 280°BTDC,
Ignition Timing: 20°BTDC, Total A/F: 12)

mal flame propagates to the zone filled with the combustion product containing radicals and CO generated by the first UV flame. This peculiar combustion phenomena of gasoline direct injection engine will promote the burning up of the soot generated in the rich mixture zone.

The luminous flame at the engine speed of 500 min⁻¹ is shown in **Figure 14**. At this engine speed, the difference between mixture formations before the combustion in the cases with and without reverse tumble can be minimized. Therefore, the direct influence of the reverse tumble on the combustion can be extracted. By adopting the reverse tumble, the luminous flame becomes feeble. This result means that the formation of soot is suppressed or the its burning is activated by the enhanced air entrainment into the combustion zone.

A MIXING CONTROL STRATEGY FOR KNOCK SUPPRESSION, "TWO-STAGE MIXING"

The authors had reported that the improvement of the full load performance caused by the charge air cooling effects and the effect of the transient knock suppression is one of the inherent characteristics of a gasoline direct injection engine [1-5]. For the further full load performance improvement, a new knock suppression method named "two-stage mixing" was invented.

Generally speaking, the most distinctive feature of the direct injection engines is the freedom of the mixture preparation. Two-stage mixing utilizes the this freedom for the knock suppression. **Figure 15** illustrates the procedure. Fuel is injected twice, a first injection is performed during the early stage of the intake stroke to prepare the premixed lean mixture and a second injection is performed during the later stage of the compression stroke to prepare the stratified mixture. In **Figure 16**, effective zone of two-stage mixing is shown with respect to second injection timing and to ratio of second injection fuel quantity to total fuel. In the figure, the effective zone is defined in which no misfire takes place, smoke density is less than 0.1 BSU and the HC emission is less than 5000 ppm. As shown in the figure, this method is effective when the major portion of the fuel is prepared by the second injection. Air to fuel ratio of the premixed mixture prepared by the first injection is from 30 to 80, which is beyond the flammability limit in usual engine condition. **Figure 17** shows the engine outputs at the trace knock conditions. Significant knock suppression effect is realized by two-stage mixing. This phenomena can be explained as follows;

(1) premixed mixture prepared by the first injection is too lean to cause the knock,

(2) stratified mixture prepared by the second injection may form the stoichiometric mixture in some location, how-

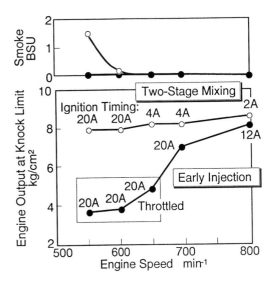

Fig. 17 Effect of Two-Stage Mixing on
Knock Suppression
(Trace Knock Condition)

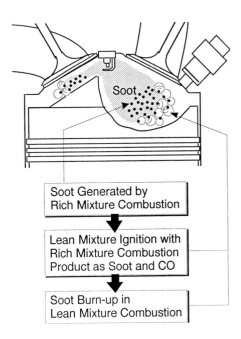

Fig. 18 Mechanism of Soot Burn-up Promotion

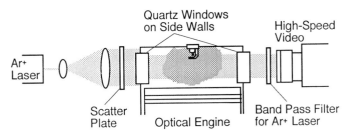

(a) Ar Ion Laser Light Extinction Method

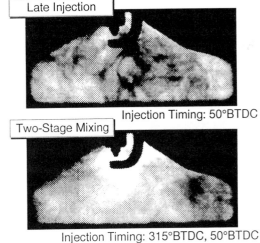

(b) Promotion of Soot Burn-up by Two-Stage Mixing
(Engine Speed: 1000min⁻¹,
Middle Load, A/F: 14,
Result in Late Stage of Expansion Stroke)

Fig. 19 Soot Visualization by
Laser Light Extinction Method

ever, the life time of that mixture before the ignition is not long enough for the precursor reaction of knock to proceed.

In the case of late injection, a large amount of soot is emitted when the average mixture strength becomes rich. In the case of two-stage mixing, however, soot emission can not be observed except for the very low engine speed condition even when the average air fuel ratio is 12. The process of soot burn-up is explained schematically in **Figure 18**:

(1) When the spark ignition takes place, only the stratified charge is ignited because the surrounding premixed mixture is too lean to be ignited.

(2) Combustion products generated in the rich stratified mixture containing CO and soot is propagated to the lean mixture.

(3) Soot plays the role of ignition site and ignite the very lean mixture beyond the flammability limit. In this process CO may assist the ignition.

(4) Soot is burned up in the combustion zone of the lean mixture.

Soot observed by the laser light extinction method is shown in **Figure 19**. In case of the late injection, unburned soot is observed even at the later stage of the expansion stroke when the average mixture strength is rich. When the two-stage mixing method is adopted, however, the generated soot is burned up during the early and the middle stage of the expansion stroke.

CONCLUSION

In order to clarify the characteristics of mixing and combustion process of GDI engine in the late injection mode, the fuel spray visualization, the mixture strength measurement, the flame spectral analysis and the flame behavior observation were performed.

Combining the results of these diagnostics, the following was clarified:

(1) The kinetic energy carried by the fuel spray is the major factor controlling the mixing in low engine speeds, the contribution of the reverse tumble to mixing becomes predominant at higher engine speeds. This mechanism supports the optimized mixing in the wide range of engine operating conditions.

(2) By changing the injection timing, local mixture strength can be controlled. The reverse tumble plays the role to expand the range of available injection timing. The wide range of available injection timing extend the freedom of mixture preparation.

(3) Gasoline with high octane number can be burned only by flame propagation. The fuel dispersed outside the piston cavity escapes from the flame propagation, resulting in the high level of HC emission. The reverse tumble suppresses the fuel dispersion to the outside of the combustion chamber, resulting in the lower HC emission in wide range of injection timing.

(4) In the combustion process under distinctively stratified charge, the flame propagation with UV chemiluminescence propagates first. Then the thermal flame propagates to the zone filled with the combustion product of the first UV flame containing radicals and CO. This peculiar combustion phenomena of gasoline direct injection engine will promote the burning up of the soot generated in the rich mixture.

Basing on these findings, a novel knocking suppression method named the "two-stage mixing" was proposed:

(5) Fuel is injected twice. A first injection during the early stage of the intake stroke prepares very lean uniform premixed mixture, and a second injection during the late stage of the compression stroke prepares the distinctively stratified mixture. Average air-fuel ratio is set to be slightly rich.

(6) Because the uniform mixture is too lean to cause the knock, and the stratified mixture does not have enough time for the precursor reactions, effective of knock suppression can be realized. Therefore, full load performance in low engine speed conditions is improved significantly.

(7) Soot generated in the rich mixture propagates towards the lean premixed mixture, then play the role of the ignition site and ignite the very lean mixture. Simultaneously, the burn-up of soot in the lean mixture zone is promoted. As a result, soot-free combustion is realized even in the rich conditions in which no excess air remains in the post flame zone.

REFERENCES

[1]   Kume, T., Iwamoto, Y., Iida, K., Murakami, N., Akishino, K. and Ando, H. : Combustion Control Technologies for Direct Injection SI Engines, SAE Paper 960600 (1996)

[2]   Kiyota, Y., Akishino, K. and Ando, H. : Combustion Control Technologies for Direct Injection SI Engines, FISITA 96 (1996)

[3]   Iwamoto, Y., Noma, K., Nakayama, O., Yamauchi, T. and Ando, H. : Development of Gasoline Direct Injection Engine, SAE Paper 970541 (1997)

[4]   Ando, H. : Combustion Control Technologies for Gasoline Engines, IMechE, International Seminar on Lean Burn Engines

[5]   Ando, H., Noma, K., Iida, K., Nakayama, O. and Yamauchi, T. : Mitsubishi GDI Engine - Strategies to meet the European requirements, AVL Engine and Environment Conference (1997)

[6]   Kuwahara, K., Watanabe, T., Shudo, T. and Ando, H. : A Study of Combustion Characteristics in a Direct Injection Gasoline Engine by High-Speed Spectroscopic Measurement (in Japanese), 13th Internal Combustion Engine Symposium, Paper No. 25 (1996)

Characteristics of Mixture Formation in a Direct Injection SI Engine with Optimized In-cylinder Swirl Air Motion

Akihiko Kakuhou, Tomonori Urushihara, Teruyuki Itoh, and Yasuo Takagi

Nissan Motor Co., Ltd.

ABSTRACT

This paper presents a study of mixture formation in the combustion chamber of a direct-injection SI engine. In-cylinder flow measurement was conducted using laser Doppler velocimetry (LDV) and particle image velocimetry (PIV), and visualization of fuel vapor behavior was done using laser-induced fluorescence (LIF). Further, fast response flame ionization detector (FID) was used to measure the hydrocarbon (HC) concentrations in the vicinity of the spark plug. Thereby mixture concentrations in the vicinity of the spark plug, within the mixture distribution observed using LIF, were quantified. Results revealed that an upward flow forms near the center of the cylinder in the latter half of the compression stroke and goes from the piston crown toward the cylinder head. This upward flow is caused by the synergistic effect of the swirl motion generated in the cylinder and the cylindrical bowl provided in the piston crown eccentrically to the central axis of the cylinder. It was confirmed that the fuel was initially trapped and vaporized in the piston bowl, then transported as a vapor cloud by the upward flow to the spark plug, thereby forming charge stratification. Performance experiments were conducted to study the ranges of fuel injection and ignition timing necessary for achieving stable stratified charge combustion. Comparisons of the data obtained with LIF visualization results and FID measurements revealed the conditions most desirable for achieving stable stratified charge combustion.

INTRODUCTION

Research efforts on direct injection SI engines have long been made. These are engines designed to provide better fuel efficiency by injecting gasoline directly into the cylinder to achieve mixture stratification. Practical application has been made of mixture stratification methods, proposed in recent times, in which the fuel is vaporized by initially spraying it onto the surface of the piston. The flow of gas within the cylinder is then used to carry an optimum fuel mixture to the vicinity of the spark plug[1] [2]. However, these mixture stratification methods required that modifies be made to the intake system used in conventional engines in order to create gas flow conditions suitable to transporting the mixture. In the previous report[3], we the authors presented a concept of direct-injection gasoline engine capable of forming charge stratification with minimal modifications made to the intake system of existing engines.

In this research, a laser Doppler velocimetry (LDV) and a particle image velocimetry (PIV) were used to measure the in-cylinder gas flow produced with the charge stratification system of that engine, and a laser induced fluorescence (LIF) was employed to visualize the fuel vapor behavior in the combustion chamber. The aim in using these techniques is to clarify the role of gas flow in transporting the mixture in the combustion chamber. By measuring mixture concentrations around the spark plug using a fast response flame ionization detector (FID) and conducting performance experiments, the fuel injection and ignition timing ranges required to attain stable combustion were established and the conditions required to produce good mixture formation in the combustion chamber were clarified.

Table1 Specification of Experimental Engine

Engine Type	4-Stroke, 4 valves/cylinder, S.I. Engine
Combustion Chamber	Pentroof Type
Number of Cylinders	1
Bore × Stroke	85 mm × 86 mm
Displacement	488 cm^3
Compression Ratio	11.0 : 1
Intake Valve Open - Close	1 deg. BTDC - 59 deg. ABDC
Exhaust Valve Open - Close	53 deg. BBDC - 7 deg. ATDC

EXPERIMENTAL ENGINE AND THE MEASURING METHODS USED

EXPERIMENTAL ENGINE

Table 1 lists the main specifications of the experimental engine used for this research. The engine has a four-valve cylinder head with pentroof type combustion chamber. As shown in Figure 1 and Figure 2,

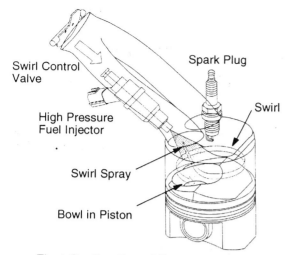

Fig.1 Configration of Experimental Engine

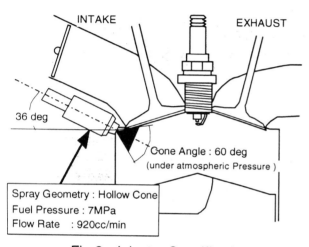

Spray Geometry : Hollow Cone
Fuel Pressure : 7MPa
Flow Rate : 920cc/min

Fig.2 Injector Specification

the fuel injector located at the lower part of the intake port is aimed toward the cylinder central axis. The injector is a swirl type producing a hollow cone spray of a cone angle of approximately 60 deg. The spray pattern and cone angle were measured by tomography based on the use of a laser light sheet. The fuel injection pressure is set at 7 MPa. Also as shown in Figure 1, a nearly cylindrical bowl is provided in the piston crown. A swirl control valve was installed at the inlet of the intake port. Three types of valves were used to generate different swirl flows of strength and thereby investigate the effect of the air motion on the mixture transportation.

METHODS OF MEASURING GAS FLOW AND MIXTURE DISTRIBUTION

As shown in Figure 3, an optical access engine, with quartz observation windows to see inside the combustion chamber both from the front and the bottom sides, was used for measuring in-cylinder gas flow and visualizing fuel behavior in this research.

In-cylinder gas flow measurement by LDV

A backscattering LDV with a rotary diffraction

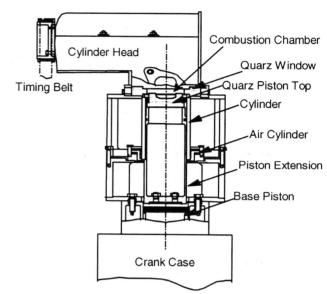

Fig.3 Optical Access Engine

gratingwas used to measure in-cylinder gas flow under motoring operation. A detailed descriptions of this LDV system was given in reference (4). The states of the in-cylinder gas flow both inside and outside the piston bowl were determined on the basis of average flow velocity patterns obtained by calculating a moving average for 10 crank angle (CA) degrees.

Measuring combustion chamber flow field by PIV

A double pulse Nd:YAG laser (532 nm) with a pulse output of 200 mJ was used as the light source. A digital CCD camera with resolution of 1000 x 1000 and a self-correlation method were used to calculate the two-dimensional flow field.

Visualization of in-cylinder mixture formation by LIF

A Kr-F excimer laser with a wavelength of 248 nm was used as the excitation light source. The fuel used was isooctane containing 0.2% dimethylaniline (DMA) as a fluorescent tracer. The fluorescence emitted by the DMA was passed through a 280 nm to 400 nm band

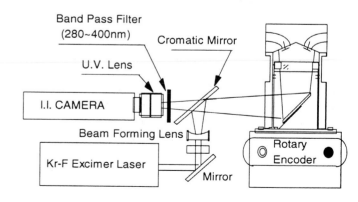

Fig.4 Coaxial LIF System for Fuel Liquid Visualization

pass filter and photographed by a high-speed shutter camera with an image-intensifier[6]. As shown in Figure 4, a coaxial LIF system was used to photograph the fluorescence in the incident direction of the laser light and thereby eliminate a dead angle within the observation range. As shown in Table 2, objects that can be visualized differ depending on the relationship between the incident direction of the laser beam and the direction of observation. For this reason, photography methods to use were chosen according to the object type to be observed.

Table2   Applicable Objects for Each LIF Method

Camera Direction	Relation between Laser Light and Camera	Fuel Spray (Liquid)	Fuel Film (Liquid)	Fuel Vapor
Front View	Coaxial	◯		◯
Bottom View	Coaxial	◯	◯	◯
	Right Angled	◯		◯

Measuring mixture density around spark plug by fast response flame ionization detector (FID)

A fast response FID system (Cambustion, HFR400) was used to measure the concentrations of mixture in the vicinity of the spark plug with a time resolution of a few degrees of crank angle. This system is illustrated in Figure 5. A special type of spark plug with a sampling tube, extended to 450 mm to minimize the sample gas transport delay, was used. Constant pressure chamber pressure was set at 47.5kPa. To reduce the effect of pressure fluctuation, the CP chamber volume was increased to 0.4 liter. In order to determine the air fuel ratio from the hydrocarbon concentrations measured, the hydrocarbon concentrations in the vicinity of the spark plug for port-injected premixed operations were also measured, as a means of calibration, under the identical volumetric charge condition. Calibration curve are shown in Figure 6.

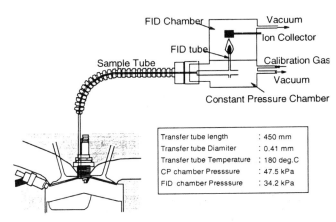

Fig.5 Fast Response FID System

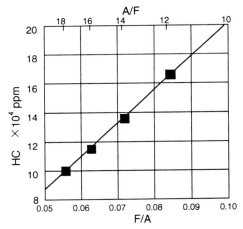

Fig.6   Calibration for HC Concentrations vs Air-Fuel Ratio

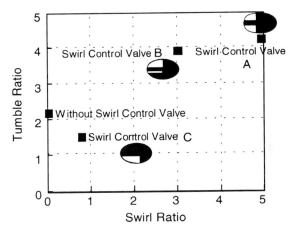

Fig.7   Steady-State Flow Characteristics of Tested Intake Systems

Table3   Conditions for Flow Measurement, Fuel Visualizaiton and Performance Experiment

	LDV and PIV Measurement (motoring)	LIF Fuel Vapor Visualization	Performance Experiment and FID Measurement
Engine Speed	1400 rpm	1400 rpm	1400 rpm
Volumetric Eff.	WOT	75 %	—
Pmi	—	—	314 kPa
A/F	—	40 : 1	40 : 1
Ignition Timing	—	TDC	—

Experimental Results and Discussion

Three types of swirl control valves with the same opening area were used for the experiments. They were designated as swirl control valves A, B and C, in the order of the largest to the smallest swirl ratio. Experiments were also conducted without a swirl control valve. Figure 7 presents the steady-state flow characteristics of the flow field produced by these four

353

types of intake systems.

RELATION BETWEEN STEADY-STATE FLOW CHARACTERISTICS AND MEAN GAS FLOW FIELD IN CYLINDER

Gas flow in the combustion chamber was measured with the above-mentioned LDV system under motoring operation for the four types of intake systems used in the steady-state flow tests. Figure 8 shows the points where the LDV measurements were made. In order to measure swirl flow in the piston bowl, velocity measurements were made in two directions in a horizontal plane at the 13 points indicated in the bowl. Moreover, to measure swirl flow outside the piston bowl, flow velocity measurements were made in the swirl direction and in the cylinder axial direction at the 35

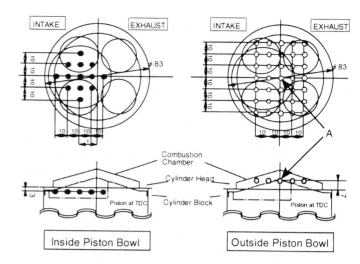

Fig.8 Measurement Points for LDV

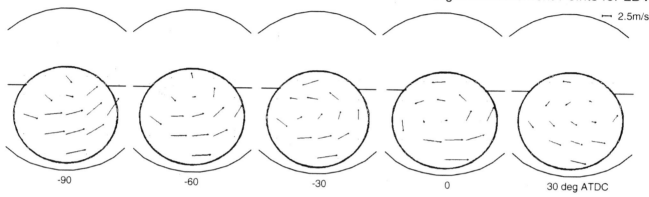

Fig.9   Mean Air Flow Field in Piston Bowl with Swirl Control Valve A

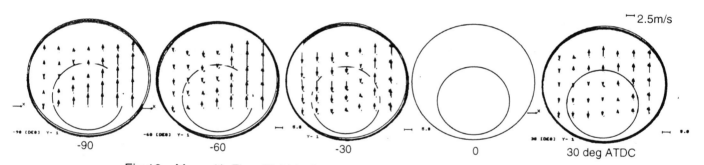

Fig.10   Mean Air Flow Field in Combustion Chamber with Swirl Control Valve A

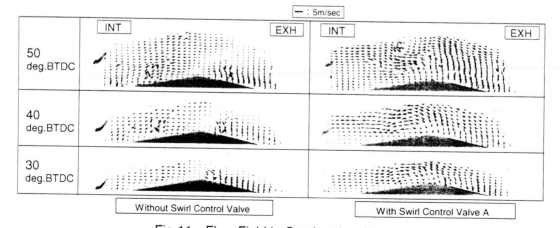

Fig.11   Flow Field in Combustion Chamber

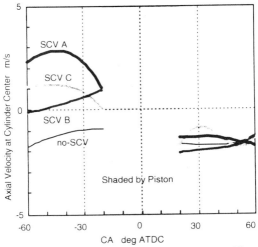

Fig.12 Time History of Axial Velocity at The Center of Combustion Chamber

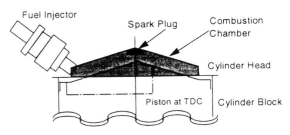

Fig.13 Fuel VIsualization Area (Front View)

points shown in the figure.

Figure 9 shows typical mean flow velocity patterns in the piston bowl when swirl control valve A was used, and Figure 10 shows the swirl flow patterns in the combustion chamber. It is seen that the swirl center in the piston bowl nearly coincided with the center of the bowl and diverged from the center of the cylinder. On the other hand, it is thought that swirl flow in the combustion chamber revolved almost around the cylinder central axis. Since the swirl center in the combustion chamber should be an area of lower pressure in relation to the peripheral portion of the swirl in the piston bowl, it is presumed that upward flow occurred from the periphery of the piston bowl toward the center of the combustion chamber. Figure 11 shows measured results, using PIV, for the cylinder flow field on the cylinder vertical plane including the spark plug, both when swirl control valve A was used and when a swirl control valve was not used. An upward flow from the piston bowl is seen on the figure when a swirl control valve A was used. Figure 12 shows the axial velocity measured by LDV at the point on the cylinder central axis, shown in Figure 8, for swirl control valves A, B and C and the case without a swirl control valve. It should be noted that the laser beam was shaded by the piston from a crank angle of 20° BTDC to 20° ATDC, preventing any measurement. It is seen that the swirl control valves with a large swirl ratio produced a

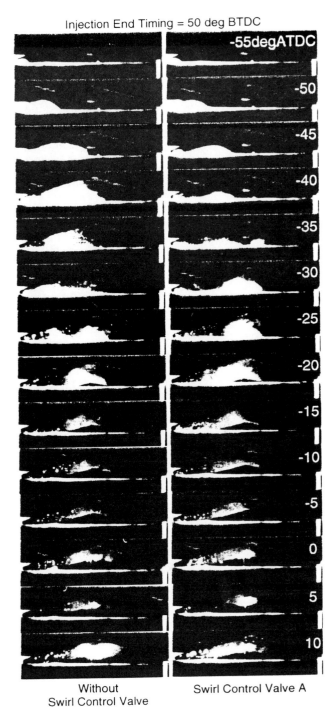

Injection End Timing = 50 deg BTDC

Without Swirl Control Valve

Swirl Control Valve A

Fig.14 Fuel Behavior in Combustion Chamber

strong upward flow from a crank angle of around 30° BTDC of the compression stroke. This confirms that the swirl flow works to generate an upward flow as mentioned earlier.

EFFECT OF SWIRL MOTION ON FUEL BEHAVIOR IN CYLINDER

Visualization of in-cylinder fuel behavior at the time of injection during the compression stroke was performed using LIF under two conditions: one using swirl control

355

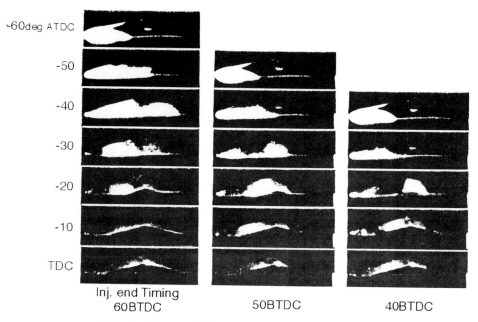

Fig.16 Effect of Fuel Injection Timing on Fuel Behavior
in Combustion Chamber

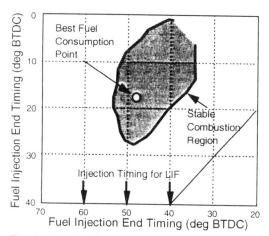

Fig.15 Stable Combustion Region in Injection
Timing - Ignition Timing Map

valve A, the valve with the highest swirl ratio, and one not using a swirl control valve. The results of coaxial LIF visualization viewed from engine front are shown in Figure 14. Figure 13 shows the area of the combustion chamber in which the fuel behavior was visualized.

The results indicate that there is less fuel spray in the 40° BTDC range when swirl control valve A is used as compared with when no swirl control valve is used. It is considered that generating a swirl motion in the cylinder suppresses tumble flow in the piston bowl during the latter half of the compression stroke and prevents the upward splashing of the fuel spray immediately after injection. This prevents droplets of liquid fuel from wetting the spark plug and reduces the volume of fuel that spills over from the piston bowl. Then, the rising of vaporized fuel from the piston bowl toward the spark plug in the 30°

BTDC to 20° BTDC range is observed, indicating that stratification of the mixture has been achieved. It is considered that the fuel, once it had entered the piston bowl, was carried by a swirl-induced upward flow, which was seen in the PIV measured results, in the central area of the cylinder.

ENGINE PERFORMANCE IN RELATION TO THE RESULTS OF MIXTURE VISUALIZATION

Results of Performance Experiments

Figure 15 shows the results of performance experiments conducted using swirl control valve A to investigate the ranges of fuel injection timing and ignition timing necessary for achieving stable stratified charge combustion. Fuel injection end timings of 50° BTDC and 40° BTDC provided a region of stable combustion, while the 60° BTDC timing did not produce any such region.

Effect of Fuel Injection Timing on Fuel behavior in Combustion Chamber

Figure 16 and Figure 17 show LIF visualization results for three fuel injection timings. Figure 16 shows a coaxial LIF image photographed from engine front, while Figure 17 shows results of mixture distribution observed on the horizontal cross-section including the spark plug. The horizontal cross-section was photographed by splitting the laser beam into two sheets of a 5 mm thickness to direct light into the combustion chamber from the opposing direction.

With a fuel injection end timing of 60° BTDC, the fuel was injected before the piston had risen to the vicinity of the fuel injector and the fuel spray was not trapped in the piston bowl and traveled as far as the exhaust side of the

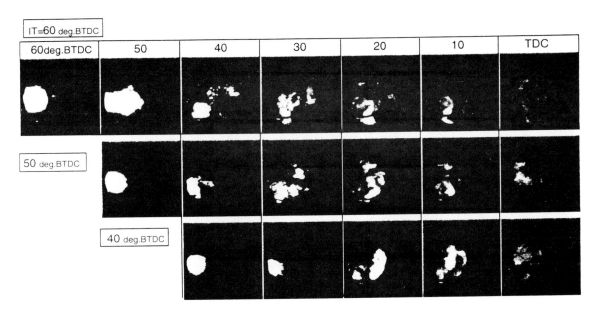

Fig.17 Fuel Behavior in Spark Plug Plane

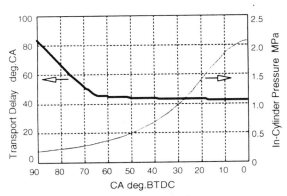

Fig.18 Effect of In-Cylinder Pressure on FID Transport Delay Time

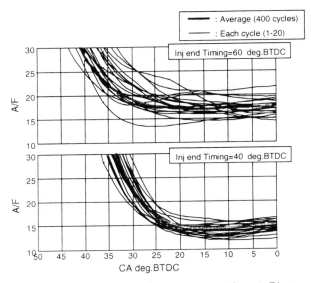

Fig.19 Fuel Concentrations near the Spark Plug

combustion chamber. As a result, the fuel vapor was not concentrated near the spark plug around a crank angle of 20° BTDC when ignition occurred. With fuel injection end timings of 50° BTDC and 40° BTDC, however, the fuel was shot into the piston bowl where it swirled, then rose in a vapor that provided effective mixture stratification in the vicinity of the spark plug. The visualized results showed good agreement with the performance experiment data.

Using Fast-response FID to Measure Mixture Concentrations Around Spark Plug

Fluorescent intensity is affected by quenching and changes in temperature. This makes it difficult to quantify mixture concentrations using LIF. The limitations of laser frequency repetition make it difficult to study the changes in concentrations that occur within a single cycle. For these reasons, the fast response FID was used to measure mixture concentrations in the vicinity of the spark plug. Figure 19 shows the measured results for two fuel injection end timings used with swirl control valve A. The figure shows the average values for 400 cycles and

the results for 20 cycle sections selected at random. The transport delay for the sample gas was calculated based on the length of the sampling tube, in-cylinder pressure and the pressure in the FID constant pressure chamber by using the software supplied with the Cambustion HFR400 [7] [8]. For this experimental condition, the transport delay was estimated at about 42° near TDC. As shown in Figure 18, the calculated result at the latter half of compression stroke was roughly constant.

With a fuel injection end timing of 40° BTDC, the average air fuel ratio in the spark plug vicinity reached the highest mixture density in the 20° to 10° BTDC range with a value of approximately 14. With an end timing of 60° BTDC, the air fuel ratio reached its highest density of approximately 17. These results indicate that the earlier the fuel injection end timing is, the poorer the mixture

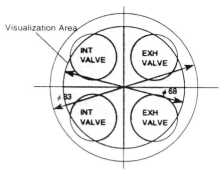

Fig.20 Visualization Area (Bottom View)

stratification will be. The results here correspond closely to the mixture visualization results by LIF in Figure 16 and 17. A fuel injection end timing of 60° BTDC produced greater cycle fluctuation of mixture concentration when compared with the 40° BTDC timing. Factors that determine the stable combustion range are not only mixture concentration. Mixture cycle fluctuation is also a contributing factor.

Visualizing Liquid Fuel Film in Piston Bowl

Next, visualization of the liquid fuel film in the piston bowl was performed using coaxial LIF with the view from the lower part of the combustion chamber. Figure 20 shows the area visualized, while Figure 21 shows the visualization results for the fuel spray and liquid fuel film. The difference of photographing condition between Figure 12 and Figure 18 is camera f number. By using large f number in Figure 18, scattered light from combustion chamber wall and fluorescence light from fuel vapor were suppressed. The large cloud shape seen in a crank angle interval of around 10° after the end of fuel injection is the fuel spray. Subsequently, the white trace remaining at the bottom of the piston bowl is the fuel film. With a fuel injection end timing of 60° BTDC, it is observed that the fuel film disappeared before the piston reached its TDC of the c P ompression stroke. However, with a fuel injection end timing of 40° BTDC, the fuel film was still present at TDC of the compression stroke. It is thought that since this fuel film does not contribute to combustion, an earlier fuel injection end timing is desirable with respect to improving combustion efficiency. It should be noted that these visualization experiments were conducted using dimethylaniline (DMA) (boiling point of 193° C) as the fluorescent tracer and that the piston surface temperature of the optical access engine is presumably lower than that of the piston surface in an actual engine. Therefore, it should be understood that the visualized results do not necessarily reproduce the exact fuel film state in an actual engine.

CONCLUSIONS

Experiments were done measuring gas flow and mixture distribution within the combustion chamber of a direct-injection, stratified-charge SI engine, of which

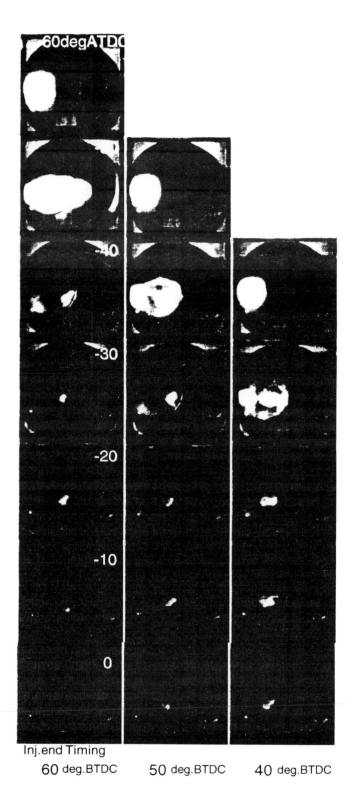

Inj.end Timing

60 deg.BTDC   50 deg.BTDC   40 deg.BTDC

Fig.21 Effect of Fuel Injection Timing
on Fuel Film in Piston Bowl

intake system was virtually identical to that of conventional gasoline engines with four valves per cylinder. The results are summarized as follows:

1. The combined effect of the swirl motion in the cylinder and the cylindrical piston bowl positioned eccentrically to the cylinder central axis, forms an upward flow that rises from the piston crown near the cylinder center area to the cylinder head in the latter half of the compression stroke.

2. The fuel directly injected into the cylinder initially enters the piston bowl. Subsequently the fuel is transported to the vicinity of the spark plug by the upward flow described above, forming mixture stratification.

Performance tests were conducted to determine the range of fuel injection timing and ignition timing that would allow stable stratified charge combustion. Comparisons of the data obtained with LIF visualization results and FID measurements showed good agreement and revealed the following points.

3. The ideal injection timings are determined from the balance between fuel trapping by the piston bowl and fuel film formation on the piston crown. The tendencies of these factors on fuel injection timing are described below.

    (1) Too early fuel injection timing: The fuel spray is not
    trapped in the piston bowl and travels as far as the exhaust side of the combustion chamber, which precludes mixture stratification and causes a large fluctuation in the mixture concentrations in the vicinity of the spark plug, thus resulting in unstable combustion.

    (2) Too late fuel injection timing: There are higher mixture concentrations in the vicinity of the spark plug. However, there is an increase in the liquid fuel film on the bottom of the piston bowl, resulting in poor combustion efficiency.

Selecting the best fuel injection timing is the key when attempting to improve fuel efficiency while achieving stable combustion using mixture stratification.

REFERENCE

1. Y. Iwamoto, K. Noma, T. Yamaguchi, H. Ando, "Development of Gasoline Direct Injection Engine", SAE Paper 970541

2. J. Harada, T. Tomita, et al,. "Development of Direct Injection Gasoline Engine", SAE Paper 970540

3. T. Itoh, A. Iiyama, S. Muranaka, Y. Takagi, "Combustion characteristics of a direct-injection stratified charge S.I engine", Japan Society of Automotive Engineers (JSAE) Trans. Vol. 19, No. 3, July 1998

4. T. Urushihara, T. Murayama, et al., "Turbulence and Cycle-by-Cycle Variation of Mean Velocity Generated by Swirl and Tumble Flow and Their Effects on Combustion" ,SAE Paper 950813

5. K. Li, T. Urushihara, et al., 10th Joint Symposium on Combustion Engines, Proceedings, p. 85, 1992

6. T. Urushihara, T. Nakada, et al., Japan Society of Automotive Engineers (JSAE), Trans. Vol. 27, No. 2, April 1996

7. Cambustion Ltd., HFR400 Fast FID User's Manual

8. T. Summers, N. Collings, "Modelling the Transit Time of a Fast Response Flame Ionization Detector During In-Cylinder Sampling ", SAE Paper 950160

1999-01-3536

Combined Catalytic Hot Wires Probe and Fuel-Air-Ratio-Laser Induced-Exciplex Fluorescence Air/Fuel Ratio Measurements at the Spark Location Prior to Ignition in a Stratified GDI Engine

B. Deschamps and V. Ricordeau
IFP, Rueil-Malmaison, France

E. Depussay and C. Mounaïm-Rousselle
L.M.E., University of Orleans

ABSTRACT

Combined Catalytic Hot Wires Probe (CHWP) and Fuel Air Ratio Laser Induced Exciplex Fluorescence (FAR-LIEF) techniques have been applied to a Gasoline Direct Injection Engine to characterize temporal and spatial evolution of the fuel/air ratio in the vicinity of the spark plug. The engine ran below stoichiometric with early injection (homogenous case taken as a reference) or late injection timing with a global equivalence ratio as low as 0.3. Under lean and late injection conditions, the temporal CHWP signal indicates that a rich vapor cloud is carried to the vicinity of the spark plug. CHWP and FARLIEF techniques show that the maximum equivalence ratio in the fuel cloud reaching the spark does not depend on the injection duration. Instead, the duration appeared to affect the size of this rich vapor pocket.

INTRODUCTION

Gasoline Direct Injection (GDI) engines promise significant advantages for improved fuel consumption, engine performance and pollution due to exhaust gases emission. They are designed to achieve stratification of the mixture so as to decrease the global equivalence ratio far below the lean operating limit. The idea is to carry a ignitable air/fuel pocket to the vicinity of the spark plug just prior to ignition by injecting the fuel very late in the compression stroke. Under the wide spacing concept, this is achieved by injecting the fuel directly on the surface of the piston bowl. This injection mode is very difficult to optimize due to the short time available for mixture preparation. Therefore, the knowledge of the local equivalence ratio in the vicinity of the spark plug is necessary in order to set the timing of the injection or/and the spark as well as possible.

Only few methods are available to estimate the local air/fuel ratio. Some are based on optical diagnostics. For example, Planar Laser Induced Fluorescence (PLIF), fully developed and used in port injection engines by Baritaud et al.[1] Neij et al.[2], and Reboux et al. [3] provides local spatial measurements by adding fluorescent tracers to iso-octane. With Infra-Red Absorption technique (Skippon et al.[4] and Kawamura et al.[5]) local temporal measurements can be obtained. But, in both techniques, the presence of liquid droplets appear to be a major problem for data analysis. Other methods use sampling probes as Flame Ionization Detector, used in SI engines by Collings [6], and the Catalytic Hot Wires Probe (Di Cocco [7], Depussay et al. [8]) : the main drawback of these techniques is the sample transit time. In the case of GDI engines, optical difficulties show up due to the complex piston shape and the enhanced tumble motion; as well, the high pressure levels made probe measurements difficult.

Recently, Mie Scattering, LIF and FID were used to evaluate the influence of the flow field on fuel vapor concentration near the spark plug in a GDI 4-valve SI Engine[9,10], and Planar Laser Induced Exciplex Fluorescence (PLIEF) imaging [11] was performed in a wide spacing optical GDI engine working under early as well as late injection conditions. This work also demonstrated that the injection/ignition timings influence the coincidence of the fuel vapor cloud with the spark. However, quantification of air-fuel ratio in the cloud using a planar laser technique was difficult at the spark timing because of the complex optical access for the laser sheet prevented it from reaching the spark location at this time.

The objective of the present work was to quantify the local equivalence ratio near the spark plug prior to ignition in the same engine. Combined Catalytic Hot Wires

Probe (CHWP) and Fuel Air Ratio Laser Induced Exciplex Fluorescence (FARLIEF) techniques have been applied to characterize temporal as well as the spatial evolution of the fuel/air ratio. The effect of the injection timing on the mixture formation has been investigated : for early injection during the suction stroke and for late injection during the compression stroke.

ENGINE DESCRIPTION

The optical GDI engine, is a research engine, developed by the Groupement Scientifique Moteurs, whose features (technical, geometrical and functional) are extremely close to those for future production model. A pent-roof, four-valve, single cylinder head is mounted on the top of the elongated crankcase. The plug is placed in the middle roof, while injection is between intake valves as shown schematically in Figure 1. The piston has a spherical section bowl, which enhances tumble and at the same time allows observation of the chamber interior via a mirror oriented at 45°. Two Suprasil windows allow lateral optical access. Geometric characteristics of the engine are listed inTable 1.

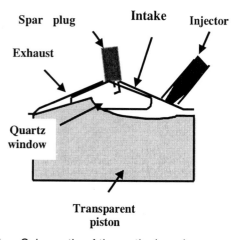

Figure 1.   Schematic of the optical engine.

Iso-octane, used as gasoline substitute, was injected at 80 bar through a swirl injector providing a 65° cone angle spray. Droplet sizing by Le Coz [12] and PLIF spray visualization [11] are reported in previous work. The behavior of this spray is very similar to these used by Ipp et al. [13]. The engine speed was fixed at 1200 rpm. The temperature of coolant and liquid oil (50°C) and intake air (30°C) are controlled before the experiments. The global equivalence ratio was controlled by modifying the injection duration while fuel distribution was varied with the injection timing.

Table 1.   Specifications of the optical GDI engine

Number of cylinders	1
Cycle	4-stroke
Number of intake valves	2
Number of exhaust valves	2
Bore	85 mm
Stroke	88 mm
Combustion chamber design	Pentroof
Displacement	500 cm3
Compression ratio	10.86

Seven stratified (late injection) engine working cases were studied with equivalence ratios varying from 0.3 to 0.6 and injection timing varying from 60 to 30 CA BTC. A homogeneous case (early injection, 32 CAD) with a global equivalence ratio of 0.9 was taken as a reference. Injection duration was adjusted for the desired global air/fuel ratio, as measured in the exhaust port with a Heated Exhaust Gases Oxygen (HEGO) sensor. Table 2 reports engine operating conditions.

Table 2.   Engine operating conditions

condition	Injection timingATD	Injection duration μs	φ
H	32	2345	0.9
S1	315	910	0.3
S2	315	1090	0.4
S3	315	1320	0.5
S4	315	1540	0.6
S2'	300	1090	0.4
S2"	320	1090	0.4

DIAGNOSTICS TOOLS

Two diagnostics have been used in this study: the Catalytic Hot Wires Probe and the Fuel Air Ratio Laser Induced Exciplex Fluorescence

CATALYTIC HOT WIRES PROBE – The feasibility of fuel concentration measurements by Catalytic Hot Wires Probe (CHWP) has been successfully tested in optical port injection S.I. engine [8, 14]. The probe consists of two platinum-iridium wires (diameter : 10 μm, length : 1 mm) heated at constant temperature via Wheatstone bridges. The first one is heated to high temperature (873 K) to favor catalytic reaction of hydrocarbon on its surface. This reaction induces an additional voltage to those usually induced by flow variations : the wire becomes sensitive to the fuel concentration in addition to the mixture flow. The second wire is maintained at lower temperature (573 K) to avoid catalytic reaction and is taken as a reference. Therefore the difference between the voltages of the two wires provides the local fuel concentration information.

The global device was fully described by Depussay [15] and Depussay et al.[8]. A Champion spark plug was modified, as it can be seen in Figure 2, by drilling a small hole through the thread and adding the capillary tube, connected to the electro-valve. As the spark plug hole in the cylinder head is very deep, this tube has 300 mm length. The electro-valve and capillary tube holder was not water-cooled to avoid iso-octane condensation.

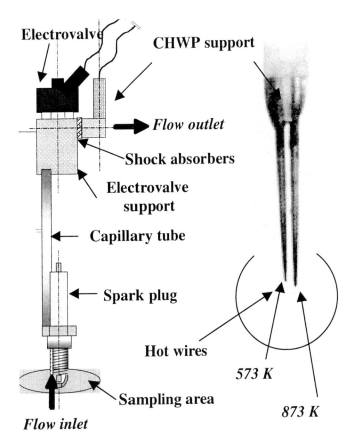

Figure 3. Transformed spark plug to provide sampling.

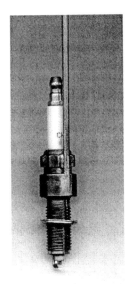

Figure 2. Transformed spark plug to provide sampling.

A schematic is shown on figure 3. The probe is located at the exit of the sampling flow in order to get constant pressure conditions (i.e. atmospheric pressure).

To limit the heat transfer between wires and also flow perturbations, the wires are offset : the gas sample first meets the "cold" wire.

An example of the response of both wires is presented in Figure 4. It can be seen that the shape of both responses is very similar until 345 CAD : in fact, from 332 CAD, the sampling begins because the pressure is sufficiently high to induce sonic regime and the sharp increase of the voltages is due to the flow velocity. If we observe the difference between both voltages, some peaks appear and certainly characterize acoustic waves or engine vibration. Indeed, both Wheatstone bridges are the same but their responses cannot be rigorously identical. Moreover, both wires are not placed at the same location in the flow. When the catalytic reaction occurs (hydrocarbons arrive in the vicinity of the probe) the voltage of the "hot" wire (maintained at 873 K) changes : the resulting difference is due to the presence of fuel.

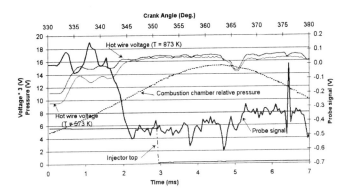

Figure 4. An example of wires responses for an air-iso-octane mixture of 0.85 equivalence ratio

As we have shown previously [8, 15], fuel concentration measurements by CHWP needs a calibration phase. The probe signal has to be tested as a function of different equivalence ratio mixtures. For this purpose a testing bench has been created : iso-octane was vaporized and injected in the airflow supplying a burner. The CHWP probe placed below the burner delivered a signal while a HEGO sensor was located in the exhaust part to measure the global equivalence ratio. An example of a calibration curve, obtained for two probes is shown in Figure 5.

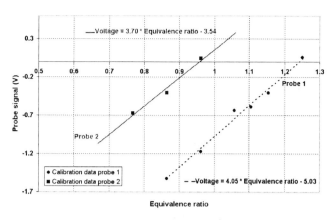

Figure 5. An example of calibration curves obtained with iso-octane.

One can point out that the voltages levels are not similar for both probes : in fact, the soldered joint between wires and pins is different for each probe and induces the voltage difference on the response. Nevertheless, the slope is conserved for each case.

As shown in Figure 6, the sampling hole was placed on a line parallel to the pentroof ridge. For this application, the electro-valve opens 30 CAD before TDC for 3 ms duration (i.e. ~ 22 CAD) in order to get concentration estimates near the ignition optimized time determined in a previous study : 20 CAD BTDC .

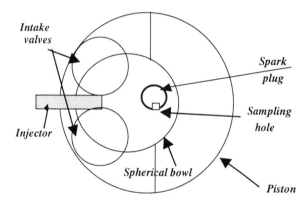

Figure 6. Sampling spatial position

A few tests were performed to evaluate the possibility of applying this technique to the GDI engine working in the late injection mode. Two types of difficulties were expected. On one hand, presence of fuel droplets can pollute the probe and on the other hand, the high pressure level encountered in these combustion chambers and soot formation can destroy the wires.

– Previous PLIEF visualizations [11] ensured the total absence of liquid phase at the spark plug prior to ignition. Effectively no peaks on the probe signals were observed.

– Under non-combustion conditions, we observed a uncontrolled electro-valve opening for sampling time duration larger than 3 ms because the cylinder pressure became too great.

– Due to our choice of combining this technique with laser beam measurements where the spark plug had to be removed (see following paragraph), the experiment was conducted without ignition so the probe would not be affected by burnt gases. Isooctane injection was activated every 5 cycles. During the next four cycles without injection, in-cylinder fuel was totally eliminated.

The sampling began at 330 CAD. Data were acquired at a sampling frequency of 20 kHz, allowing 60 data records during the 3ms sampling duration for each injection cycle.

Probe signal description – Instantaneous CHWP signals obtained (with Probe 2) for early injection case (H, global $\phi = 0.9$), late injection case (S2, global $\phi = 0.4$), no injection case, and pressure signals are plotted in Figure 7.

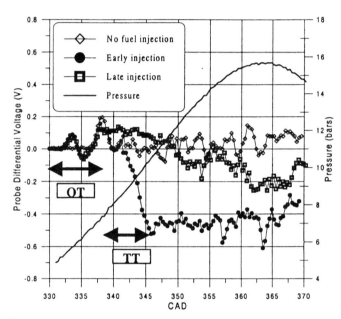

Figure 7. Instantaneous signals : no injection case, early injection case global $\phi = 0.9$ (H), and late injection case global $\phi = 0.4$ (S2)

The pressure record shows that pressure range is between 5 and 15 bars. No fuel injection and early injection temporal signals were needed to evaluate the time response of the CHWP system. This delay includes the time necessary to open the electro-valve, "OT" (about 1 ms); as well as the transit time, "TT", necessary for the extracted mixture to reach the 2 hot wires (1ms). Total delay is about 2 ms (15 CAD). Results presented below are corrected for this delay.

In the late injection case, the analysis of the signal is difficult due to its complex shape. As it can be seen in Figure 7, the voltage signal does not decrease until 5 CAD after the transit time, when it reaches a plateau before achieving the zero value again and decreases until an other plateau, 20 CAD after the transit time. With iso-octane as fuel, the response of the probe is close to zero in the case of no fuel presence (so no catalysis) as well as in a stoichiometric mixture. In the example of the figure, two hypothesis are possible :

1. until 335 CAD (corrected angle), no fuel arrives in the vincinity of the spark plug and fuel "pockets" appear later;
2. just after the transit time, the air-fuel mixture is near stoichiometric.

Previous studies performed at the same engine conditions with PLIEF [11] showed that a rich cloud with no other fuel nearby reached the spark plug at 340 CAD and confirmed preliminary combustion tests conducted to determine the best ignition timing. This result allows the good choice of the interpretation of the CHWP signal in the case of "late injection". This example shows that both techniques can allow a good analysis of the mixing process in GDI engines.

FUEL AIR RATIO LASER INDUCED EXCIPLEX FLUORESCENCE – Laser-Induced Exciplex Fluorescence was also used to quantify air/fuel ratio. In this work, isooctane which does not fluoresce when excited above 200 nm was seeded with benzene and triethylamine (TEA). It has been demonstrated [16,17,18] that the combination of those tracers leads to an exciplex formation in the liquid phase under 248 and 266 nm excitation. With 2.9% benzene and 4.1% TEA a complete spectral separation of TEA/Benzene (tracing the vapor) and exciplex (tracing the liquid) occurs, enabling isolation of vapor from liquid fluorescence by the use of an optical filter centered around 290nm. Characteristics of those tracers are reported inTable 3.

Table 3. Characteristics of tracers

| | BP. | Excitation | Fluorescence | |
			vapor	Liquid
Iso-octane	98°	266 nm	No	
Benzene	81°	266 nm	290	350 nm <exciplex
TEA	89°	266 nm	290	

Several works [19,18] also demonstrated the linear dependence of the vapor fluorescence signal of both tracers with the air/fuel ratio due to important quenching with oxygen molecules. Figure 8 exhibits an example of a linear dependency of the signal versus the equivalence ratio obtained with 2.9%Benzene/4.1%TEA tracing iso-octane in a previous work (266 nm excitation).

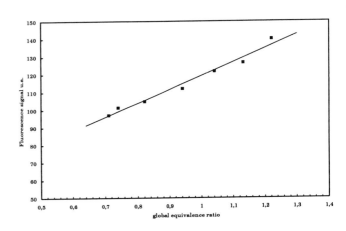

Figure 8. Vapor signal versus the equivalence ratio. 2.9% Benzene / 4.1% TEA tracing iso-octane (from Ortolan and Deschamps[18])

Optical set up – A quadrupled ND:YAG laser (PIV200 Spectra Physics) acting as a UV source of excitation delivered about 80 mJ per pulse. A short laser pulse (7ns), 4 mm diameter beam was directed into the chamber horizontally through a lateral window. It crossed the CHWP extraction location at the inter electrode space. To avoid beam reflection, the spark plug was removed in this step of the experiment. Figure 9 shows a schematic of the laser beam passing through the combustion chamber. The fluorescence signal along the beam was collected through the quartz piston via a 45° mirror through a combination of a 270-305 nm band pass filter and a 60 mm Cerco UV lens by an ICCD. The intensifier gate was set to 25 ns.

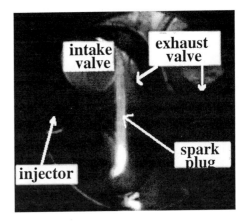
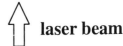

Figure 9. Laser beam passing through the combustion chamber (bottom view through the transparent piston bowl)

Some 266nm residual light passed through the band pass filter. We placed an additional 45° 266 nm dichroic mirror in front of the collection and applied a few nanoseconds delay to the intensifier gate in order to avoid the maximum laser pulse power. The flash lamp and Q-switch were both synchronized to the engine for the best temporal match. For each engine condition, 800 images were recorded.

FARLIEF Calibration under early injection conditions — Previous work [11] demonstrated a homogeneous mixture under early injection condition H. We decided to study the possibility of using this configuration as a calibration case.

For ten injection durations (from 910ms to 3280ms) corresponding to equivalence ratios varying from 0.3 to 1.2, 800 fluorescence images were acquired at 340 CAD (expected ignition time). The images were not corrected from optical distortions, spatial response of the camera or laser intensity profile. Only correction from laser pulse energy and from background images were applied. Individual images are shown on Figure 10. The signal is almost constant along the beam in each condition. One can notice the fluorescence intensity increasing. Nevertheless for conditions close to stoichiometry, the signal does not seem to increase anymore. This result is different from the expected behavior of FARLIEF signal shown on Figure 8.

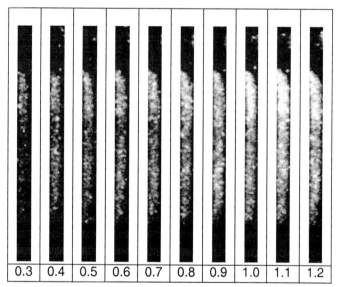

| 0.3 | 0.4 | 0.5 | 0.6 | 0.7 | 0.8 | 0.9 | 1.0 | 1.1 | 1.2 |

Figure 10. Instantaneous visualization of the fluorescence of the vapor along a beam passing horizontally through inter-electrodes location (for different global equivalence rations

The averaged fluorescence signals near the spark plug at 340 CAD are plotted in Figure 11, for the ten injection duration conditions. The values are obtained over 800 cycles and into a 4x9.5 mm rectangle surrounding the spark plug.

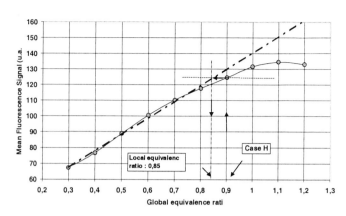

Figure 11. Evolution of the fluorescence signal versus equivalence ratio measured in the exhaust pipe. 1200 rpm, injection 32 CAD. (Equivalence ratio is varied by adjusting the injection duration)

This result confirms the linear dependence of FARLIEF signal to the global equivalence ratios up to 0.8. But above this value, the slope of the "calibration curve" decreases. The most simple explanation for this phenomenon is that the injection duration necessary to get high equivalence ratio becomes too great. An equivalence ratio of 0.8 corresponds to an injection duration of ~2 ms (~14,5 CAD) while an equivalence ratio of 1.2 corresponds to ~3.3 ms (~23,5 CAD). For large ϕ the injection duration may exceed the time limit where the spray is completely impacting the piston bowl. Then a part of the liquid spray deposits on the cold cylinder wall and is kept trapped during the compression stroke. This fuel amount does not take part to in vapor mixture formation which is visualized at the spark timing (340 CAD) but concurs with late combustion after the top dead center. Then the equivalence ratio at the spark is lower than the equivalence ratio measured with the lambda probe.

The slope of the linear part of the curve in Figure 11 was used to define the actual FARLIEF calibration curve. As an example, the local equivalence ratio of the case H ($\phi=0.9$) can be read on the curve equal to 0.84.

Local equivalence ratios for late injection conditions were determined on the basis of the above calibration method. By this means, we assumed that the temperature difference at the spark between late and early injection does not lead to a major difference in the benzene and TEA fluorescence.

EACH CYCLE AND 1/5 CYCLE INJECTION MODE COMPARISON – Since the engine is not fired, significant amounts of residual fuel remained in the cylinder. That is why iso-octane was injected every 5 cycles. Therefore additional tests were performed with both techniques to estimate the equivalence ratio when injecting every cycle, 1/5 and 1/10 cycles.

In Figure 12, the local equivalence ratio measured with the CHWP is plotted for each cycle. The averaged signal between 336 and 342 CAD exhibits an equivalence ratio, 8% larger than when injecting only 1/5 cycles. Moreover, fuel accumulation observed in the combustion chamber does not seem to be repeatable : with the fuel injection at each cycle, cyclic fluctuations increase. One can notice the 0.5% cyclic variation in the 1/5 injection case which will be discussed further

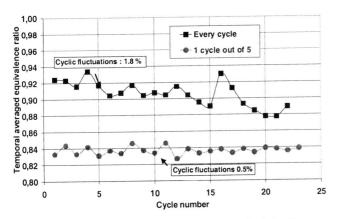

Figure 12. Comparison between each cycle injection and a injection one cycle/5 cycles.

With the FARLIEF technique, the same 8% variation was measured. One injection every 10 cycles resulted in the same equivalence ratio than as injecting every 5 cycles. So injecting every 5 cycles is enough to simulate the mixture process behavior with combustion.

DISCUSSION

Results on the effect of injection timing and duration on the equivalence ratio obtained at the spark with both techniques are discussed in this part.

EARLY INJECTION CASE (H) – Figure 13 presents a local equivalence ratio history averaged over 25 cycles obtained with CHWP. Around the spark timing (338-342 CAD) its value of 0.837 is really close to the FARLIEF measurement already presented in the diagnostic part. This is a first encouraging result for both techniques applied to GDI engine. The curve in Figure 13 displays minimal variation from 0.835 to 0.852 showing a really good homogeneity along the extracted volume, which is actually the mixture carried by the tumble and reacting with the spark. The relatively constant equivalence ratio obtained with FARLIEF along the beam (Figure 14, case H) displays the homogeneity of the charge in the direction perpendicular to the tumble motion. Also the 0.5% cyclic variation of the equivalence ratio of the mixture

reacting with the spark (measured with the CHWP, 4 CAD around the arrival time, Figure 12) shows the stability of the early injection case as well as the quality of the injection. One can estimate the air/fuel mixture is homogeneous. 10% cyclic fluctuations (including 4% shot to shot laser pulse energy variations) were obtained with the high temporal resolution of our FARLIEF diagnostic (several ns) in a 4x9.5 mm area surrounding the spark plug. This is much higher than fluctuations measured by the CHWP during 4 CAD which is more representative of the spark duration.

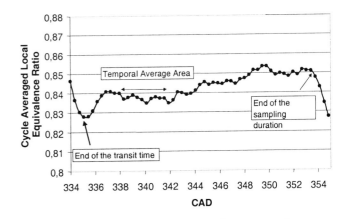

Figure 13. Temporal evolution of cycle averaged equivalence ratio for case H.

LATE INJECTION (CASES S) – Figure 14 shows 5 single shot FARLIEF vapor signal images along the beam for a late injection case (S1) compared to single shot images obtained in the early injection case taken as a reference. One can see on this figure the heterogeneous character of the FARLIEF signal along the beam at the optimized spark timing (340 CAD). Occasionally, no fuel is present at the spark location. This can explain 100% fluctuations measured in a rectangle 4x9.5 mm surrounding the spark plug. But this does not mean that the vapor cloud is heterogeneous. Cyclic variations of the equivalence ratio measured during 4 CAD with CHWP exhibits only 1 to 1.5 % variation showing a relatively stable equivalence ratio in the vapor pocket. Actually, the optimized spark timing has been chosen from previous combustion stability analysis [11]. This timing corresponds to the mean CAD where the vapor cloud starts to reach the spark. FARLIEF measurements were acquired at this CAD with a high temporal resolution. Swirl injectors were of good quality and tumble motion very stable. The turbulence flow field affects the perimeter of the vapor cloud resulting in a slight variation of its time of arrival at the spark location, which is detectable with the nanoseconds duration FARLIEF images but not with 4 CAD CHWP measurements and so not by the spark.

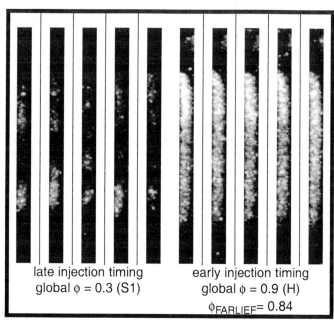

late injection timing global φ = 0.3 (S1) early injection timing global φ = 0.9 (H)

$\phi_{FARLIEF}$= 0.84

Figure 14. Instantaneous FARLIEF images (along a beam passing horizontally through electrodes) at the optimized spark timing 340 CAD

Injection duration effect – Figure 15 presents the local equivalence ratio history, averaged over 25 cycles obtained with CHWP for 4 injection durations (S1,S2,S3,S4).

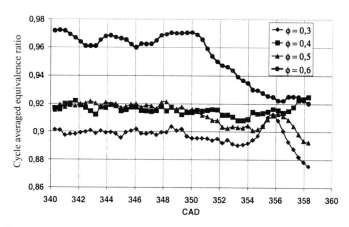

Figure 15. Temporal evolution of cycle averaged equivalence ratio

On average, this injection duration does not seem to affect the evolution. In Table 4, we report the crank angle at which a detectable air-fuel mixture appears : it can be confirmed that the optimised spark timing (20 BTDC) is in fact due to the presence of an ignitable mixture. The equivalence ratio averaged between 340 – 345 CAD is also given : the level of the equivalence ratios are quasi-similar for the four injection durations.

Table 4. Arrival time of fuel pockets and averaged equivalence ratios for late injection cases measured with CHWP.

Injection Duration	910 μs	1090 μs	1320 μs	1540 μs
φ : Global equivalence ratio (HEGO probe)	0.3	0.4	0.5	0.6
Arrival time of fuel pockets (CAD)	340,3	341,8	340,7	340,9
ϕ_{CHWP} : Local mean equivalence ratio, averaged between 340 – 344 CAD	0,9	0,913	0,919	0,966

With the FARLIEF technique, if we use the same spark surrounding area (9,5 mm) to spatially average the fluorescence signal, the values are also found constant but very low (80 a.u.) which corresponds to 0.45 equivalence ratio. In fact, this is an average over 800 samples and the instantaneous images presented on Figure 14 show that at 20 CAD BTDC, the air-fuel pocket may or may not have arrived in the measuring volume. When the pocket is detectable, the equivalence ratio is about 0.9. So fluctuations of 100% provide an average measurement of 0.45.

For the analysis, the PLIEF images, obtained in a previous study [11] are needed. In Figure 16, an example of instantaneous PLIEF images are presented for cases H (φ = 1), S1, S2, S3. These images have been obtained at the spark timing with an inclined laser sheet which could not exactly cross the spark plug position. They show that for stratified cases, S2 and S3, the spatial fluorescence intensity is lower than the homogeneous case and for S1 is similar to H. So, with the fluorescence calibration curve (Figure 11) we can deduce a local equivalence ratio value at the spark plug of 0.91, for the global stoichiometric mixture case, in early injection mode. Therefore, the results obtained with CHWP are in a good agreement with those obtained with FARLIEF.

In Figure 17, we can see that the choice of the length value for spatial averages is crucial. Indeed, for the smallest injection duration, the averaged fluorescence intensity is nearly the same : it means that the air-fuel mixture, averaged for 800 cycles, is homogeneous along the laser beam. On the other hand, for injection durations > 1 ms , the mean signal increases with the length : the mixture is leaner at the center than at bowl boundaries.

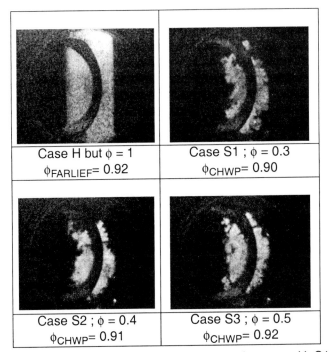

| Case H but φ = 1 $\phi_{FARLIEF}$= 0.92 | Case S1 ; φ = 0.3 ϕ_{CHWP}= 0.90 |
| Case S2 ; φ = 0.4 ϕ_{CHWP}= 0.91 | Case S3 ; φ = 0.5 ϕ_{CHWP}= 0.92 |

Figure 16. Instantaneous PLIEF images for cases H, S1, S2, S3. HEGO equivalence ratio and local FARLIEF or CHWP equivalence ratios are indicated (from [11]).

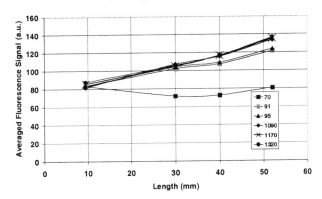

Figure 17. Effect of the area for spatial averaging of fluorescence intensity

<u>Injection timing effect</u> – To study the effect of the injection timing on local equivalence ratio at the spark plug, we present, in Figure 18, 3 cyclic averaged signals obtained with the CHWP. The injection duration was fixed at 1090 μs (Case S2). It can be seen that the air-fuel pocket arrival time is a function of the injection timing. For the injection timing nearest the TDC, the equivalence ratio reaches, at the arrival time, a value near stoichiometric and after a small plateau of 2 CAD, continues to decrease. This behavior is really similar to the results obtained by Kawamura et al. [5], provided by IR absorption technique.

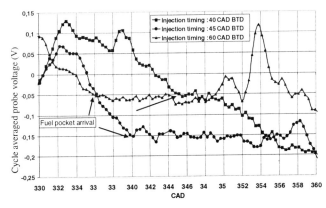

Figure 18. Effect of injection timing on probe signals

The decrease observed, for example in the case of an injection timing of 45 BTDC, between 330 CAD and 340 CAD, indicates the evolution of fuel concentration inside the sampling system itself. It means that hydrocarbons are present in the combustion chamber at the vicinity of the spark plug at this crank-angle. But due to dilution phenomenon inside the sampling systems we can not say if the fuel concentration changes in the combustion chamber itself or near the wires. In fact, we can notify that the signals rise a minimum indicating the presence of homogeneous fuel pockets. For example, in the case of an injection timing of 45 BTDC, the equivalence ratio remains constant around 0.9. Thus, one can estimate at what crank-angle, the air-fuel mixture is constant in the vicinity of the spark plug.

In Table 5, these arrival times and equivalence ratios at these times are given. No real effect on the cycle averaged equivalence ratios can be seen. Therefore it can be concluded that the injection timing affects the pockets arrival at the spark plug, inducing possible misfires but not the air-fuel ratio itself.

Table 5. Injection timing effects.

Injection timing	60 CAD BTDC	45 CAD BTDC	40 CAD BTDC
Arrival time	24 CAD BTDC	20 CAD BTDC	15 CAD BTDC
Equivalence ratio : ϕ_{CHWP}	0.935	0.92	0.94

CONCLUSION

Combined Catalytic Hot Wires Probe (CHWP) and Fuel Air Ratio Laser Induced Exciplex Fluorescence (FARLIEF) techniques have been applied to a Gasoline Direct Injection engine to characterize the temporal and spatial evolution of the fuel/air ratio in the vicinity of the spark plug.

The global equivalence ratio was controlled by modifying the injection duration while fuel distribution was varied with the injection timing : The engine ran below stoichiometric with early injection (homogenous case taken as a reference) or late injection timing with global equivalence ratios as low as 0.3. Preliminary combustion stability tests were conducted to determine the best ignition timing associated with late injection timing.

Some of the fuel/air mixture was extracted near the spark plug and acted on the CHWP, inducing a temporal fuel/air ratio signal. A short pulsed UV laser beam crossed the extraction location at spark timing, exciting the fluorescence of benzene and TEA. tracing the iso-octane used as a fuel. Those combined tracers enabled local FAR-LIEF measurements along the beam, avoiding liquid deposit fluorescence interference. Calibration of both techniques were performed independently. Both techniques yield similar equivalence ratios for the reference stoichiometric case characterized by a slowly varying time evolution.

Under lean and late injection conditions, the temporal CHWP signal indicates that a rich vapor cloud is carried to the vicinity of the spark plug. This result is consistent with vertical PLIEF visualizations. The vapor reaches the spark location at the optimum spark timing. CHWP and FARLIEF techniques show that the maximum equivalence ratio in the fuel cloud reaching the spark does not depend on the injection duration, which instead seemed to act on the size of this rich vapor pocket. FARLIEF measurements exhibit fluctuations of fuel/air ratio about ten time larger than for the homogeneous case. These fluctuations are mainly due to cyclic variations in the arrival time of fuel vapor in the measuring volume. These results illustrate the sensitivity of combustion stability to the injection timing which controls the coincidence of the rich fuel vapor with the spark.

This work confirmed that CHWP is a low cost fuel/air ratio diagnostic well adapted to stratified gasoline direct injection engine studies.

ACKNOWLEDGEMENT

The Authors express their gratitude to the Groupement Scientifique Moteurs for making an optical GDI engine available for this study.

REFERENCES

1. BaritaudT. and Heinze T, "Gasoline Distribution Measurements with PLIF in a SI Engine" SAE Paper 922355, 1992

2. Neij H. Johannsson B. and Aldén M. "Development and Demonstration of 2D-LIF, for Studies of Mixture Preparation in SI Engines",25th Symp (Int) on Combustion, Irvine, July 31st-August 5th , 1994

3. Reboux J., Puechberty D., Dionnet F.; Study of mixture inhomogeneities and combustion development in a S.I. engine using a new approach to laser induced fluorescence, SAE Paper 961205

4. Skippon S.M., Nattrass S.R., Kitching J.S., Hardiman L., Miller H., "Effects of fuel Composition on In-Cylinder Air/Fuel ratio During Transients in a SI Engine, Measured using Differential Infra-Red Absorption, SAE Paper 961204, 1996l

5. Kawamura K., Suzuoki T., Saito A, Tomoda T., Kanda M., "Development of instrument for measurement of fuel-air ratio in vicinity of spark plug : application to DI gasoline engine", JSAE review 19 p 305-306, 1998

6. Collings N., "A new technique for measuring HC concentration in real Time, in a Running Engine, SAE Paper 880517, 1988Collings

7. Di Cocco E., PhD thesis – Université de Versailles Saint Cyr L'Ecole, 1998

8. Depussay E., Mounaïm-Rousselle C., Burnel S., Ricordeau V., Deschamps B., "Comparative measurements of local iso-octane concentrations by Planar Laser Induced Fluorescence and Catalytic Hot Wires Probe in SI Engines, SAE Paper 982474, 1998

9. Kakuhou A., Urushihara T., Itoh T. and Takagi Y. « Characteristics of mixture formation in a Direct injection SI Engine with optimized in-cylinder Swirl Air motion » SAE Paper 1999-01-0505

10. Alger T., Hall M. and Matthews R . «Fuel spray dynamics and fuel vapo concentration near the spark plug in DI 4-valve SI engi ne» SAE 1999-01-0497

11. Deschamps B., Ricordeau V. Visualisation du carburant et de la combustion dans un moteur IDE à géométrie réelle. IFP report n° 44964"

12. Le Coz JF., "Comparison of different Drop sizing techniques on direct injection gasoline sprays", 9Th International Symposium on Applications of Laser Techniques to Fluid Mechanics, Lisbon, 1998

13. Ipp W, Wagner V., Krämer H., Wensing M. and Leiperz A., Arnt S. and Jain A. "Spray formation of high pressure swirl gasoline injectors investigated by two Dimentional Mie and LIEF Techniques" SAE paper 1999-01-0 4 9

14. Depussay E., Mounaïm-Rousselle C., Brunel S., "Experimental study of a new technique for the estimate of the local fuel ratio in engine environment", International Colloquium of Engines (ICE97), Capri, Sept. 97

15. Depussay E., PHD thesis, Université d'Orléans, 1998,

16. Knibbe H., Rehm D., Weller A.; Ber. Bunsenges, Physik. Chem. 73:839

17. Münch K.-U., Kramer H., Leipertz A.; Investigation of fuel evaporation inside the intake of a S.I. Engine using L.I.E.F. with a new seed, SAE Paper 961930 (1996)

18. Ortolan G., Deschamps B., "Separate Vapor/Liquid visualization with Laser-Induced Exciplex Fluorescence applied to Gasoline Direct Injection engines." Poster presented at the 10th Gordon Research Conference on "Laser Diagnostics" Il Ciocco Italy june 20th to 25th 1999

19. Fröba A.P., Rabenstein F., Münch K.-U., Leipertz A.; Mixture of T.E.A. and benzene as a new seeding material for the quantitative two-dimensional L.IE.F. imaging of Vapor and liquid fuel inside S.I. engines, Comb. & Flame 112: 199-209 (1998).

1999-01-3688

Optical Investigations of a Gasoline Direct Injection Engine

J. Reissing, J.M. Kech, K. Mayer, J. Gindele, H. Kubach, U. Spicher
Institut für Kolbenmaschinen, Universität Karlsruhe (TH)

ABSTRACT

In this paper optical investigations of a gasoline direct injection engine with narrow spacing arrangement of spark plug and injector are presented. For the combustion analysis spectroscopy techniques based on the fiber technique are used. With this measurement technique information about soot formation and temperature progression in the combustion chamber is obtained. Furthermore a validation of numerical simulation of the stratified combustion with data obtained experimentally, is performed and discussed.

INTRODUCTION

The gasoline direct injection concept provides a high potential concerning a lowering of fuel consumption. This has been presented in several recent publications [7, 11, 14, 15, 19, 28]. Higher fuel economy is mainly achieved by lower throttling and lower heat losses during stratified combustion, as well as a higher compression ratio. Furthermore, with the GDI concept no wall wetting and build up of wall films inside the intake ports occurs. This leads to combustion with lean mixtures and reduced HC- and CO-emissions during cold start, warm-up, and transient mode.

The renewed interest in the technology has been made possible by advances in high-pressure fuel injection, powerful electronic control systems in conjunction with the development of new measurement techniques and improved numerical simulation of the combustion process. In spite of recent intensified development activities recently, no combustion concept was established because direct injection gasoline engines require a strongly controlled in-cylinder gas motion, fuel injection and ignition time. The tuning of these parameters to generate a stable charge stratification has to be realized over a wide range of operating conditions. The exhaust after-treatment represents a further problem especially at part load under stratified charge conditions.

Powerful tools such as optical diagnostics and 3D-CFD simulation are essential for efficient combustion analysis. Modern measurement techniques allow investigations of in-cylinder flow conditions and the physical and chemical processes during mixture formation and combustion. Especially the knowledge of temperature distribution in the combustion chamber of GDI engines due to the formation of nitric oxide is of eminent importance.

The most common technique of measuring temperature in diesel engines is the two-color method [2, 20]. In SI engines the laser induced fluorescence (LIF) [5, 22] as well as coherent anti stokes-raman-spectroscopy [6] are applied. However, both techniques are very complex as an optically accessible engine is required and expensive laser equipment is necessary.

Therefore, in this work a spectroscopic determination of temperature using the emission lines of sodium (Na) and potassium (K) is used in combination with the well-known optical fiber technique.

Beside the experimental approach for analysis and optimization of the stratified combustion, numerical simulation were performed using a model for partially premixed flames. The measured data allowed validation of the stratified-charge mode combustion-process simulation.

TEMPERATURE DETERMINATION

An atom rises from its ground state to one of its low-energy excited states due to energy feeding in form of temperature. Once, an atom's electrons are in the excited state, a certain number will relax to the ground state and emit radiation corresponding to the difference in energy levels, $h\nu$. The intensity of this emitted radiation can be used to determine the temperature in an observed volume.

Spectroscopic temperature measurements using emission spectroscopy in optically thin layers require the premise of local thermodynamic equilibrium. Following assumptions have to be made [8, 12, 18]:

- Maxwellian velocity distribution of all particles.

- The distribution of the different energy levels is consistent with the Boltzmann distribution.

- The rate of ionization is according to the Eggert-Saha-equation.

- Instead of the blackbody radiation, the distribution of radiation can be described by different atomic or molecular variables because of the strong energy exchange between the particles themselves.

If the relations mentioned before are valid, the emission of an atomic or molecular spectral line is:

$$I_L = \int_0^l \frac{1}{4\pi} A_{nm} n_n h\nu \, dx \qquad \left[\frac{W}{m^2 sr}\right] \qquad (1)$$

Herein are:

A_{nm} = transition probability for spontaneous emission from energy level n to energy level m, also called Einstein coefficient

n_n = particle density of the upper level n

ν = line frequency

l = layer thickness

h = Planck's constant

With the Boltzmann distribution:

$$n_n = n \frac{g_n}{Z} \exp\left[-\frac{E_n}{kT}\right] \qquad (2)$$

with:

E_n = energy level of the state n

n = total particle density

k = Boltzmann's constant

g_n = $(2J_n+1)$ = statistical weight of the upper level

$$Z = \sum_i g_i \exp\left[-\frac{E_i}{kT}\right] = \text{state sum} \qquad (3)$$

it is obtained:

$$I_L = \frac{1}{4\pi} A_{nm} \frac{ng_n}{Z} h\nu l \exp\left[-\frac{E_n}{kT}\right] \qquad \left[\frac{W}{m^2 sr}\right] \qquad (4)$$

The quotient of the line intensities of two different lines with the subscripts x and y gives:

$$\frac{I_{Ly}}{I_{Lx}} = \frac{A_{nm}^y n_y Z_x g_n^y \nu_y}{A_{nm}^x n_x Z_y g_n^x \nu_x} \exp\left[-\frac{E_n^y - E_n^x}{kT}\right] \qquad (5)$$

After a re-arrangement the following expression for the temperature is obtained:

$$T = \frac{E_n^x - E_n^y}{k \ln\left(\frac{I_{Ly} A_{nm}^x n_x Z_y g_n^x \nu_x}{I_{Lx} A_{nm}^y n_y Z_x g_n^y \nu_y}\right)} \qquad [K] \qquad (6)$$

As this equation shows, knowing the atomic or molecular constants, the temperature is only a function of measured intensities.

A further optical measurement technique to determine flame temperature in optical thick layers is the well established two-color method. In Diesel engines and in diffusive flames respectively this technique allows measurement of temperature as well as soot concentration. The principle of the two-color method is based on the detection of radiation at two different wavelengths of incandescent soot particles arising during combustion. This has been reported in several publications [10, 29]. Consideration of Planck's radiation law, the spectral emissivity rate of any temperature radiation emitter and Kirchhoff's law results in an equation which allows one to calculate the soot temperature iteratively:

$$T = \frac{1}{\frac{1}{T_{S1}} + \frac{\lambda_1}{C_2} \ln\left\{1 - \left[1 - \exp\left(\frac{C_2}{\lambda_2}\left(\frac{1}{T} - \frac{1}{T_{S2}}\right)\right)\right]^{A_{av}}\right\}} \qquad [K] \qquad (7)$$

Herein:

C_1 = the first Planck's constant

C_2 = the second Planck's constant

λ_1, λ_2 = wavelength

T_{S1}, T_{S2} = temperature of blackbody which will emit the same radiation intensity as a non-blackbody at temperature T and wavelength λ_1 and λ_2

Aav = quotient of mass-absorption cross-section at the wavelength λ_1 and λ_2

Soot concentration is calculated according to Beer-Lambert's law of absorption, Mie's theory of light scattering and the temperature obtained by equation (7):

$$c = \frac{1}{A_{a2}s} \cdot \ln\left\{1 - \exp\left[\frac{C_2}{\lambda_2}\left(\frac{1}{T} - \frac{1}{T_{S2}}\right)\right]\right\} \quad \left[\frac{g}{m^3}\right] \quad (8)$$

Herein:

A_{a2} = mass-absorption cross-section at the wavelength λ_2

s = optical path length

For this calculation the diameter of soot particles were assumed to *50 nm* and the soot density to *2 g/cm³*.

EXPERIMENTAL SETUP

APPLICATION OF MEASUREMENT TECHNIQUE

The presented measurement technique for temperature determination is based on the well-known optical fiber technique [25]. In contrast to laser methods, the optical fiber technique is more adaptable to production engines. The spectroscopic investigations were performed using the experimental configuration shown in Fig. 1.

The radiation due to temperature excitation of gas is monitored through an optical fiber mounted in an optical probe. The optical probe operates like an aperture with a detection diameter of 1.0 *mm*. Therefore the observed cone-shaped volume element is limited by a 30 degree field of view. The flame radiation is detected by the probe and transmitted to a monochromator. The spectrum of the dispersed light is detected by an intensified CCD-camera. Analogue to the approach described by Kech et al. [16] and Reissing et al. [24] for temperature determination, the emission lines of the alkalis sodium (Na) and potassium (K) are applied. For temperature estimation the resonance transition 3^2S-3^2P of sodium and 4^2S-4^2P of potassium were detected. The values of energy-levels and Einstein coefficients were taken from [4] and [27]. The alkalis were admixed to the fuel in the form of additives based on salt. A chemical analysis of the fuel resulted in a concentration of *17 mg Na/kg fuel* and *33 mg K/kg fuel*.

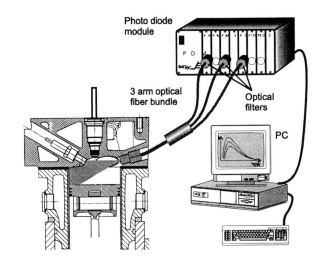

Fig. 2: Experimental configuration for the two-color method

The experimental apparatus used for the two-color method is shown in Fig. 2. For these investigations special optical fiber bundles were applied, which bifurcate into three arms. To minimize the influence of unequally distributed radiation from the observed volume, the

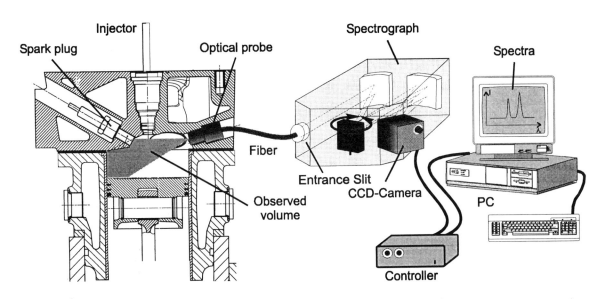

Fig. 1: Experimental configuration

optical fibers are stochastically bifurcated. Three different wavelengths in the visible and near infrared spectrum (λ_1 = 600 nm, λ_2 = 750 nm, λ_3 = 900 nm) were chosen. The bandpass filters have a bandwidth of about $\Delta\lambda$ = 3 nm. A tungsten band lamp calibrated on a blackbody temperature of T_s = 2395 K at the wavelength λ = 650 nm was used to calibration the optical setup.

Additionally, a pressure transducer was installed to allow simultaneous measurement of cylinder pressure data.

GDI ENGINE

The measurements were performed with a single cylinder GDI engine with narrow spacing arrangement of spark plug and injector. With this configuration, contact between fuel and combustion chamber walls can be avoided. In contrast to other GDI concepts this signifies a considerable advantage concerning emission of unburned hydrocarbons. A well-homogenized mixture at full load is possible with a centrally-located injector. A leaner mixture can be achieved due to a much better stratification of fuel at part load.

However, the turbulent in-cylinder gas motion interacts strongly with the spray motion and, in cases of strong variations in the mixing process, the system reacts sensitively implicating misfires. Consequently an accurate coordination of location of spark plug and injector, in-cylinder gas motion, injection parameters and ignition time is required. Transient conditions during injection period influence spray characteristics. Spray angle, penetration as well as fuel distribution in the spray cone varies with different in-cylinder pressure. The controllability of spray presents a serious problem over a wide range of engine operating conditions. Low engine speed and low flow velocity could generate large fuel droplets promoting soot formation. High engine speed and high turbulence supports the mixture formation. However, high flow velocity and small droplet size complicates a realization of stabilized ignitable mixture close to the spark plug.

The engine has a pentroof cylinder head with a central position of the injector. To generate different in-cylinder flow motion, one inlet valve can be throttled. For fuel injection a common-rail system was used in combination with a hollow-cone swirl nozzle with a cone angle of 90° at atmospheric conditions. The main geometrical details of the engine were as follows:

Bore:	89.6 mm
Stroke:	86.6 mm
Compression ratio:	10
Inlet valve:	2
Outlet valve:	1
Combustion chamber:	pentroof

The experimental investigations were performed using the following operation parameters for homogeneous and stratified charge mode listed in Table 1 and 2:

Table 1: Homogeneous charge

Speed	1000 rpm		1500 rpm
IMEP	2 bar	4 bar	2 bar
Ignition	22°BTDC	18°BTDC	25°BTDC
Injection pressure	85 bar	85 bar	85 bar
Start of injection	280°BTDC		
A/F-ratio	1.0	1.0	1.0

Table 2: Stratified charge

Speed	1000 rpm		1500 rpm
IMEP	2 bar	4 bar	2 bar
Ignition	42°BTDC	38°BTDC	42°BTDC
Injection pressure	85 bar	85 bar	85 bar
Start of injection	52°BTDC	60°BTDC	58°BTDC
A/F-ratio	0.29	0.43	0.25

To consider the varying intensity of light during one cycle and the cyclic variation of the process, the CCD-camera was exposed over 40 to 60 cycles accumulating the spectra of Na and K. Analogue to this approach, the soot radiation detected by the photo diode module was recorded over 100 cycles. The curves of soot temperature and soot concentration shown in the following chapter are averaged values of 100 cycles.

MEASUREMENTS AND DISCUSSION

EVALUATION

Preliminary investigations with emission of different molecules have shown that for temperature determination the radiation of the OH-radical at 306.4 nm ($A^2\Sigma$-$X^2\Pi$-system) can be used [3]. If the OH-emission is applied, the temperature evaluation will be simplified because the particle density n_x and n_y (Eq. 6) can be neglected. However, the concentration of OH molecules decrease in the end gas region. Consequently estimation of temperature during expansion period is difficult due to the weak radiation. Furthermore as soon as the flame front reaches the observed volume, the temperature curves are characterized by considerable fluctuations. This behavior allows the conclusion that the assumed thermodynamical equilibrium of the radiating molecules does not apply in the reaction zone. Therefore in this work the very strong emission lines of sodium and potassium are used.

As mentioned before the temperature calculated with the quotient of two line intensities depends on physical constants which are known, the measured intensities and the density of radiating particles, respectively, and the quotient of particle density. However, the concentrations of the alkalis are only known in the fuel. During the combustion process the concentration distribution of free atoms of sodium and potassium are unknown. Expandable laser methods based on absorption are required to determine the concentration behavior of sodium and potassium. Because there is no suitable optical access in the combustion chamber to adapt laser methods, the quotient of particle concentration is obtained using thermodynamical calculations. An average temperature curve of the burned gas for homogeneous charge is calculated using a two-zone model [9], plotted in Fig. 3. Assuming homogenous temperature distribution at 120°CA ATDC in the combustion chamber, the measured temperature value is fitted to the calculated temperature value. On the basis of this assumption a theoretical quotient of particle concentration is obtained. This quotient is assumed to be constant and is used to calculate the temperature values at crank angle positions before. The measured temperature curve is plotted in Fig. 3 as well.

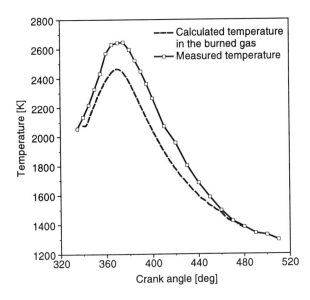

Fig. 3: Measured and calculated temperature curves for homogeneous charge, n = 1500 rpm, IMEP = 2 bar

Furthermore soot formation is expected due to the occurrence of premixed as well as diffusive controlled flames and the short time for mixture preparation with stratified combustion. This behavior can be seen in Fig. 4 which shows the visible emission spectra of a homogeneous stoichiometric charge in contrast to

stratified charge for the operating point n = 1000 rpm and IMEP = 4 bar. Both line emissions of sodium (589.5 nm) and potassium (766.5 nm) predominate in the recorded spectra. Nevertheless, the radiation intensity increases significantly with stratified engine mode. Soot, as a blackbody emitter, radiates in the visible spectrum described in Planck's law. Therefore increasing intensities in the observed emission spectrum indicate soot formation with stratified combustion.

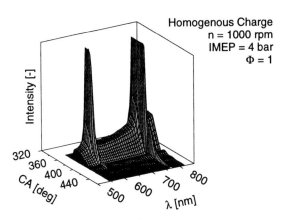

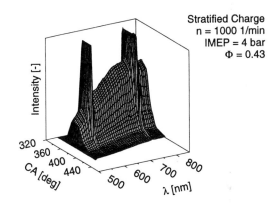

Fig. 4: Visible emission spectra of homogeneous and stratified combustion, n = 1000 rpm, IMEP = 4 bar

The absorption of soot particles has to be taken into account when calculating the temperature with the intensities of alkali emission. Therefore the absorption coefficients of soot at the wavelength of alkali-emission were computed in regard to the assumptions of soot density and diameter of particles mentioned before. However, the soot concentration, which was also needed for determination of absorption behavior, is unknown. In order to avoid this problem, a correlation between soot concentration and soot radiation was set up. For this a standardized ratio of the line intensities and the soot intensities close to the wavelength of Na and K was used for this. The temperature evaluation was performed additionally using the soot concentration obtained with the two-color-method to verify the correlation. Fig. 5

shows two temperature curves evaluated with the different approaches for stratified combustion. As one can see, the differences between the temperature values can be disregarded.

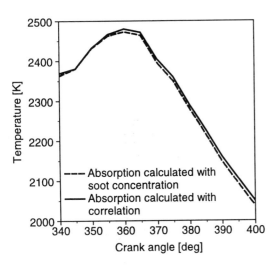

Fig. 5: Comparison of temperature curves obtained with different approaches

RESULTS

Fig. 6 illustrates the temperature behavior at stratified charge mode obtained with the detection of line emission intensities of sodium and potassium and the two-color method. The temperature progression estimated with the two-color method starts at $40°CA$ BTDC with values about 2200 K corresponding to the values measured with the radiation of the alkalis. Basically, the soot temperature is lower than the temperature measured with the radiation of the alkalis. Especially in the range

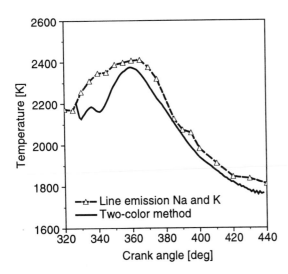

Fig. 6: Comparison of temperature curves obtained with the two-color method and the emission lines of Na and K, n = 1000 rpm, IMEP = 2 bar

between $30°CA$ BTDC to $10°CA$ BTDC differences of about 100 K exist. During the expansion period the two temperature curves show similar behavior.

However, the evaluation of soot temperature includes a lot of assumptions about soot and temperature distribution, size of soot particles or flame shape and reacts very sensitively to calibration [10]. Especially inhomogeneous distribution of temperature and soot particles in the observed volume can cause lower estimated soot temperatures. At the beginning of the combustion the flame does not extend over the hole observed volume. Therefore the temperature values measured with the two-color method are obviously lower due to inhomogeneous soot distribution. Both temperature curves correlate well until the reaction zone extends into the predominantly part of the observed volume.

Compared to the two-color method the assumption for temperature determination using the emission lines of sodium and potassium mentioned before seems to be justifiable. Therefore, in the following plots the temperature curves obtained with the line emissions of the alkalis are pictured.

Although the combustion characteristic at full load does not essentially differ from homogeneous working SI engines, the combustion characteristics of GDI engines change significantly with the combustion control strategy at part load. After combustion has started around the spark plug the flame develops rapidly into the stoichiometric mixture region. However, in the lean outer region of the stratified charge, the speed of flame propagation is reduced. This early combustion may be characterized by flames as they occur in typical premixed lean or stoichiometric mixtures. The significantly lower combustion rate near the end of the combustion process is caused by high contribution of diffusively controlled combustion to the global combustion rate. Flame propagation in partially premixed flows, such as the stratified charge in GDI engines, is determined by the movement of triple flames, which consist of two premixed wings and a trailing diffusion flame [17, 23]. The partially premixed combustion is essentially determined by the mixture fraction, because the laminar burning velocity depends on the local equivalence ratio ϕ and for that reason on the mixture fraction. The maximum of laminar burning velocity is close to $\phi = 1.0$. Therefore, flames will propagate fastest along the surface $\phi = 1.0$ in a mixture field. The leading edge of the flame, called the triple point, propagates along the surface of stoichiometric mixture. On the lean side of that surface there is a lean premixed flame branch and on the rich side a rich premixed flame branch both propagating with lower burning velocity. A diffusion flame develops behind the triple point on the surface of stoichiometric mixture where the unburned fuel from the rich premixed branch burns together with the remaining oxygen from the lean premixed branch.

Compared to a conventional engine, the duration of the combustion process increases because of the lower local burning velocity in the rich and lean premixed flames. This behavior is demonstrated in Fig. 7. Here standardized light signals of different cycles are compared as well as the averaged light signal of 100 cycles detected by a photomultiplier. Differing from the optical approach for temperature determination for this investigations, the optical access is achieved using a modified spark plug with an implemented optical probe with a field of view of 8°. The used photomultiplier is sensitive in a range of wavelength between 200 *nm* and 600 *nm*. The engine was operated with constant speed *n* = 1500 *rpm* and load *IMEP* = 2 *bar*.

The homogeneous combustion is characterized by a spherical flame propagation. As soon as the reaction zone reaches the observed volume the light signals increase obviously in consequence of the radiation of reacting radicals. The light signals decrease in the burned gas due to decreasing concentration of radiating molecules and decreasing temperature. It is remarkable, that all detected signals show a similar behavior which confirms the numerical calculated spherical flame propagation shown in Fig. 12.

A completely different progression of radiation intensity is observed compared to the homogeneous combustion with the stratified charge mode. The signals show irregular behavior and are characterized by a second or third maximum of radiation intensity. Due to inhomogeneous fuel distribution the flame does not propagate in a spherical manner. A local bordered combustion with ragged flames and moderate reaction rate in consequence of fuel burning diffusive controlled is observed in the experiment as well as in the numerical simulation presented in the next chapter.

Gathering from the inhomogeneous fuel distribution and in consequence of statistical distribution of reaction, an inhomogeneous distribution of temperature in the combustion chamber is expected.

Fig. 8 shows plots of temperature progression at stratified operating points with constant speed *n* = 1000 *rpm* and *IMEP* = 2 *bar* as well as 4 *bar* obtained by spectroscopic determination using the emission lines of sodium and potassium. The maximum temperature at *IMEP* = 2 *bar* in the observed volume reaches values of about 2400 *K* and was detected at *TDC*. The observed temperature plateau over 2200 *K* is caused by slow burning mixture followed by a strong temperature decrease during expansion. This strong decrease is the consequence of the mixing of burned gas and cold air coming from the bordered zone of the combustion chamber. Due to the close timing of injection and ignition

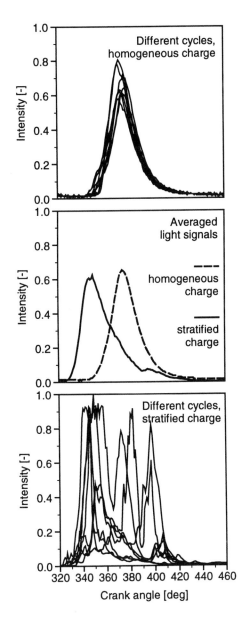

Fig. 7: Light signals of different cycles, *n = 1500 rpm,*
IMEP = 2 bar

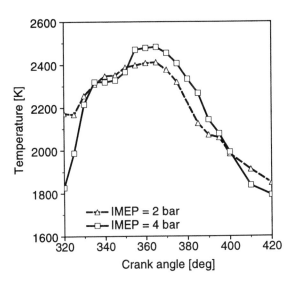

Fig. 8: Temperature progression at load variation, *n*
= *1000 rpm, IMEP = 2 bar, 4 bar*

time mentioned before, the temperature progression conforms to maximum heat release characterized by timing before *TDC* [7, 15].

Increasing load effects increasing temperatures. Correspondingly, the maximum temperature increases at *IMEP* = 4 *bar* up to values of 2480 *K* and a temperature plateau with a strong decrease is found as well. Due to higher mass of injected fuel the temperature stays longer on this level compared to at lower load. Conforming to this behavior, the exhaust temperature measured with a thermocouple in the exhaust pipe increases from 500 *K* up to 620 *K* as does the average emission of nitric oxide (NO) increases from 360 *ppm* up to 1200 *ppm*. This behavior emphasizes the strong influence of temperature on NO formation exceeding the influence of higher oxygen concentration at lower load [13].

In Fig. 9 the temperature progression is plotted with speed variation and constant load. With increasing speed the level of temperature in the observed volume increases. In consequence of more fuel burning with the A/F-ratio close to 1 and an increasing turbulence intensity due to higher inlet flow velocity at *n* = 1500 *rpm* the maximum temperature reaches values of about 2470 *K*. As mentioned before, characteristic temperature plateaus are found. Both operating points show similar gradients in temperature progression during compression and expansion because of similar charge in the combustion chamber. However, the NO concentration at both operating points measured in the exhaust pipe are on the same level.

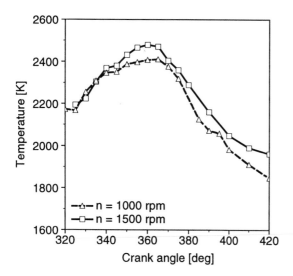

Fig. 9: Temperature progression at speed variation, *IMEP* = 2 bar, n = 1000 rpm, 1500 rpm

Fig. 10 shows a comparison of temperature curves obtained with homogeneous and inhomogeneous combustion at a constant speed of *n = 1500 rpm* and load of *IMEP = 2 bar*. The maximum detected temperature with stratified charge mode are about 140 K lower than with homogeneous charge. This behavior

corresponds to the measured NO emission which is significantly lower at inhomogeneous combustion. Due to areas burning very rich, NO-formation is affected by less oxygen. On the other hand, in the outer region burning lean the temperatures are too low for strongly increasing NO-formation. Only a small zone burning with the A/F-ratio close to 1 is dominated by high burning temperature and sufficient oxygen concentration, respectively.

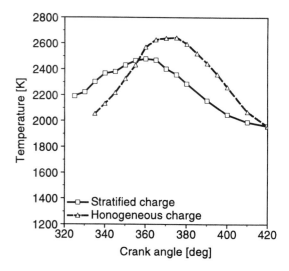

Fig. 10: Temperature progression with stratified charge and homogeneous charge, *n = 1500 rpm, IMEP = 2 bar*

As mentioned before, the short time for mixture preparation with stratified charge mode due to the short time between injection and ignition causes soot formation depending on load and speed. Fig. 11 illustrates the soot formation during the inhomogeneous combustion process at a constant speed of *n* = 1000 *rpm* and *IMEP* = 2 *bar* and 4 *bar* obtained with the two-color method. As a consequence of short injection duration at low load, the remaining time between injection time and ignition time

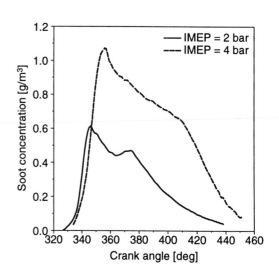

Fig. 11: Soot concentration with stratified charge mode, *n = 1000 rpm*

suffices to evaporate nearly all fuel droplets. Therefore maximal soot concentrations of about 0.6 g/m^3 are observed at 20 °CA BTDC. During the expansion period the soot is oxidized which causes decreasing soot concentrations. A Bosch smoke number of 1.3 is determined in the exhaust pipe. Increasing load effects an imperfect mixture preparation and rich burning regions with less oxygen resulting in soot concentrations of about 1.1 g/m^3 at 5 °CA BTDC. Soot oxidation occurs as well. However, due to decreasing temperature during the expansion period, the oxidation process is stopped before the soot is burned completely. Therefore a Bosch smoke number of 3 is measured in the exhaust pipe.

COMPARISON WITH 3D SIMULATION

In this chapter numerical investigations of the GDI engine are presented and compared to experimental results.

The numerical simulations were performed using KIVA-3V, a CFD code for transient reactive flows with spray. KIVA-3V uses a block-structured mesh and permits flexibility for modeling complex geometries with various boundary conditions. The equations and the numerical method are discussed in detail by Amsden et al. [1] In the presented study, the original KIVA submodel for gas turbulence, hollow-cone spray atomization and vaporization was used. A flame area evolution model for homogeneous-charge and turbulent combustion presented by Weller et al. [26] has been implemented and amplified for partially premixed turbulent combustion. The model expansion is based on the discussion of the burning velocity in partially premixed flames described by Peters [23] and Müller et al. [21].

The used mesh comprises about 120,000 cells. The KIVA-3V calculation started at the crank angle position representing the inlet valves opening. The boundary conditions were defined as pressures at the intake flow and outflow boundaries, which were associated with experimental values. For the initial condition, a zero velocity field and homogeneous pressure and temperature field in the combustion chamber were assumed. For further study of the numerical approach of partially premixed flame propagation please refer to [17].

First of all, a comparison of numerical calculated temperature distribution for homogeneous and stratified combustion clarifies the differences in temporal progression of flame propagation. Fig. 12 shows cross-sections of the calculated temperature distributions at both operating conditions (n= 2000 rpm, IMEP = 3 bar) at 10°CA ATDC. The homogeneous combustion is characterized by a spherical propagation of a thin reaction zone separating the burned and the unburned region. Conforming to this behavior temperature distribution with maximum values close to the spark plug and radial temperature gradients are observed. The flame propagation with stratified charge mode shows inhomogeneous distribution of reaction zones inducing inhomogeneous temperature distribution. As mentioned before these results correspond to the light signals measured with the photomultiplier.

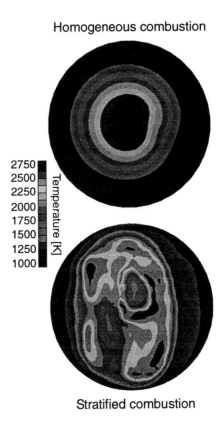

Homogeneous combustion

Stratified combustion

Fig. 12: Comparison of temperature distribution at homogeneous and stratified combustion

Additionally in Fig. 13 shows calculated temperature distributions in horizontal planes at six different crank angle positions. The area of maximum temperature propagates according to the fuel distribution and encloses the region with rich mixture. The temperature reaches values of about 2750 K. The position of the temperature maximum belongs to a quasi stationary flame.

Compared to the experimental data, the numerical calculated temperature values are locally higher. Especially shortcomings at the beginning of the combustion, illustrated in Fig. 13 at 20 °CA BTDC, seem to be caused by incorrect simulation of the fuel evaporation. However, in temporal progression the comparison of the results correlate well.

In the next step of the experimental investigations a more detailed spatial resolution of temperature distribution using a multitude of optical probes with small detection volumes are planned. This further approach should enable a more exact verification of calculated temperature data. Further on, an analysis of radiation behavior of molecules, i.e. OH, C_2 or CH with the stratified charge mode should provide a detailed understanding of the combustion process of a direct injection engine.

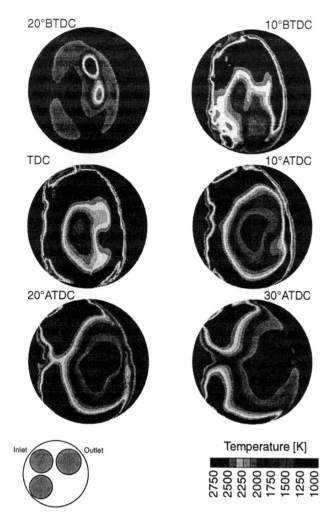

20°BTDC   10°BTDC

TDC   10°ATDC

20°ATDC   30°ATDC

Inlet   Outlet

Temperature [K]

2750 2500 2250 2000 1750 1500 1250 1000

Fig. 13: Calculated temperature distribution, n = 1500 rpm, IMEP = 2 bar

CONCLUSION

The GDI concept provides a high potential concerning fuel economy and pollutant emission. Direct injection technology is therefore becoming of increasing interest to the present and future development and optimization of SI engines. This technology differs over a wide range of engine operating conditions from the conventional mixing formation and combustion process. In consequence, new experimental methods as well as numerical models are required to analyze the processes in such engines.

This paper presents optical investigations of a single cylinder GDI engine with narrow spacing arrangement of spark plug and injector. Spectroscopic measurements allow investigations of physical and chemical processes as well as the determination of temperature development in the combustion chamber using the emission lines of the alkalis sodium and potassium. Additionally, the two-color method was applied for estimating soot formation.

The presented results of temperature measurement show realistic values. The maximum temperatures with stratified charge mode are lower than with homogeneous

charge mode. The process of stratified combustion with fast flame propagation in the first stage of combustion and slow flame propagation in the second stage of combustion leads to temperature progression characterized by temperature plateaus. The in-cylinder measurements of soot formation correspond to the smoke numbers measured in the exhaust pipe.

Furthermore, the results of numerical simulation of the inhomogeneous combustion at part load was described. Therefore a model for partially premixed combustion was applied.

The comparison of the calculated results with first experimental data shows good correlation in temporal progression. Although there are currently shortcomings especially at the beginning of the combustion, the discussion of the flame structure provides an insight into the combustion process of GDI engines. The combined application of optical diagnostics and numerical methods provide the opportunity to realize the potential benefits of the GDI engines.

REFERENCES

[1] Amsden, A.A.: KIVA-3: A KIVA Program with Block-Structured Mesh for Complex Geometries, Los Alamos National Lab., LA 12503-MS, 1993

[2] Arcoumanis, C.; Bae, C.; Nagwaney, A.; Whitelaw, J.H.: Effect of EGR on Combustion Development in a 1.9L DI Diesel Optical Engine, SAE-Paper 950850, 1995

[3] Bach, M.; Reissing, J.; Spicher, U.: Temperature Measurement and NO Determination in SI Engines Using Optical Fiber Sensors, SAE-Paper 961922, 1996

[4] Bashkin, S.; Stoner, J.O.: Atomic Energy Level & Grotrian Diagrams, Volume 2, Department of Physics, Tuscon, 1978

[5] Becker, H.; Monkhouse, P.B. Wolfrum, J.; Cant, R.S.; Bray, K.N.C.; Maly, R.R.; Pfister, W.; Stahl, G.; Warnatz, J.: Investigation of Extinction in Unsteady Flames in Turbulent Combustion by 2D-LIF of OH Radicals and Flamelet Analysis, 23. Symposium on Combustion/The Combustion Institute, pp. 817 – 823, 1990

[6] Furono, S.; Akihama, K.; Hanabusa, M.; Iguchi, S.; Inoue, T.: Nitrogen CARS Thermometry for a Study of Temperature Profiles through Flame Fronts, Combustion and Flame 54, pp. 149 – 154, 1983

[7] Fraidl, G.K.; Piock, W.F.; Wirth, M.: Gasoline Direct Injection: Actual Trends and Future Strategies for Injection and Combustion Systems, SAE-Paper 960465, 1996

[8] Gaydon, A.G.; Wolfhard, H.G.: Flames -Their Structure, Radiation and Temperature, 4.Edition, Chapman and Hall, London, 1979

[9] Gorenflo, E.: Einfluß der Luftverhältnisstreuung auf die zyklischen Schwankungen beim Ottomotor, Universität Karlsruhe (TH), 1997

[10] Gstrein, W.: Ein Beitrag zur spektroskopischen Flammentemperaturmessung bei Dieselmotoren, Technische Universität Graz, 1986

[11] Harada, J.; Tomita, T.; Mizuno, H.; Mashiki, Z.; Ito, Y.: Development of Direct Injection Gasoline Engine, SAE-Paper 970540, 1997

[12] Herzberg, G.: Atomic Spectra and Atomic Structure, Dover Publications, New York, 1944

[13] Heywood, J.B.: Internal Combustion Engine Fundamentals, McGraw-Hill, New York, 1988

[14] Jackson, N.S.; Stokes, J.; Whitaker, P.A.; Lake, T.H.: Stratified and Homogeneous Charge Operation for the Direct Injection Gasoline Engine – High Power with Low Fuel Consumption and Emissions, SAE-Paper 970543, 1997

[15] Karl, G.; Kemmler, R.; Bargende, M.; Abthoff, J.: Analysis of a Direct Injected Gasoline Engine, SAE-Paper 970624, 1997

[16] Kech, J.; Reissing, J.; Bach, M.; Spicher, U.: Determination of Local Temperature-Distribution in the Combustion Chamber of SI Engines, Internal Combustion Engines: Experiments and Modeling, ICE97, Capri, 17.-19.09.1997

[17] Kech, J.M.; Reissing, J.; Gindele, J.; Spicher, U.: Analyses of the Combustion Process in a Direct Injection Gasoline Engine, International Symposium COMODIA, Japan, 1998

[18] Kovács, I.: Rotational Structure in the Spectra of Diatomic Molecules, Adam Hilger LTD, London, 1969

[19] Kume, T.; Iwamoto, Y.; Lida, K.; Murakami, M.; Akishino, K.; Ando, H.: Combustion Control Technologies for Direct Injection SI Engine, SAE-Paper 960600, 1996

[20] Li, X.; Wallace, J.S.: In-cylinder Measurement of Temperature and Soot Concentration Using the Two-Color Method, SAE-Paper 950848, 1995

[21] Müller, C.M.; Breitbach, H.; Peters, N.: Partially Premixed Turbulent Flame Propagation in Jet Flames, 25. Symposium on Combustion/The Combustion Institute, pp. 1099 – 1106, 1994

[22] Orth, A.; Sick, V.; Wolfrum, J.; Maly, R.R.; Zahn, M.: Simultaneous 2D-Single Shot Imaging of OH Concentrations and Temperature Fields in a SI Engine Simulator, 25. Symposium on Combustion/The combustion Institut, pp. 143 – 150, 1994

[23] Peters, N.: Four Lectures on Turbulent Combustion, ERCOFTAC Summer School, Aachen, 1997

[24] Reissing, J.; Kech, J.M.; Gindele, J.; Spicher, U.: Optische und spektroskopische Untersuchungen im Brennraum eines Ottomotors mit Benzin-Direkteinspritzung, Tagung „Optisches Indizieren in der Motorenentwicklung", Haus der Technik, Essen, 1998

[25] Spicher , U.; Krebs, R.: Optical Fibre Technique as a Tool to Improve Combustion Efficiency, SAE-Paper 902138, 1990

[26] Weller, H.G.; Uslu, S.; Gosman, A.D.; Maly, R.R.; Herweg, R.; Heel, B.: Prediction of Combustion in Homogenous-Charge Spark-Ignition Engines, International Symposium COMODIA, 1994

[27] Wiese, W.L.; Smith, M. W.; Miles, B. M.: Atomic Transition Probabilities NSRDS-NBS 22, Washington, 1969

[28] Wirth, M.; Piock, W.F.; Fraidl, G.K.; Schoeggl, P.; Winklhofer, E.: Gasoline DI Engines: The Complete System Approach by Interaction of Advanced Development Tools, SAE-Paper 980492, 1998

[29] Zhao, H.; Ladommatos, N.: Optical Diagnostics for Soot and Temperature Measurement in Diesel Engines, Prog. Energy Combust. Sci., Vol. 24, pp. 221 – 225, 1998

1998-01-0497

Fuel Spray Dynamics and Fuel Vapor Concentration Near the Spark Plug in a Direct-Injected 4-Valve SI Engine

Terrence Alger, Matthew Hall, Ronald Matthews
University of Texas at Austin

ABSTRACT

The mixture preparation process was investigated in a direct-injected, 4-valve, SI engine under motored conditions. The engine had a transparent cylinder liner that allowed the fuel spray to be imaged using laser sheet Mie scattering. A fiber optic probe was used to measure the vapor phase fuel concentration history at the spark plug location between the two intake valves. The fuel injector was located on the cylinder axis. Two flow fields were examined; the stock configuration (tumble index 1.4) and a high tumble (tumble index 3.4) case created using shrouded intake valves. The fuel spray was visualized with the engine motored at 750 and 1500 RPM. Start of injection timings of 90°, 180° and 270° after TDC of intake were examined. The imaging showed that the fuel jet is greatly distorted for the high tumble condition, particularly at higher engine speeds. The tumble was large enough to cause significant cylinder wall wetting under the exhaust valves for some conditions. The fuel vapor concentration history near the spark plug was very different for the two flow fields; for the high tumble case a richer mixture quickly reached the spark plug and was then diluted before reaching a steady concentration near the end of the compression stroke. For the stock valve configuration the fuel vapor concentration increased gradually throughout compression. The fuel spray imaging provided evidence linking the fuel-spray/charge-motion behavior with the fuel vapor measurements.

INTRODUCTION

Direct injection of gasoline is emerging as one of the most viable technologies in the effort to develop more efficient, cleaner running SI engines. Relative to a port-fuel injected SI engine, the direct-injected SI engine potentially offers several significant advantages, especially in the areas of fuel economy, power output and emissions [1,2]. The two primary strategies for obtaining optimum performance are early injection timing at high load and late injection timing at low load and idle operating conditions. The early injection timing allows time for significant mixing and homogenous charge formation. Late injection timing allows for operation at an overall lean condition while maintaining a combustible mixture in the vicinity of the spark plug. A combination of injection timings has been found to be successful in production engines [3,4,5].

In addition to varying the injection timing in order to optimize the engine's performance, several combinations of swirl, tumble and piston configuration have been used to influence the mixture preparation process. One production engine uses a combination of high reverse tumble and spherical piston cavity to direct the fuel spray to a centrally located spark [3]. Another scenario involves high swirl and an involute shaped piston cavity [5].

Several studies have visualized the fuel spray dynamics in fired and motored engines [6,7,8,9]. There have also been several studies using numerical methods to model the fuel spray dynamics in a direct-injected SI engine [10,11]. A fiber optic spark plug probe has been successfully used to measure fuel vapor concentration history in both a fired and motored port-fuel injected SI engine [12,13,14].

The purpose of this study was to investigate the interaction between the intake flow and the fuel jet. There is a significant difference between the spray pattern of an injector in a quiescent chamber and the pattern in a running engine. The goal of this study is to determine the effect of varying the injection timing and tumble ratio on the fuel spray dynamics and the fuel vapor concentration near the spark plug.

METHODOLOGY

ENGINE CHARACTERISTICS The engine system used in this study was used in several previous studies [6,12]. The experimental setup for the imaging experiments is shown schematically in Figure 1. The experimental setup for the fuel vapor concentration is shown schematically in Figure 2.

The engine is a modified CLR engine with a GM Quad-4, 4-valve head that has been modified for direct injection. The cylinder liner is 22-mm thick quartz. The injector is located along the axis of the cylinder and the spark plug is located on the fire deck between the intake valves at an angle of 33.5 degrees from the horizontal. The piston crown was modified by machining a shallow bowl (12.5 mm in diameter and 4 mm deep) to allow clearance for the fiber optic probe. Access to the spark plug is through the fuel injector port in the intake manifold. A schematic indicating the injector and spark plug location is shown in Figure 3. Engine specifications are shown below in Table 1.

A Zexel hollow-cone swirl injector was used in this study. The injector forms a hollow, 60° cone when used with a fuel pressure of 5 MPa and produces droplets of 15-30 μm SMD [3,15]. Wide-angle swirl injectors are used to decrease penetration and promote air/fuel mixing. High fuel pressure promotes increased vaporization and mixing. Howell EEE gasoline was used in every test and the engine was motored in an unthrottled condition which resulted in a manifold air pressure of 1 atm (101.1 kPa) and a manifold air temperature of approximately 23° C (300 K).

The injection timings for this study were 90°, 180° and 270° aTDC of the intake stroke. The 90° injection scenario allowed us to compare the results with the results of a previous study [6], where the spray was visualized in this engine with a standard valve configuration. The 270° timing provides a late injection scenario and the 180° case is a median case that will allow study of the transition from early to late injection.

Stroke	95.25 mm (3.750 in)
Bore	92.0 mm (3.625 in)
Compression Ratio	7.5 : 1
Piston	Modified flat top, AL alloy
Cylinder Head	Modified GM Quad-4, 4-valve
Fuel Injector Location	Central, aimed axially
Fuel Injector	Zexel 60° hollow-cone swirl pattern
Fuel Pressure	5 MPa
Fuel	Howell EEE
Intake Valve Opening	22° bTDC
Intake Valve Closing	45° aBDC
Exhaust Valve Opening	120° aTDC
Exhaust Valve Closing	20° aTDC

Table 1. Engine Specifications

Injection timing is controlled by a PC equipped with a software based synchronization system [16] and a shaft encoder connected to the engine. The encoder and signal conditioning system allow for 0.125 crank angle degree resolution.

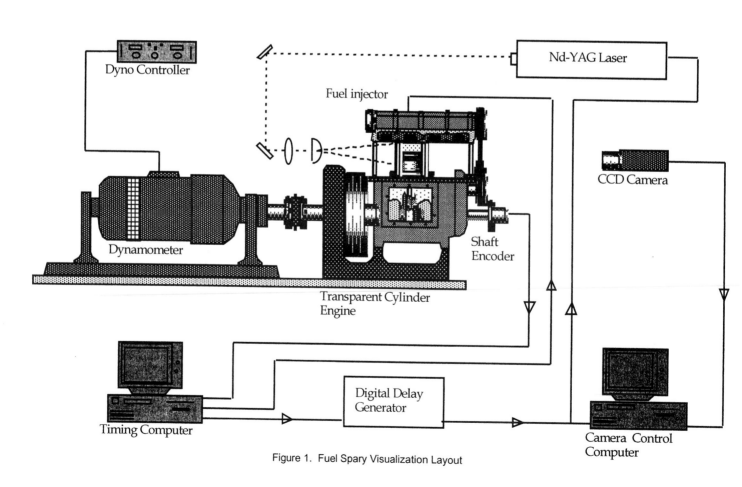

Figure 1. Fuel Spary Visualization Layout

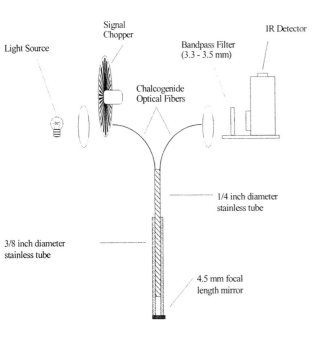

Figure 2. Fiber Optic Experimental Layout

Light Source

Signal Chopper

IR Detector

Bandpass Filter (3.3 - 3.5 mm)

Chalcogenide Optical Fibers

1/4 inch diameter stainless tube

3/8 inch diameter stainless tube

4.5 mm focal length mirror

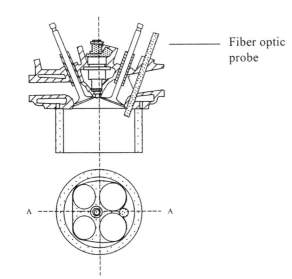

Fiber optic probe

A ---- A

Figure 3. Head Elevation and Plan View

DIAGNOSTICS

Mie Scattering To observe the effect of injection timing and intake flow on the fuel jet, the spray was visualized at different times after injection using laser sheet droplet imaging. Light from a Quanta Ray DCR-11 Nd-YAG laser (λ=532 nm) passed through a 100 cm focal length spherical lens and a 7.5 cm focal length cylindrical lens to form a 3 mm thick laser sheet. The sheet was passed

through a plane on the cylinder centerline. The plane passed in-between both the intake and exhaust valves (Plane A in Figure 3). The cylinder liner was illuminated throughout the entire range of motion of the piston and the field of view is shown in Figure 4. A Cidtec Model 2250 CCD camera with 512 x 512 pixel resolution with a dedicated PC equipped with a Data Translation Model DT2862 framegrabber was used to capture the images. A 10 ns laser pulse was used at the desired crank angle to illuminate the droplets. The PC controlling the injection timing also was used to control the laser and camera. Each image corresponds to a different engine cycle, since the acquisition equipment can only acquire one image per engine cycle. This experimental setup was identical to the one used in a previous study [6].

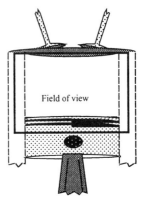

Field of view

Figure 4. Field of View

Fiber Optic Measurements The setup for measuring the fuel vapor concentration using a fiber optic probe was very similar to that used in several previous experiments involving port-fuel injection [12,13,14]. The quartz cylinder liner used in the imaging experiments was replaced with an aluminum cylinder liner fitted with a pressure transducer. The chalcogenide fibers are transparent (losses between 0.5 – 4.0 dB m^{-1}) in the IR range from approximately 2.5 to 12.5 μm. The fibers had a core diameter of 500 μm and were glass clad and plastic coated.

A 9.5 mm diameter stainless steel tube was fitted with a 4.5 mm focal length stainless steel mirror at one end. The sides of the tube were machined away to allow gas to flow across the mirror. The fibers were placed in a 6.4 mm diameter stainless steel tube with epoxy. The larger tube was threaded into the engine head and the end projected approximately 5.5 mm into the combustion chamber. The smaller tube was inserted into the larger and held in place with a compression fitting.

Light from a 75 W halogen bulb was focused onto the end of a chalcogenide fiber using an ellipsoidal reflector and a 25 mm focal length chalcogenide lens. A chopper wheel was used to modulate the light, producing a square wave. The speed of the chopper was adjusted according to the engine speed so that the period was equal to approximately 2.5 crank angle degrees.

The radiation traveled through the first fiber and was reflected from the mirror into the second fiber. Of the radiation that passed from the first fiber into the second fiber, a portion was absorbed by the fuel vapor in the gap. The spatial resolution was equal to the distance from the fibers to the mirror, which was 4.5 mm. The radiation exited the second fiber and passed through a second 25 mm focal length chalcogenide lens and a 3.3 – 3.5 μm band pass filter onto a liquid N_2 cooled InSb IR detector. The detector output was recorded, along with the corresponding crank angle from the shaft encoder and cylinder pressure using a DSP Technology Inc. engine data acquisition system. The signals from the engine were sampled with 1/8 crank angle degree resolution. The raw data files were evaluated using a Fortran code to convert differences in the signal peak height to fuel vapor concentration.

RESULTS AND DISCUSSION

FUEL SPRAY IMAGING The development of the fuel spray was visualized using 2-D laser sheet droplet imaging. Three injection scenarios were investigated: i) Start of Injection (SOI) at 90° aTDC, ii) SOI at 180° aTDC and iii) SOI at 270° aTDC. The injector used in this study is of a different generation than the injector used in the previous visualization study [6] and had a mass flow rate of approximately one half that of the previous injector. The injector was pulsed for the same duration, 2 ms, in order to compare the results to those published in a previous study [6]. Thus, the mass observed in this study was one half that of the previous study [6]. When comparing images for two different injectors, the injection duration is more important to keep constant than the mass injected. For the equivalence ratio comparisons presented later in this paper, the mass injected with the present injector was increased to match that with the prior injector.

To observe the effect of an increase in intake flow momentum, three engine speeds were investigated. They were: i) engine still with piston at BDC, ii) engine motored at 750 RPM, iii) engine motored at 1500 RPM. The resulting images for the shrouded valve configuration are shown in Figures 5-11. The bright vertical lines on the sides of each image are the result of unavoidable reflections from the cylinder liner and head support columns.

Figure 5 illustrates the development of the fuel spray when the injector is fired into a quiescent chamber. The spray development takes on a pattern that has been outlined and discussed in previous experiments [6]. The spray is characterized by an initial slug of fuel inside a hollow cone of fuel. As the pattern develops and fuel penetrates further into the cylinder, the cone enlarges and then disappears as toroidal vorticies along the sides of the cone entrain fuel back towards the top of the cylinder.

Figures 6-8 are images of the fuel spray taken at an engine speed of 750 RPM using the shrouded valves. At these speeds, varying the injection timing changes the amount of mixing and the amount of wall and piston wetting. The final fuel concentration is heavily dependant on the amount of mixing and, to a large extent, on how much fuel is deposited on the wall of the cylinder and the piston top. The amount of mixing and fuel deposition are determined by the inducted air velocity during and following injection and by piston position, and, thus, depends on injection timing and engine speed.

In the instance of injection while the intake valves are fully opened, 90° SOI (Figure 6), the fuel jet is immediately influenced by the intake flow. The intake side of the cone is suppressed and the exhaust side is entrained down towards the piston in the direction of the tumble. By 2.4 msec after SOI, the toroidal vorticies have begun to form but are inhibited by the intake flow. The bottom of the cone has rotated in the tumble direction and, by 3.4 msec after SOI, the exhaust side of the cone has impacted the top of the piston. As time progresses, the fuel remains concentrated near the center of the piston, close to the piston top. In this case, little or no fuel is convected toward the top of cylinder by toroidal vorticies. The position of the fuel droplets at 4.4 msec after SOI indicates that a significant amount of fuel was deposited on top of the piston.

In Figure 7, in which the SOI was 180°, the fuel was injected during the period of maximum tumble in the cylinder. Since the intake valves are almost closed, the intake flow is minimal and the cone is not effected during the initial 2 msec of the injection period. By 2.4 msec after injection, the jet penetrates to the location of maximum tumble and the exhaust side of the spray is entrained under the jet in the direction of tumble. The toroidal vorticies, which began to form at 1.9 msec after SOI, are suppressed by the tumble. By 3.4 msec after injection, the jet has lost its shape and fuel droplets are moving with the tumble in the center of the cylinder. The toroidal vorticies have been suppressed and the fuel, by 5.4 msec after SOI, is concentrated in the center of the cylinder, well above the piston top.

In Figure 8, the injection occurs at 270°, very late in the compression stroke. The higher pressures in the cylinder inhibit the jet penetration and fuel vaporization. The bulk motion in the cylinder still inhibits the intake side of the spray while convecting the exhaust side down and in the tumble direction. Toroidal vorticies begin to form by 2.4 msec after SOI, but by the time the cone impacts the piston top at 3.4 msec after SOI, these vorticies have disappeared. Little or no fuel is transported upward by the vorticies but fuel does appear

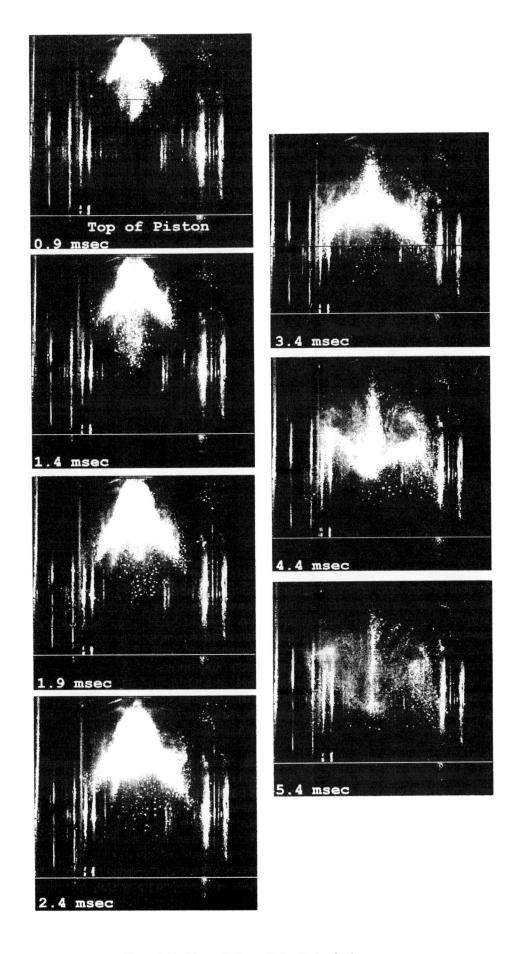

Figure 5. Fuel Spray Pattern with the Engine Static

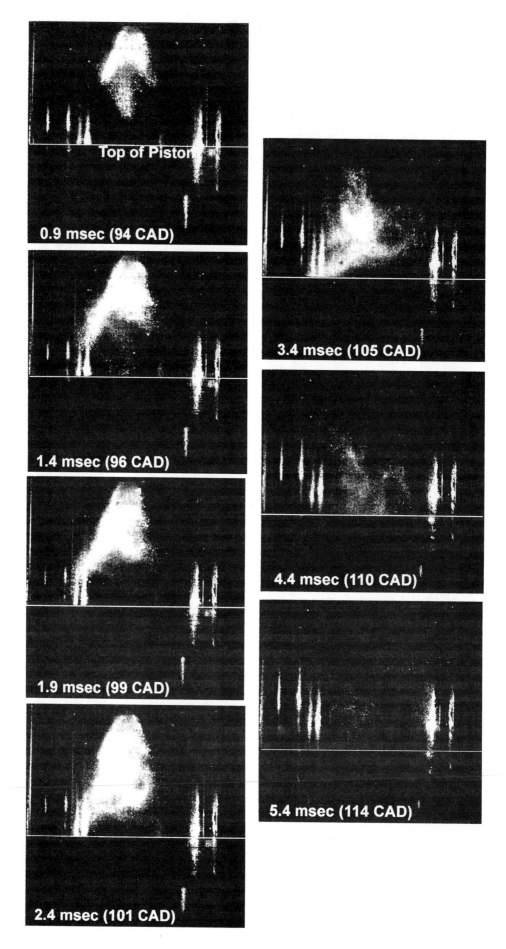

Figure 6. Fuel Spray Pattern at 750 RPM with SOI at 90°

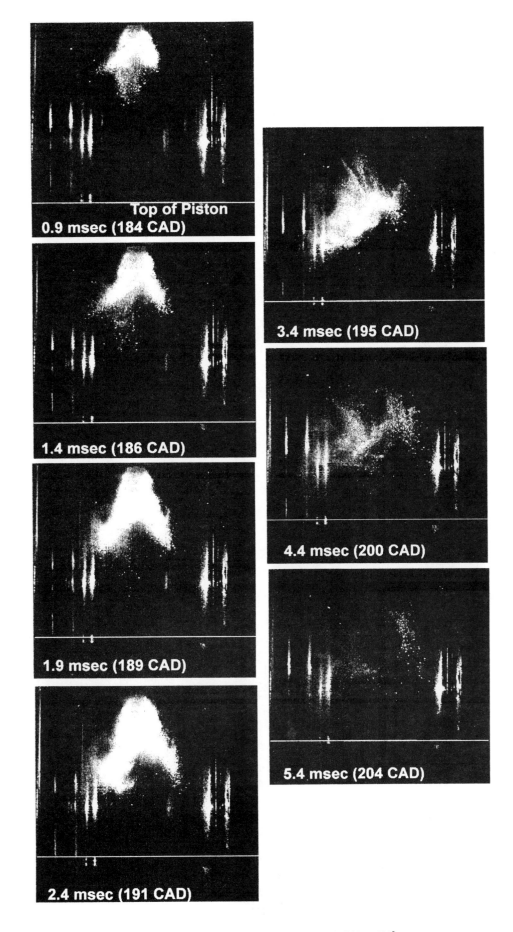

Figure 7. Fuel Spray Pattern at 750 RPM with SOI at 180°

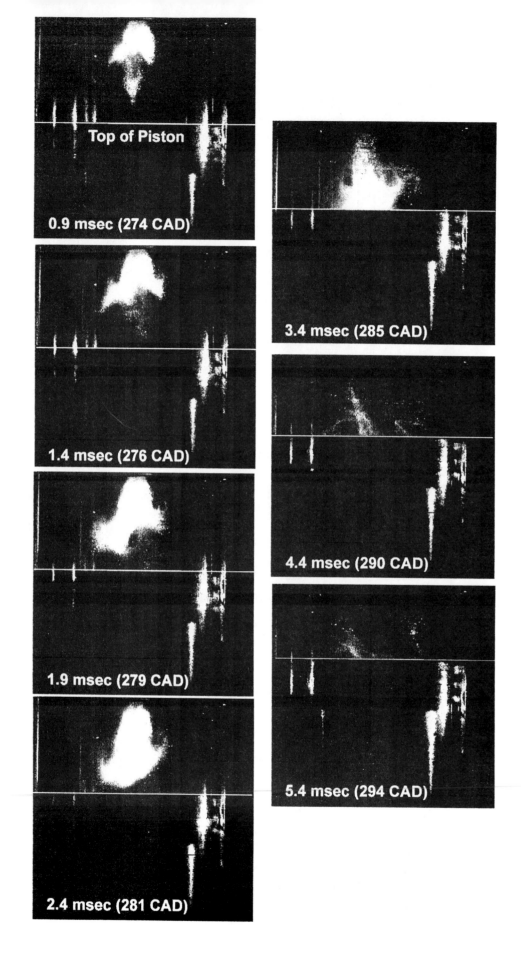

Figure 8. Fuel Spray Pattern at 750 RPM with SOI at 270°

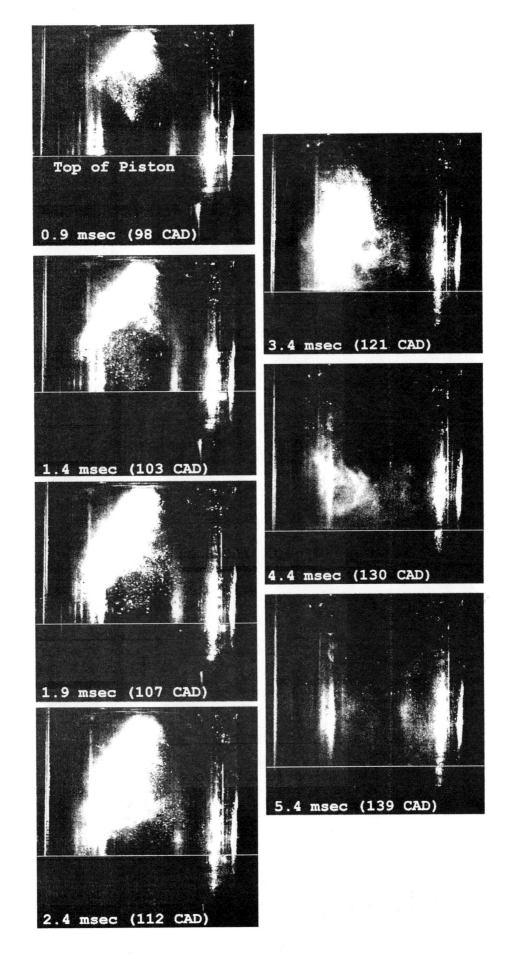

Figure 9. Fuel Spray Pattern at 1500 RPM with SOI at 90°

393

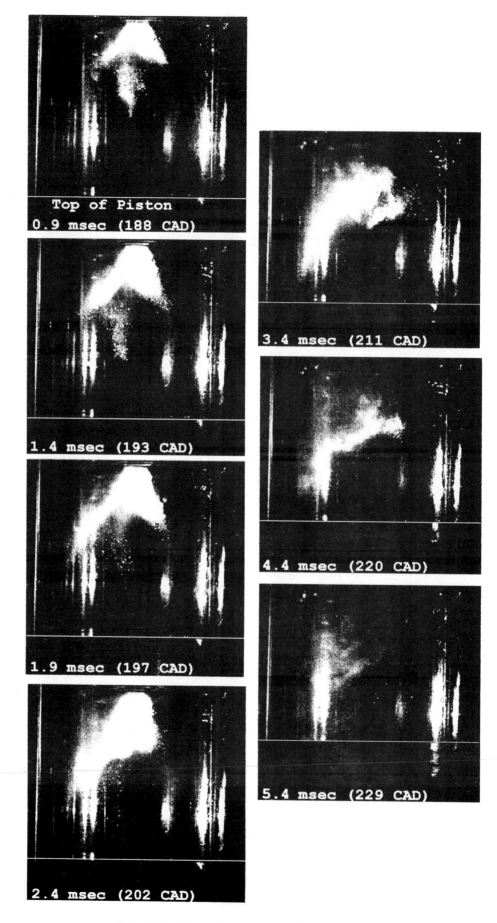

Figure 10. Fuel Spray Pattern at 1500 RPM with SOI at 180°

394

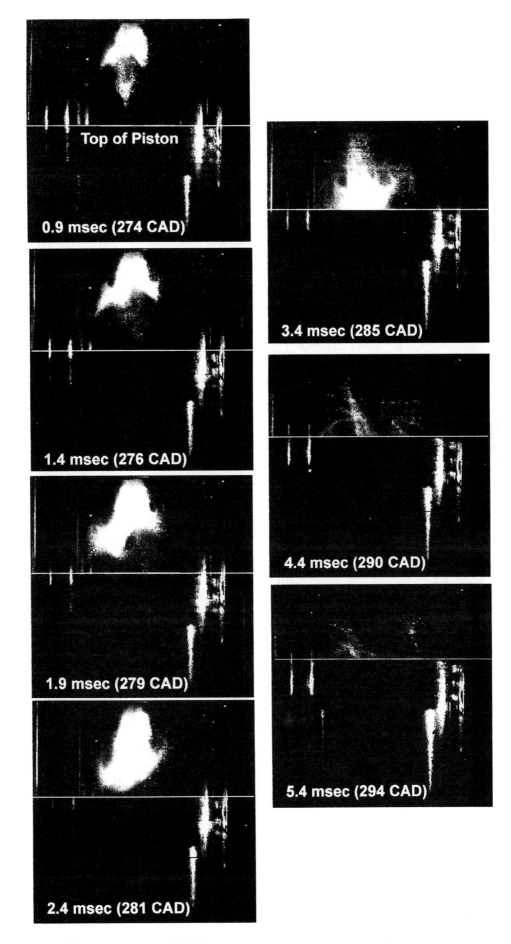

Figure 11. Fuel Spray Pattern at 1500 RPM with SOI at 270°

to be splashing off the piston. At 5.4 msec after SOI, clouds of fuel are located on the intake and exhaust sides of the cylinder, with very little fuel in the center of the cylinder. The positions of the fuel cloud indicate that a lot of fuel remains on the top of the piston.

Figures 9-11 illustrate the fuel spray development at an engine speed of 1500 RPM with the shrouded valves. At a higher engine speed, the in-cylinder flow has a much greater momentum and a corresponding effect on the fuel jet. While mixing is increased by the increased turbulence, the amount of wall and piston wetting also increases. The fuel jet is rapidly entrained and swept towards the piston top soon after SOI, and, as compared to the slower engine speed, much more wetting occurs at every injection timing.

Figure 9 shows the fuel spray development for SOI at 90°. Unlike at the slower engine speed, the formation of the hollow cone is immediately inhibited by the intake flow. By 1.9 msec after SOI, the jet begins to contact the cylinder liner under the exhaust valves. In this case, the toroidal vortices on the sides of the cone are not present at all and are replaced by a vortex that is aligned parallel to the axis of the tumble and located near the top of the piston. By 3.4 msec after injection, much more fuel has impacted the piston top than in the 750 RPM case. The jet is so greatly deflected by the intake flow that by 3.4 msec after SOI, the fuel is concentrated close to the cylinder wall on the exhaust side and near the piston top. After 3.4 msec, the fuel is swept along the top of the piston and is entrained by the tumble. By 5.4 msec after SOI, there is fuel located near the intake side of the cylinder slightly above the piston as the tumble carries it in a counter-clockwise direction.

In Figure 10, the SOI was at 180°. Unlike the case illustrated in Figure 9, the greatly slowed intake flow initially seems to only inhibit the jet slightly on the intake valve side. By 1.9 msec after SOI, the tumble in the cylinder is entraining the fuel downward on the exhaust valve side while pushing the fuel on the intake valve side closer to the cylinder centerline. By 2.4 msec after SOI, the intake side of the cone has almost completely collapsed while the exhaust side has elongated and has impacted the cylinder liner. From 3.4 msec to 5.4 msec after SOI, the fuel remains in the center of the cylinder and slightly toward the side of the exhaust valves. In this case, no roll up vortices formed at all and the fuel appears to only minimally impact the piston by 5.4 msec after SOI.

Figure 11 is the last image at 1500 RPM. In this case, SOI was at 270°. The fuel jet is immediately impacted by the in-cylinder flow, with the spray at 0.9 msec after SOI showing significant distortion. The initial slug of fuel has been pushed over to the exhaust valve side of the cone. By 1.4 msec after SOI, some fuel has already impacted the piston top and roll up vortices have begun to develop on either side of the hollow cone. By 1.9 msec after SOI, the roll up vortex on the exhaust

side has entrained some fuel back to the top of the cylinder and the vortex on the intake side is beginning to develop. By 2.4 msec after SOI, the exhaust side of the cone has rolled under the original axis of the spray and a large amount of fuel has impacted the piston. At the same time, the vortex on the exhaust side is beginning to disappear. On the intake side, the roll up vortex continues to develop. By 3.4 msec after SOI, the fuel jet is strongly impacting the piston and some fuel appears to be splashing off the piston top. The intake side roll up vortex is still visible and is transporting fuel up away from the piston. By 5.4 msec after injection, all fuel is evaporated or on the piston top, except for a small amount visible almost directly under the intake valves near the piston. The late injection scenario shown in Figures 8 and 11 were the only cases where the bulk flow transported the fuel droplets to the intake side of the cylinder.

At higher engine speeds, it appears that more fuel impacts the piston top and cylinder walls than at lower speeds. This can be attributed to a combination of two effects. The first is that the stronger bulk motion of the tumble convects the fuel jet more. The second is that the piston is moving much faster, and thus covers a greater range of motion during the injection event. The greater amount of piston wetting will cause the equivalence ratio in the gas phase to be lower than expected, even as the higher engine speed promotes more mixing.

FUEL VAPOR CONCENTRATION MEASUREMENTS
While the laser sheet droplet images provide valuable information about the fuel distribution close to the injection event, substantial mixing occurs after the time the last images were taken. To study fuel vapor distribution after that time, a fiber optic probe was used. While the probe only yields information from within a small volume, it does provide a basis for examining the effects of changing parameters on the fuel/air mixing. In this case, since the volume is located within what would be the spark plug gap, this is arguably the location of the most importance in the engine.

The fiber optic probe was located in the engine where the spark plug would be if the engine were to be fired. For these experiments, we used the same injection timings of SOI at 90°, 180° and 270°. To examine the effect of changing engine speed on the local fuel vapor concentration and the local equivalence ratio, engine speeds of 750 and 1500 RPM were investigated. Since the engine was not fired, the injector was triggered only every sixth cycle to help reduce the amount of residual fuel remaining in the cylinder at the beginning of the sixth cycle. To yield an overall equivalence ratio of $\phi=1.0$, the injector was pulsed for 4.3 ms, which delivered 46 mg of fuel to the cylinder.

Figures 12 and 13 show the fuel vapor concentration history for the high tumble valve configuration with a SOI of 180° for 1500 RPM and

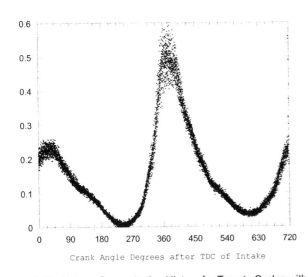

Figure 12. Fuel Vapor Concentration History for Twenty Cycles with High Tumble Valves at 1500 RPM and SOI at 180°

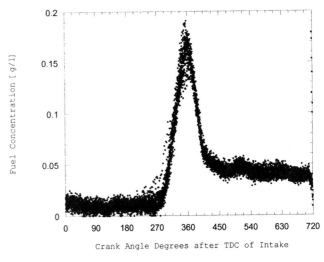

Figure 13. Fuel Vapor Concentration History for Twenty Cycles with High Tumble Valves at 750 RPM and SOI at 180°

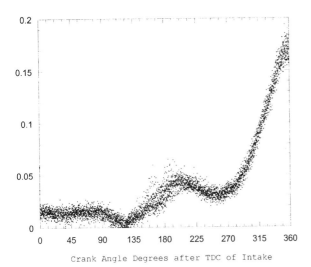

Figure 14. Fuel Vapor Concentration History for Twenty Cycles with High Tumble Valves at 750 RPM and SOI at 90°

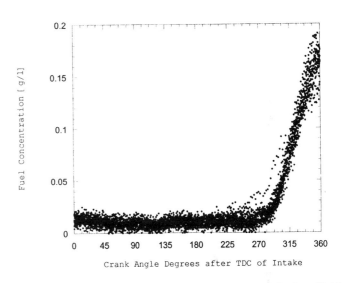

Figure 15. Fuel Vapor Concentration History for Twenty Cycles with High Tumble Valves at 750 RPM and SOI at 180°

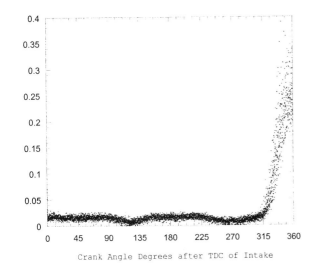

Figure 16. Fuel Vapor Concentration History for Twenty Cycles with High Tumble Valves at 750 RPM and SOI at 270°

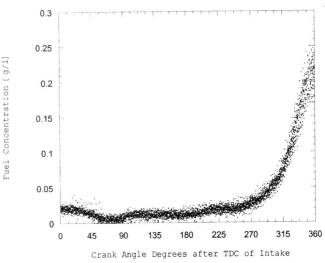

Figure 17. Fuel Vapor Concentration History for Twenty Cycles with Standard Valves at 750 RPM and SOI at 90°

750 RPM respectively measured in 20 individual cycles. The high speed case shows a larger amount of residual fuel in the cylinder, as shown by the high concentrations at the beginning of the intake stroke. The low speed case, however, does not show this extreme amount of residual fuel. In this case, the fuel has more time to vaporize before being exhausted. Also, in the high tumble case, the laser sheet images suggest that more fuel impacts the piston crown and cylinder liner (Figure 10) at the higher engine speed than at the lower engine speed (Figure 7).

The fuel on the piston vaporizes slowly, resulting in high concentrations of fuel vapor remaining in the engine after several cycles.

Figures 14, 15 and 16 show the fuel vapor concentration histories during the intake and compression strokes for the low speed, high tumble case for each injection timing. An early injection timing, SOI of 90° (Figure 14), results in an increase in concentration about 25° after SOI, with the fuel concentration peaking and falling back before beginning to rise again. An increase in fuel vapor concentration can be caused by the in-cylinder flow transporting fuel to the probe volume or an increase in pressure, which causes the density of both the air and fuel in the probe volume to increase. In the case of the first peak in the 90° SOI case, the increase is caused by the tumble transporting the fuel vapor up to and past the probe volume.

This peak is not evident in the 180° (Figure 15) and 270° (Figure 16) SOI cases. In all cases, the fuel vapor concentration during the initial intake stroke shows an oscillating pattern. The frequency of these oscillations corresponds to the tumble frequency. At a tumble ratio of 3.4, the bulk flow will rotate approximately once every 106°. In the 180° case the pattern is difficult to see due to lower levels of residual fuel and the cycle-to-cycle variation, but it is clearly evident in the 90° and 270° SOI cases. These oscillations in the fuel vapor concentration during the intake stroke in Figures 14 and 16 suggest that as the intake valves open, the shrouds on the lower half of the valves block the fresh air from immediately diluting the air around the probe volume. As the fresh air enters the cylinder, it mixes with the residual fuel vapor and, as the tumble brings the fresh air up to the probe, the fuel concentration gradually decreases. The same tumble, however, also brings the displaced, richer fuel vapor mixture back up to the probe on the next rotation. In the case of the 90° SOI (Figure 14) case, the tumble also carries injected fuel along with it.

Figure 17 is the fuel vapor concentration for the standard valve configuration with the engine run at 750 RPM and a SOI of 90°. In this case, the fuel vapor oscillations are not seen since the tumble effects are much smaller, with a tumble index of approximately 1.4. The initial fuel concentration around the probe is diluted by the intake flow, but the concentration then slowly increases after SOI.

The fuel vapor concentration is affected by pressure in that a higher pressure results in higher densities for both fuel and air in the cylinder. At different engine speeds, the in-cylinder pressure varies, increasing with engine speed. To present comparable results from different engine speeds, the fuel vapor concentration was converted to an equivalence ratio. Using the fuel concentration information and cylinder pressure measurements, the equivalence ratio was calculated in the manner illustrated below.

The density of the fuel is measured by the probe as ρ_{fuel} in grams per liter. The temperature was calculated by assuming isentropic compression and using the formula:

$$T = T_o \left(\frac{P}{P_o} \right)^{\frac{(k-1)}{k}} \tag{1}$$

T_o and P_o are taken as the intake conditions, 101.1 kPa and 300K and k is the specific heat ratio, which is taken as 1.3 for the mixture and P is the cylinder pressure measured by the transducer.

Using ρ_{fuel} and T, the partial pressure of the fuel is calculated by:

$$P_f = \rho_f R_f T \tag{2}$$

where R_f is the gas constant for octane, which is 0.0729 kJ/kg-K.

To find the equivalence ratio, the density of the air in the probe volume needed to be determined. The partial pressure of the air is calculated by:

$$P_a = P - P_f \tag{3}$$

Using the result from equation three, the density of the air is then calculated as:

$$\rho_a = \frac{P_a}{R_a T} \tag{4}$$

where R_a is the gas constant for air, 0.2867 kJ/kg-K.

The equivalence ratio is then determined by using the relationship:

$$\phi = \frac{\rho_f / \rho_a}{FAR_{stoich}} \tag{5}$$

FAR_{stoich} is the fuel-air ratio at stoichiometric conditions, which is 0.068.

Figure 18 shows the results of transforming the fuel concentration data into an equivalence ratio for the high tumble valves at 750 RPM and 90° SOI. The steep rise in the equivalence ratio following injection is clearly present. This, again, is an effect of the flow bringing fuel to the probe volume. After correcting for pressure effects, the local equivalence ratio then decreases and stabilizes around TDC. The plot also shows that the cycle-to-cycle variation is larger during the early part of the intake stroke, where the residual vapor and the injected fuel are mixing with the intake flow. The variation decreases significantly later in the compression stroke and the equivalence ratio levels off near TDC.

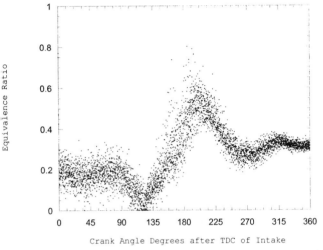

Figure 18. Equivalence Ratio History for Twenty Cycles with High Tumble Valves at 750 RPM and SOI of 90°.

Since the engine is not fired, large amounts of residual fuel remain in the cylinder even after several cycles with no fuel injection (Figures 12 and 13). In order to observe the effect that injection has on the local equivalence ratio, the fuel concentration from the cycle immediately before the cycle with injection was subtracted from the cycle with the injection. This resulted in an overall decrease in fuel vapor concentration and local equivalence ratio, but allows the observation of the changes in the cylinder due solely to the injection event.

Figure 19 shows the ensemble average over 20 cycles of both the cycle with injection and the cycle immediately preceding it for the 750 rpm and 90° SOI case. By subtracting the concentration of the preceding cycle, the fuel vapor concentration prior to injection goes to zero, while the increases in the fuel vapor concentration after injection are still apparent. Figure 20 shows the equivalence ratio when the results of Figure 19 are converted to an equivalence ratio. These figures show that the effect of the previous cycle near the point of ignition (300°-360°) is negligible.

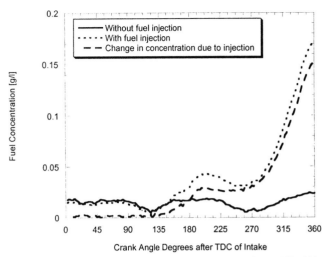

Figure 19. Ensemble Average of Twenty Cycles' Fuel Vapor Concentration With and Without Fuel Injection at 750 RPM and SOI of 90° for High Tumble Valves

Figure 21 and 22 show the effect of injection timing on the local equivalence ratio at 750 RPM. Figure 21 is the high tumble valve configuration while Figure 22 is the standard valve configuration. In Figure 21, the equivalence ratio at TDC is the highest for the 270° SOI case, with the 180° SOI case being the next highest and the 90° SOI case having the lowest value. The pattern of

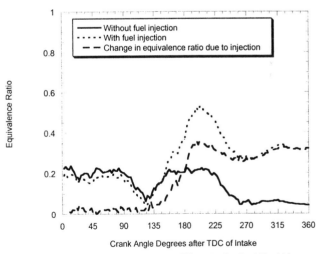

Figure 20. Ensemble Average of Twenty Cycles' Fuel Vapor Concentration With and Without Fuel Injection at 750 RPM and SOI of 90° for High Tumble Valves

the equivalence ratio increase is much different for the high tumble case from that of the standard valves. With a 90° SOI, the high tumble valves show a sharp increase in the equivalence ratio starting approximately 45° after SOI. The 180° SOI case shows a gradual rise commencing at 45° after SOI and the 270° SOI case

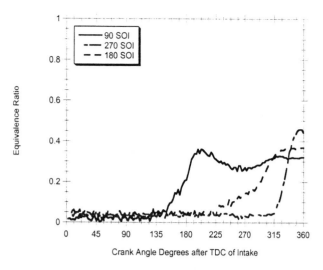

Figure 21. Ensemble Average of Twenty Cycles' Equivalence Ratio History
With High Tumble Valves at 750 RPM

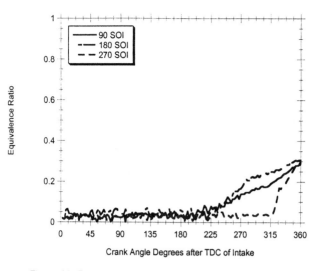

Figure 22. Ensemble Average of Twenty Cycles' Equivalence Ratio
History With Standard Valves at 750 RPM

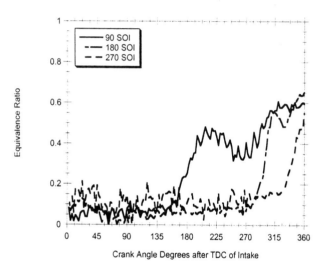

Figure 23. Ensemble Average of Twenty Cycles' Equivalence Ratio History
With High Tumble Valves at 1500 RPM

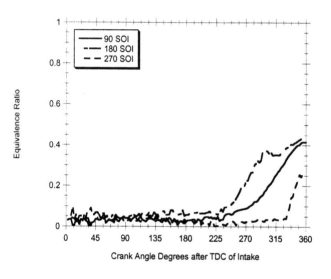

Figure 24. Ensemble Average of Twenty Cycles' Equivalence Ratio
History With Standard Valves at 1500 RPM

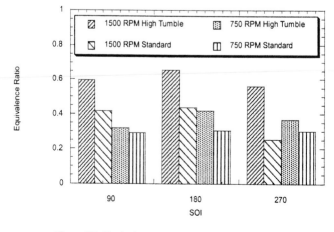

Figure 25. Equivalence Ratio at Top Dead Center for All Cases

shows a very sharp rise in equivalence ratio at 45° after SOI. The higher equivalence ratio in the 270° SOI case is consistent with Figure 8, which shows the remnants of the tumble pushing the fuel droplets towards the probe location. In every case, the equivalence ratios are lower than expected. This is partly due to fuel impacting the top of the piston and partly because of the high intake manifold pressure, which inhibits fuel vaporization. The fuel on top of the piston not only leads to large amounts of residual fuel in the cylinder (Figures 12 and 13) but also decreases the local equivalence ratio.

In Figure 22, which shows the equivalence ratios for the standard valve case at 750 RPM, the equivalence ratio for each injection timing does not show the same increase at approximately 45° after SOI that the high tumble valves demonstrate. In addition, the local equivalence ratios at TDC are roughly the same, regardless of the injection timing. These levels are up to 20% lower than those of the high tumble cases. This suggests that the fuel jet did not experience as much mixing in the cylinder and that the lower amount of tumble does not promote as much vaporization and mixing as the high tumble.

Figures 23 and 24 show the effect of injection timing on the equivalence ratio at 1500 RPM. Figure 23 shows the information for high tumble valves and Figure 24 is the same information for standard valves. Comparing the two figures to Figures 21 and 22, the equivalence ratio at the lower engine speeds is lower than at higher engine speeds for the same injection timing, with the exception of the 270° case with standard valves. In this instance, without the high tumble to deflect the fuel spray towards the probe, the fuel does not have time to diffuse enough to yield a high concentration around the probe. The difference in equivalence ratio between engine speeds, with the exception of the 270° case, may be the result of increased mixing due to higher turbulence in the cylinder at higher speeds.

The effects of injection timing vary with the engine speed. In Figure 23, the high tumble case, the equivalence ratio shows the same pattern for much of the cycle. Due to greater turbulence and cycle-to-cycle variations, the equivalence ratio at the beginning of the stroke shows some erratic behavior. The same pattern of a rise in equivalence ratio is present occurring approximately 50°-90° after SOI. At TDC, the 180° SOI case has the highest equivalence ratio, followed by the 90° SOI and then the 270° SOI case. While the 180° and 90° SOI cases do not show a great difference in equivalence ratio, they are both significantly greater than that of the 270° SOI case. This indicates that while the mixing and vaporization of the fuel spray may be enhanced by the higher engine speeds, the lack of time for diffusion is depressing the late injection (270° SOI) TDC equivalence ratio. For the 180° SOI and 90° SOI cases, the images indicate a higher amount of piston wetting at a SOI of 90° than at a SOI of 180°. The

amount of fuel on the piston top could be the difference between the TDC equivalence ratios in these cases.

The standard valve case shown in Figure 24 has the equivalence ratio for the 180° and 90° SOI cases both beginning to increase at the same time, approximately 225°. This shows that the standard valve case does not mix the intake air and fuel vapor as well as the high tumble case. The gradual increase in the equivalence ratio up to TDC by both the 180° SOI and 90° SOI cases may be caused by a rich region close to the piston, which follows the piston throughout the compression stroke. Without high levels of tumble to mix the gasses, the equivalence ratio will remain low and increase as the piston approaches the probe. Like the high tumble case, the 90° SOI case has a slightly lower equivalence ratio at TDC than the 180° SOI case. This is again the result of the increased piston wetting at 90° SOI. This difference in equivalence ratio between the two cases could be due to increased wetting of the piston top and/or less time for the fuel to be convected to the probe.

Figure 25 shows the top dead center equivalence ratios for all cases investigated in this study. As can be seen in the chart, the TDC equivalence ratios are much smaller than expected, especially in the early injection cases. While most of this can be attributed to piston wetting, the possibility exists that the mixture, even when early injection is used, is far from homogeneous. The present results agree with those of a prior multi-dimensional modeling study [17] that focused on DISI engines with early injection. The model demonstrated that the air/fuel ratio is non-uniform at TDC even with early injection. The model also showed that increasing swirl or tumble made the air/fuel distribution at TDC less non-uniform. The relatively low equivalence ratios at TDC found in the present study imply that a combination of an injection scenario that minimizes piston wetting and an intake flow that maximizes mixing would increase the TDC equivalence ratio near the spark plug significantly.

CONCLUSIONS

1. The intake flow configuration has a significant effect on fuel distribution. Increasing the amount of tumble in the cylinder gases severely distorts the fuel jet. As the engine speed is increased, the effect of the intake flow upon the fuel spray is increased. Images taken late in the compression stroke suggest that bulk motion due to tumble exists very late into the compression stroke.

2. The amount of tumble in the cylinder has a significant effect on the local equivalence ratio at TDC. At high engine speeds, an increase in tumble at 1500 RPM results in a factor of 2 increase in local equivalence ratios at TDC with a late injection timing (270° SOI). The other injection timings resulted in an approximately 40% increase over the standard valve configuration. At 750

RPM, the increase in tumble results in a 10% increase in the equivalence ratio at TDC for the 90° SOI case, a 20% increase for the 180° SOI case and a 55% increase for the 270° SOI case.

4.   Piston wetting has a significant effect on the equivalence ratio at TDC. In the cases that the laser sheet imaging showed to have more piston wetting, the equivalence ratio at TDC was lower. At 1500 RPM, the 90° SOI shows more piston wetting than the 180° SOI case and consequently the 90° SOI case has a lower TDC equivalence ratio than the 180° SOI case. The 270° SOI case at 1500 RPM has a much lower TDC equivalence ratio than the other two injection timings, which corresponds to more piston wetting as well as less time to diffuse.

ACKNOWLEDGEMENTS

This research was funded by General Motors and conducted using the facilities of the General Motors Foundation Combustion Sciences and Automotive Research Labs on the UT Campus and the fuel injectors were provided by General Motors Corporation and Zexel. The Ford Motor Company provided funding for the development of the fiber optic spark plug. The authors would also like to thank Dr. Rudolf Stanglmaier of Southwest Research Institute and Dr. Paul Najt of General Motors for their assistance and guidance with this work. Support was provided to Mr. T. Alger in the form of a Thrust 2000 Fellowship at UT as well as a National Defense Science and Engineering Graduate Fellowship.

REFERENCES

1. Takagi, Y., T. Itoh, S. Muranaka, A. Iiyama, Y. Iwakiri, T.Urushihara and K. Naitoh (1998), "Simultaneous Attainment of Low Fuel Consumption, High Output Power and Low Exhaust Emissions in Direct Injection SI Engines", SAE Paper 980149.

2. Anderson, R.W., D. D. Brehob, J. Yang, J.K. Vallance and R. M. Whiteaker (1996), "Understanding the Thermodynamics Of Direct Injection Spark Ignition (DISI) Combustion Systems: An Analytical and Experimental Investigation", SAE Paper 962018.

3. Iwamoto, Y., K. Noma, T. Yamauchi and O. Nakayama (1997), "Development of the Direct Injection Gasoline Engine", SAE Paper 970541.

4. Noma, K., Y. Iwamoto, N. Murakami, K. Iida and O. Nakayama (1998), "Optimized Gasoline Direct Injection Engine for the European Market", SAE Paper 980150.

5. Harada, J., T. Tomodita, H. Mizuno, Z. Mashiki and Y. Ito (1997), "Development of the Direct Injection Gasoline Engine", SAE Paper 970540.

6. Stanglmaier, R., M. Hall and R. Matthews (1998), "Fuel-Spray/Charge-Motion Interaction within the Cylinder of a Direct-Injected, 4-Valve, SI Engine", SAE Paper 980155.

7. Shiraishi, T., Y. Nakayama, T. Nogi and M. Ohsuga (1998), "Effect of Spray Characteristics on Combustion in a Direct Injection Spark Ignition Engine", SAE Paper 980156.

8. Kano, M., K. Saito, M. Basaki, S. Matsushita and T. Gohno (1998), "Analysis of Mixture Formation of Direct Injection Gasoline Engine", SAE Paper 980157.

9. Davy, M.H., P. Williams and R.W. Anderson (1998) "Effects of Injection Timing on Liquid-Phase Fuel Distributions in a Centrally-Injected Four-Valve Direct-Injection Spark-Ignition Engine", SAE Paper 98FL-375.

10. Han, Z., R.D. Reitz, J. Yang and R.W. Anderson (1997), "Effects of Injection Timing on Air-Fuel Mixing in a Direct-Injection Spark-Ignition Engine", SAE Paper 970625.

11. Preussner, C., C. Doring, S. Fehler and S. Kampmann (1998), "GDI: Interaction Between Mixture Preparation, Combustion System and Injector Performance", SAE Paper 980498.

12. Koenig, M. and M. Hall (1996), "A Fiber Optic Probe to Measure Pre-Combustion In-Cylinder Fuel-Air Ratio Fluctuations in Production Engines", presented at the 26th International Combustion Symposium, Naples, Italy 8/96.

13. Koenig, M. and M. Hall (1997), "Measurements of Local In-Cylinder Fuel Concentration Fluctuations in a Firing SI Engine", SAE Paper 971644

14. Koenig, M. and M. Hall (1998), "Cycle-Resolved Measurements Of Pre-Combustion Fuel Concentration near the Spark Plug in a Gasoline SI Engine", SAE Paper 981053.

15. Kume, T., Y. Iwamoto, K. Iida, M. Murakami, K. Akishino, and H. Ando (1996), "Combustion Control Technologies for Direct Injection SI Engine", SAE Paper 960600

16. Zur Loye, A.O., "A Software-Based Controller for Generation of Synchronized Pulses for Use with Rotating Machinery", Sandia National Laboratories Unlimited Release SAND89-8247, 1989.

17. Han, K., R. Reitz, P. Claybaker, C. Rutland, J. Yang, R. Anderson (1996), "Modeling the Effects of Intake Flow Structures on Fuel/Air Mixing in a Direct-Injected Spark-Ignition Engine", SAE Paper 961192

1999-01-0502

The Effect of In-Cylinder Wall Wetting Location on the HC Emissions from SI Engines

Rudolf H. Stanglmaier
Southwest Research Institute

Jianwen Li and Ronald D. Matthews
The University of Texas at Austin

ABSTRACT

The effect of combustion chamber wall-wetting on the emissions of unburned and partially-burned hydrocarbons (HCs) from gasoline-fueled SI engines was investigated experimentally. A spark-plug mounted directional injection probe was developed to study the fate of liquid fuel which impinges on different surfaces of the combustion chamber, and to quantify its contribution to the HC emissions from direct-injected (DI) and port-fuel injected (PFI) engines. With this probe, a controlled amount of liquid fuel was deposited on a given location within the combustion chamber at a desired crank angle while the engine was operated on pre-mixed LPG. Thus, with this technique, the HC emissions due to in-cylinder wall wetting were studied independently of all other HC sources.

Results from these tests show that the location where liquid fuel impinges on the combustion chamber has a very important effect on the resulting HC emissions. The highest HC emissions resulted from fuel impingement on the exhaust-side of the cylinder liner, and wetting on the piston-top was second in importance. The HC emissions due to this mechanism were found to be insensitive to the timing of the injection and to the coolant temperature. This surprising result is indicative of low vaporization rates for the liquid gasoline on the combustion chamber surfaces. The results presented here indicate that the impingement of liquid fuel on the piston top and cylinder liner is an important source of HC emissions from direct-injection spark-ignition (DISI) engines, and that this mechanism is at least partially responsible for their high HC emissions.

INTRODUCTION AND BACKGROUND

The dominant sources of engine-out hydrocarbon emissions from homogeneous charge spark ignition engines are generally accepted to include:

i) fuel or fuel/air mixture protected from the combustion process in:
   a) crevices (dominated by the piston top-land crevice),
   b) oil films (especially when the oil is cold), and
   c) deposits.
ii) exhaust valve leakage.
iii) liquid fuel effects (especially during cold start and warm-up).
iv) misfires, especially during cranking, cold operation, and decelerations.

For direct injection, stratified charge gasoline engines, the crevice, oil film, and deposit effects may be minimized by the stratification, but may be replaced by the following sources: i) bulk flame quenching, ii) decreased in-cylinder and exhaust oxidation resulting from lower average temperatures due to the overall lean operation, iii) reduced oxidation in the catalyst due to lower exhaust temperatures, and iv) liquid fuel impingement on in-cylinder surfaces. The present study is focused upon liquid fuel effects on HC emissions from both port fuel injection (PFI) and direct injection, spark ignition (DISI) engines.

In PFI engines, some of the fuel that does not vaporize completely in the port is deposited on the walls of the combustion chamber, where it can escape combustion. This behavior occurs predominantly under cold-start and warm-up conditions and is generally referred to as wall wetting. A number of experiments have demonstrated that wall wetting in PFI engines is a significant mechanism for engine-out HC emissions during cold operation (e.g. Cheng et al., 1993; Alkidas, 1994; Alkidas and Drews, 1996; Fox et al., 1992; Fulcher et al., 1995). In 4-valve PFI engines, wall wetting occurs on one of two locations within the combustion chamber: i) open intake valve (OIV) injection results predominantly in fuel droplet impingement on the cylinder liner, directly under the exhaust valves; and ii) closed intake valve (CIV) injection results mainly in droplet impingement on the piston top (e.g. Stanglmaier et al., 1997a; Kelly-Zion *et al.*, 1997). In DISI engines, wall wetting can occur on several locations within the combustion chamber, depending on the engine geometry, spray cone angle, injection timing and fuel pressure. Fuel droplet impingement on the walls may occur to some extent over all engine operating conditions, but is most likely during late injection (stratified charge) mode.

The location within the combustion chamber where wetting occurs has a large effect on the fate of this fuel. Liquid fuel which impinges on the cylinder liner can be scraped-up by the piston, forced into the top-land crevice, and protected from combustion (Takeda *et al.*, 1995; Saito *et al.*, 1995). In contrast, liquid fuel which is deposited on the piston top cannot be easily collected into any crevices and it is generally believed that liquid fuel deposited on the piston top is more likely to be consumed during the combustion and post-flame oxidation processes than liquid fuel which was deposited on the cylinder liner. The work presented in this paper is part of an effort to gain a better understanding of the relationship between in-cylinder wall wetting and HC emissions from spark ignition engines, and the focus of the present study was to investigate the effects of wall wetting location and injection timing.

A procedure was developed to allow the liquid fuel source to be essentially isolated from the other HC sources. A 4-valve engine was operated on a gaseous fuel, thereby eliminating liquid fuel sources except for that imposed by injecting a controlled amount of liquid fuel on a chosen surface within the cylinder. The gaseous fuel used was liquefied petroleum gas (LPG) composed of ~ 95% propane, which is insoluble in engine oil (Kaiser *et al.*, 1994). While liquid fuel was injected, the equivalence ratio was adjusted to the same value as during LPG-only operation, thereby maintaining post-flame oxidation temperatures nearly constant. Although the overall fueling was dominated by LPG, the flow of LPG was necessarily decreased somewhat (by ~ 15 %) during liquid fuel injection to maintain the same overall lean stoichiometry. In turn, crevice packing by LPG was also decreased somewhat. Therefore, any increases in HC emissions observed during liquid fuel injection is the net effect of the HC increase via the liquid fuel effects and the relatively slight decrease in HCs due to lower LPG loading.

METHODOLOGY

An innovative technique for studying wall wetting was developed as part of this study. This technique makes it possible to isolate the effects of in-cylinder wall wetting from other sources of emissions of unburned and partially-burned HCs in gasoline fueled engines, and is described in detail in this section.

DIRECTIONAL INJECTION PROBE

A prototype spark-plug-mounted directional injection probe was developed as a tool for conducting

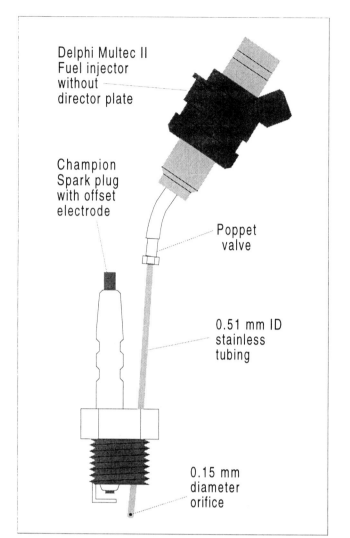

Figure 1. Spark-plug mounted directional injection probe.

wall-wetting experiments in unmodified production SI engines, and is shown schematically in Fig. 1. This injector probe uses a conventional port fuel injector (Delphi Multec II), with the director plate removed, for metering the fuel quantity and timing of the injection event. The injector is connected with flexible tubing to a small poppet valve with a cracking pressure of 280 kPa (Zizelman *et al.*, 1992). The bottom of the poppet valve is attached to a short section of 1.6 mm O.D, 0.51mm I.D. (1/16" O.D., 0.020" I.D.) stainless-steel tubing which passes through an orifice in an offset-electrode spark-plug designed for cylinder pressure measurements[*] (Champion Model 304-802). This directional injection probe is designed to introduce a small stream of fuel directly into the combustion chamber, where it can be aimed at a desired wall location. Two different tube-end designs were required to perform the wall-wetting location studies described herein: i) a side hole location (depicted in Fig. 1) was used to deposit fuel on the cylinder liner, and ii) an axial hole location was used to

---

[*] The ground electrode on this spark plug was relocated so as to not interfere with the tubing.

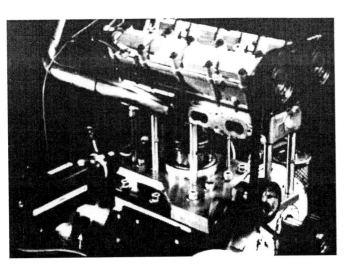

Figure 2. Optically accessible engine used to verify the targeting of the injection probe.

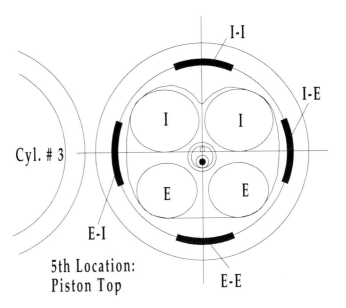

Figure 3. Schematic of the Quad-4 combustion chamber, showing the four wall-wetting target areas around the periphery of the cylinder liner.

I-I

I-E

Cyl. # 3

E-I

5th Location:
Piston Top

E-E

deposit fuel on the piston top. Both tube-end holes had a diameter of 0.15 mm (0.006").

INJECTION PROBE VERIFICATION TESTS

Prior to using this new tool to gather data in a firing engine, a number of tests were performed to verify its operation. First, to verify the dynamic response of the injector probe, a pulsed strobe-light was used to observe the fuel jet emerging from the injection probe at an equivalent firing rate of 1200 rpm. This observation verified that there was a distinct beginning and end to the injection event, and that a stream of liquid fuel emerged perpendicular from the injection probe tip (with side-hole configuration) for a distance of at least 15 cm. Only an occasional droplet of fuel was observed to dribble off the

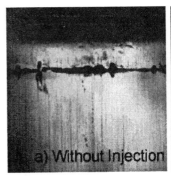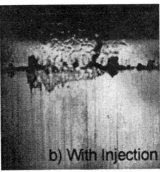

Figure 4. Back-lit images of the cylinder liner under the exhaust port, demonstrating the liner wetting achieved with the directional injection probe. Both images were taken at 226 °ATDC intake, with the engine motored at 750 rpm. No injection was performed for Fig. 4a, whereas injection from 31 - 40 °ATDC (2 ms pulsewidth) was performed for Fig. 4b. Note: the dark horizontal line on these pictures is generated by oil accumulation by the top compression ring.

tip of the probe every few seconds (see Appendix A for a discussion of the significance). A volumetric flow calibration of the injection probe was obtained to verify its linearity, and this calibration was repeated at different simulated engine speeds to ensure no frequency dependency of the injector-probe cavity.

Finally, the most crucial test of the operation of this injector probe was to verify that liquid fuel was actually deposited on the desired location within the combustion chamber of an actual engine. This test was performed with the use of an optically accessible engine which has been described previously (Stanglmaier, 1997a; Stanglmaier et al., 1997b), and is shown in Figure 2. The optical access engine is geometrically identical to the production engine used to make the HC measurements reported here (GM Quad-4). The injector probe was aimed at four different locations around the periphery of the cylinder liner (see Figure 3): i) under the exhaust port, ii) under the intake port, and iii) midway between the intake and exhaust ports. The optical access engine was motored at different speeds, and fuel was injected through the probe at various crank angles during the intake stroke to verify that the liquid fuel stream impacted at the desired liner location. Figure 4 shows an example of these tests. The pictures in Fig. 4 are back-lit digital photographs of a portion of the cylinder liner, directly under the exhaust port (viewing area approx. 5 cm by 5 cm). The dark horizontal line on these images is caused by oil accumulation by the top compression ring at the top of its stroke. These images were taken at 226 °ATDC intake, with the engine motored at 750 rpm. No injection was performed for Fig. 4a, whereas injection from 31° - 40° (2 ms pulsewidth) was performed for Fig. 4b. Liquid fuel impingement on the upper portion of the liner, just above the dark oil line, is visible in Fig. 4b.

The axial-hole configuration of the injection probe, for depositing fuel on the piston top, could not be checked with this optical arrangement. However, it is reasonable to assume that a fuel jet injected directly downward from the spark-plug will impinge on the piston top.

EXPERIMENTAL PROCEDURE

These experiments were performed on one cylinder (Cyl. # 4) of a production GM Quad-4 engine. The exhaust manifold of the engine was modified so that the exhaust stream from cylinder # 4 was isolated from the other three cylinders. A wide-range oxygen sensor (Horiba Model 101λ) was used to measure the equivalence ratio of this cylinder, and a Flame Ionization Detector (Horiba FIA-34A-2) was used to measure the hydrocarbon (HC) concentration in the exhaust stream.

The aim of this study was to examine the influence of in-cylinder wall wetting on the HC emissions. In order to minimize changes to the overall combustion process and the burned gas temperatures (which affect the post-flame HC oxidation process), the engine was fueled with LPG (introduced upstream of the throttle via a venturi mixer) and only a small amount of liquid fuel (California Phase 2 RFG) was injected into the cylinder. For all cases studied, about 85% of the fuel was LPG, and about 15% liquid fuel (1.45 mg/injection $\pm$ 3%), although this ratio varied slightly depending on the wetting location. The air/fuel equivalence ratio was maintained slightly lean at $\phi = 0.9$ ($\lambda = 1.1$) for all tests, to minimize the sensitivity of the HC emissions to minor variations in the excess air ratio (λ). All measurements were conducted at 1000 rpm and at either MBT (minimum advance for best torque) or advanced (44 °BTDC) ignition timing. To simulate both warm idle and cold idle, two coolant temperatures were used: 90 ± 2 °C and 36 ± 1 °C. The manifold air pressure (MAP) was 32.5 kPa for the warm idle case and 34.5 kPa for the cold idle measurements. Typically, the engine speed and load were set, the LPG mixer was adjusted so that the exhaust oxygen sensor read $\phi_{LPG} = 0.9$, and a baseline HC measurement was taken. The liquid fuel injection was then enabled and the resulting equivalence ratio was recorded (LPG plus liquid fuel, ϕ_{LPG+LF}). Then, the LPG flow was reduced until ϕ_{LPG+LF} reached a value of 0.9 once again and a steady-state HC measurement was recorded. Finally, the liquid fuel injector was turned off and the exhaust oxygen sensor value was recorded again as ϕ_{LPG-LF} (LPG minus liquid fuel).

RESULTS AND DISCUSSION

EFFECT OF WETTING LOCATION

The HC emissions for all wetting locations are shown in Fig. 5, for an injection timing of 160 °ATDC, MBT ignition timing, and a coolant temperature of 90 °C. The baseline HC emissions on LPG, at the same

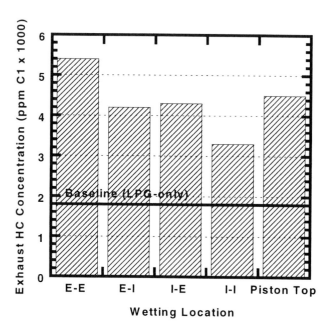

Figure 5. Effects of wall wetting location on HC emissions. Injection timing = 160 °ATDC, MBT ignition timing, coolant temperature 90 °C. The baseline emissions with LPG-only were obtained at the same equivalence ratio (ϕ_{LPG} = 0.9). All experiments performed on same day.

equivalence ratio ($\phi_{LPG} = 0.9$), are shown for reference on the same plot. Since engine-out HC emissions measurements are known to vary (by a few hundred ppm) from day-to-day, all of the data shown in Fig. 5 were collected on the same day. It is obvious from this plot that the location of wetting within the combustion chamber has a very significant impact on the resulting HC emissions, and that all locations result in substantial increases in HC emissions from the baseline (LPG-only) levels. The smallest increase in HC emissions is observed for the wetting location I-I on the cylinder liner, which is located directly under the intake ports. The location E-E (under the exhaust ports) results in roughly twice the HC increase as the I-I location, and the 'side' liner locations E-I and I-E result in HC increases somewhere between the I-I and E-E locations. Surprisingly, wetting of the piston-top resulted in an HC increase second in importance to the worst liner wetting location. This result contradicts the findings of a previous study (Frank and Heywood,1991), which concluded that fuel wetting of the piston top was not a significant source of HCs in gasoline-fueled DI engines. This contradiction in findings is discussed in more detail later.

Differences in HC emissions for the four liner locations can be explained by considering the physical distance that hydrocarbons deposited on these locations must travel to reach the exhaust port. Much of the liquid fuel impinging on the liner will get scraped into the topland during the compression stroke. Some of this liquid may escape to the crankcase past the rings, but

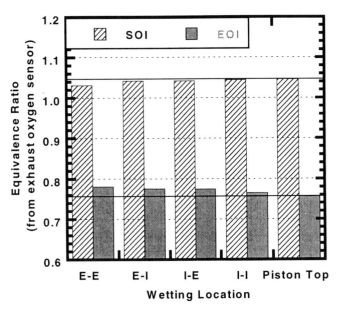

Figure 6a. Equivalence ratio measured via a wide-range oxygen sensor in the exhaust stream. Measurements were made immediately after starting injection (SOI, before ϕ was adjusted back to a value of 0.9), and immediately after ceasing injection from a ϕ value of 0.9 (EOI).

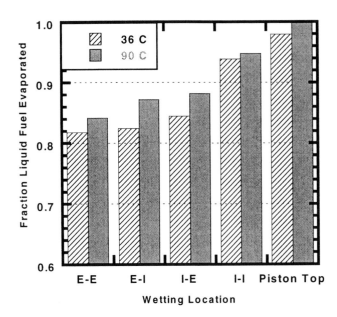

Figure 6b. Fraction of the liquid fuel injected that vaporizes for all wetting locations, coolant temperatures of 36 and 90 °C. This is calculated from the measured equivalence ratio (i.e. Fig. 6a), according to the procedure described in Appendix B.

the remainder will be laid back along the liner as a film after TDC. Any vapors from this film that survive post-flame oxidation (e.g., due to evaporation too late, when core temperatures are low) will merge into the roll-up vortex during the exhaust stroke. When this vortex reaches the open exhaust valves, the portions of the vortex that are physically closer to the exhaust valves have a higher probability of leaving through the exhaust ports than those portions of the vortex that are farther from the exhaust valves. The portion of the vapor that does not escape through the exhaust valve will be retained within the cylinder and consumed during the following cycle. The liquid remaining within the cylinder is available for evaporation in subsequent cycles. Thus, it should be expected that wetting the liner farthest from the exhaust port (between the intake valves) will yield lower HC emissions than wetting directly under the exhaust ports, and that intermediate liner wetting locations will yield HC emissions between these two extremes.

Further insight into the difference between wetting on the cylinder liner and the piston top can be gained by examining the measurements taken with a wide-range oxygen sensor, as described in the experimental procedure. Figure 6a shows the exhaust equivalence ratios for all wetting locations for injection timing of 160 °ATDC, MBT timing, and coolant temperature of 90 °C. Beginning with $\phi_{LPG} = 0.9$ for LPG/air with no liquid injection, injection onto the

piston yields and equivalence ratio that is slightly higher than for all of the liner wetting locations. Thus, the LPG flow had to be decreased more to achieve $\phi_{LPG+LF} = 0.9$ for injection onto the piston than for injection onto the liner. Therefore, it is not surprising that when the liquid injection was turned off, the mixture was somewhat leaner (ϕ_{LPG-LF}) for the piston case.

These results indicate that a larger fraction of the liquid fuel evaporates, and/or less is lost into the crankcase, when the liquid is deposited on the piston top. These differences in equivalence ratio can be used to estimate the fraction of liquid fuel evaporated (Appendix B). As shown in Figure 6b, it is estimated that 100% of the liquid fuel evaporates in the warm coolant case, for wetting on the piston top. For all wetting locations, decreasing the coolant temperature decreases the percent evaporated by a few percentage points. The worst case is for injection onto the liner underneath the exhaust valves with cold coolant, for which only ~ 82% of the liquid evaporates. These results are reasonable in light that the piston is hotter than any other surface within the combustion chamber (except the exhaust valve faces), and that little, if any, of the liquid fuel that impacts the piston top can escape into the crankcase. The high HC emissions for piston impingement indicate not only that more fuel evaporates from the piston top (compared to the liner), but also that some of the fuel that evaporates off the piston top vaporizes too late in the cycle to be oxidized.

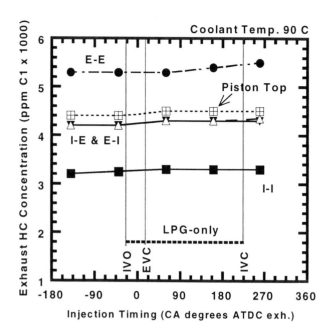

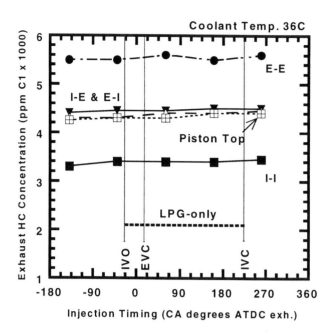

Figure 7. Effects of injection timing on HC emissions for all five wall wetting locations with 90 °C coolant, MBT ignition timing. All experiments performed on the same day.

Figure 8. Effects of injection timing on HC emissions for all four liner wetting locations with 36 °C coolant, MBT ignition timing. All experiments performed on the same day.

EFFECT OF INJECTION TIMING

The effect of injection timing on the HC emissions for all locations is shown in Fig. 7 for MBT ignition timing and a coolant temperature of 90 °C; and in Fig. 8 for MBT ignition timing and a coolant temperature of 36 °C. The baseline HC emissions on LPG, at the same equivalence ratio (ϕ_{LPG} = 0.9), are shown for reference on these plots. The HC emissions for all wetting locations display a very slight upward trend as the injection timing is retarded relative to IVO. This increase in HC emissions with retarded injection timing may be attributed to the reduced time available for vaporization of the liquid fuel.

It is interesting to note that the HC dependence on injection timing is extremely weak, and that both coolant temperatures display very similar trends. This behavior indicates that vaporization of liquid fuel off the liner surface is a slow process relative to the engine cycle. There is only a 100-200 ppm difference in the HC emissions for injection timing at 140 °BTDC (min HCs) and at 260 °ATDC (max HCs). This 400 ° injection timing differential corresponds to an additional 67 msec for vaporization at this engine speed.

The very weak dependence of HC emissions on injection timing suggests that only some of the fuel that is injected on a given cycle evaporates during that cycle, and that some of the liquid may evaporate on subsequent cycles. Furthermore, the increase in HC emissions with liquid injection indicates that some of the vaporization occurs late in the cycle, when cylinder pressures are low (lower pressures enhance

vaporization). Several studies have shown that liquid fuel deposited on the cylinder surfaces is not significantly affected by flame passage - the rate of vaporization is predominantly dictated by heat transfer from the surface, not the gas (Shin *et al.*, 1994; Witze and Green, 1997). Because the vaporization rate appears to be limited by the heat transfer process from the surface to the liquid, fuel impacting an in-cylinder surface on any given cycle may not contribute to HC emissions until the following cycle or even later.

Injection timing is known to have a significant impact on HC emissions from PFI engines (e.g. Alkidas, 1994; Alkidas and Drews, 1996; Fox *et al.*, 1992; Fulcher *et al.*, 1995) and DISI engines (e.g. Karl *et al.*, 1997). The present study, however, essentially isolated the effects of wall wetting from other HC sources, and the results obtained in this study show that fuel residence time on the walls has a weak effect on the resulting HC emissions. As noted previously, for cold start and warm-up of PFI engines open intake valve injection (OIV) results in liner wetting under the exhaust valves whereas closed intake valve (CIV) injection results in fuel impingement on the piston. The present results indicate that wetting the liner underneath the exhaust valves is worse, from an HC emissions perspective, than for liquid fuel on the piston. Additionally, it was previously shown that a larger mass of liquid impinges on the liner during OIV injection than impinges on the piston during CIV injection (Takeda *et al.*, 1997). Similarly, at least one of the effects of injection timing on HC emissions from DISI engines may due to the amount of liquid impinging on the surfaces. Brehob *et al.* (1998) showed that the HC

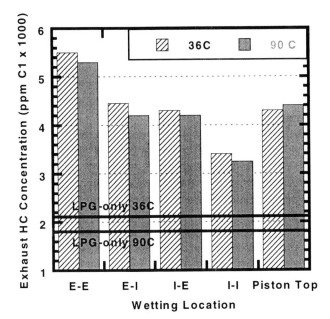

Figure 9. Effect of coolant temperature on HC emissions, for injection timing = 160 °ATDC, MBT ignition timing. The baseline emissions with LPG-only were obtained at the same equivalence ratio (ϕ_{LPG} = 0.9) and coolant temperature.

increase in HC emissions is observed for wall wetting on the piston top.

The dependence of HC emissions on combustion chamber surface temperature in SI engines is well documented. (e.g. Meyers and Alkidas, 1978; Russ et al., 1995). Excluding mixture preparation effects (which may lead to wall wetting), increased coolant temperatures result in smaller crevice volumes, reduced crevice content density, and thinner quench layers; all of which result in lower HC emissions.

As shown in Figure 9, the expected increase in HC emissions with decreased coolant temperature is observed for all of the liner wetting locations, but not for liquid fuel impingement on the piston top. The temperature of the piston increases as the coolant temperature increases, which should enhance vaporization of the liquid fuel off the piston top[*]. However, as noted previously, evaporation appears to be a slow process. The present results suggest that some of the additional fuel vaporization that should occur when the piston is hotter (see Fig. 6b) happens when cylinder pressures and core temperatures are low - conditions for which the probability of in-cylinder oxidation is also low. Unlike the results presented in prior sections, the comparisons between the warm and cold coolant were from data acquired on two different days. The day-to-day repeatability for the HC emissions when operating solely on LPG was ± 170 ppm about the mean. The repeatability with injection depended upon wetting location and was worst for liner wetting under the intake valves (I-I) at approximately ± 310 ppm about the mean. For all other wetting locations, the repeatability was ± 200 ppm or better. For any specific wetting location, data acquired on the same day with the two coolant temperatures showed the same trend as presented in Figure 9: increased HCs with decreasing coolant temperature for all liner wetting locations but increased HCs with increasing temperature for piston wetting. Given the uncertainties in the data, it appears that - for the liner wetting locations - the observed increase in HC emissions with decreasing coolant temperature may be due solely to the crevice effect for the LPG. However, the crevice effect for the LPG must also occur when the liquid is injected onto the piston, but the data consistently revealed decreased HC emissions with decreasing coolant temperature for piston wetting. Therefore, as the coolant temperature decreases, the HC emissions resulting from piston wetting must decrease more than the crevice HCs increase.

The apparently weak dependence of HC emissions on coolant, and therefore piston surface, temperature led Frank and Heywood (1991) to conclude that piston top wetting was not an important source of HC emissions from DISI engines. In their paper,

emissions from a production stratified-charge DISI engine increased rapidly as injection was performed closer to TDC, where wall wetting is to be expected. The present results show that piston wetting results in increased HC emissions and imply that conditions that produce more piston wetting will produce increased HC emissions. It should be noted that wall wetting in DISI engines may also occur for early injection timings (near BDC), when bulk motion (tumble or swirl) may lead to droplet impingement on the liner. The present results indicate that, at least from the HC emissions perspective, wetting the liner under the intake valves (e.g., by reverse tumble) is preferable to other liner wetting locations. Of course, from the perspectives of fuel economy and engine durability, all liner wetting locations should be avoided.

EFFECT OF COOLANT TEMPERATURE

Figure 9 compares the exhaust HC concentration for all wetting locations for coolant temperatures of 90 and 36 °C, for injection timing of 160 °ATDC and MBT ignition timing. The baseline HC emissions on LPG, at the same equivalence ratio (ϕ_{LPG} = 0.9) and both coolant temperatures are shown for reference on the same plot. It is evident from Fig. 9 that the coolant temperature only has a modest effect on the HC emissions due to wall wetting; roughly equal in magnitude to the effect it has on the LPG-only tests. A decrease in HC emissions is observed for increased coolant temperature for all liner locations, but an

---

[*] An exception to this occurs if the excess temperature between the surface and liquid lies in the region between the critical heat flux and the Leidenfrost point on the boiling curve (e.g. Incropera and DeWitt, 1990)

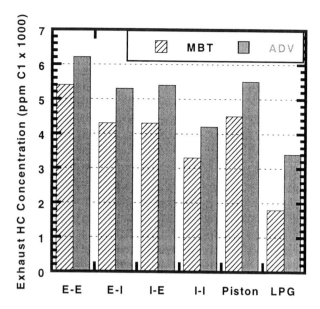

Figure 10. Effect of ignition timing on HC emissions for all wetting locations and LPG-only. Injection timing = 160 °ATDC, MBT ignition timing, coolant temperature of 90 °C. (MBT timing ~ 32 °BTDC, ADV timing = 44 °BTDC).

Frank and Heywood cite several previous studies that indicated that piston wetting is an important source of HC emissions from DISI engines but also several prior studies indicating that flame quenching near the surfaces dominates the HC emissions from these engines. All of those prior studies, including that by Frank and Heywood, were for engine configurations similar to the Texaco Controlled Combustion System (TCCS) design. In the TCCS concept, the injector is aimed toward the spark plug, which is sparked throughout the injection process. The intention is to burn the resulting stratified charge during the injection process. In contrast, current DISI engines do not aim the fuel spray directly toward the spark plug, the spark plug is located much farther from the injector, a single spark is used, and there is a significant delay between the end of injection and the ignition event. These fundamental differences may explain the opposite conclusion resulting from the present work and that by Frank and Heywood.

EFFECT OF IGNITION TIMING

Additional experiments were performed with the ignition timing advanced to 44 °BTDC relative to MBT (~32 °BTDC). For homogeneous charge SI engines, advancing the ignition timing decreases the burned gas temperatures during the expansion and exhaust strokes (Heywood, 1988; Kaiser *et al.*, 1984). The peak pressure also increases when the ignition timing is advanced (Matekunas, 1983) and thus more unburned fuel is trapped in the crevices. The net effect is that HC emissions increase with advanced ignition timing. Kaiser *et al.* (1984) showed that advanced ignition timings result

in increased emissions of unburned fuel and decreased emissions of partially-burned fuel, which is indicative of reduced post-flame oxidation.

The exhaust HC concentration for all wetting locations and LPG-only are shown in Figure 10, comparing MBT and advanced ignition timings, and for injection timing of 160 °ATDC and coolant temperature of 90 °C. The results provided in Figure 10 display the expected increasing HC emissions trend with advanced ignition timing for the LPG-only case and all wall wetting cases. However, the HC emissions are most sensitive to ignition timing for the LPG-only case, increasing by about 1600 ppm compared to about 1000 ppm for all wetting cases. For all of the liquid injection cases, the measured increase in HC emissions is less than that expected due to the crevice effect alone (approximately the LPG-only value). Advancing the ignition timing increases both the surface heat fluxes and surface temperatures, except for the exhaust valve (French and Atkins, 1973; Watts and Heywood, 1980). Therefore, the higher surface temperatures for advanced timing may result in higher liquid evaporation rates and more evaporation early in the expansion stroke when the probability of in-cylinder oxidation is highest. Increased early evaporation with advanced timing explains why the HC emissions are less sensitive to timing variations when some of the fuel is provided via surface wetting compared to the crevice effect alone.

CONCLUSIONS AND IMPLICATIONS

The results presented in this paper not sufficient to fully explain the complex relationship between in-cylinder wall wetting and HC emissions from SI engines. However, a number of important conclusions relevant to port injected and direct injected SI engines can be drawn from these results:

1) The results presented in this paper show that liquid fuel wall wetting on all surfaces of the combustion chamber is detrimental to HC emissions.

2) The residence time of liquid fuel on the walls appears to have only a small effect on the observed HC emissions due to wall wetting. This suggests that the vaporization process off the walls is slow relative to the engine cycle (even at 1000 rpm). Evaporation is a slow process that is controlled by the surface temperature and is enhanced at lower cylinder pressures. It appears that a significant fraction of the liquid fuel that impinges on in-cylinder surfaces evaporates late in the cycle, when the probability of in-cylinder oxidation is low.

3) It is evident from these results that, from the standpoint of HC emissions, wetting of the cylinder liner under the exhaust ports and the piston top is to be avoided. Interestingly, these are the two predominant locations where wall wetting occurs in

current production engines. In 4-valve PFI engines, fuel injection during the open intake valve period results in wall wetting on the cylinder liner, directly under the exhaust ports; and injection during the closed intake valve period results in wall wetting of the piston top. In direct injected engines, however, it may be possible to change the location where wall wetting occurs (given that wall wetting cannot be avoided all together) to reduce the negative impact on HC emissions.

The results presented in this paper provide some new insights into the relationship between in-cylinder wall wetting and HC emissions from SI engines. However, these limited experiments are not sufficient to fully characterize this complex interaction. Further studies in this area are planned, including the use of fuels with varying volatility and resistance to oxidation, as well as fast-FID and speciated emissions measurements.

ACKNOWLEDGMENTS

This research was funded by the General Motors Corp. and conducted using the facilities of the General Motors Foundation Combustion Sciences and Automotive Research Labs on the UT Campus. The authors would like to thank Mr. Curtis Johnson of the University of Texas at Austin for his contributions in designing and fabricating the injector probe used for this study, and Prof. Matt Hall for his assistance with the optical engine tests. We would like to gratefully acknowledge Mr. Dan Tribble of Champion for supplying us with a modified spark plug; Mr. Lee Markle of Delphi and Ms. Lisa Nemecek, formerly of Delphi, for supplying the prototype fuel injectors; and Mr. Kevin Whitney of Southwest Research Institute for providing the RFG used in this study. We are also grateful to Mr. Lee Dodge and Dr. Rick Anderson for providing feedback on this manuscript. Any opinions, findings, and suggestions expressed herein are those of the authors and do not necessarily reflect the views of the General Motors Corporation.

CONTACT

Rudolf H. Stanglmaier is affiliated with the Engine Research Department at the Southwest Research Institute. He can be contacted at:

6220 Culebra Rd., Bldg. 151
San Antonio, TX 78238

Tel: (210) 522-5505
Fax: (210) 522-2019
E-mail: rstanglmaier@swri.org

REFERENCES

Alkidas, A. C. (1994), "The Effects of Fuel Preparation on Hydrocarbon Emissions of a S.I. Engine Operating Under Steady-State Conditions," SAE Paper No. 941959.

Alkidas, A. C., and R. J. Drews (1996), "Effects of Mixture Preparation on HC Emissions of a S.I. Engine Operating Under Stead-State Cold Conditions," SAE Paper No. 961958.

Brehob, D. D., J. Fleming, M. Haghgooie, and R. A. Stein (1998), "Stratified-Charge Engine Fuel Economy and Emissions Characteristics," SAE Paper No. 982704.

Cheng, W. K., D. Hamrin, J. B. Heywood, S. Hochgreb, K. Min, and M. Norris (1993), "An Overview of Hydrocarbon Emission Mechanisms in Spark-Ignition Engines," SAE Paper No. 932708.

Fox, J. W., K. D. Min, W. K. Cheng, and J. B. Heywood (1992), "Mixture Preparation in a SI Engine With Port Injection During Starting and Warm-Up," SAE Paper No. 922170.

Frank, R. M. and J. B. Heywood (1991), "The Effect of Piston Temperature on Hydrocarbon Emissions from a Spark-Ignited Direct-Injection Engine," SAE Paper No. 910558.

French, C.C.J., and K.A. Atkins (1973), "Thermal Loading of a Petrol Engine," Proceedings of the Institution of Mechanical Engineers 189(49/73): 561-573.

Fulcher, S. K., B. F. Gajdeczko, P. G. Felton, and F. V. Bracco (1995), "The Effects of fuel Atomization, Vaporization, and Mixing on the Cold-Start HC Emissions of a Contemporary S.I. Engine With Intake Manifold Injection," SAE Paper No. 952482.

Heywood, J. B. (1988), Internal Combustion Engine Fundamentals, Published by McGraw-Hill, Inc., New York, 1988.

Incropera, F. P, and D. P. DeWitt (1990), Fundamentals of Heat and Mass Transfer, Published by John Wiley and Sons, New York, 1990.

Kaiser, E. W., W. G. Rothschild, and G. Lavoie (1984), "Storage and Partial Oxidation of Unburned Hydrocarbons in Spark-Ignited Engines - Effect of Compression Ratio and Spark Timing," Combustion Science and Technology, Vol. 36, pp. 171-189, 1984.

Kaiser E. W., W. O. Seigl, and R. W. Anderson (1994), "Fuel Structure and the Nature of Engine-Out Emissions," SAE Paper No. 941690.

Karl, G., R. Kemmler, M. Bargende, and J. Abthoff (1994), "Analysis of Direct Injected Gasoline Engine," SAE Paper No. 970624.

Kelly-Zion, P., J, Styron, C. Lee, R. Lucht, J. Peters, and R. White (1997), "In-Cylinder Measurement of Liquid Fuel During the Intake Stroke of a Port-Injected Spark Ignition Engine," SAE Paper No. 972945.

Matekunas, F.A. (1983), "Modes and Measures of Cyclic Combustion Variability," SAE Paper No. 830337; also in: SAE Trans., Vol. 92.

Meyers, J. P., and A. C. Alkidas (1978), "Effects of Combustion Chamber Surface Temperature on the Exhaust Emissions of a Single-Cylinder Spark-Ignition Engine," SAE Paper No. 780642.

Roberts, C. E., and R. D. Matthews (1996), "Development and Application of an Improved Ring Pack Model for Hydrocarbon Emission Studies," SAE Paper No. 961966.

Russ, S., E. W. Kaiser, and W. O. Seigl (1995), "Effect of Cylinder Head and Engine Block Temperature on HC Emissions from a Single Cylinder Spark Ignition Engine," SAE Paper No. 952536.

Saito, K., K. Sekiguchi, N. Imatake, K. Takeda, and T. Yaegashi (1995), "A New Method to Analyze Fuel Behavior in a Spark Ignition Engine," SAE Paper No. 950044.

Shin, Y., W.K. Cheng, and J.B. Heywood (1994), "Liquid Gasoline Behavior in the Engine Cylinder of a SI Engine," SAE Paper No. 941872.

Stanglmaier, R. H. (1997a), "Fuel Transport, Mixture Preparation, and Hydrocarbon Emissions from Gasoline-Fueled Spark-Ignition Engines," Ph. D. Dissertation, Dept. of Mechanical Engineering, The University of Texas at Austin; Austin, TX 1997.

Stanglmaier, R. H., M. J. Hall, and R. D. Matthews (1997b), "In-Cylinder Fuel Transport During the First Cranking Cycles in a Port Injected 4-Valve Engine," SAE Paper No. 970043.

Stanglmaier, R. H., C. E. Roberts, O. A. Ezekoye, and R. D. Matthews (1997c), "Condensation of Fuel on Combustion Chamber Surfaces as a Mechanism for Increased HC Emissions from SI Engines During Cold Start," SAE Paper No. 972884.

Stanglmaier, R. H., M. J. Hall, and R. D. Matthews (1998), "Fuel-Spray/Charge-Motion Interaction within the Cylinder of a Direct-Injected, 4-Valve, SI Engine," SAE Paper No. 980155.

Takeda, K., T. Yaegashi, K. Sekiguchi, K. Saito, and N. Imatake (1995), "Mixture Preparation and HC Emissions of a 4-Valve Engine With Port Injection During Cold Start and Warm-Up," SAE Paper No. 950074.

Watts, P.A., and J.B. Heywood (1980), "Simulation Studies of the Effects of Turbocharging and Reduced Heat Transfer on Spark-Ignition Engine Operation," SAE Paper No. 800289.

Witze, P.O., and R.L. Green (1997), "LIF and Flame-Emission Imaging of Liquid Fuel Films and Pool Fires in an SI Engine During a Simulated Cold Start," SAE Paper No. 970866.

Zizelman, J., M. J. Seino, M.C. Graves, and J. Manz (1992), "Central Port Fuel Injection," SAE Paper No. 920295.

APPENDICES

APPENDIX A: SIGNIFICANCE OF OCCASIONAL DROPLETS WHICH DRIBBLE OFF INJECTOR TIP.

As noted in the section entitled Injection Probe Verification Tests, an occasional droplet of fuel was observed to dribble off the probe tip every few seconds. This serves as an additional source of liquid fuel, the significance of which is explored in this appendix. The measured air, LPG, and liquid fuel mass flow rates are equal to the exhaust mass flow rate. If all of the liquid fuel evaporates but escapes combustion, the mass fraction of HCs in the exhaust due to this source is the ratio of the liquid fuel mass flow to the total exhaust mass flow. This can be converted to a mole fraction given the molecular weight of the liquid fuel (as C1) and an estimate of the molecular weight of the exhaust (30). Thus, it is estimated that, if all of the liquid fuel evaporates but escapes combustion, the result would be an increase of HC emissions of about 20,000 ppmC1 for both the cold and warm coolant cases. Variation of the estimated molecular weight of the exhaust produces a 10% variation in this value. Because this is much higher than the observed increase (1300-3500 ppmC1 depending upon location), it is concluded that most of the liquid fuel both evaporates and burns. To determine the significance of the droplets that are occasionally formed, the injector probe was fired 150 times and the droplets were collected on a slide. The mass was determined to be ~ 0.02 mg per injection. Applying the same procedure as discussed above to this mass yields an estimated increase in HC emissions of 250 ppmC1 if none of it burns. Because this is much smaller than the observed increase, it must be concluded that this occasional drop cannot be responsible for the observed effects.

APPENDIX B: ESTIMATES OF PERCENT LIQUID FUEL EVAPORATED.

The fraction of liquid fuel that vaporized (m_{LF_EVAP} / m_{LF_INJ}), shown in Fig. 6b, was calculated by the procedure described in this appendix. See section on Methodology for description of experimental procedure.

Measured quantities:

ϕ_{LPG+LF} = Equiv. ratio with LPG and liquid fuel injection.
ϕ_{LPG-LF} = Equiv. ratio without liquid fuel injection.
m_{AIR} = trapped air mass (mg/cylinder/cycle).

Eq.1a
$$\phi_{LPG+LF} = \frac{AF_S}{AF} = \frac{AF_{S\_MIX}}{\dfrac{m_{AIR}}{m_{LPG} + m_{LF\_EVAP}}}$$

Eq.1b
$$m_{LF\_EVAP} = \frac{\phi_{LPG+LF}}{AF_{S\_MIX}}\left(m_{AIR}\right) - m_{LPG}$$

Eq.2a
$$\phi_{LPG-LF} = \frac{AF_S}{AF} = \frac{AF_{S\_LPG}}{\dfrac{m_{AIR}}{m_{LPG}}}$$

Eq.2b
$$m_{LPG} = \frac{\phi_{LPG-LF}}{AF_{S\_LPG}}\left(m_{AIR}\right)$$

combining Eqs. 1b and 2b yields:

Eq.3
$$m_{LF\_EVAP} = m_{AIR}\left(\frac{\phi_{LPG+LF}}{AF_{S\_MIX}} - \frac{\phi_{LPG-LF}}{AF_{S\_LPG}}\right)$$

where AF_{S_MIX} depends upon the mass of liquid fuel evaporated. A value for AF_{S_MIX} is assumed (between AF_{S_LPG} and AF_{S_LF}). Equation 3 is solved for m_{LF_EVAP}, then AF_{S_MIX} is calculated from:

Eq.4
$$AF_{S\_MIX} = \frac{AF_{S\_LPG} - AF_{S\_LF}}{1 + \dfrac{m_{LF\_EVAP}}{m_{AIR}}} - AF_{S\_LF}$$

Equations 3 and 4 are iterated to convergence.

413

982703

Stability Improvement of Direct Fuel Injection Engine under Lean Combustion Operation

Toshiharu Nogi, Takashi Shiraishi, Youko Nakayama, Minoru Ohsuga and Nobuo Kurihara
Hitachi Ltd.

ABSTRACT

Meeting future exhaust emission and fuel consumption standards for passenger cars will require refinements in how the combustion process is carried out in spark ignition engines. A direct injection system reduces fuel consumption under road load cruising conditions, and stratified charge of the air-fuel mixture is particularly effective for lean combustion.

This paper describes an approach to improve combustion stability for direct fuel injection gasoline engines. Effects of spray characteristics (spray pattern and diameter) and air flow motion on the combustion stability were investigated. Spray patterns were observed by the laser sheet scattering method and 3-dimensional laser doppler velocimetry. Mixture behavior in the combustion chamber was observed by the laser-induced fluorescence method using an excimer laser and single cylinder optical engine. It was found that the spray pattern for a pressurized condition affects the combustion stability and smoke generation. Spray having both high and low speed particles leads to mixture stratification around the spark plug and reduced the fuel impingement on the piston cavity. The engine test results showed improvement of combustion stability and smoke reduction under lean operation.

INTRODUCTION

Many methods to get a stratified charge mixture have been developed for direct-injected spark ignition (SI) engines. A side injection type was proposed which has an injector located between two intake valves.[1-4] This type has a cavity in the top of the piston. Fuel is injected into the cavity to form the stratified charge late in the compression stroke. A center injection type was also proposed which has an injector located at the center and top of the cylinder. [3-12] This type also has a cavity in the top of the piston. An ignition plug is also located at the center of the cylinder.

This paper reports the combustion stability improvement approach for direct fuel injection gasoline engines. Effects of spray characteristics (spray pattern and diameter) and air flow motion on the combustion stability were investigated. Spray patterns were observed by the laser sheet scattering method and 3-dimensional laser doppler velocimetry. Mixture behavior in the combustion chamber was observed by the laser-induced fluorescence method using an excimer laser and single cylinder optical engine. Relation between the spray pattern and the combustion stability was studied. The spray patterns to achieve mixture stratification around the spark plug and reduce the fuel impingement on the piston cavity were proposed.

SYSTEM CONFIGURATION

Fig.1 shows the system configuration for this study. The engine studied is a 4 cycle, 4 cylinder, 2.0 liter engine remade from a port injection (PI) engine. The fuel injectors located in the space between the two intake ports of the cylinder at a 36 degrees decline to get the horizontal line to give side injection. The piston has a cavity and the compression ratio is 10. The fuel pressure is variable up to 10 MPa by adjusting the pressure regulator. A high pressure fuel

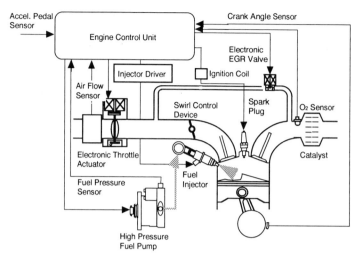

Fig.1   System Configuration

Table1   Engine Specifications

Engine Type	4-stroke, 4 cylinder 4-valve DOHC
Displacement	1998cc
Bore-Stroke	86mm × 86mm
Compression Ratio	10.0
Combustion Chamber	Pent-roof
Spark Plug Location	Center of Chamber
Fuel Injector Location	Below Intake Port

pump is driven by the engine cam-shaft. The engine control unit (ECU) calculates the injection pulse signal based on the air flow sensor. The injector is driven by the injector drive circuit to get a wide dynamic range of fuel metering. The electronic throttle is used for the engine torque control, even if the air fuel ratio changes from rich to lean, in accordance with the accelerator pedal angle. A swirl control device is included in the intake manifold which generates swirl air motion in the cylinders. The electric EGR valve and NOx catalyst are used to reduce NOx emissions during operation at a lean air fuel ratio. The crank angle signal is sent to the ECU to detect the engine stability. Table 1 lists the engine specifications.

CONCEPT OF MIXTURE FORMATION

Table 2 lists features of side and center injection type systems. For the former, the fuel injector is mounted between the intake valves, and the spark plug is located at the top of the pent-roof. The piston has a cavity. Fuel is injected into this cavity late in the compression stroke to get the stratified charge mixture. Fuel is vaporized at the surface of the piston, and vaporized fuel is transported to the spark plug and

ignited. In this case, vaporized fuel is introduced by in-cylinder air flow and fuel spray momentum. The injector is easily installed and ignition is good because of the vaporized fuel movement. But there are possibilities for a fuel film to form on the piston surface and smoke and unburned HC to be emitted. This can be solved by spray optimization.

For the center injection type, the injector and spark plug are located at the center of the cylinder. Fuel is injected downward from the top of the cylinder. Fuel is vaporized during the compression stroke. This type of engine is potentially less dependent of the piston shape, but combustibility is affected by injector performance. In this paper, the side injection type is selected and optimized.

Fig.2 shows the mixture formation and combustion process of the direct injection engine. Fuel spray and in-cylinder air flow are important for air/fuel mixture formation. Spray (droplet size, pattern) is determined by the fuel injector design, the fuel

Table 2 Feaures of Side and Center Injection Type

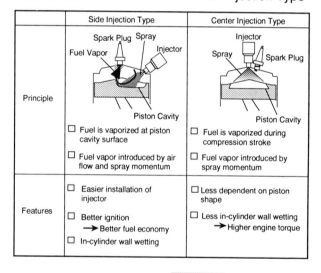

	Side Injection Type	Center Injection Type
Principle	☐ Fuel is vaporized at piston cavity surface ☐ Fuel vapor introduced by air flow and spray momentum	☐ Fuel is vaporized during compression stroke ☐ Fuel vapor introduced by spray momentum
Features	☐ Easier installation of injector ☐ Better ignition → Better fuel economy ☐ In-cylinder wall wetting	☐ Less dependent on piston shape ☐ Less in-cylinder wall wetting → Higher engine torque

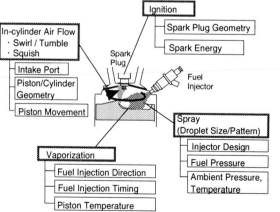

Fig.2   Mixture Formation and Combustion Process

416

pressure, the ambient temperature and pressure. The air flow motion are affected by the intake port, piston, cylinder geometry, piston movement. In this study, the effects of the spray pattern and in-cylinder air flow motion on the combustion stability and smoke emissions are studied.

COMBUSTION STABILITY IMPROVEMENTS

EXPERIMENTAL APPARATUS AND METHOD- Fig. 3 shows the experimental apparatus to visualize a section of the fuel spray in the chamber by using an argon laser and high speed camera. The chamber (inside diameter 200mm, height 300 mm) has two acrylic windows on the side walls and the injector is set at the chamber center. The argon laser beam passes through one window after being made into a sheet, 2 mm thick, by a cylindrical lens. The sheet beam irradiates onto the fuel spray. Scattered light from the spray is observed through the other side windows by the high speed camera. Fuel is pressurized by nitrogen gas and the injector is operated intermittently by an injector driver.

Fig. 4 shows the setup with the 3-dimensional laser doppler velocimetry (LDV). The beam from an argon laser (4 watts.) is split and distributed through optical fibers to the spray. The doppler burst signals are detected by photodiodes and the burst signal analyzer. The velocity distributions of the injected spray at the horizontal and vertical sections are measured.

Fig. 5 shows the transparent engine, which is the same dimension as the engine studied. The cylinder and piston of this engine are made of quartz glass. To analyze interaction of fuel spray and intake air flow, the argon laser light sheet is sent into the cylinder through the piston, and scattered light from the fuel spray is observed though the cylinder wall with a high speed camera.

Fig.6 shows the experimental apparatus for laser induced fluorescence (LIF). The light source is a KrF excimer laser, the wavelength is 248 nm and the output energy is 380 mJ/pulse. The laser beam is formed into two sheet beams, with a beam size 90mm wide-1mm thick, through the use of mirrors and cylindrical lenses. The Cylinder is irradiated from both sides. The control unit sends injection signals to the engine and a synchronized signal to the PC for the laser and the camera on the basis of the crank angle signals. The CCD camera, with an image intensifier (I.I.) and an optical filter (250-400 nm and 75% efficiency), takes photographs of the mixture distribution in the cross section using the gating signal from the P.C. The fuel is n-octane with 0.2% N,N-dimethyl aniline (DMA). The wavelength of fluorescent light is in the range of 300-400 nm.

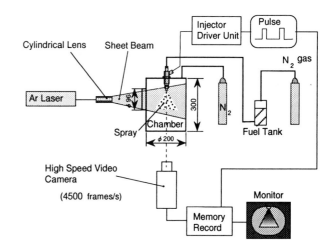

Fig.3 Apparatus for Spray Measurement using Laser Light Scattering

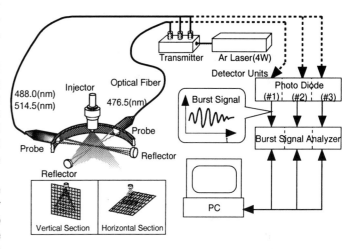

Fig.4 Setup with the 3-D Laser Doppler Velocimetry

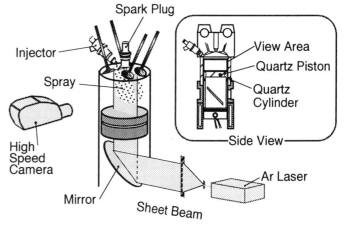

Fig. 5 Transparent Engine Layout

SPRAY CHARACTERISTICS - Fig.7 shows the simulation results of the evaporation rate and droplet diameter. Droplets are vaporized during the compression stroke. The evaporation rate at TDC is 100% when the droplet diameter is less than 20 μm and fuel is injected at BTDC.

Fig.8 shows an example of the atomization characteristics of the injector. The swirl type injector is used for this study. Spray diameter is measured by the diffraction method. Fuel pressure is set at 7 MPa. Sauter mean diameter of the spray is about 15μm. Droplet diameter at mass fraction 90% is 22 μm. This is effective to minimize impingement of fuel on the piston.

Fig.9 shows the spray pattern observed by the argon laser sheet. Nozzle A is a swirl type fuel injector with minimized initial spray. The surrounding pressure is set at atmospheric pressure (0.1MPa) and a pressurized condition (0.5MPa). The injection period is varied from 1ms to 5 ms. The spray pattern is a hollow cone shape and affected by the surrounding pressure. The spray shape is more compact than the atmospheric condition, this helps the spray remain in the piston cavity. Spray angle can be changed by fuel swirl intensity.

VISUALIZATION ANALYSIS - Fig. 10 shows the fuel behavior around the piston cavity. The model piston is set inside of the pressurized chamber shown in Fig.3. The spray angle is measured at a pressure 0.5MPa. The injected fuel impinges on the piston cavity surface, is guided by the cavity and is introduced to the spark plug area. When the spray angle is narrow, example 55 degrees, the spray impinges on the cavity surface and forms a film which causes more smoke and deteriorates combustion stability. When the spray angle is wide, the impingement is reduced, but the spray does not have enough momentum to reach the spark area, so combustion stability deteriorates. When the spray angle is optimized, the fuel impingement on the cavity is reduced and the spray is introduced to the spark plug area, and combustion stability is maintained.

Fig. 11 shows the mixture distribution in the cylinder observed by the LIF method. The swirl air motion is applied. When the spray angle is narrow, the mixture can not remain in the cavity area because of the higher spray momentum. When the swirl air motion intensity is higher, the mixture can not remain in the cavity. The spray pattern and air flow motion optimization are very important for stratified mixture formation.

ENGINE EVALUATION -Fig.12 shows the effects of spray angle on combustion stability and smoke. End of fuel injection timing and end of ignition timing were changed. Combustion stability is improved and smoke

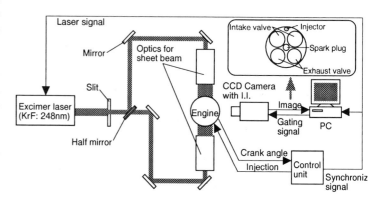

Fig. 6   Experimental Apparatus for LIF

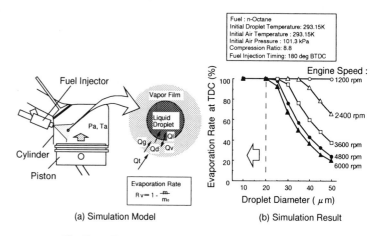

(a) Simulation Model   (b) Simulation Result

Fig.7   Evaporation Rate vs. Droplet Diameter

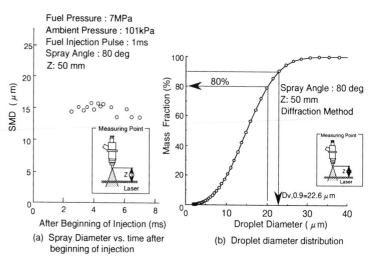

(a) Spray Diameter vs. time after beginning of injection

(b) Droplet diameter distribution

Fig.8   Atomization Characteristics

418

reduction is achieved by spray optimization.

Fig. 13 shows the spray angle effects on combustion stability and smoke. The smoke amount is reduced by the increase of the spray angle because of less spray impingement on the cavity. In this condition, combustion stability is improved and smoke are cut when the spray angle is 65 -85 degrees.

SPRAY PATTERN CONTROL SPRAY DURING COMPRESSION STROKE - The higher torque operation condition requires a homogeneous mixture. Fig. 14 shows behavior of the fuel injected in the cylinder at WOT. The fuel pressure is set at 9 MPa. The fuel sprays are dispersed in the cylinder by the interaction of the spray momentum and the intake air flow. In the 65 degrees spray case, the fuel spray flows towards the exhaust side of the cylinder because the spray has greater momentum than the intake air flow. In the 85 degrees angle spray case, the fuel spray flows towards the center of the cylinder and is turns around there. The wider spray angle causes the fuel spray to have less momentum. So the fuel spray flows toward the intake side of the cylinder. Fig. 15 shows the effects of spray angle on engine torque at an end of injection timing of 210 degrees BTDC. Engine torque at 9 MPa is influenced by spray angle.

The spray pattern in a pressurized condition is important for the stratified mixture and that in a atmospheric condition is important for the homogenous mixture. The spray of the injector must meet both requirements. The swirl intensity of the injector and the fuel pressure are key parameters. Other parameters are considered to control the spray pattern to meet a variety of engine bore sizes and operation conditions.

Fig. 16 shows spray patterns and velocity vectors measured by the 3-dimensional LDV. After injection, fuel has its vector toward the inside of the hollow cone. The pressure inside of the hollow cone is lower than the ambient pressure. Then spray is forced to direct inside of the spray. The pressure inside of the hollow cone spray is important to control the spray pattern

Fig. 17 shows the effects of the initial spray on spray patterns. Nozzle B type has longer penetration than nozzle A. The initial spray entrains the air inside the hollow cone and decreases the pressure. This pressure reduces the spray angle, especially under pressurized conditions.

Fig. 18 shows the effects of initial spray amount on spray angle. Spray angle is basically determined by the swirl intensity of fuel. When the swirl number is increased, the spray angle is increased. The initial spray amount affects the spray angle at surrounding pressure of 0.5 MPa. When the initial amount of spray is increased, the spray angle is reduced. So the spray angle of the atmospheric conditions and pressurized

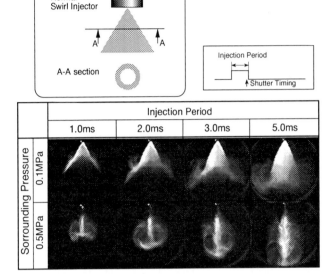

Fig.9 Spray Patterns (Nozzle A)

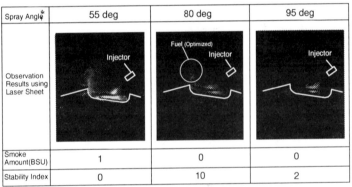

Fig.10 Fuel Spray Behavior around Piston Cavity

Fuel is introduced by the cavity surface and swirl air motion

Swirl Air Motion

Swirl Concept

Narrow Spray Pattern

Higher Swirl Air Intensity

Optimization of Sray Pattern and Swirl Air Motion

Fig.11 Mixture Distribution in the Cylinder

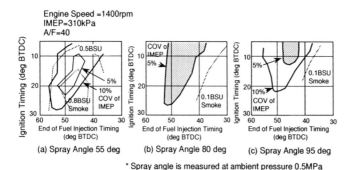

Engine Speed =1400rpm
IMEP=310kPa
A/F=40

(a) Spray Angle 55 deg
(b) Spray Angle 80 deg
(c) Spray Angle 95 deg

* Spray angle is measured at ambient pressure 0.5MPa

Fig. 12 Effects of Spray Angle on
Combustion Stability and Smoke

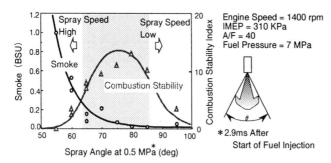

Engine Speed = 1400 rpm
IMEP = 310 KPa
A/F = 40
Fuel Pressure = 7 MPa

* 2.9ms After
Start of Fuel Injection

Fig.13   Effects of Spray Angle to
Combustion Stability and Smoke

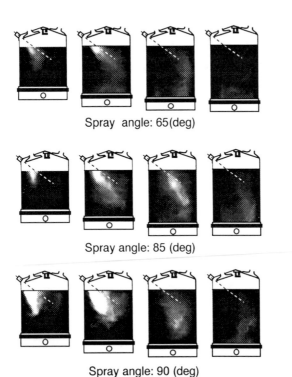

Spray angle: 65(deg)

Spray angle: 85 (deg)

Spray angle: 90 (deg)

Fig.14   Behavior of fuel injected in cylinder

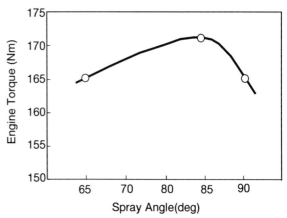

Fig. 15   Effect of Spray Angle on Engine
Torque (End of Injection: Optimized)

conditions can be controlled by the swirl intensity and the initial spray amount.

CONCLUSION

An approach to combustion stability improvement approach for direct fuel injection gasoline engines has been studied. Effects of spray characteristics and air flow motion on the combustion stability and smoke emissions were investigated. The major conclusions are as follows.

(1) The spray patterns at the pressurized conditions affected the combustion stability and smoke emissions. A narrow spray angle degree (spray having higher speed particles) increases smoke amount because of more impingement of the spray on the cavity surface. The wider spray angle degree (spray having lower speed particles) could not reach the spark plug area because of less fuel momentum.

(2) After spray angle optimization, the engine test results and the visualization analysis demonstrate the improvement of combustion stability was better and smoke amount was cut under air fuel ratio 40.

(3) Spray angle at atmospheric conditions affects homogenous mixture formation and WOT engine torque performance. Engine torque on spray angle 85 degree is increased 3.5% compared to that of 65 degree.

(4) Spray angle on the swirl type injector was narrowed when the fuel injection period was longer and surrounding pressure was higher. The pressure difference between the inside and outside of the hollow cone spray caused the spray pattern to change.

(5) Spray angle at the pressurized conditions were controlled by the initial spray amount and the swirl intensity of the injector.

REFERENCES

(1) T. Kume, Y. Iwamoto, K. Iida, M. Murakami,K. Akishino, H. Ando, Combustion control technologies for direct injection SI engine, SAE paper No. 960600 (1996)

(2) S. Kono , Development of the stratified charge and stable combustion method in DI gasoline engines, SAE paper No.950688 (1995)

(3) Y. Iriya, T. Noda, A. Iiyama, H. Fujii, Engine Performance and the effects of fuel spray characteristics on direct injection S.I. engines, JSAE paper No.9638031(1996)

(4) S. Matsushita, K. Nakanishi, T. Gohno, D. Sawada, Mixture formation process and combustion process of direct injection S.I. engine, JSAE paper No.9638022 (1996)

(5) T. Shiraisihi, M. Fujieda, M. Ohsuga, Y. Ohyama, A study of the mixture preparation process on spark ignited direct injection engine, SAE IPC-8 paper No. 9530409

(6) M. Fujieda, T. Shiraishi, M. Fujieda, M. Ohsuga, Influence of the spray pattern on combustion characteristics in a direct injection engine, JSAE paper No.9631911 (1996)

(7) R. W. Anderson, J. Yi, Z. Han, R.D. Reiz, Challenges of Stratified Charge Combustion, Direkteinspritzung im Ottomotor, Haus Der Technik E.V (1997)

(8) K. Shimotani, K. Oikawa, Y. Tashiro, O. Horada, Characteristics of exhaust emission on gasoline in-cylinder direct injection engine, 13th Symposium of Internal Combustion Engine No.20 (1996)

(9) G. Karl, M. Bargende, R. Kemmler, M. Kuhn, G. Bubeck, Thermodynamic analysis of a stratified direct injected gasoline engine, VDI 17th International Wiener Motoren Symposium No.267(1996)

(10) H. Stuzenberger, C. Preussner, J. Gerhardt, Gasoline direct injection for S.I. engines-development status and outlook, VDI 17th International Wiener Motoren Symposium No.267(1996)

(11) G.K. Fraidl, W.F. Piock, M. Wirth, Gasoline direct injection, Actual and future strategies for injection and combustion system, SAE paper No. 960465 (1996)

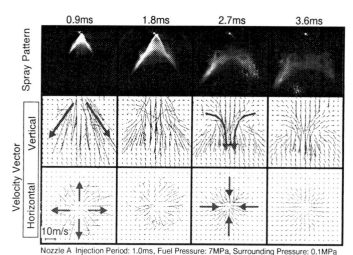

Nozzle A Injection Period: 1.0ms, Fuel Pressure: 7MPa, Surrounding Pressure: 0.1MPa

Fig.16 Spray Patterns and Velocity Vectors

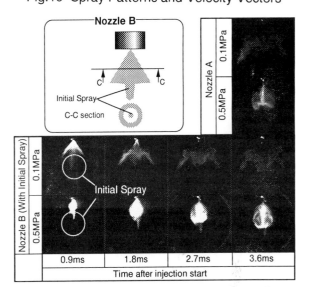

Fig.17 Effects of Initial Spray on Spray Patterns

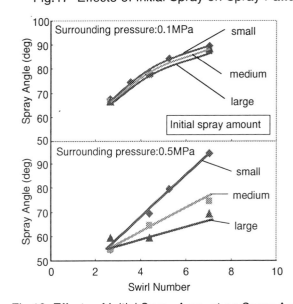

Fig.18 Effects of Initial Spray Amount on Spray Angle

421

950110

Fuel Distributions in a Firing Direct-Injection Spark-Ignition Engine Using Laser-Induced Fluorescence Imaging

Todd D. Fansler and Donald T. French
Thermosciences Dept.
General Motors Research and Development Center
Warren, MI

Michael C. Drake
Physical Chemistry Dept.
General Motors Research and Development Center
Warren, MI

ABSTRACT

Two- and three-dimensional images of fuel distributions in a continuously firing direct-injection stratified-charge engine have been recorded under moderate-load conditions using planar laser-induced fluorescence (LIF) from commercial gasoline. Cyclic variations in the fuel concentration at the spark gap (deduced from individual-cycle two-dimensional images) appear sufficient to account for the observed incidence of misfires and partial burns. Tomographic three-dimensional LIF images of the average fuel distribution at the time of spark indicate that ignitable mixture is present only in a thin shell around the periphery of the fuel cloud. Differences in power output and combustion stability during engine warm-up observed with two injectors of the same type are reflected in systematic differences in the fuel concentration near the spark gap as inferred from LIF data.

INTRODUCTION

Fuel-air mixture preparation is important for all spark-ignited engines. Problems associated with unfavorable mixture preparation include misfires, roughness, poor fuel economy, and high hydrocarbon emissions. Direct in-cylinder imaging of fuel distributions in firing engines should therefore be a great aid to engine design and optimization.

Controlling the in-cylinder fuel distribution is particularly critical to maximizing the performance and minimizing the emissions of direct-injection stratified-charge (DISC) engines [1--7][1], which have long been of interest as an alternative to premixed-charge engines because of their potentially superior fuel economy (Ref. 1 includes a review of early work). Over the last decade, two-stroke DISC engines have been the subject of intense research and development activity [8--10]. Recently there has also been a revival of interest in four-stroke DISC engines [10--13].

At light load, DISC engines typically operate with a highly stratified fuel-air mixture, which is achieved by igniting the charge during or just after the end of fuel injection. Improper fuel-air concentration (too rich or too lean) near the spark gap at the time of ignition is the most likely source of misfires and partial burns, although the flow field may also be important. Over-mixing of the fuel and air, leading to regions too lean to burn, is believed to be a major source of light-load hydrocarbon emissions. Variation in the fuel distribution from one engine cycle to the next is believed to be a major cause of cyclic variability. Currently, the locations of the spark plug and fuel injector and the optimization of injection and spark timing in DISC engines are primarily determined by traditional "black-box" testing.

Gas sampling [14] and line-of-sight infrared absorption [15] have been used previously to characterize fuel distributions in four-stroke premixed-charge engines, but these techniques are of much more limited utility than 2-D imaging. Rayleigh scattering from freon [16] has been used in single-point measurements of "fuel" concentration in a *nonfiring* port-injected four-stroke-engine model. Smoke tracer techniques [17,18] have been used to obtain 2-D images of the flame front in firing optical engines and engine simulators, although there is some concern as to how well particle flows mimic fuel vapor transport. Fuel-spray imaging using Mie scattering from liquid fuel droplets can be done readily in both motored and fired engines [19]. More experimentally-complex exciplex laser-induced fluorescence imaging [20], which separately analyzes the liquid and vapor components of the fuel distribution, has been used in motored DISC engines [21], premixed-charge engines [22] and diesel sprays [23]. Exciplex fluorescence has also been applied to an earlier generation of the DISC engine used in the present study, and the experimental results have been compared with detailed computer modeling [24]. Exciplex imaging requires the addition of special dopants to the fuel, however, and is of very limited value in fired engines because the fluorescence is strongly quenched by oxygen.

---

[1]The references cited in this brief literature survey are illustrative rather than exhaustive.

Laser induced fluorescence (LIF) of non-exciplex-forming fuel dopants (such as acetone, acetaldehyde, other ketones and aldehydes, and pyrene, toluene and other polyaromatics) has been used to image fuel distributions in flames and in premixed-charge four-stroke and diesel engines [25--35]. Fluorescence from these species (particularly aldehydes and ketones) is not significantly quenched by oxygen, and the fluorescence intensity is believed to be proportional to the dopant concentration whether in the liquid or vapor state. LIF from undoped, commercial gasoline has been used to visualize fuel distributions in a port-fuel-injected engine [36]. LIF imaging in a firing DISC engine has not been reported, however.

The purposes of the present study are to further the development of laser-induced fluorescence as a tool to image the fuel distribution (both liquid and vapor) in firing engines, to use the technique to measure fuel distributions in a continuously fired DISC engine under realistic speed/load conditions, and to use the results to understand better the causes of intermittent partial burns and misfires.

FUEL LIF METHOD

Fuel LIF, which is sensitive to both liquid- and vapor-phase fuel, is a useful complement to laser Mie scattering, which responds only to liquid fuel. Furthermore, the LIF images under certain conditions can be interpreted quantitatively, with the fluorescence intensity proportional to the mass of fuel present in the laser light sheet [31--34]. In contrast, the Mie-scattering intensity is always a complex function of drop size, shape, and number density, none of which is usually known with any certainty in an in-cylinder experiment.

Laser induced fluorescence involves absorption of light followed by spontaneous emission. As noted above, in previous LIF studies of fuel distributions in flames or engines, other researchers have used fuels of pure ketones or acetaldehyde; conventional fuels doped with a ketone, with acetaldehyde, or with toluene; or undoped commercial-grade fuels. The fluorescing component(s) of commercial gasoline has (have) not yet been identified, however. Likely candidates include polycyclic aromatics, substituted benzenes (e.g., toluene), aldehydes and ketones. In the present study, we use Amoco commercial-grade unleaded gasoline (RON 91) with no special additives.

Absorption and emission spectra for a range of organic ketones and aldehyde vapors have been studied as a function of temperature, pressure, and air dilution [25,31]. Absorptions usually maximize near 290 nm and fluorescence emissions extend from 350 to 500 nm with a maximum near 420 to 450 nm. Liquid- and vapor-phase absorption spectra are similar. The fluorescence quantum yield (photons emitted/photons absorbed) is typically in the range of 10^{-5} to 10^{-3} and is limited by a rapid intramolecular process of intersystem crossing between electronic excited states. This internal de-excitation mechanism is fast compared to intermolecular collisional de-excitation, however. The fluorescence intensity is therefore essentially independent of collisional processes (e.g., quenching by O_2), and the fluorescence intensity should be proportional to the number den-

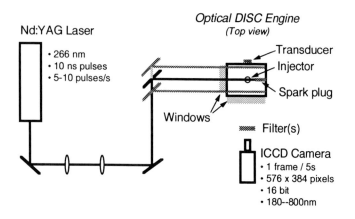

Figure 1. Schematic diagram of fuel LIF imaging experiment, including DISC engine modified for optical access. Translating the final beam-steering mirror moves the laser sheet through the combustion chamber.

sity of the fuel molecules, whether they are present as liquid or as vapor. The quantum yield declines when the excitation wavelength is shorter than 280 nm because of photo-dissociation.

Polycyclic aromatics and substituted benzenes are generally less volatile than small ketones and aldehydes, but have roughly similar absorption and fluorescence spectra. For example, dimethyl aniline has a boiling point of 195°C, absorbs strongly in the near ultraviolet, and fluoresces strongly (quantum yield 0.1 vs. ~0.001 for aldehydes and ketones) from 310--400 nm [37]. Much less is known about the quenching characteristics of these species, however.

APPARATUS

FUEL LIF IMAGING --- A schematic of the experimental apparatus is shown in Figure 1. A pulsed Nd:YAG laser, frequency quadrupled to 266 nm, is the excitation source, with a power of 10 mJ/pulse at a maximum repetition rate of 10 Hz. This excitation wavelength may be shorter than optimum, but it was more convenient than a Nd:YAG-pumped frequency-doubled dye laser at 280 nm. The laser beam is formed by two quartz lenses into a sheet approximately 0.5 mm thick by 50 mm high, cropped by a iris to a height of 25 mm, and directed into the engine through an ultraviolet-grade quartz window. Light scattered perpendicular to the laser beam exits a second ultraviolet-grade quartz window. Radiation at the laser wavelength (from reflections off solid surfaces, Mie scattering and Rayleigh scattering) was strongly attenuated with a long-pass glass filter (WG-345). The fuel fluorescence was imaged by a quartz camera lens (Nikkor UV lens, 105-mm, f/4) onto a digital computer-controlled camera system (Princeton Instruments ICCD-576) that is equipped with a gated image intensifier.

The Princeton Instruments camera has high dynamic range (16-bit digitization), low noise (cooled detector to minimize dark counts and slow readout to minimize readout noise), wide spectral sensitivity (180--800 nm), high quantum efficiency (≥12% at 300 nm), adjustable gain (≤1 to 80 counts/photoelectron), and adjustable electronic gating (3.5 ns to 80 ms). One disadvantage of the camera system is its relatively modest spatial resolution, which is limited by the 576 x 384-pixel detector and further reduced by a factor of 2 by the fiber-optic-coupled intensifier. In our application, each pixel in the image corresponds to a 0.25-mm square. Another disadvantage is the long image-readout time (about 5 s for an unbinned full frame), so no more than one image can be obtained per engine cycle. The camera produces monochrome (gray-scale) images, but false-color processing (assigning a color to correspond to an intensity range) is often used for image enhancement.

For in-cylinder fuel LIF imaging, the engine, laser and camera gate are all electronically synchronized to acquire the image at a specified crank angle. The camera gate duration was 0.2 μs, and camera gain was adjusted for each engine condition. The engine was typically fired continuously for 30--60 seconds before starting data acquisition, and typically either three 16-laser-shot average images or up to 50 individual images were taken, transferred to computer memory, and written to disk. Image acquisition typically required about 40 seconds, but one experiment spanned a six-minute period of continuous firing.

Some auxiliary experiments were performed with a uniform vapor distribution in the engine combustion chamber. The relative fuel fluorescence intensity with 266 nm laser excitation from different fuels was compared to Amoco commercial-grade unleaded gasoline (RON 91). Relative intensities of $\leq$ 0.1, 0.9, 1.0 and 5 were found for reagent-grade isooctane, California certification gasoline, Amoco 91 gasoline, and acetone, respectively. Amoco 91 was used in the present fuel LIF images because it gave high-intensity signals and avoided possible changes in combustion if additives had been used. Because the specific molecular species in the multicomponent gasoline giving rise to the fuel LIF is (are) not known, the measured fuel LIF spatial intensity distribution may not represent an appropriately weighted average over the various fuel constituents. It is important to note, however, that large LIF intensities were measured when a drop of gasoline was vaporized on an ~85° C surface, implying that the LIF is not completely dominated by high-boiling-point components.

A uniform vapor distribution was also used to measure the intensity distribution across the laser sheet, which was found to be constant within ±5% in the bottom 80% of the image, with the intensity decreasing to 60% of the maximum intensity at the top of the chamber. Corrections for pulse-to-pulse variations in laser power were not measured, and window fouling led to decreases in measured LIF intensities and sometimes to clouding and streaking of the images. Calibrations of LIF intensity to absolute fuel concentration were not attempted, and effects of temperature and pressure on LIF intensities were not quantified.

We have, however, carried out a preliminary comparison of undoped-gasoline LIF images to LIF images recorded with a mixture of 20% 3-pentanone and 80% spectroscopic-purity isooctane. This fluorescent-marker/fuel mixture provides a single-component fuel, a marker with nearly the same boiling point as the fuel, and a fluorescence yield that should be largely unaffected by collisonal quenching. The fluorescence intensity should therefore be proportional to the fuel concentration [31,33,34]. (A disadvantage of this system is that the fluorescence is significantly weaker than that from gasoline.) Although LIF images acquired at corresponding crank angles during injection and at 1° before ignition with gasoline and with the pentanone/isooctane mixture are not identical, they do agree in terms of their major features. The smaller differences between the two sets of images may be associated with the fluorescence or vaporization characteristics of the fluorescing component(s) of gasoline; alternatively, the differences between the two sets of images may reflect genuine differences in the fuel distributions that are related to observed differences in engine performance with the two fuels. In either case, we conclude that *the gasoline fluorescence imaging adequately reveals the major features of the fuel distribution which are the focus of the present study.*

HIGH-SPEED FILM AND VIDEO IMAGING --- We have also applied high-speed imaging techniques to examine the fuel spray and combustion processes in this engine. The experimental arrangement is the same as shown in Figure 1 except that (1) the Nd:YAG laser is replaced by a copper vapor laser (CVL), which produces pulses of 2--3-mJ energy and ~30-ns duration at rates up to 20000 pulses/s, and (2) the Princeton Instruments camera is replaced by either a HyCam high-speed color film camera or a pair of Kodak EktaPro image-intensified high-speed video cameras. The CVL output beam is formed into a thin (~0.5-mm) sheet, and the laser is synchronized to fire once during each frame of the film or video cameras, which we operate at 4000 frames/s. A beamsplitter permits the two video cameras to record the same view of the combustion chamber. Interference filters (10-nm bandpass) placed in front of each video camera isolate combustion luminosity from important radical species, such as OH (306-nm wavelength). These techniques provide a continuous record over many consecutive engine cycles of the liquid fuel spray (Mie scattering), of ignition and combustion (spectrally resolved flame emission), and of condensed-phase constituents of crevice flows (Mie scattering). The high-speed imaging is complemented by simultaneously-acquired data on cylinder pressure.

ENGINE --- The combustion chamber of the direct-injection two-stroke-cycle engine used in this study is shown schematically in Figure 2, together with a Mie-scattering cross-sectional image of the fuel spray digitized from one frame of a high-speed movie. The engine block is taken from one bank of a prototype GM Mod-3 V6 DI two-stroke engine, and the optical engine uses the prototype engine's piston, rings, and lubrication. The air-assist fuel injection system, which is similar to that described previously in Ref. [24], operates with fuel- and air-rail pressures of 620 and 540 kPa, respectively. Fuel-injection and spark timings are established with a computer-based engine set-point controller that is synchronized to the engine using a shaft encoder with 0.5° crank-angle resolution.

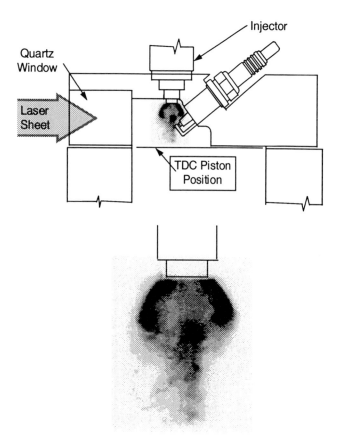

Figure 2. Cross-sectional diagram of the combustion chamber of the optical DISC engine. Superimposed on the diagram (and shown enlarged beneath it) is a 2-D Mie-scattering image of the fuel spray digitized from a high-speed film.

Table 1: Engine Specifications and Operating Conditions

Number of cylinders	1
Displacement	0.5 liter
Bore	86 mm
Stroke	86 mm
Piston height above deck at TDC	0.95 mm
Compression Ratio (trapped)	6.5
Engine Speed	1500 RPM
Delivery Ratio	0.23
Avg. Fuel Injected	6 mg/cycle
Fuel Injection Duration	5.0 ms
Injector Poppet Opens	83° BTDC
Injector Poppet Closes	37° BTDC
Spark Timing	39° BTDC
Coolant Temperature	90° C

In this paper, we present results for a single test condition that represents realistic (cruise) operation at a moderate speed (1500 RPM) and load [180 kPa indicated mean effective pressure (IMEP)]. Engine specifications and operating conditions are summarized in Table 1. Note that the engine is *not* skip fired.

To permit high-quality laser imaging, the rounded contours of the combustion chamber in the Mod-3 engine's cylinder head were modified to a nearly rectangular shape and fitted with flat ultraviolet-grade quartz windows. As indicated in Figure 2, the vertical laser sheet enters from the left, typically traverses along the centerline of the engine, and strikes the right metal sidewall. Laser scatter from the side wall and the spark plug electrodes was not a serious problem, but laser scatter from the fuel injector sometimes had to be reduced by clipping the laser sheet near the top of the chamber. At TDC, essentially all of the combustion-chamber volume can be seen because the piston crown comes within 1 mm of the bottom of the front window. A pressure transducer (not shown) is mounted on the rear wall. Cylinder, crankcase and exhaust-port pressures are all digitized with a resolution of 0.5° crank angle.

Systematic experiments revealed only minor differences in operation between the optical and unmodified engine configurations. Specifically, *when fully warmed-up* at the speed/load condition used in this paper, both the optical and unmodified engines operated with the same combustion

stability [coefficient of variation (COV) of IMEP: 2.5--3%] and with engine-out hydrocarbon emissions that agreed within 15%. However, *during engine start-up* (done here with the engine block and cylinder head heated to their normal 90°C operating temperature and with injection and ignition timings that are optimized for fully warmed-up operation), both the optical and unmodified engines often ran more roughly for the first minute or two. (In an actual vehicle application, of course, the engine control system could adjust injection and ignition timing and fueling level to ensure smooth operation during starting and warm-up.)

In order to minimize fouling of the quartz windows, the engine was usually fired for only 30--60 seconds before recording images, and occasional (~1--2%) partial burns or misfires were encountered, as illustrated explicitly in the next section. Near the end of the paper, we present LIF results recorded during six minutes of continuous firing following such a warm engine start.

RESULTS AND DISCUSSION

ENGINE OPERATION AND CYLINDER-PRESSURE ANALYSIS --- Before presenting our fuel-LIF imaging results, we summarize the understanding of engine performance and combustion at our specific test condition that emerges from cylinder-pressure analysis and high-speed film and video imaging.

Figure 3 shows an average cylinder-pressure curve which represents the ensemble average of 100 consecutive individual cycles of pressure-time data acquired simultaneously with a high-speed film discussed below. Data acquisition began 30 s after a "warm start," i.e., the engine block and head had been heated to operating temperature (90°C), but the engine had not been fired for at least 5 minutes. Superimposed are the timings and durations for exhaust and transfer port opening and closing, fuel injection, and spark. Note that the injector and spark timings used here had been optimized for minimum exhaust-hydrocarbon emissions and minimum COV(IMEP) under fully warmed-up, steady-state operation. Also shown in Figure 3 is a mass-burning-rate curve (%/deg)

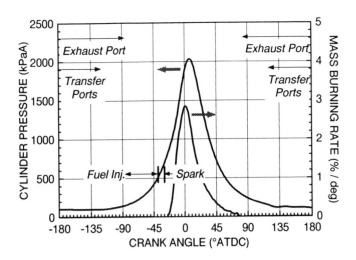

Figure 3. Ensemble-averaged cylinder pressure and mass-burning rate vs. crank angle. Also shown are port, fuel-injection and spark timings.

which represents the average of the burning-rate curves for all cycles which burned at least 80% of the fuel in the cylinder, as determined by individual-cycle heat-release analysis.

Results of the individual-cycle cylinder-pressure analysis are summarized in Figure 4, where the IMEP, burn durations and final mass fraction of fuel burned are plotted against cycle number. In four cycles (cycles 80, 87, 88, and 92), notably substandard combustion occurred: IMEP values below 140 kPa, long 0-10% burn durations, and final mass-burned fractions less than 70%. Such poor-burn cycles lead to engine roughness, high COV(IMEP), and high HC emissions; understanding their causes is a major motivation for the present study.

Plots of pressure (solid lines) and mass-burning rate (dashed lines) vs. crank angle are presented in Figure 5 for cycles 82-90, where the 100-cycle-average results are shown as thin lines and the individual-cycle results are shown as thick lines. Some cycles have a relatively slow early burn rate (e.g., cycles 83 and 85) and still have near-average peak burn rates, IMEP and final mass fraction burned. Other cycles which initially burn slowly (e.g., cycles 87 and 88) continue to burn slowly, resulting in low values for peak pressure, IMEP and final mass fraction burned. No complete misfires were observed in this data set.

HIGH-SPEED IMAGING --- We now summarize pertinent features of high-speed films and videos that reveal the nature of the combustion process and discuss how these features relate to the results of the cylinder-pressure analysis presented in the previous section. We begin with high-speed films which were used to visualize *liquid* fuel spray using Mie scattering, ignition and combustion luminosity, and early flame-kernel growth.

High-speed color films show that liquid fuel first emerges from the injector at about 85°BTDC, as expected, and reaches the spark plug some 10--12 crank-angle degrees later. (Figure 2 includes a spray image digitized from a high-speed film.) The spray initially leaves the injector as a hollow cone, as expected from the poppet geometry, but the cone sub-

sequently collapses and fills in with liquid.[2] The spray penetrates to the bottom of the field of view by about 70--65° BTDC. The Mie-scattering intensity then diminishes, and by about 50°BTDC, the liquid spray extends only as far as the spark plug. At ignition (39°BTDC), Mie scattering is no longer visible anywhere in the interior of the chamber for most engine cycles, suggesting that essentially all of the fuel within the field of view has vaporized.[3] For some cycles, however, a small amount of liquid appears briefly on the left side of the injector tip. The spark usually appears as an intense blue streak lasting about 1 ms. No combustion luminosity is evident on the color films until typically 5--15° ATDC, when yellow, orange and red emission appear at the top right corner of the chamber between the fuel injector and the slanted wall. This luminosity is typically visible for 20--40 crank-angle degrees. Still later, around 40°ATDC, some Mie scattering is often observed near the injector, apparently due to liquid emerging from the injector-tip or injector-mounting crevices.

We have also used high-speed films and videos of Mie scattering from minute silicone-oil droplets seeded into the engine intake air supply to visualize early flame-kernel growth. With this technique, Mie-scattered light from the oil droplets marks the *un*burned gas [17,18]. (In these tests, the fuel injector was raised to be flush with the top wall of the combustion chamber.) After ignition, the flame kernel initially moves rapidly downward and slightly to the right. Much of the early combustion probably occurs below our observation area. Later, combustion appears in the bottom left part of the chamber and moves upward. In general, combustion from the spark plug moves in a clockwise direction through the combustion chamber.

The absence of visible luminosity during flame-kernel growth is a strong indication that this stage of combustion occurs as (partially) premixed flame propagation at lean or stoichiometric equivalence ratio. This identification is reinforced by high-speed spectrally-resolved videos recorded

---

[2]The fact that initially hollow-cone sprays collapse and fill in is well known and has been understood for more than forty years. The earliest explanation that we know of is given in Giffen and Muraszew's 1953 book [Ref. 38, p. 50]. Both air-assist and single-fluid hollow-cone sprays collapse because a low-pressure region forms downstream of the injector as a result of air entrainment into the spray. Collapsing hollow-cone sprays of both varieties have been studied extensively by Bracco's group; see for example Refs. [21,39,40]. Ref. [24] describes our earlier exciplex-fluorescence visualization of this process under motored conditions in an early DI two-stroke engine.

[3] For reasons mentioned earlier, the Mie-scattering images do not permit the amount of liquid present to be evaluated. We estimate, however, that in our previous DISC-engine exciplex-fluorescence experiments [24], less than 1% of the total amount of fuel injected per cycle remained in the liquid phase after the end of injection. Because the exciplex visualizations were performed in a motored engine with a lower trapped compression ratio (4.0 vs. 6.5) and with an injector that produced a less well atomized spray, it is likely that substantially less than 1% of the fuel injected (i.e., <0.06 mg/cycle) is present as liquid within the field of view at the time of ignition in the present work.

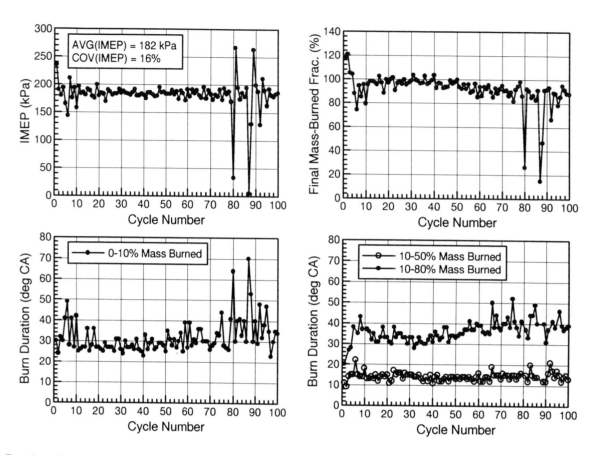

Figure 4. Results of individual-cycle heat-release analysis for 100 consecutive engine cycles. Note the large fluctuations between cycles 80 and 92. Burn-duration values are not plotted for cycles that failed to burn the specified mass fraction (10, 50, or 80%).

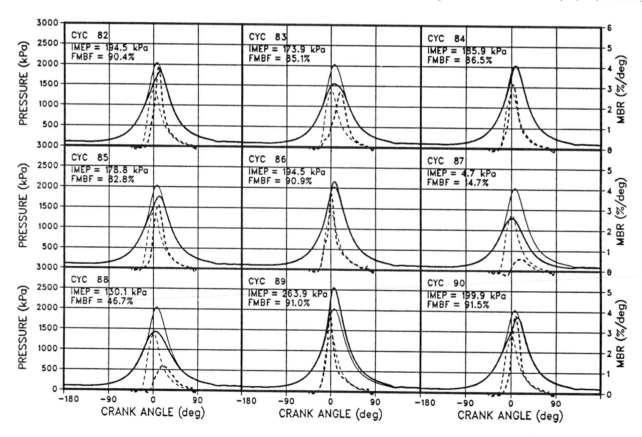

Figure 5. Cylinder-pressure measurements (*solid lines*) and mass-burning rate (MBR) results (*dashed lines*) for cycles 82--90. *Thick lines:* Individual-cycle results. *Thin lines:* ensemble-average results. Each plot is also labeled with individual-cycle IMEP and final mass-burned fraction (FMBF).

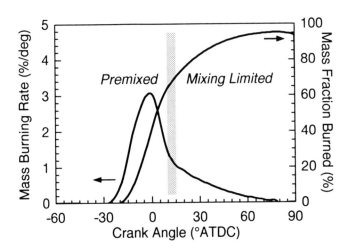

Figure 6. Mass burning rate and cumulative mass-fraction burned vs. crank angle for 100 cycles of cylinder-pressure data acquired after five minutes of continuous firing. The vertical gray bar indicates the transition from partially premixed to mixing-controlled combustion regimes.

under these test conditions with a pair of image-intensified high-speed cameras, as described above. The 4000-frame/s videos show that the flame-kernel growth is accompanied by ultraviolet (306-nm) emission from the OH radical and deep-blue visible (430-nm) emission from the CH radical, which characteristically are formed in lean and stoichiometric combustion, respectively. (Color film has poor sensitivity at these wavelengths.) Longer-wavelength emission, in part from the C_2 radical (516 nm), is observed during the spark itself, but then disappears until the partially premixed flame has propagated across the combustion chamber. The reappearance of the longer-wavelength continuum emission shortly after TDC corresponds to the time at which yellow, orange and red luminosity appear in the color movies. These observations strongly imply that the later stage of combustion proceeds as partially premixed or "diffusion" flames whose combustion rate is limited by mixing.

This scenario of an initial rapid premixed-burning phase followed by a slow mixing-controlled-burning phase has been suggested in earlier studies of light-load combustion in direct-injection stratified-charge engines [2,6,7]. To illustrate how our imaging and heat-release results support this scenario, Figure 6 contains another example of heat-release data under our test conditions, derived in this case from 100 cycles of cylinder-pressure data that were collected after the engine had been firing continuously for five minutes. As in Figure 3, the instantaneous mass-burning-rate and cumulative mass-fraction-burned curves represent averages over all the cycles (99 here) that burned at least 80% of the fuel in the cylinder. To interpret these data, note that completely premixed combustion leads to a burning-rate curve that is much more symmetric about its peak than the examples in Figures 3 and 6. On the expanded crank-angle scale in Figure 6, the burning-rate curve is seen to have a long tail that indicates slow combustion (<1%/degree) that continues until ~84°ATDC, shortly before nominal exhaust-port opening. The knee on the burning-rate curve at 10-15°ATDC coincides with the onset of yellow/orange/red visible luminosity and C_2 emission

in our high-speed films and spectrally-resolved videos, respectively. It is reasonable to consider that the knee identifies the transition between the premixed and mixing-controlled combustion stages. The cumulative mass-fraction-burned curve in Figure 6 indicates that roughly two-thirds of the fuel burns premixed at this test condition, while the remaining third experiences slower mixing-controlled combustion.

TWO-DIMENSIONAL FUEL LIF IMAGING --- Figure 7 shows representative examples of fuel LIF images recorded as a function of crank angle from shortly after the start of injection (80° BTDC) to just before spark (40° BTDC). Each image is a 16-shot average obtained with the laser sheet passing through the combustion-chamber centerline. The combustion-chamber cross section is superimposed on the second and seventh frames of the figure to show the locations of the fuel injector and spark plug. As we will demonstrate later, the fluorescence-intensity data recorded by the digital camera have a much wider dynamic range (ratio of maximum to minimum detectable intensity) than can be reproduced in printed gray-scale and false-color images, which are limited to <256 distinguishable levels. For best results with black-and white printing, gray-scale images are presented as negatives, i.e., black represents bright fluorescence and white represents dim fluorescence.

Images acquired just before the start of fuel injection with the same camera gain (not shown here) detect weak fluorescence from fuel vapor in the residual gases that remain from previous cycles. Figure 7 shows that at the start of fuel injection, the fuel fluorescence exhibits the expected hollow-cone shape of the fuel spray. The spray images are usually asymmetric, with the higher intensity occurring on the left side of the spray, although the 75°BTDC image is an exception to this trend. The fuel distribution fills in along the spray centerline due both to vaporization and gas-phase turbulent diffusion and to entrainment of air within the spray cone, which causes the spray to collapse [21,24,38--40].

The high-speed Mie-scattering films and the LIF images present a generally consistent picture of the fuel-spray behavior. However, because of film-saturation effects, the films give an exaggerated impression of the amount of liquid that has penetrated to the bottom of the field of view. As noted earlier, the Mie-scattering results show that liquid fuel disappears soon after the end of injection, indicating that vaporization is essentially complete prior to ignition at this test condition. The LIF images at 40°BTDC (i.e., at 1° before spark) therefore represent predominantly the fuel vapor distribution. In many LIF images at 40°BTDC, a small, very bright spot appears on the left side of the injector tip. This probably corresponds to the last liquid to exit the injector. The gradients in fuel concentration at the outer edges of the spray at this time are less steep than the gradients observed earlier in the spray, due to vaporization and turbulent mixing. The highest fuel vapor concentrations occur on the fuel injector centerline and extend to the bottom of the field of view. It is likely that substantial fuel vapor exists below the field of view. The high fuel vapor concentration in the lower left part of the image in Figure 7 may represent early injected

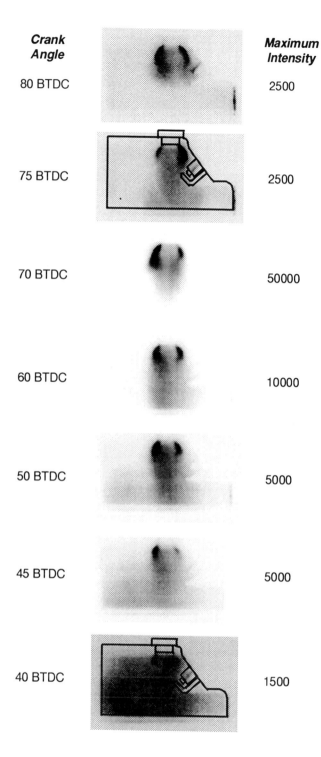

Crank Angle		Maximum Intensity
80 BTDC		2500
75 BTDC		2500
70 BTDC		50000
60 BTDC		10000
50 BTDC		5000
45 BTDC		5000
40 BTDC		1500

Figure 7. LIF images of average fuel distribution (liquid and vapor) during fuel injection. The image plane passed through the centerlines of the combustion chamber, fuel injector and spark plug.

fuel which has hit the piston surface and is moving upward back into the field of view.

Several images taken after ignition are shown in Figure 8. (LIF images were not acquired during ignition in order to avoid the intense spark emission.) The top row displays 16-shot average images for crank angles of 40, 30, 20, and 10° BTDC. Representative instantaneous (i.e., single-laser-shot) images at the same crank angles are shown in the bottom

three rows. The average images show a gradual decrease in fuel concentration with time after ignition and a maximum fuel concentration in the lower left side of the chamber. The intense signal on the right of the fuel injector seen in the average 10°BTDC image is definitely combustion luminosity, not fuel fluorescence, because it does not disappear when the ultraviolet laser illumination is blocked. Recall, furthermore, that strong luminosity in the corner between the fuel injector and the slanted combustion-chamber side wall is a common feature of our high-speed films and spectrally resolved videos. The average fuel LIF images in Figure 8 show no indication of the steep fuel gradients that one expects from flamefronts because the flame propagation varies significantly from one cycle to another.

All the instantaneous images in Figure 8 vary markedly from cycle to cycle. In all three examples at 40°BTDC, however, the fuel is concentrated on the centerline below the fuel injector and in a region to the lower left, as discussed earlier for the averaged image in Figure 7. Little direct evidence for combustion can be seen in the images taken at 30°BTDC. However, the sharp gradients that are expected from propagating flame fronts do appear mainly in the bottom right corner in the 20°BTDC images and throughout the right side in the 10°BTDC images.

THREE-DIMENSIONAL FUEL LIF IMAGING --- A CAT-scan-like tomographic representation of the *average* fuel distribution can be assembled from many two-dimensional images derived from parallel laser-sheet illumination. The laser sheet is stepped across the chamber from back to front by translating the last laser turning mirror shown in Figure 1. Figure 9 displays a set of 2-D liquid and vapor fuel images recorded during injection (at 70°BTDC) by stepping the laser sheet through eleven parallel planes separated by 2.5 mm from z = -12.5 mm (closer to the back wall of the combustion chamber) to z = +12.5 mm (closer to the front quartz window). Note that each 2-D image is an average of 16 laser shots and that the images were taken sequentially, *not* simultaneously. Images recorded in the back (front) half of the chamber are shown in the left (right) two columns of the figure. To demonstrate the excellent dynamic range of the camera, all the images are shown on two intensity scales (0--10000 and 0--50000 counts). To bring out the extent to which the spray is symmetric, images taken at equal and opposite offsets from the injector centerline (z=0) are paired in the same row, and the centerline images are shown in both groups. The images from the back half of the chamber (-12.5 mm to -2.5 mm) show the shadow of the spark plug and to a lesser extent a shadow of the fuel injector. Most of the fuel is near the injector centerline, and the distribution appears reasonably symmetric from front to back.

A similar series of tomographic images of the average fuel vapor distribution at 1° before spark (40° BTDC) is shown in Figure 10. Here the fuel is much more widely distributed, with substantial fuel in nearly all parts of the chamber. The horizontal streaks in the images are an artifact due to fouling of the laser entrance window. The strong signals from the right-hand side wall are caused by

Laser-Induced Fluorescence Imaging of Fuel Distribution in Firing Engine

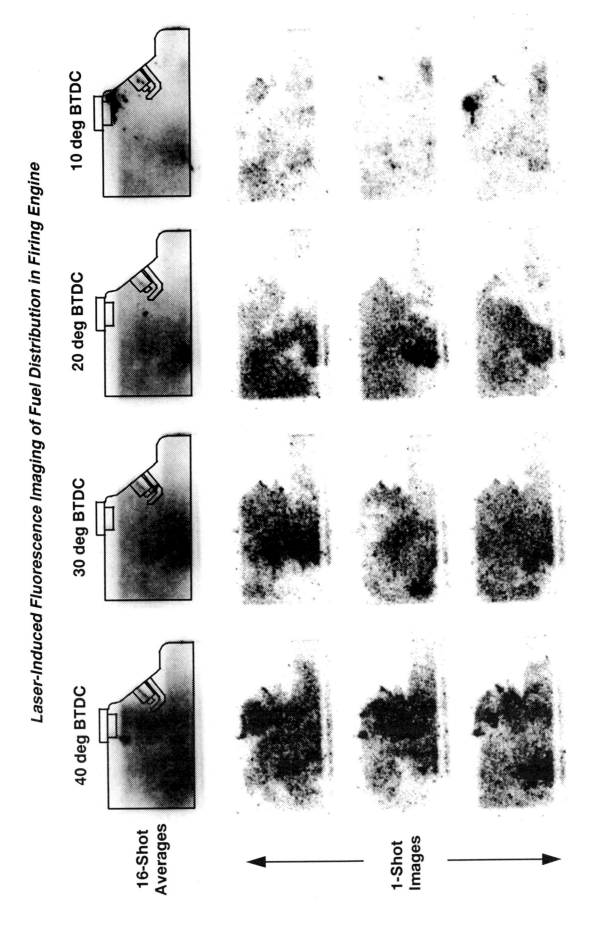

Figure 8. Average and representative instantaneous LIF 2-D images of fuel distribution before and after ignition (39° BTDC). The intense signal near the injector in the top and bottom images at 10° BTDC is flame luminosity, not fuel LIF.

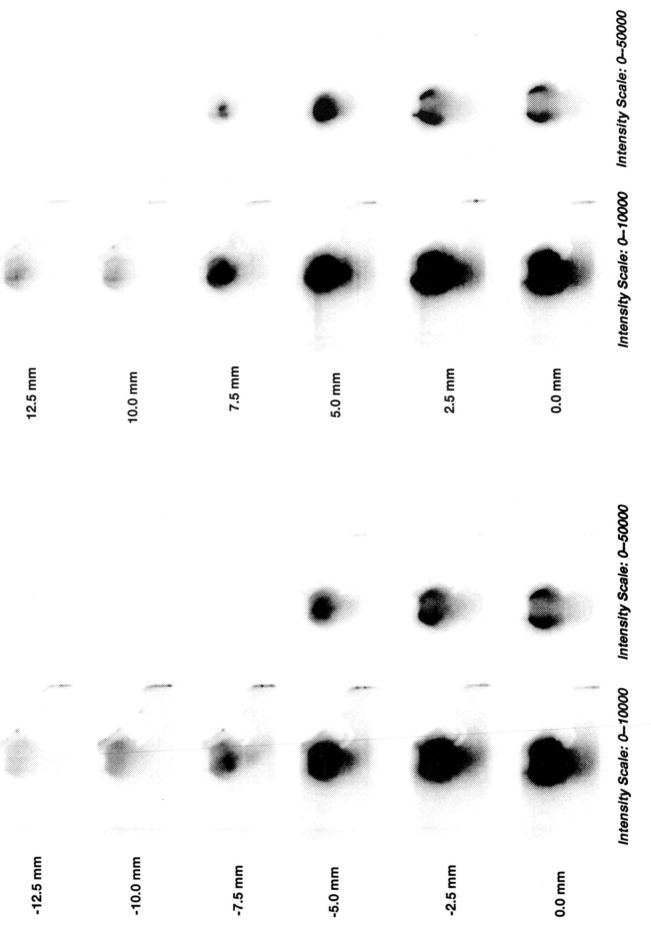

12.5 mm

10.0 mm

7.5 mm

5.0 mm

2.5 mm

0.0 mm

Intensity Scale: 0–50000

Intensity Scale: 0–10000

-12.5 mm

-10.0 mm

-7.5 mm

-5.0 mm

-2.5 mm

0.0 mm

Intensity Scale: 0–10000

Intensity Scale: 0–50000

Figure 9. Tomographic series of 2-D images of average liquid and vapor fuel distribution at 70°BTDC. Images were recorded by offsetting laser sheet from injector axis by distances indicated at left (positive distances are closer to the front window). Different intensity scales illustrate the high dynamic range in LIF image intensity.

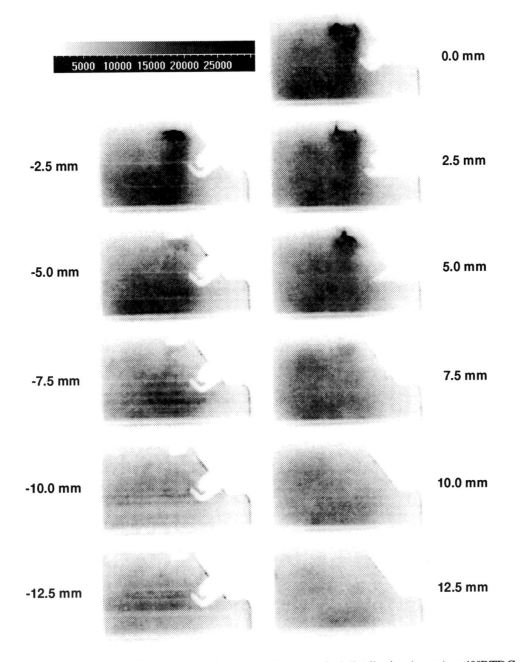

Figure 10. Tomographic series of 2-D images of average fuel distribution (vapor) at 40°BTDC (1° before spark).

fluorescence from oil deposits or condensed fuel. The highest fuel vapor concentration is close to the fuel injector.

Fully three-dimensional representations of the average fuel distributions at these two crank angles have been generated using commercial scientific-visualization software to stack together the tomographic images in Figures 9 and 10, respectively, to interpolate linearly between the images, and to render the results as 3-D geometrical objects. In Figure 11, the fuel distributions at 70° and 40° BTDC are each given two distinct 3-D renderings. In all the examples in Figure 11, a simple wireframe drawing indicates the edges of the combustion chamber, and a cross marks the spark-gap position. Fluorescence intensity is color-coded using a rainbow scale from dark blue (low intensity, low fuel concentration) to

bright red (high intensity, high fuel concentration), as shown by the color bars.

In the left column of Figure 11, the two fuel distributions are each rendered as a 3-D solid. The exterior of each solid is a blue isosurface that represents a relatively low fuel concentration (different for each crank angle), and the isosurface has been "lighted" so as to cast shadows and bring out its 3-D character. To reveal the interior structure of each fuel distribution, a 3-D section has been removed from the isosurface (the top-right octant from the strawberry-shaped spray at 70° BTDC and the top-front quadrant from the more dispersed fuel distribution at 40° BTDC). The 3-D renderings of the fuel distribution can also be sliced along arbitrary planes and rotated arbitrarily in three dimensions, e.g., they can be examined from the back or from the bottom.

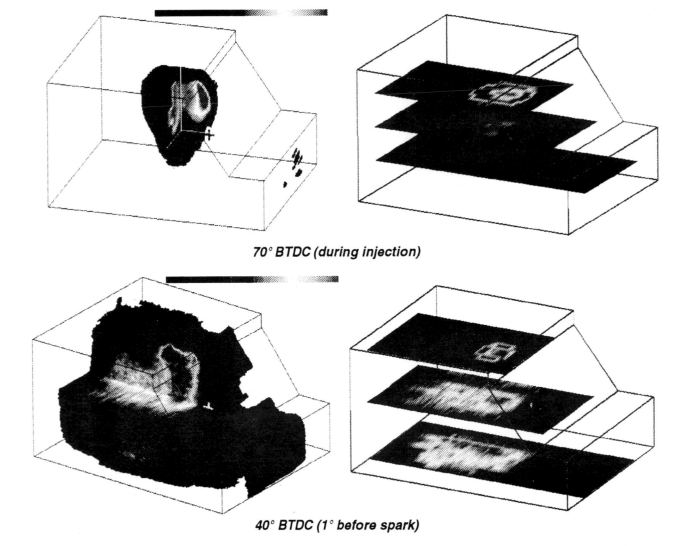

70° BTDC (during injection)

40° BTDC (1° before spark)

Figure 11. 3-D fuel distributions constructed from sets of 2-D LIF images. Wireframe drawings show combustion-chamber boundaries. Cross marks spark-gap location. *Left:* blue shaded isosurface of constant fuel concentration with section removed to show interior structure of fuel distribution. *Right:* three horizontal cross sections through fuel distribution.

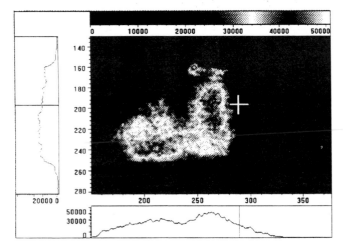

Figure 12. False-color image of (16-laser-shot) *average* fuel distribution along chamber centerline at 1° before ignition. Color scale for fuel LIF intensity is shown at top. Vertical and horizontal cross sections through cursor location (just above spark gap) are shown at left and bottom of image.

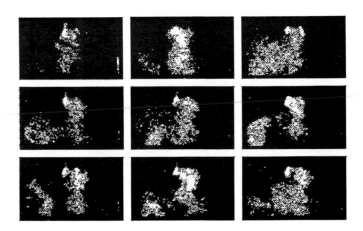

Figure 14. Cyclic variablility of fuel distribution at 1° before ignition illustrated by group of *instantaneous* 2-D LIF images selected from a series of 50 such images. Results in top, middle, and bottom rows exhibit the lowest, average, and highest fuel LIF intensities near the spark gap, respectively.

The right column of Figure 11 shows three horizontal slices through each fuel distribution. Despite the coarser resolution in the third dimension than in the vertical planes of the original images, the horizontal slices vividly depict aspects of the fuel distributions (e.g., degree of symmetry about the injector axis) that are much less apparent in the original images (Figures 9 and 10).

In comparing the fluorescence intensities at 70° BTDC and 40° BTDC, it is important to note that the images at these two crank angles were recorded with different camera gain settings. The camera sensitivity was about 10 times higher at 40° BTDC than at 70° BTDC, i.e., the same image intensity (same gray level in Figures 9 and 10, same color in Figure 11) represents about 10 times *lower* fuel concentration in the 40° BTDC images than it does in the 70° BTDC spray images.

FUEL LIF IMAGE ANALYSIS --- Strictly speaking, our LIF images constitute *qualitative* measures of the fuel distribution. As mentioned earlier, corrections to the images have not been made for the spatial nonuniformity of the laser-sheet intensity. Corrections for pulse-to-pulse variations in laser power were not measured, and window fouling sometimes led to decreases in measured LIF intensities and to clouding and streaking of the images. Direct calibrations of gasoline LIF intensity to absolute fuel concentration were not attempted, and effects of temperature and pressure on LIF intensities were not quantified. In addition, the measured fuel LIF intensities may not represent a properly weighted average over the various components of the Amoco 91 fuel blend because the molecular species giving rise to the fluorescence is (are) not known. However, we believe that these corrections are not severe, and quantitative analysis of the LIF images using reasonable assumptions and approximations is found to give useful insight, as discussed next.

Average Fuel Distributions --- Perhaps the most important time to measure fuel distributions is just prior to spark, because the fuel distribution (together with the flow field) in the spark gap and its immediate neighborhood controls ignition and early flame growth. Figure 12 shows the average fuel vapor distribution determined from a 16-laser-shot-average fuel LIF image obtained on the chamber centerline 1° before spark. The intensities are color coded as shown at the top of the image, and the white cursor marks a spot just above the spark gap. The curves at the side and bottom of the image are vertical and horizontal intensity profiles, which make clear that although the vertical fluorescence-intensity gradients at the spark gap are small, the horizontal gradients are large.

If we assume that the measured fuel LIF intensity is linearly proportional to fuel concentration and that the residual is well mixed at this time, then the fuel LIF intensity should also be proportional to the fuel/air equivalence ratio. It seems reasonable that the optimum spark-gap position is where a stoichiometric fuel-air mixture is found on average, so we further assume that 20000 counts in Figure 12 corresponds to a stoichiometric equivalence ratio Φ=1. This implies that the maximum equivalence ratio at this time is ~3,

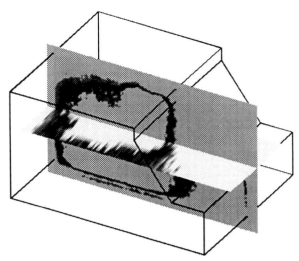

Figure 13. Region of ignitable mixture (Φ ≈ 0.6--1.6) at 1° before ignition estimated roughly from LIF image analysis. Vertical and horizontal cutting planes (gray) intersect just above the spark gap. Dark regions show the comparatively thin zone of ignitable mixture in each cutting plane.

which occurs at one corner of the fuel injector and near the lower left part of the chamber. This result appears reasonable considering that the overall equivalence ratio for the fuel-air mixture emerging from the air-assist injector is ~14 (the masses of fuel and *injected* air are approximately equal at our test condition) and that considerable time for mixing with the residual has elapsed since the start of injection.

Using the rough scale of equivalence ratio vs. fluorescence intensity provided by these estimates, the regions of flammability in Figure 12 can be estimated from laminar flame speeds as a function of Φ. For gasoline [41] at atmospheric pressure and room temperature, the laminar flame speed peaks near Φ=1 and falls to 50% of its maximum at Φ =0.6 and Φ=1.6. [Certainly, the elevated temperature (~550 K) and pressure (480 kPa) at ignition will enhance flame speeds, but dilution by residual gases -- which is substantial in this engine at light load -- will considerably narrow the flammability range.] To a first approximation, this suggests that in Figure 12, the highest flame speeds occur in the relatively narrow blue-green zones, with substantially slower flame propagation in yellow or deep blue regions, and no propagation in red-white (too rich) and deep blue-black (too lean) areas. This interpretation is qualitatively consistent with the high-speed videos and films in which Mie-scattering from silicone oil is used to mark the flame front. These show that the early flame kernel burns downward in the chamber and does not initially propagate towards the center of the chamber. Later combustion occurs on the lower left before propagating clockwise into the center of the chamber.

Figure 13 is a 3-D rendering of the estimated ignitable-mixture zone at 1° before ignition. The procedure of the preceding paragraph is used to relate LIF image intensity roughly to equivalence ratio, and the range Φ=0.6 to 1.6 is taken to define ignitable mixture. The region of space occupied by such mixture is visualized by passing two orthogonal cutting planes (shown in light gray) through the combustion chamber so that they intersect just above the spark gap. Within each plane, pixels whose fuel LIF intensity corresponds to the range Φ=0.6--1.6 are colored dark gray or black. Although the rough calibration may be somewhat inaccurate, Figure 13 illustrates clearly that, *at spark, ignitable mixture is found only within a relatively thin "shell" near the periphery of the fuel cloud.* Initially, the flame kernel can propagate only along this shell: fuel located outside the shell is too lean to burn, while fuel within the large central gray region is too rich to burn. The predominantly downward motion typically observed for the flame kernel almost certainly results from the combined effects of the initially thin shell of ignitable mixture (which broadens rapidly due to turbulent transport) and the in-cylinder air motion established as a result of scavenging and fuel injection. Note that the thickness of the estimated ignitable-mixture region (~2--6 mm) is consistent with the observation that, for fixed injection and ignition timings, the spark-plug electrode length could not be changed by more than a few mm before combustion stability deteriorated.

Instantaneous Fuel Distributions and Cyclic Variability --- Having a stoichiometric mixture at the spark gap just prior to ignition *on the average* is not sufficient to avoid ignition problems and even misfires if the fuel-air mixture at the spark gap *for an individual cycle* is beyond the ignitability limits. To help assess cycle-to-cycle variability, Figure 14 (located on same page as other color figures) shows nine instantaneous fuel LIF images selected from a larger set of 50 images taken 1° before spark. The top row of images has the least fuel at the spark gap, the middle row has an average amount of fuel, and the bottom row has the most fuel at the spark gap. The shape of the fuel cloud and the total amount of fuel within the cloud appear very roughly the same in each image, but the amount of fuel at the spark gap varies substantially because the spark gap is located in a region of strong fuel-concentration gradient.

Cyclic variation in the fuel concentration near the spark gap was quantified by calculating a histogram of fuel LIF intensity from a small region of interest (0.75 mm x 1 mm image area by ~0.5 mm light-sheet thickness) located just above the spark gap. The spark gap itself was not chosen because of laser-sheet blockage by the ground electrode. The histogram, shown in Figure 15a, displays a symmetric distribution centered at approximately 1800 counts with two cases of high intensity. Assuming again that the LIF intensity is proportional to equivalence ratio and that the average equivalence ratio corresponds to Φ=1 at the spark gap, we estimate from Figure 15 that the instantaneous fuel-air equivalence ratio at the spark gap varies from 0.45 to 2.0. This implies a very large variation in laminar flame speed in the early flame kernel, from which some misfires and partial burns are likely. For example, eight of the 50 images have a fuel LIF intensity which lies outside the 0.6--1.6 equivalence-ratio range, which is roughly consistent with the ~5% incomplete-combustion (<80% final mass burned fraction) cycles determined from cylinder-pressure analysis. This approach to analyzing the fuel LIF images probably over-estimates the incidence of misfires and partial burns because

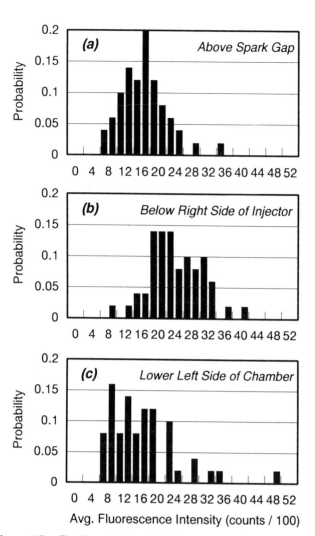

Figure 15. Cyclic variation in fuel concentration shown by histograms of fuel LIF intensity from small regions (0.75 x 1 x 0.5 mm³) at three different locations in combustion chamber.

the spark can enflame gases from a region somewhat larger than the ~0.4 mm³ volume examined here. However, the present LIF data certainly suggest that *fuel concentration fluctuations can account for the observed cyclic variability in early burn rates.*

Histograms from two other regions were also calculated in order to see whether a narrower histogram distribution could be found. The histogram in Figure 15b is from a region at approximately half the distance from the fuel injector tip to the spark gap, and the histogram in Figure 15c is from the lower left part of the chamber. These locations were chosen to have roughly the same average fuel LIF intensity as in Figure 15a. Neither of the two lower histograms in Figure 15 is narrower than the top histogram, suggesting that neither of the corresponding locations is preferable in terms of ignition stability.

Fuel Distribution during Engine Warmup --- With the fuel injector (designated here as "injector A") used for all of the experiments reported to this point, both the optical and unmodified versions of the engine ran more roughly for the first minute or two after a warm start (i.e., having not been run for at least five minutes, and with the engine head and block heated to 90°C, the engine was started using injection and spark timings that had been optimized for fully warmed-up, steady-state operation). Consistent changes in fuel LIF images were also observed over a corresponding period of time under these circumstances. Figure 16 plots the LIF intensity vs. time after a warm start for the 0.4 mm³ area just above the spark gap for a set of 50 images taken at uniformly spaced intervals over a six-minute period. All the images were 16-shot averages acquired at 1° before spark. These results imply that the fuel concentration near the spark plug gradually increased by a factor of about 1.5 over the first three minutes of the test. Although window fouling might explain the slow decay over the second three minutes, the slow increase in fuel concentration after starting the engine is surprising because separate experiments on several injectors of the type used here have established that the injectors normally reach full fuel delivery within just a few engine cycles after starting.

While acquiring the images for Figure 16, we also monitored a real-time moving-average (100-cycle) readout of the engine's IMEP and COV(IMEP). As the fuel LIF intensity at the spark gap increased, the IMEP increased from 155 kPa to 177 kPa, COV(IMEP) decreased from about 20% to 3.7%, and misfires decreased from 1% to zero. Together with the LIF images, these observations suggest that the fuel-air mixture at the spark gap was initially too lean for reliable ignition. If we assume that the average equivalence ratio was Φ=1 at the spark gap for the smoothly-operating engine, then Figure 16 implies that Φ=0.65 *on average* at the spark gap when the engine was started. Given even modest cyclic variability in Φ, misfires or slow burns are therefore to be expected in some engine cycles during warm-up under these circumstances.

Different warm-up behavior was observed with a second (nominally identical) injector (designated B) which was tested more than a year after injector A. The LIF intensity for injector B in Figure 16 is essentially constant over the three-minute period of observation. Correspondingly, engine operation with injector B stabilized much more rapidly (i.e., immediately after the start of fuel injection, the engine fired steadily with a moving-average IMEP of 175 kPa, a COV(IMEP) of 14%, and no misfires. Within 30 s after starting, IMEP and COV(IMEP) stabilized at 183 kPa and 2.5--3%, respectively.)

We have not conclusively identified why these two nominally identical injectors behaved differently. Nevertheless, the differences observed in engine performance during warm-up can be explained plausibly by the differences in fuel concentration near the spark gap as inferred from the measured fuel fluorescence intensity.

SUMMARY AND CONCLUSIONS

LIF imaging of fuel distributions has been accomplished in a firing direct-injection stratified-charge engine. Fluorescence from one or more components of commercial unleaded gasoline was excited by a thin sheet of ultraviolet light (266-nm wavelength) from a pulsed laser and detected by an

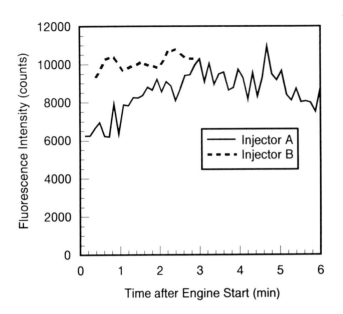

Figure 16. Fuel fluorescence intensity near spark gap observed during engine warm-up with two nominally identical fuel injectors.

intensified CCD camera with low noise and high dynamic range. The LIF signal intensity is expected to be approximately proportional to the fluorescing-species concentration (whether liquid or vapor).

Instantaneous and multi-cycle-averaged 2-D (cross-sectional) images of the fuel distribution in the combustion chamber were obtained at selected crank angles during fuel injection, ignition and combustion. Tomographic 3-D images of the fuel distribution were constructed from sets of averaged 2-D LIF images taken while stepping the laser light sheet across the combustion chamber.

For the direct-injection stratified-charge engine studied here, the assumption that the fluorescence intensity is proportional to the fuel concentration leads to the following conclusions:

1. *Multi-cycle-averaged 2-D and 3-D images show that the average fuel distribution is highly stratified at the time of ignition, with steep fuel-concentration gradients existing near the spark gap.* Ignitable mixture appears to occur only in a relatively thin shell around the periphery of the fuel cloud. Such images can help optimize spark-plug location and ignition timing for a given injector, combustion-chamber geometry, and fuel.

2. *Instantaneous 2-D fuel LIF image intensities demonstrate that there are large cycle-to-cycle variations in the fuel distribution (a factor of five has been observed near the spark gap at 1° crank angle before ignition).* The cyclic variation in fuel LIF intensity is consistent with the incidence of poor burns and misfires obtained from cylinder pressure analysis and is also consistent with high-speed videos of fuel sprays and combustion in this engine.

3. *The fuel LIF results imply that the early flame fronts are very irregular, and flame propagation is highly non-uniform.* However, the flame kernel generally appears to move through the ignitable-mixture region (deduced from the measured average fuel distribution) in the direction to be expected from the in-cylinder flow field.

4. *Comparison of LIF images acquired with two nominally identical injectors imply systematic differences in fuel concentration near the spark gap that correspond to observed differences in engine performance during a warm start-up.*

The assumption that the fluorescence intensity is proportional to fuel concentration has not been established quantitatively here because the identity of the fluorescing constituent(s) of commercial gasoline has (have) not been determined yet. Quantitative interpretation of the fluorescence images would be compromised if the fluorescence is collisionally quenched, if the absorption is temperature-dependent, or if the fluorescing constituent(s) do not co-evaporate with representative gasoline components. More reliable quantitative interpretation of fuel LIF imaging should be possible by doping a single-component fuel (such as isooctane) with an appropriately matched fluorescent tracer (such as 3-pentanone).

Ensemble-averaged fuel LIF images can be compared directly to computational-fluid-dynamics calculations of mixture preparation and combustion in order to validate or improve modeling approaches. In addition, instantaneous fuel LIF images can provide direct information about important engine-development problems by quantifying fuel distributions on a cycle-to-cycle basis during both steady and transient engine operation. For example, fuel LIF imaging -- especially if combined with simultaneous heat-release and exhaust-hydrocarbon measurements on a cycle-resolved basis -- should be especially helpful in addressing sources of misfires, partial burns, and hydrocarbon emissions during cold start and highly dilute operation for a wide range of spark-ignition engines.

ACKNOWLEDGMENTS

We are grateful to A.S.P. Solomon for providing his two-stroke version of P.M. Najt's individual-cycle pressure-analysis code and for tutoring us in its use, and to E.G. Groff and M.R. Galasso for helpful discussions and information. T.E. Zak helped with use of the visualization software and devised a routine to assemble our 2-D images into a 3-D stack.

REFERENCES

1. C.D. Wood, "Unthrottled Open-Chamber Stratified-Charge Engines," SAE Paper 780341, 1978.

2. D.R. Lancaster, "Diagnostic Investigation of Hydrocarbon Emissions from a Direct-Injection Stratified-Charge Engine with Early Injection," I. Mech. E. Paper C397-80, 1980.

3. R. Diwakar, "Direct-Injection Stratified-Charge Engine Computations with Improved Submodels for Turbulence and Wall Heat Transfer," SAE Paper 820039, 1982.

4. A.J. Giovanetti, et al., "Analysis of Hydrocarbon Emissions in a Direct-Injection Spark-Ignition Engine," SAE Paper 830587, 1983.

5. J.M. Lewis, "UPS Multifuel Stratified Charge Engine Development Program -- Field Test," SAE Paper 860067, 1986.

6. R.M. Frank and J.B. Heywood, "Combustion Characteristics in a Direct-Injection Stratified-Charge Engine and Implications on Hydrocarbon Emissions," SAE Paper 892058, 1989.

7. R.M. Frank and J.B. Heywood, "The Importance of Injection System Characteristics on Hydrocarbon Emissions from a Direct-Injection Stratified-Charge Engine," SAE Paper 900609, 1990.

8. D. Scott and J. Yamaguchi, "Pneumatic Fuel Injection Spurs Two-Stroke Revival," *Automotive Engineering* 94(8), 74, 1986.

9. P. Duret, A. Ecomard, and M. Audinet, "A New Two-Stroke Engine with Compressed-Air Fuel Injection for High Efficiency Low Emissions Applications," SAE Paper 880176, 1988.

10. "Clean Air Fuels Engine Evolution, Not Revolution," *Ward's Auto World*, June 1991, p. 33.

11. H. Schäpertöns, K.-D. Emmenthal, H.-J. Grabe, and W. Oppermann, "VW's Gasoline Direct Injection (GDI) Research Engine," SAE Paper 910054, 1991.

12. J. Yamaguchi, "Global Viewpoints," *Automotive Engineering*, February 1994, p. 39.

13. N. Miyamoto, H. Ogawa, T. Shudo and F. Takeyama, "Combustion and Emissions in a New Concept DI Stratified Charge Engine with Two-Stage Fuel Injection," SAE Paper 940675, 1994.

14. F. Galliot, W.K. Cheng, M. Sztenderowicz, J.B. Heywood, and N. Collings, "In-cylinder Measurements of Residual Gas Concentration in a Spark Ignition Engine," SAE Paper 900485, 1990.

15. E. Winklhofer, G.K. Fraidl, and A. Plimon, "Monitoring of Gasoline Fuel Distribution in a Research Engine," *Proc. I. Mech. E.* D206, 107, 1992.

16. C. Arcoumanis and A.C. Enotiadis, "In-Cylinder Fuel Distribution in a Port-Injected Model Engine using Rayleigh Scattering," *Experiments in Fluids* 11, 375, 1991.

17. T.A. Baritaud and R.M. Green, "A 2-D Flame Visualization Technique Applied to the I.C. Engine," SAE 860025, 1986.

18. A.O. zur Loye and F.V. Bracco, "Two-Dimensional Visualization of Premixed-Charge Combustion Flame Structure in an IC Engine," SAE 870454, 1987.

19. P.G. Felton, J. Mantzaras, M.E.A. Bardsley, and F.V. Bracco, "2-D Visualization of Liquid Fuel Injection in an Internal Combustion Engine," SAE Paper 872074, 1987.

20. L.A. Melton, "Spectrally Separated Fluorescence Emissions for Diesel Fuel Droplets and Vapor," *Applied Optics* 22, 2224-2226, 1983.

21. M.E.A. Bardsley, P.G. Felton, and F.V. Bracco, "2-D Visualization of a Hollow-Cone Spray in a Cup-in-Head, Ported, I.C. Engine," SAE Paper 890315, 1989.

22. R. Shimizu, S. Matumoto, S. Furono, M. Murayama and S. Kojima, "Measurement of Air-Fuel Mixture Distribution in a Gasoline Engine Using LIEF Technique," SAE Paper 922356, 1992.

23. J. Senda, Y. Fukami, Y. Tanabe and H. Fujimoto, "Visualization of Evaporative Diesel Spray Impinging Upon Wall Surface by Exciplex Fluorescence Method," SAE Paper 920578, 1992.

24. R. Diwakar, T.D. Fansler, D.T. French, J.B. Ghandhi, C.J. Dasch and D.M. Heffelfinger, "Liquid and Vapor Fuel Distributions from an Air-Assist Injector -- An Experimental and Computational Study," SAE Paper 920422, 1992.

25. D.A. Hansen and E.K.C. Lee, "Radiative and Nonradiative Transitions in the First Excited Singlet State of Simple Linear Aldehydes," *J. Chem. Phys.* 63, 3272, 1975.

26. F. Beretta, V. Cincotti, A. D'Alessio, and P. Menna, "Ultraviolet and Visible Fluorescence in the Fuel Pyrolysis Regions of Gaseous Diffusion Flames," *Combust. Flame* 61, 211, 1985.

27. A. Lozano, B. Yip, and R.K. Hanson, "Acetone: a tracer for concentration measurements in gaseous flows by planar laser-induced fluorescence," *Experiments in Fluids* 13, 369, 1992.

28. P. H. Paul, I. van Cruyningen, R.K. Hanson, and G. Kychakoff, "High Resolution Digital Flowfield Imaging of Jets," *Experiments in Fluids* 9, 241 1990.

29. P. Andresen, G. Meier, H. Schluter, H. Voges, A. Koch, W. Hentschel, W. Opperman, and E. Rothe, "Fluorescence Imaging Inside an Internal Combustion Engine Using Tunable Excimer Lasers," *Applied Optics* 29, 2392, 1990.

30. A. Arnold, H. Becker, R. Suntz, P. Monkhouse, J. Wolfrum, R. Maly, and W. Pfister, "Flamefront imaging in an internal-combustion engine by laser-induced fluorescence of acetaldehyde," *Optics Letters* 15, 831, 1990.

31. W. Lawrenz, J. Kohler, F. Meier, W. Stolz, R. Wirth, W.H. Bloss, R.R. Maly, E. Wagner, and M. Zahn, "Quantitative 2D LIF Measurements of Air/Fuel Ratios During the Intake Stroke in a Transparent SI Engine," SAE 922320, 1990.

32. T.A. Baritaud and T.A. Heinze, "Gasoline Distribution Measurements with PLIF in a SI Engine," SAE Paper 922355, 1992.

33. A. Arnold, A. Buschmann, B. Cousyn, M. Decker, F. Vannobel, V. Sick, and J. Wolfrum, "Simultaneous Imaging of Fuel and Hydroxyl Radicals in an In-line Four-Clinder SI Engine," SAE Paper 932696, 1993.

34. H. Neij, B. Johansson and M. Aldén, "Development and Demonstration of 2D-LIF for Studies of Mixture Preparation in SI Engines," *Combust. Flame* 99, 449, 1994.

35. J. Reboux, D. Puechberty and F. Dionnet, "A New Approach of Planar Laser Induced Fluorescence Applied to Fuel/Air Ratio Measurement in the Compression Stroke of an Optical S.I. Engine," SAE Paper 941988, 1994.

36. E. Winklhofer, H. Phillipp, G. Fraidl, and H. Fuchs, "Fuel and Flame Imaging in SI Engines," SAE Paper 930871, 1993.

37. Isadore B. Berlman, *Handbook of fluorescence spectra of aromatic molecules*, 2d ed., Academic Press, New York, 1971.

38. E. Giffen and A. Muraszew, *The Atomisation of Liquid Fuels*, Wiley, New York, 1953.

39. B. Chehroudhi and F.V. Bracco, "Structure of a Transient Hollow-Cone Spray," SAE Paper 880522, 1988.

40. J. Emerson, P.G. Felton and F.V. Bracco, "Structure of Sprays from Fuel Injectors Part III: The Ford Air-Assisted Fuel Injector," SAE Paper 900478, 1990.

41. J.B. Heywood, *Internal Combustion Engine Fundamentals*, McGraw-Hill Book Company, New York, 1988, p. 403- 406.

Analysis of Mixture Formation of Direct Injection Gasoline Engine

Masao Kano, Kimitaka Saito and Masatoshi Basaki
Nippon Soken, Inc.

Souichi Matsushita and Takeshi Gohno
Toyota Motor Corp.

ABSTRACT

Direct injection gasoline engines require extremely advanced control of air-fuel mixture in order to achieve good stratified combustion. The method of examining quality of mixture formation in combustion chambers is essential for the achievement. In this research, air-fuel mixture in combustion chamber of the TOYOTA D-4 engine was analyzed in space and time by visualization as well as Air/Fuel ratio measurement by multi-point and high response techniques. Thus the effects that injection timing, swirl and fuel pressure exerted to mixture formation were elucidated. For realizing good stratified combustion, the following were identified as necessary:

(1) to select the optimized timing interval between injection and ignition at which the Air/Fuel ratio near the ignition plug at ignition can be maintained at 10 to 20

(2) to control swirl intensity accurately corresponding to engine operating conditions, in order to guide air-fuel mixture to ignition plug without dispersing the mixture

(3) to allow spray to follow swirl flow by 15 μm droplet size with 12 MPa fuel pressure

1. INTRODUCTION

To cope with energy and environmental issues in recent years, there has been extensive research conducted to achieve lower fuel consumption for the engines of motor vehicles. As a result of this need for low fuel consumption technology, there has been considerable attention given to a direct injection gasoline engine. The reason is the direct injection gasoline engine is capable of creating high power while maintaining low fuel consumption. Based on the promise of the DI Engine, there has been considerable research and development work conducted in the recent years.[1-21]

The direct injection gasoline engine performs stratified charge combustion with an extremely sophisticated air-fuel mixture control and thus a method for examining the quality of the air-fuel mixture formation in the combustion chamber is necessary. Therefore, in this research we analyzed the air-fuel mixture in the combustion chamber of the direct injection gasoline engine by the following means:

1) Observation of in-cylinder fuel spray through the spark plug hole.
2) Measurement of the in-cylinder Air/Fuel ratio distribution by multi-point timed gas sampling.
3) High response measurement of Air/Fuel ratio close to the spark plug.

This research will examine of the effects of injection timing, swirl intensity, and fuel pressure upon the formation of air-fuel mixture in the cylinder as well as injection(combustion) cycle-to-cycle fluctuation will be reported hereinafter.

2. TEST ENGINE

Table 1 D-4 Engine Specification

Displacement	1998cc
Bore * Stroke	86 * 86 mm
Type	In-line , 4-Stroke
Number of Valve	4
Compression Ratio	10.0 : 1
Intake Port	Helical and Straight with Swirl Control Valve
Fuel Pressure	12MPa
Fuel Supply	High Pressure Swirl Injector
Fuel	91RON

Table 1 shows the direct injection gasoline engine (hereinafter called the "D-4" engine) specifications used in the experiments. Figure 1 shows the concept of the stratified air-fuel mixture control in the D-4 engine. Fuel injected in the last half of the compression stroke enters the piston cavity and flows toward a spark plug along the cavity wall by means of swirl air and penetrating force of the fuel spray. Fuel injected with high pressure obtains heat from the wall surface of the piston and space in this process, and it

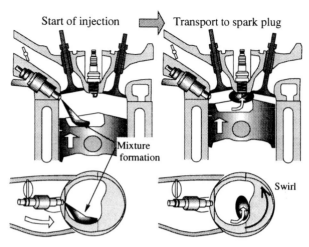

Figure 1 Fuel spray behavior in the piston cavity at compression stroke in D-4 engine

forms combustible air-fuel mixture near the spark plug while evaporating. At the moment of spark, rich air-fuel mixture which is easily ignited exists near the spark plug and, as a whole, extremely lean air-fuel ignition mixture is created. As a result, stable combustion is realized.

3. EXAMINATION OF AIR-FUEL MIXTURE FORMATION

For understanding the air-fuel mixture formation in the combustion chamber of a gasoline engine, a single cylinder long piston engine is often utilized. However, this single cylinder engine does not provide a direct correspondence with an actual DI engine because of the following:

1) High revolution measurement is difficult.
2) Phenomenon, such as intake pulsation, typical of a multiple cylinder engine cannot be reproduced.
3) Application to complicated piston design is not possible.

Therefore, in this research, actual engines were utilized for spatial and time analysis with in-cylinder visualization of air-fuel mixture, multi-point Air/Fuel ratio measurement, and high response Air/Fuel ratio measurement near spark plug.

3.1 IN-CYLINDER VISUALIZATION OF BEHAVIOR OF AIR-FUEL MIXTURE

3.1.1 CONVENTIONAL VISUALIZATION METHOD

Analyzing a series of phenomena such as gas flow, air-fuel mixture formation, combustion and exhaust in the combustion chamber has the most important role in the research and the development work for the engine.

There are many methods utilized for the

phenomena in the combustion chamber. Visualization techniques have been utilized for many years because these methods can provide a large amount of information and simultaneous occurrence of spatial data. Thus, several types of visualized engines have been proposed. The configuration of visualization engines can be classified into three types such as Bottom View, Side View, and Top View as shown in Figure 2.

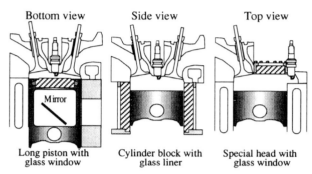

Figure 2 Conventional visualization engines

For the Bottom View Method, the cylinder liner and the piston are extended, an observation window is arranged at the piston top, and the combustion chamber is observed from the bottom through a mirror located inside the extended piston. This method is called the Bowditch system, and was developed by General Motors. [22] In the Bottom View Method, the combustion chamber roof does not have to be modified at all, and thus, it is the most widely used at present. However, its application is difficult when there is a cavity on the piston, as in the case of the D-4 engine.

For the Side View Method, a transparent glass cylinder is buried in a cylinder block and the combustion chamber is observed from the side. The author and his group have had much experience in visualizing many engines with this method [23, 24]. But its operation for many hours is a problem because of the difference in thermal expansion, as well as, the sealing between the piston and the glass cylinder.

For the Top View Method, an observation window is provided in part of the cylinder head, and the combustion chamber is observed from the top. However, many gasoline engines have overhead four-valve designs, and thus the application of the Top View Method is difficult.

In addition, the items required for the visualization engine are:

1) Configuration / shape of the combustion chamber, valve layout, and the like, must strictly conform to the actual engine (conformity).
2) Stable observation must be available for a

long period of time (operationability).
3) Visual field of observation must be wide, and clear observation must be available (observability).
4) Handling must be easy and operation must be safe (maintainability).

However, all the visualization engines proposed in the past rarely satisfied all requirements described above. Moreover, for the visualization engines, the time required from scheme establishment of the visualization to test including design fabrication and adjustment is typically three to six months, and this is a problem. Therefore, for this visualization of the direct injection engine, we decided to develop an approach to visualize the combustion chamber quickly without modifying the actual engine.

3.1.2 NEW VISUALIZATION APPROACH

For visualizing the inside of the combustion chamber without changing the construction of the actual engine at all, methods are limited to those utilizing a spark plug hole. An M14 thread fixes a spark plug on the D-4 engine. Although outer diameter of the M14 thread is 14mm, the available diameter is 12mm due to the thread dimensions. In the new visualizing method developed by the authors, the device can be packaged within the space of 12mm in diameter inside the spark plug.

One device for visualizing the inside of the cylinder through a limited hole diameter is a fiber scope. The authors tried a fiber scope with illuminating function, made by Olympus at first. But any clear images fulfilling the requirements (several hundred thousand picture elements) were not obtained because the resolution was restricted by the number of fibers (several ten thousand fibers for this try), and thus was abandoned.

Figure 3 shows an ultra compact built-in camera which has been developed. Normally, an insulator and a central electrode is located in a housing body of the spark plug. Instead of the insulator and the central electrode, we placed a CCD video camera, fibers for stroboscopic lighting, a group of lenses, and a quartz observation window.

During the observation of the combustion chamber, the ignition function is lost by installing the visualization equipment in place of the spark plug. However, this was not a problem because the analysis of the air-fuel mixture formation from fuel injection to ignition was the main objective.

Figure 4 shows the visualizing devices on an actual engine. The stroboscopic lighting fibers are made of about 40,000 fibers surrounding the video camera. Since the piston travels up and down in an 86mm stroke, a group of lenses at the front of the camera were selected as large focus depth and wide view are obtained to observe the whole combustion chamber around the top dead center. We achieved fine visualization with this idea. A quartz observation window, with its surface facing the combustion chamber, was formed in a concave shape. Thus a lens effect was obtained and the view angle was widened. We had to make extra effort for taking countermeasures against reflection, at the end surface of the observation window, due to the stroboscopic light radiated from the fibers and also against piston halation. As a result of trial and error, we resolved the problems by providing adequate optic diffusion. The camera has a casing of 7mm in diameter, a number of effective picture elements is 410,000, required illuminance of object is 200/20 luxes (standard/minimum), and 1/4 inch color single plate CCD (Toshiba) is used. Strobe light source is a xenon type (Sugawara). Its energy per flash is 8J (170 luxes/sec.). A flash time of 12 to 25 μ sec. is transmitted with the optical fibers.

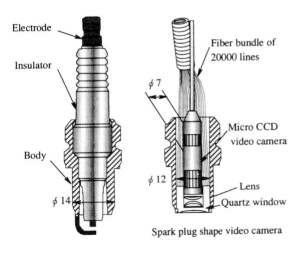

Figure 3 Packaging video camera in spark plug space

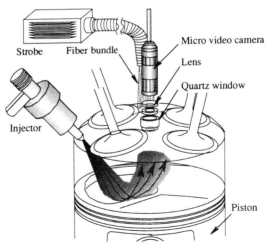

Figure 4 Video system to observe air-fuel behavior in cylinder

Since the shooting direction of the camera and lighting direction from the fiber are the same in the visualization device, the forward scattered light of spray particles is photographed. Here, the Mie's forward scattered light intensity, E, can be given by the following equation when the lighting intensity is E_0.

$$E = NK \pi a^2 E_0$$

where,
N: Number of particles
K: Scattering coefficient
a: Radius of particle

This equation expresses that the forward scattering intensity decreases as the particle diameter becomes smaller. The minimum particle diameter of the spray that can be photographed by the visualization is about 0.3 μm, estimating based on the light source intensity of the xenon minimum illuminance of object for the CCD video camera, and light transmission loss of the fibers. Therefore fuel particles or those vaporized components smaller than that cannot be observed.

A block diagram of the video system is shown in Figure 5. An injection signal is branched from an engine control computer to a strobe controller and a strobe emits light with a predetermined time delay. The video signal from the ultra compact video camera is stored in a digital frame memory in synchronization with the strobe light emission, and the flicker of the image due to the strobe light emission was eliminated. The behavior of the spray at the target timing can be observed with constantly delayed strobe lighting. Slowmotion the photographing of the spray behavior can be observed with swept delay of strobe lighting. The picture image stored by strobe synchronization can be recorded and played back by a VCR. The spray behavior in the combustion chamber can be clearly visualized in real time with this visualization system.

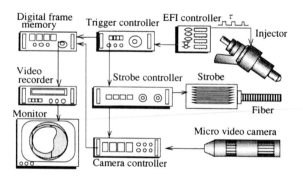

Figure 5   Block diagram of the strobe-synchronized video system

3.2   IN-CYLINDER AIR/FUEL RATIO DISTRIBUTION MEASUREMENT

Figure 6 shows an outline of an in-cylinder timed gas sampling apparatus for measuring in-cylinder

Air/Fuel ratio distribution. Figure 7 shows the features of an electromagnetic valve for the gas sampling. The in-cylinder air-fuel mixture is sampled by six sampling tubes of 1.7 mm inner diameter: one tube at spark plug, four tubes projected out by 4.5 mm from the combustion chamber surface and one tube mounted flush on the combustion chamber surface in a squish area. Timed sampling is applied to the air-fuel mixture with predetermined timing by using electromagnetic valves installed upstream of the sampling tubes, and Air/Fuel ratio was calculated for each sampling tube by HC concentration measured with HORIBA gas analyzer after collection of the gas in a bag.

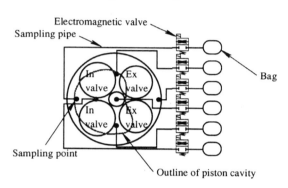

Figure 6   In-cylinder, multi-point timed gas sampling apparatus

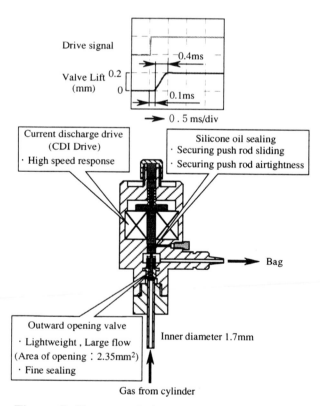

Figure 7   Feature of electromagnetic valve for in-cylinder multi-point timed gas sampling

Features of the electromagnetic valve used here are as follows:
   1)   High speed response with 0.5 msec. for full

lift (0.2 mm) by using a current discharge driver (a CDI driver).

2) Lightweight and large flow (opening area: 2.35 mm^2) of the valve realized by an outward opening valve, and fine sealing is also realized.
3) Silicon oil assures both smooth sliding and air tightness of the push rod.

Iso-Air/Fuel ratio diagram in cylinder was calculated by using the Air/Fuel ratio measured at six points. Air/Fuel ratio in the outer periphery of combustion chamber and in piston cavity were used as

3.3 HIGH RESPONSE MEASUREMENT OF AIR/FUEL RATIO BEHAVIOR NEAR SPARK PLUG

Figure 8 shows an outline of the high response measurement of Air/Fuel ratio behavior near spark plug. Air-fuel mixture near the spark plug was measured with a sampling tube, near the spark plug, connected to an ultrahigh response FID (HFR-400/T90, Cambustion Co.; 4msec. response).[25-28] The FID used for the research had a compact and lightweight detector, with the following features:

1) Mounting close to the sampling point for both ultra quick response and transit time of several msec.
2) Easy calibration in dynamic state
3) A wide measuring range from low concentration HC in combustion gas to high concentration uHC in unburned air-fuel mixture.

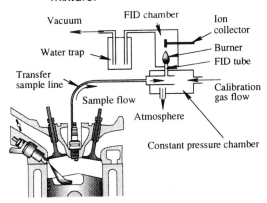

Figure 8 FID in-cylinder sampling apparatus

This FID is installed just above the engine, with a 400mm length sampling tube, and therefore a gas transport delay in the tube is minimized.

Because the collector output of the FID is affected by cylinder pressure, as shown in Figure 9, the collector output and calibration characteristics of exhaust Air/Fuel ratio for premixed combustion (injection in intake stroke) were measured in each operating condition. Then the measurement for

stratified combustion (injection in compression stroke) was carried out. In Figure 9, the air-fuel mixture injected during intake stroke is detected from the last half of the intake stroke. The FID collector output increases toward the compression stroke and becomes a constant value up to ignition. The FID collector output suddenly drops during combustion when the combustion gas is sampled. The constant value of the collector output depends on the injection quantity, so that we have developed the correspondence between the constant value of collector output and the exhaust Air/Fuel ratio for calibration curves. We also adjusted the engine crank angle phase to correlate the FID collector output, considering the 4 msec. response

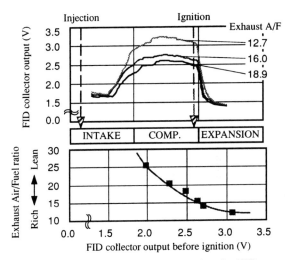

Figure 9 Characteristics of calibration for FID collector output vs. exhaust Air/Fuel ratio

4. RESULTS AND REVIEW

4.1 STATE OF STRATIFIED AIR-FUEL MIXTURE FORMATION

Figure 10 shows visualized spray behavior between BTDC70° CA and BTDC30° CA. It is observed that the spray injected upstream of swirl, relative to the spark plug, is induced in the direction of the spark plug along the cavity side wall. Figure 11 shows Air/Fuel ratio distribution immediately before ignition (BTDC16° CA). It is confirmed that rich air-fuel mixture of Air/Fuel ratio 10 to 20 is present near the spark plug.

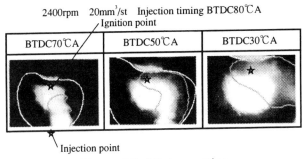

Figure 10 Mixture motion

445

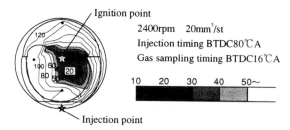

Figure 11 Mixture distribution

Its surrounding is ultra-lean with Air/Fuel ratio 50 to 100. The air-fuel mixture in the combustion chamber is stratified in compact space as originally designed.

4.2 EFFECT OF INTERVAL BETWEEN INJECTION AND IGNITION TIMING

Figure 12 shows the effect of timing intervals between injection and ignition upon the combustion by varying the injection timing. Under the stratified combustion, an optimum interval (B in the figure) exists. If the interval is longer (A in the figure) or shorter (C in the figure) than that, then the combustion becomes worse and indicated mean effective pressure (IMEP) drops. Therefore, we performed visualization and Air/Fuel ratio distribution measurement at A and C where IMEP is low as well as B where it is high.

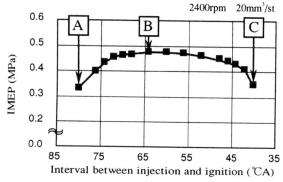

Figure 12 Effect of interval on IMEP

Figure 13 shows the visualization at BTDC50° CA. Figure 14 shows the Air/Fuel ratio distribution immediately before ignition (BTDC16° CA). In the case of A, where the interval between injection and ignition is long, the spray diffuses out of the cavity, scatter of air-fuel mixture progresses and, as a result, Air/Fuel ratio exceeds 20 near the spark plug. On the other hand, in the case of short interval C, the air-fuel mixture is largely concentrated in the cavity close to the injector, and Air/Fuel ratio near the spark plug is about 40, which is too lean to be ignited. In the interval B, where the combustion is good, Air/Fuel ratio is 10 to 20 near the spark plug. This leads to the conclusion that the Air/Fuel ratio surrounding the spark plug must be controlled within 10 to 20 for good stratified combustion.

The interval between injection and ignition is

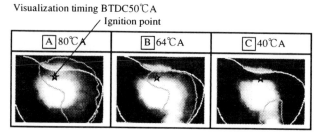

Figure 13 Effect of interval on mixture motion

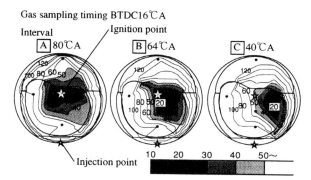

Figure 14 Effect of interval on mixture distribution

identified as an important factor for the stratified combustion.

4.3 EFFECT OF SWIRL

Figure 15 shows the effect of swirl intensity upon the combustion by varying opening of a swirl control valve (SCV). An optimum swirl intensity exists for the stratified combustion, and the IMEP drops when the SCV opening is shifted from the optimum to closing side or opening side. Therefore, we performed visualization and Air/Fuel ratio distribution measurement to analyze air-fuel mixture at A and C, where the IMEP is low as well as at B where it is high, as shown in Figure 15.

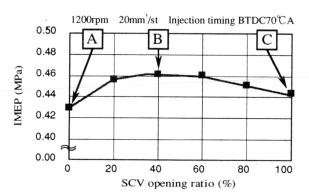

Figure 15 Effect of swirl on IMEP

Figure 16 shows the visualization at BTDC55° CA. Figure 17 shows the Air/Fuel ratio distribution immediately before ignition (BTDC16° CA). The results indicate that the swirl is too strong at A (with

Visualization timing BTDC55℃A

Ignition point

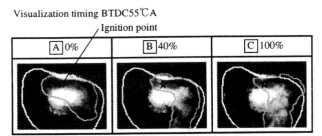

Figure 16 Effect of swirl on mixture motion

Gas sampling timing BTDC18℃A

SCV opening ratio　Ignition point

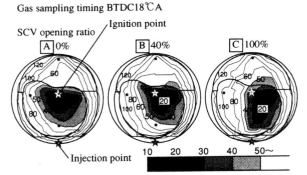

Injection point

Figure 17 Effect of swirl on mixture distribution

0% SCV opening) so that the spray moves quickly. In this case, most of the spray passes the spark plug, the air-fuel mixture disperses and, as a result, the stratification weakens. The results also visualizes that no swirl at C (with 100% SCV opening) results in spray diffusion out of the piston cavity after its impingement to piston top, then the spray disperses in the combustion chamber.

The required air-fuel mixture can be directed toward the spark plug by adequate swirl flow, without dispersing the mixture.

4.4   EFFECT OF ATOMIZATION

Figure 18 shows the relationship between spray droplet size and injection pressure. The Sauter Mean Diameter (SMD) of the spray at 5MPa is about 20 μ m, while that at 12MPa is about 15 μ m. The atomization is improved by increasing the injection pressure. Figure 19 shows the effect of fuel pressure on stratification. By comparing the IMEP between 12 MPa and 5 MPa fuel pressure, the IMEP of 5 MPa is 15.5% lower than that of 12 MPa. Figure 20 shows the visualization at BTDC55° CA. Figure 21 shows the Air/Fuel ratio distribution immediately before ignition (BTDC18° CA). Since the SMD is about 15 μ m at 12 MPa, spray particles follow the swirl flow and are guided to the spark plug; however, at 5 MPa, spray particles are larger and are not directed by the swirl flow, so that the spray flows as injected, and the stratification weakens.

Droplet size around 15 μ m is identified for spray to follow the swirl flow and to be guided to around the spark plug.

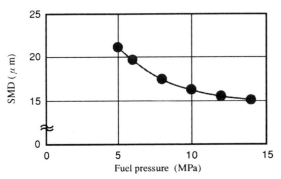

Figure 18 Droplet sizes of injection spray on fuel pressure

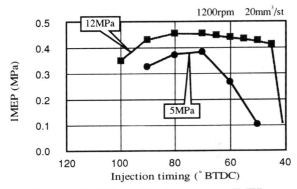

Figure 19 Effect of fuel pressure on IMEP

Visualization timing BTDC55℃A

Ignition point

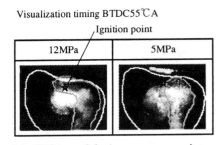

Figure 20 Effect of fuel pressure on mixture motion

Gas sampling timing BTDC18℃A
SCV opening ratio 40%

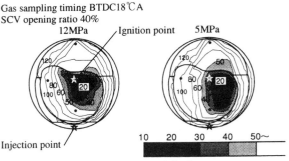

Injection point

Figure 21 Effect of fuel pressure on mixture distribution

447

4.5 CYCLE-TO-CYCLE FLUCTUATION IN THE AIR-FUEL MIXTURE FORMATION

Cycle-to-cycle fluctuation in the air-fuel mixture formation in cylinder is an unavoidable problem in the internal combustion engine. It is often discussed, particularly from the relations with torque fluctuation or lean limit. The fluctuation in air-fuel mixture is also an important issue related to the combustion stability in the direct injection, and the issues have to be clarified.

commonly used for evaluating injector spray in a test chamber under predetermined pressure and temperature. Figure 23 compares cycle-to-cycle fluctuation of spray between injector test bench and actual engine, with 1.5 msec. delay after injection. The same injector was used in the tests, and injection quantity, injection pressure and drive frequency were all the same. While the spray in the test chamber is stable in each cycle, the actual engine indicates cycle-to-cycle fluctuation of spray. The authors think that disturbance, such as engine vibration and cycle-to-cycle fluctuation in fuel pressure as well as in air flow to the injector, causes the fluctuation of spray in the actual engine.

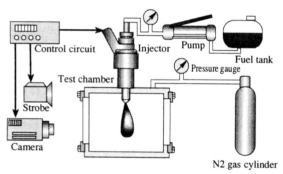

Figure 22 Schematic diagram of fuel injector test bench

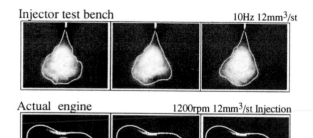

Figure 23 Comparison of fuel spray form between injector test bench and actual engine

Figure 24 shows cycle-to-cycle fluctuation of Air/Fuel ratio at the spark plug. The cycle-to-cycle fluctuation of the engine was also verified by rising Air/Fuel ratio profile at the spark plug, as shown in the figure.

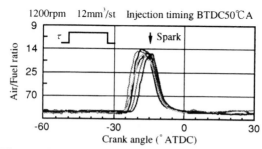

Figure 24 Cycle-to-cycle fluctuation of Air/Fuel ratio nearby spark plug

It has been clarified that even though a stable spray is obtained in a test chamber, its stability is not always guaranteed in the combustion chamber of the actual engine.

However, the cycle-to-cycle fluctuation in the air-fuel mixture mentioned above is too small to cause engine torque fluctuation. The stable combustion is achieved in gasoline direct injection engine because angle, direction, distribution, penetration and others of the spray are optimized to minimize the mixture fluctuation.

5. CONCLUSION

By analyzing the air-fuel mixture in the D-4 combustion chamber, we have clarified that the following items are required:
(1) Interval between injection and ignition is an important parameter for realizing good stratified combustion. Air/Fuel ratio around spark plug at ignition should be maintained between 10 to 20 with the suitable interval.
(2) Swirl intensity should be controlled accurately depending on the operating conditions in order to guide the air-fuel mixture to the spark plug without dispersing the mixture.
(3) Fine spray with about 15 μm SMD allows itself to follow the swirl flow.

REFERENCES

[1] A. J. Giovanetti et. al. "Analysis of Hydrocarbon Emissions Mechanisms in a Direct Injection Spark-Ignition Engine" SAE830587,1983
[2] E. N. Balles et. al. "Fuel Injection Characteristics and Combustion Behavior of a Direct-Injection Stratified-Charge Engine" SAE841379,1984
[3] J. M. Lewis "UPS Multifuel Stratified Charge Engine Development Program-Field Test" SAE860067,1986
[4] R. M. Siewert et. al. "Unassisted Cold Starts to

-29℃ and Steady-State Tests of a Direct-Injection Stratified-Charge (DISC) Engine Operated on Neat Alcohols" SAE872066,1987

[5] R. M. Frank et. al. "Combustion Characterization in a Direct-Injection Stratified-Charge Engine and Implications on Hydrocarbon Emissions" SAE892058,1989

[6] H. Schapertons et. al. "VW's Gasoline Direct Injection (GDI) Research Engine" SAE910054,1991

[7] L. Spiegel et. al. "Mixture Formation and Combustion in a Spark Ignition Engine with Direct Fuel Injection" SAE920521,1992

[8] R. Kagawa et. al. "A Study of a Direct-Injection Stratified-Charge Rotary Engine for Motor Vehicle Application" SAE930677,1993

[9] T. H. Lake et. al. "Preliminary Investigation of Solenoid Activated In-cylinder Injection in Stoichiometric S.I. Engines" SAE940483,1994

[10] Y. Ohyama et. al. "A New Engine Control System Using Direct Fuel Injection and Variable Valve Timing" SAE950973,1995

[11] G. K. Fraidl et. al. "Gasoline Direct Injection : Actual Trends and Future Strategies for Injection and Combustion Systems" SAE960465,1996

[12] T. Kume et. al. "Combustion Control Technologies for Direct Injection SI Engine" SAE960600,1996

[13] J. Harada et. al. "Development of Direct Injection Gasoline Engine" SAE970540,1997

[14] Y. Iwamoto et. al. "Development of Gasoline Direct Injection Engine" SAE970541,1997

[15] M. Ohsuga et. al. "Mixture Preparation for Direct-Injection SI Engines" SAE970542,1997

[16] N. S. Jackson et. al. "Stratified and Homogeneous Charge Operation for Direct Injection Gasoline Engine-High Power with Low Fuel Consumption and Emissions" SAE970543,1997

[17] J. Meyer et. al. "Spray Visualization of Air-Assisted Fuel Injection Nozzles for Direct Injection SI-Engines" SAE970623,1997

[18] G. Karl et. al. "Analysis of a Direct Injected Gasoline Engine" SAE970624,1997

[19] Z. Han et. al. "Effects of Injection Timing on Air-Fuel Mixing in a Direct-Injection Spark-Ignition Engine" SAE970625,1997

[20] J. Miok et. al. "Numerical Prediction of Charge Distribution in a Lean Burn Direct-Injection Spark Ignition Engine" SAE970626,1997

[21] M. Pontoppidan et. al. "Direct Fuel Injection -A Study of Injector Requirements for Different Mixture Preparation Concepts" SAE970628,1997

[22] Bowditch F. W. "A New Tool for Combustion Research A Quartz Piston Engine" SAE Transactions, Vol. 69, pp. 17-23,1961

[23] K. Saito et. al. "Analysis of Oil Consumption by Observing Oil Behavior Using a Glass Cylinder Engine" SAE982107,1989

[24] K. Saito et. al. "A New Method to Analyze Fuel Behavior in a Spark Ignition Engine" SAE950044,1995

[25] D. Rose et. al. "In-Cylinder Mixture Excursions in a Port-Injected Engine During Fast Throttle Opening" SAE940382,1994

[26] T. Summers et. al. "Modeling the Transit Time of a Fast Response Flame Ionization Detector During In-Cylinder Sampling" SAE950160,1995

[27] N. A. Henein et. al. "Cycle-by-Cycle Analysis of HC Emissions During Cold Start of Gasoline Engines" SAE952402,1995

[28] T. Summers et. al. "Signal Reconstruction Applied to a Fast Response Flame Ionization Detector" SAE952541,1995

Recommended Reading

The following papers have been published by SAE and are available in original or photocopy form. For ordering information, contact SAE's Customer Sales and Support Department, SAE International, 400 Commonwealth Drive, Warrendale, PA 15096-0001, USA; Telephone: 724/776-4970; Fax: 724/776-0790; e-mail: publications@sae.org; web site: www.sae.org.

Combustion System Design and Development of Direct-Injection Gasoline Engines

Tomoda, T.; Sasaki, S.; Sawada, D.; Saito, A.; Sami, H.; "Development of Direct-Injection Gasoline Engine—Study of Stratified Mixture Formation," SAE Technical Paper 970539

Effects of spray characteristics for stratified combustion of a direct-injection gasoline engine have been researched. The highly functional piezoelectric (PZT) injector was selected for this research. A hole and swirl nozzle were examined in a wide range of fuel pressures. The hole nozzle aims to make stratified mixture formation by vaporizing fuel on the piston, and the swirl nozzle aims to do so in the air above the piston by utilizing the spray characteristic of lower penetration and higher dispersibility. Both sprays could realize stable stratified combustion. The stability mainly depends on the combination of spray characteristic and piston cavity shape, and the swirl air motion which strength changes corresponding to engine operating conditions. The hole nozzle requires high, and the swirl nozzle less fuel pressure. Even by a large amount of EGR, stratified combustion has the advantage of combustion stability, and is useful to reduce exhaust emissions, especially NOx emissions.

Jackson, N.S.; Stokes, J.; Whitaker, P.A.; Lake, T.H.; "Stratified and Homogeneous Charge Operation for the Direct-Injection Gasoline Engine—High Power with Low Fuel Consumption and Emissions," SAE Technical Paper 970543

This paper describes an experimental investigation to explore and optimize the performance, economy and emissions of a direct-injection gasoline engine. Building on previous experimental direct-injection investigations at Ricardo, a single-cylinder engine has been designed to accommodate common rail electronically controlled fuel injection equipment together with appropriate port configuration and combustion chamber geometry. Experimental data is presented on the effects of chamber geometry, charge motion and fuel injection characteristics on octane requirement, lean limit, fuel consumption and exhaust emissions at typical automotive engine operating conditions. The configuration is shown to achieve stable combustion at air/fuel ratios in excess of 50:1 enabling unthrottled operation over a wide operating range. Strategies are demonstrated to control engine out emissions to levels approaching conventional port-injected gasoline engines. Overall operating strategies and alternative exhaust aftertreatment approaches are demonstrated with the potential to meet future automotive emissions requirements.

Lake, T.H.; Stokes, J.; Whitaker, P.A.; Crump, J.V.; "Comparison of Direct-Injection Gasoline Combustion Systems," SAE Technical Paper 980154

The methods of operation of four of the leading combustion system designs for fuel-only gasoline direct-injection (G-DI) engines have been compared by applying a classical analysis procedure for defining fuel transport. The fuel spray requirements for the different systems are discussed in relation to results obtained from a Phase Doppler Anemometry (PDA) rig for different injectors. The combustion systems have then been considered regarding the functional requirements of future G-DI engines. These include power potential, stratified and homogeneous performance, variable air motion requirements, OBD component function monitoring, packaging and manufacturing issues, and calibration effort. The paper concludes that there are at least four main approaches capable of producing acceptable combustion and that the choice of system will depend on packaging, cost, and manufacturing constraints.

Fuel Spray Characteristics for Direct-Injection Gasoline Engines

Miyamoto, T.; Kobayashi, T.; Matsumoto, Y.; "Structure of Sprays from an Air-Assist Hollow-Cone Injector," SAE Technical Paper 960771

An Eulerian model of evaporating transient sprays and a new method to describe air-atomization near the injector exit to predict the mean size and velocity of droplets have been developed to study the influence of operating conditions of an air-assist hollow-cone injector and the influence of fuel atomization on the spray structure. Good agreement between the results of the computation and experiment in terms of spray shape has been achieved. The numerical results show the typical structure of sprays from the air-assisted fuel injector and show the influence of atomization on the structure.

Tatsuta, H.; Matsumura, M.; Yajima, J.; Nishide, H.; "Mixture Formation and Combustion Performance in a New Direct-Injection SI V-6 Engine," SAE Technical Paper 981435

One advantage of a direct-injection SI engine is lower fuel consumption due to the use of lean stratified charge combustion. Another advantage is greater power output resulting from evaporation of the fuel in the cylinder. A critical factor in making the most of these advantages is to achieve optimum mixture formation for both stratified and homogeneous charge combustion. To achieve the optimum mixture, the new direct-injection SI V-6 engine adopts a piston with a shallow bowl, a valve that changes in-cylinder air motion between swirl and tumble by opening and closing one side of separated air intake port, an air intake port that has optimized inward and port angle to induce swirl in the piston bowl, and a CASTING NET injector that injects the hollow cone spray in a deflected pattern toward the spark plug.

Shelby, M.H.; VanDerWege, B.A.; Hochgreb, S.; "Early Spray Development in Gasoline Direct-Injected, Spark-Ignition Engines," SAE Technical Paper 980160

The characteristics of the early development of fuel sprays from pressure swirl atomizer injectors of the type used in direct-injection gasoline engines is investigated. Planar laser-induced fluorescence (PLIF) was used to visualize the f distribution inside a firing optical engine. The early spray development of three different injectors at three different fuel pressures (3, 5, and 7 MPa) was followed as a function of time in 30 μsec intervals. Four phases could be identified: 1) a delay phase between the rising edge of the injection pulse and the first occurrence of fuel in the combustion chamber, 2) a solid jet of prespray phase, in which a poorly atomized stream of liquid fuel during the first 150 μsec of the injection, 3) a wide hollow cone phase, separation of the liquid jet into a hollow cone spray once sufficient tangential velocity has been established, and 4) a fully developed spray, in which the spray cone angle is narrowed due to a low pressure zone at the center. The spray penetration, quality of atomization and the specific fuel distribution during the duration of injection (about 2 ms or 12 CAD) are functions of fuel delivery pressure as well as ambient pressure. The experiments are intended for future comparison with computational results, as well as for guidance in the design of GDI combustion systems.

Preussner, C.; Döring, C.; Fehler, S.; Kampmann, S.; "GDI: Interaction Between Mixture Preparation, Combustion System and Injector Performance," SAE Technical Paper 980498

The development of future engine generations for gasoline direct injection requires sophisticated combustion systems to reach reduced fuel consumption and future emission standards. The design process of these combustion systems has to be based on a fundamental knowledge of the interacting mixture preparation mechanisms. Beside the air motion inside the cylinder, mixture preparation is mainly fed by the fuel spray quality, injector performance, respectively. The article therefore presents a fundamental analysis of the GDI mixture preparation and affords an insight into the injector development. Comprehensive experimental studies were performed in high pressure/temperature vessels using phase doppler anemometry, laser-induced fluorescence and video techniques to define the significant fuel spray features for GDI. CFD calculations were additionally applied to study the temporal behavior of the mixture preparation under injection parameter variation. The adjustment of the spray and injection parameters is described under the view of engine application requirements. Using the results from these fundamental studies, an appropriate injector design could be developed very efficiently.

Ren, W.M.; Nally Jr., J.F.; "Computations of Hollow-Cone Sprays from a Pressure-Swirl Injector," SAE Technical Paper 982610

A computational model is proposed and analysis is carried out to study the atomization processes of hollow-cone fuel sprays from pressure-swirl injectors for a gasoline direct-injection (GDI), spark-ignition (SI) engine. The flow field inside a swirl injector is numerically analyzed, and characteristics of the liquid sheet at the nozzle exit are predicted. The intact length (i.e., breakup length) of the sheet is calculated from a semi-empirical correlation and a sauter mean diameter (SMD) at breakup location is estimated based on the classical wave instability theory. The spray dynamics that address the interactions between liquid drops and surrounding gas phase are simulated using FIRE code with modified spray models. The objective is to understand the effects of nozzle geometry and engine operating conditions on spray characteristics so that the spray structure can be optimized through the injector design to meet the fundamental requirements of GDI engines. Experimental measurements are performed to provide the global spray images, drop size distribution, and mean particle size (i.e., SMD). Computed and measured spray characteristics, such as spray width, cone angle, and tip penetration, are compared, and good levels of agreement are achieved. The results show that spray width and penetration decrease significantly as the ambient pressure increases. However, they are relatively insensitive to the injection pressures evaluated.

Schmidt, D.P.; Nouar, I.; Senecal, P.K.; Rutland, C.J.; Martin, J.K.; Reitz, R.D.; Hoffman, J.A., "Pressure-Swirl Atomization in the Near Field," SAE Technical Paper 1999-01-0496

To model sprays from pressure-swirl atomizers, the connection between the injector and the downstream spray must be considered. A new model for pressure-swirl atomizers is presented which assumes little knowledge of the internal details of the injector, but instead uses available observations of external spray characteristics. First, a correlation for the exit velocity at the injector exit is used to define the liquid film thickness. Next, the film must be modeled as it becomes a thin, liquid sheet and breaks up, forming ligaments and droplets. A linearized instability analysis of the breakup of a viscous, liquid sheet is used as part of the spray boundary condition. The spray angle is estimated from spray photographs and patternator data. A mass-averaged spray angle is calculated from the patternator data and used in some of the calculations. This new model is referred to as the Linearized Instability Sheet Atomization (LISA) model, and is a complete set of equations which provide the required boundary data for the multi-dimensional, transient code, KIVA-3V. KIVA-3V is used to model the development of the spray, including spray collision, coalescence, breakup, and drag. The calculated results are compared to experimental data for two injectors. The spray penetration is compared to measured values from the literature. The spray calculations are also compared qualitatively to photographs of sprays. Further comparisons to transient patternator data are used to check the predicted mass flux distribution.

Park, J.; Xie, X.; Im, K.-S.; Kim, H.; Lai, M.-C.; Yang, J.; Han, Z.; Anderson, R.W.; "Characteristics of Direct-Injection Gasoline Spray Wall Impingement at Elevated Temperature Conditions," SAE Technical Paper 1999-01-3662

The direct-injection gasoline spray-wall interaction was characterized inside a heated pressurized chamber using various visualization techniques, including high-speed, laser-sheet macroscopic and microscopic movies up to 25,000 frames per second, shadowgraph, and double-spark particle image velocimetry. Two hollow-cone, high-pressure swirl injectors having different cone angles were used to inject gasoline onto a heated plate at two different impingement angles. Based on the visualization results, the overall transient spray impingement structure, fuel film formation, and preliminary droplet size and velocity were analyzed. Results show that upward spray vortex inside the spray is more obvious at elevated temperature condition, particularly for the wide-cone-angle injector, due to the vaporization of small droplets and decreased air density. Film build-up on the surface is clearly observed at both ambient and elevated temperature, especially for narrow cone spray. Vapor phase appears at both ambient and elevated temperature conditions, particularly in the toroidal vortex and impingement plume. More rapid impingement and faster horizontal spread after impingement are observed for elevated temperature conditions. Droplet rebounding and film break-up are clearly observed. Post-impingement droplets are significantly smaller than pre-impingement droplets with a more velocity component horizontally regardless of the wall temperature and impingement angle condition.

Multi-Dimensional Modeling of Direct-Injection Gasoline Engine Phenomena

Abraham, J.; Bracco, F.V.; "3-D Computations to Improve Combustion in a Stratified-Charge Rotary Engine Part II—A Better Spray Pattern for the Pilot Injector," SAE Technical Paper 892057

A three-dimensional combustion model of a direct-injection stratified-charge rotary engine is used to identify modifications that might lead to better indicated efficiency. The engine, which has a five-hole main injector and a pilot injector, is predicted to achieve better indicated efficiency if a two-hole 'rabbit-ear' pilot injector is used instead of its present single-hole pilot injector. This rabbit-ear arrangement is predicted to increase the surface area of the early flame (on account of better distribution of the fuel), and thereby result in an increased overall burning rate. Computations were made at high and low engine speeds and loads, encompassing the practical operating range. It is concluded that the modified pilot injector will increase indicated efficiency by about 5% within the computed operating range.

Han, Z.; Reitz, R.D.; Claybaker, P.J.; Rutland, C.J.; Yang, J.; Anderson, R.W.; "Modeling the Effects of Intake Flow Structures on Fuel/Air Mixing in a Direct-Injected Spark-Ignition Engine," SAE Technical Paper 961192

Multidimensional computations were carried out to simulate the in-cylinder fuel/air mixing process of a direct-injection spark-ignition engine using a modified version of the KIVA-3 code. A hollow cone spray was modeled using a Lagrangian stochastic approach with an empirical initial atomization treatment which is based on experimental data. Improved Spalding-type evaporation and drag models were used to calculate drop vaporization and drop dynamic drag. Spray/wall impingement hydrodynamics was accounted for by using a phenomenological model. Intake flows were computed using a simple approach in which a prescribed velocity profile is specified at the two intake valve openings. This allowed three intake flow patterns, namely, swirl, tumble and non-tumble, to be considered. It was shown that fuel vaporization was completed at the end of compression stroke with early injection timing under the chosen engine operating conditions. The mixing process and the in-cylinder fuel distribution were found to be significantly affected by the flow structures which are dominated by the intake flow details. More uniform distributions of air-fuel ratio and mixture temperature in the combustion chamber were obtained at the end of compression in the cases using tumble and swirl flow patterns.

Laek, T.M.; Sapsford, S.M.; Stokes, J.; Jackson, N.S.; "Simulation and Development Experience of a Stratified-Charge Gasoline Direct-Injection Engine," SAE Technical Paper 962014

Computational fluid dynamics (CFD) simulation has been used to investigate the fuel air mixing regimes of an open-chamber gasoline direct-injection (GDI) engine. Acceptable homogeneous stoichiometric charge operation was predicted by the CFD simulation and confirmed by data from engine experiments with early injection timing. The simulation also predicted that late injection timing would be inoperable with the open chamber geometry employed. This was confirmed by injection timing experiments on the test engine. Subsequent initial engine development using a different engine geometry with top-entry inlet ports and a piston containing a spherical bowl has demonstrated very stable combustion with an unthrottled late injection strategy. The use of recycled exhaust gas (EGR) is demonstrated to produce better emissions and fuel consumption than purely lean operation. The effect of throttling is found to provide emissions improvements at the expense of fuel economy.

Choi, K.H.; Daisho, Y.; Saito, T., "A Study on Process of Direct Injection Stratified Charge Combustion in a Constant-Volume Vessel," SAE Technical Paper 891223

A numerical simulation model has been developed to predict the direct injection stratified charge combustion in a constant-volume vessel. Important factors such as local fuel concentration, their fluctuation and turbulent flow characteristics were measured throughout the vessel as function of time. These data were utilized to estimate the buring rate composed of the turbulent fuel-air mixing rate and chemical reaction rate. The model can predict the combustion pressures and heat release rates measured for different ignition timings and spark location.

Corcione, F.E.; Rotondi, R.; Gentile, R.; Migliaccio, M.; "Modeling the Mixture Formation in a Small Direct-Injected Two-Stroke Spark-Ignition Engine," SAE Technical Paper 970364

Computations were carried out to simulate in-cylinder flow field and mixture preparation of a small port scavenged direct-injection two-stroke spark-ignition engine using a modified version of KIVA-3 code. Simulations of the interaction between air flow and fuel were performed on a commercial Piaggio (125 cc) motorcycle engine modified to operate with a hollow-cone injector located in different positions of the dome-shaped combustion chamber. The engine has a large exhaust port and five smaller transfer ports connecting the cylinder to the crankcase. The numerical grid of this complex geometry was obtained using an IBM grid generator based on the output of engine design by CATIA solution. To take into account the rapid distortion of flow, the standard k-ϵ turbulence model in KIVA-3 was replaced by the RNG k-ϵ model. Three cases of hollow-cone injector locations were explored: one, located on the head, injected the fuel along the cylinder axis; another one, also located on the head, injected the fuel in counter flow; the third one, located on cylinder wall, injected the fuel towards the combustion chamber. It was found that the most important parameters that strongly influence in-cylinder droplet vaporization process and spatial vapor distribution are: fluid flow pattern, injector location, injection pressure and injection timing.

Pontoppidan, M.; Gaviani, G.; Bella, G.; Rocco, V.; "Improvements of GDI-Injector Optimization Tools for Enhanced SI-Engine Combustion Chamber Layout," SAE Technical Paper 980494

The suggestions for upcoming Euro 2000 clean air act puts an increasing legislative pressure for lower specific fuel consumption in order to reduce the emission of CO2, and thereby, decrease the impact of the "greenhouse" effect. One of the possible suggestions to meet these requirements for SI engines is the gasoline direct-injected (GDI) power unit. One of the key points of the success of a layout of a GDI system is the optimization of the fuel injector and combustion chamber charge formation parameters. A brief description of the basic GDI system used during the study is given. Hereafter are outlined the computational and experimental optimization tools which have been used to produce, on a reasonable industrial time scale, the main indications to optimize the design of a given injector/chamber configuration. The paper discusses in detail the results produced by the latest enhancements introduced into the 3D multiphase computational approach, NCF-3D. These are fully implemented combustion sub-model as well as a submodel for an autoignition mechanism. Three different injector nozzle geometries are evaluated in the same combustion chamber. The computational and experimental results, obtained with different injector positions, are compared for the given combustion chamber of a 4-

cylinder, 16-valve, 2-liter engine. An alternative ignition strategy, the multispark concept, is discussed.

Pontoppidan, M.; Gaviani, G.; Bella, G.; de Maio, A.; Rocco, V.; "Experimental and Numerical Approach to Injection and Ignition Optimization of Lean GDI-Combustion Behavior," SAE Technical Paper 1999-01-0173

The first part of the paper gives an overview of the current development status of the GDI system layout for the middle displacement engine, typically 2 liter, using the stoichiometric or weak lean concept. Hereafter are discussed the particular requirements for the transition to a small displacement/small bore engine working in stratified lean conditions. The paper continues with a description of the application of the different steps of the optimization methodology for a 1.2 liter, small bore 4 cylinder engine from its original base line MPI version towards the lean stratified operation mode. The latest changes in the combustion model, used in the numerical simulation software applied to the combustion chamber design, are discussed and comparison made with the previous model. The redesign of the combustion chamber geometry, the proper choice of injector atomizer type and location and the use of two-stage injection and multi-spark strategies are discussed in detail. The paper concludes with a discussion of the impact on the component industrialization layout of the specification requirements necessary to fulfill an optimized lean combustion behavior for the small displacement/small bore engine.

Fan, L.; Li, G.; Han, Z.; Reitz, R.D.; "Modeling Fuel Preparation and Stratified Combustion in a Gasoline Direct Injection Engine," SAE Technical Paper 1999-01-0175

Fuel preparation and stratified combustion were studied for a conceptual gasoline direct-injection spark-ignition (GDI or DISI) engine by computer simulations. The primary interest was on the effects of different injector orientations and the effects of tumble ratio for late injection cases at a partial load operation condition. A modified KIVA-3V code that includes improved spray breakup and wall impingement and combustion models was used. A new ignition kernel model, called DPIK, was developed to describe the early flame growth process. The model uses Lagrangian marker particles to describe the flame positions. The computational results reveal that spray wall impingement is important and the fuel distribution is controlled by the spray momentum and the combustion chamber shape. The injector orientation significantly influences the fuel stratification pattern, which results in different combustion characteristics. The gas tumble also affects the fuel distribution and the ignition process. Under certain conditions, the mixing is characterized by the existence of many lean regions in the cylinder and the burning speed is very low; hence the combustion can be poor in these cases.

Kim, Y.-J.; Lee, Sang H.; Cho, N.-H.; "Effect of Air Motion on Fuel Spray Characteristics in a Gasoline Direct Injection Engine," SAE Technical Paper 1999-01-0177

Numerical simulation was carried out to investigate the effect of transient in-cylinder air motion on fuel spray characteristics in a side-injection gasoline direct injection engine. KIVA-3V code with a fuel spray impingement model was used to simulate a swirl flow driven stratification for a late injection mode. For better understanding of in-cylinder air motion during the induction and compression strokes a flat piston and a bowled piston are compared with each other. Also a simplified simulation considering only the compression stroke was compared with the full simulation. As the high-pressure fuel spray jet flow is much stronger than ambient swirl and tumble flow in the present combustion system the spray development shows similar behavior in both simulations.

Joh, M.; Huh, K.Y.; Noh, S.H.; Choi, K.H.; "Numerical Prediction of Stratified Charge Distribution in a Gasoline Direct-Injection Engine—Parametric Studies," SAE Technical Paper 1999-01-0178

Numerical analysis of the flow field and fuel spray in a gasoline direct-injection (GDI) engine is performed by a modified version of the KIVA code. A simple valve treatment technique is employed to handle multiple moving valves without difficulties in generation of a body-fitted grid. The swirl motion of a hollow-cone spray is simulated by injection droplets with initial angular momentum around the nozzle periphery. The model for spray-wall impingement is based on single droplet experiments with the droplet behaviors after impingement determined by experimental correlations. Different behaviors of an impinging droplet depend on the wall temperature and the critical temperature of fuel with the fuel film taken into account. The test engine is a 4-stroke 4-valve gasoline engine with a pent-roof head and vertical ports to form a reverse tumble flow during the intake stroke. A hollow-cone spray by a high-pressure swirl injector is employed to enhance mixture preparation and mixing. A piston bowl is implemented to trap the mixture around the spark plug and to enhance evaporation by retarding impingement of the spray. Parametric study is performed for two different chamber geometries with respect to the speed/load, the fuel injection timing, the spray cone angle and the incidence angle of the fuel injector. Results show that poor mixture distribution can be improved by adjusting some of these parameters.

Suh, E.S.; Rutland, C.J.; "Numerical Study of Fuel/Air Mixture Preparation in a GDI Engine," SAE Technical Paper 1999-01-3657

Numerical simulations are performed to investigate the fuel/air mixing preparation in a gasoline direct-injection (GDI) engine. A two-valve OHV engine with wedge combustion chamber is investigated since automobiles equipped with this type of engine are readily available in the U.S. market. Modifying and retrofitting these engines for GDI operation could become a viable scenario for some engine manufacturers. A pressure-swirl injector and wide spacing injection layout are adapted to enhance mixture preparation. The primary interest is on preparing the mixture with adequate equivalence ratio at the spark plug under a wide range of engine operating conditions. Two different engine operating conditions are investigated with respect to engine speed and load. A modified version of the KIVA-3V multi-dimensional CFD code is used. The modified code includes the Linearized Instability Sheet Atomization (LISA) model to simulate the development of the hollow cone spray. The model includes spray collision, coalescence, breakup and drop drag. Mixing for this engine is difficult due to the non-symmetric engine geometry and complicated flow structures in the cylinder. The results reveal that knowledge of the optimal timing interval between start of injection and ignition is important. Also, a flat piston and a radically designed piston with bowl cavity are compared in order to gain additional understanding of the in-cylinder air motion that affects fuel preparation.

Diagnostic Methods for Direct-Injection Gasoline Engines

Salters, D.; Williams, P.; Greig, A.; Brehob, D.; "Fuel Spray Characterization within an Optically Accessed Gasoline Direct Injection Engine Using a CCD Imaging System," SAE Technical Paper 961149

A test facility was constructed at University College London to study fuel spray structure within a gasoline direct injection engine. The facility consisted of a single cylinder research engine with extensive optical access and a novel video imaging and analysis system. The engine used an experimental prototype 4-valve cylinder head with direct in-cylinder high pressure fuel injection provided by a major automotive manufacturer. The fuel spray was illuminated using a pulsed copper-vapor laser. Results are presented that illustrate the spray behavior within the fired research engine. A laser light sheet provided an insight into the inner spray cone behavior.

Shiraishi, T.; Nakayama, Y.; Nogi, T.; Ohsuga, M.; "Effect of Spray Characteristics on Combustion in a Direct-Injection, Spark-Ignition Engine," SAE Technical Paper 980156

Meeting the future exhaust emission and fuel consumption standards for passenger cars will require refinements in how the combustion process is carried out in spark-ignition engines. A direct-injection system decreases fuel consumption under road load cruising conditions, and stratified charge of the fuel mixture is particularly effective for ultra-lean combustion. On the other hand, there are requirements for higher output power of gasoline engines. A direct-injection system for a spark-ignition engine is seen as a promising technique to meet these requirements. To get higher output power at wide open throttle conditions, spray characteristics and in-cylinder air flow must be optimized. In this paper, the engine system, which has a side-injection-type engine and flat piston, was investigated. We tried some injectors, which have different spray characteristics, and examined effects of spray characteristics on combustion of the direct-injection gasoline engine.

Ozasa, T.; Kozuka, K.; Fujikawa, T.; "Schlieren Observations of In-Cylinder Phenomena Concerning a Direct-Injection Gasoline Engine," SAE Technical Paper 982696

The Schlieren visualization of in-cylinder processes from the side of an engine cylinder is useful to understand the phenomena which change along the cylinder axis. A transparent collimating cylinder, TCC, permits Schlieren observation inside the cylinder through its transparent wall. In this study, a single-cylinder visualization engine with the TCC was applied to a direct-injection gasoline engine. A fuel spray, mixture formation and combustion were observed with a simultaneous measurement of in-cylinder pressure. The shape of the fuel spray and subsequent mixture formation process are drastically changed with the injection timing. The images of luminous flame were also taken with the Schlieren images during the combustion period. Stable combustion, misfire and abnormal combustion are discussed with the comparison between the observed results and in-cylinder pressure analysis

Choi, K.H.; Park, J.H.; Lee, N.H.; Yu, C.H.; Noh, S.H.; "A Research on Fuel Spray and Air Flow Fields for Spark-Ignited Direct Injection Using Laser Measurement Technology," SAE Technical Paper 1999-01-0503

The change of spray shape under different ambient pressure and the size distribution of fuel droplets from swirl injector were investigated. And, for the stratified-mixture operation in the vicinity of spark plug, fuel distribution after impingement on piston in motoring engine was examined in the cases of two different injector positions. Also, the influence of intake port geometry and piston top shape on in-cylinder air flow during intake to compression stroke was investigated. This paper concludes that, in tumble-based combustion strategy, locating swirl injector at intake side of cylinder head and making counter-rotating tumble increased in intake port geometry and conserved on bowl piston surface would be very effective in stratified combustion of spark-ignited, direct-injection engine.

Weimar, H.-J.; Töpfer, G.; Spicher, U.; "Optical Investigations on a Mitsubishi GDI Engine in the Driving Mode," SAE Technical Paper 1999-01-0504

Optical investigations using optical fibers were carried out in the first available direct-injection, SI engine, the Mitsubishi GDI, in the driving mode. The optical access to the combustion chamber was realized by 8 optical sensors evenly distributed in a ring on the ground electrode of the standard spark plug. All investigations, steady state (constant load and velocity) and unsteady state (engine starts), show that there is preferred flame propagation to the intake valves, caused by a reverse tumble in-cylinder flow. As the inflammation depends on thermodynamic conditions, flow characteristics and the actual air/fuel ratio at the spark plug, the optical sensors can be used to describe the quality of stratification. The time between spark and the first flame detection at the sensors is short in case of high velocity and high load (induction stroke injection) and nearly comparable to conventional SI engines, and in case of low velocity and low load (compression stroke injection) the time duration until flame detection is long. The behavior during start of the GDI engine was tested, too. The measurements using a fast response FID show very high emissions of hydrocarbons, both during cold-start and during warm-start. In any case, the engine starts in the homogeneous mode to assure a safe driving behavior and changes depending on oil and coolant temperature to the stratified mode.

Richter, M.; Axelsson, B.; Aldén, M.; Josefsson, G.; Carlsson, L.-O.; Dahlberg, M.; Nisbet, J.; Simonsen, H.; "Investigation of the Fuel Distribution and the In-Cylinder Flow Field in a Stratified Charge Engine Using Laser Techniques and Comparison with CFD-Modeling," SAE Technical Paper 1999-01-3540

This paper presents an investigation of a Volvo direct-injection, spark-ignition (DISI) engine, where the fuel distribution and the in-cylinder flow field have been mapped by the use of laser techniques in an engine with optical access. Along with the experimental work, CFD modeling of flow and fuel distribution has been performed. Laser-Induced fluorescence (LIF) visualization of the fuel distribution in a DI engine has been performed using an endoscopic detection system. Due to the complex piston crown geometry it was not possible to monitor the critical area around the sparkplug with conventional, through-the-piston, detection. Therefore, an endoscope inserted in the spark plug hole was used. This approach gave an unrestricted view over the desired area. In addition, the in-cylinder flow fields have been monitored by particle image velocimetry (PIV) through cylinder and piston. The results from both the LIF and the PIV measurements have been compared with CFD-modeling at Volvo. The validation was made at part load when the engine was operating in stratified mode, i.e., late injection during the compression phase. Qualitative agreement was found between the calculated and measured fuel distribution around the spark plug prior to ignition. Also the PIV measurements showed a promising agreement with the flow fields obtained by CFD-modeling. In addition, the transportation properties of the fuel distribution that was monitored by LIF could to a great extent be explained by the results from the PIV measurement and the CFD-modeling. All three techniques showed promising agreements with each other and the measured properties could be used to further increase the accuracy of the CFD-modeling. The close collaboration and comparison between different techniques described in this paper increased the understanding of the processes going on in the combustion chamber.

Hildenbrand, F.; Schulz, C.; Hartmann, M.; Puchner, F.; Wawrschin, G.; "In-Cylinder NO-LIF Imaging in a Realistic GDI Engine Using KrF Excimer Laser Excitation," SAE Technical Paper 1999-01-3545

The formation of nitric oxide in a transparent direct-injection gasoline engine was studied experimentally using two different schemes of laser-induced fluorescence (LIF) with KrF excimer (248 nm) excitation. With detection of the fluorescence shifted towards the red, strong interference from fluorescence of partially burned fuel was found. With blue-shifted fluorescence, interference was minimized allowing selective detection of NO. Possibilities of quantifying NO fluorescence intensities in inhomogeneous combustion are discussed.

Wagner, V.; Ipp, W.; Wensing, M.; Leipertz, A.; "Fuel Distribution and Mixture Formation Inside a Direct-Injection SI Engine Investigated by 2D Mie and LIEF Techniques," SAE Technical Paper 1999-01-3659

Two-dimensional Mie and LIEF techniques were applied to investigate the spray propagation, mixture formation and charge distribution at ignition time inside the combustion chamber of a direct-injection SI engine. The results obtained provide the propagation of liquid fuel relative to the piston motion and visualize the charge distribution (liquid fuel and fuel vapor) throughout the engine process. Special emphasis was laid on the charge distribution at ignition time for stratified charge operation. By means of a LIEF technique it was possible to measure cyclic fluctuations in the fuel vapor distributions which explain the occurrence of misfiring.

Hentschel, W.; Homburg, A.; Ohmstede, G.; Müller, T.; Grünefeld, G., "Investigation of Spray Formation of DI Gasoline Hollow-Cone Injectors Inside a Pressure Chamber and a Glass Ring Engine by Multiple Optical Techniques," SAE Technical Paper 1999-01-3660

The paper describes detailed studies about the spray formation of a direct-injection, high-pressure gasoline injector and the interaction of the droplets with the surrounding compressed air in pressure chamber experiments and inside an optically accessible research engine. Different optical techniques, like stroboscopic video technique, high-speed filming with flood-light illumination or with light-sheet illumination by a copper vapor laser, particle image velocimetry of the droplets, laser-induced fluorescence of the liquid phase, and spontaneous Raman spectroscopy for the measurement of the fuel/air ratio are used. From the recorded images spray characteristics such as spray penetration and spray cone angle are evaluated for different settings of the chamber pressure and temperature and for different rail pressures. The results show that all techniques are suitable to derive the quantities mentioned above. Their differences are not larger than the shot-to-shot variations, but for a direct comparison of results it is advisable to keep the same optical technique. Measurements performed in the combustion chamber of a DI gasoline engine show stronger cycle-to-cycle variations compared to pressure chamber experiments. Both systematic pressure chamber experiments under well-defined conditions as well as measurements in optically accessible engines highlight special phenomena of the spray formation process and thus support the development of the combustion process for DI gasoline engines.

Related Reading

This appendix is a collection of papers for related reading, suggested by the individuals who assisted with the development of PT-80. These papers are in critical related fields. The list is by no means exhaustive but should serve as a starting point. Due to space constraints, abstracts for these papers were not included.

The following papers have been published by SAE and are available in original or photocopy form. For ordering information, contact SAE's Customer Sales and Support Department, SAE International, 400 Commonwealth Drive, Warrendale, PA 15096-0001, USA; Telephone: 724/776-4970; Fax: 724/776-0790; e-mail: publications@sae.org; web site: www.sae.org.

Gasoline Direct-Injection Engine – Calibration and Control

970540   *Development of Direct-Injection Gasoline Engine*; Harada, J.; Tomita, T.; Mizuno, H.; Mashiki, Z.; Ito, Y.

980150   *Optimized Gasoline Direct-Injection Engine for the European Market*; Noma, K.; Iwamoto, Y.; Murakami, N.; Iida, K.; Nakayama, O.

1999-01-1281   *Development of the Control and Aftertreatment System for a Very Low Emission G-DI Vehicle*; Lake, T.H.; Bending, R.G.; Williams, G.P.; Beaumont, A.J.; Warburton, A.; Andersson, J.

1999-01-1282   *Gaseous and Particulate Emissions from a Vehicle with a Spark-Ignition, Direct-Injection Engine*; Cole, R.L.; Poola, R.B.; Sekar, R.R.

1999-01-1284   *Motronic MED7 for Gasoline Direct-Injection Engines: Engine Management System and Calibration Procedures*; Küsell, M.; Moser, W.; Philipp, M.

Gasoline Direct-Injection Engine – Aftertreatment

1999-01-1279   *Development of NOx Storage-Reduction, Three-Way Catalyst for D-4 Engines*; Ikeda, Y.; Sobue, K.; Tsuji, S.; Matsumoto, S.

1999-01-1285   *Durability Aspects of NOx Storage Catalysts for Direct Injection Gasoline Vehicles*; Müller, W.; Strehlau, W.; Hoehne, J.; Okumura, A.; Göbel, U.; Lox, E.; Hori, M.

1999-01-3499   *Optimising the Aftertreatment Configuration for NOx Regeneration on a Lean-NOx Trap*; Marshall, R.A.; Gregory, D.; Eves, B.; Peirce, G.; Taylor, T.; Cornish, S.; Dearth, M.; Hepburn, J.

1999-01-3501   *Influence of Sulfur Concentration in Gasoline on NOx Storage—Reduction Catalyst*; Asanuma, T.; Takeshima, S.; Yamashita, T.; Tanaka, T.; Murai, T.; Iguchi, S.

1999-01-3504   *Sulphur Poisoning and Regeneration of NOx Trap Catalyst for Direct-Injected Gasoline Engines*; Erkfeldt, S.; Larsson, M.; Hedblom, H.; Skoglundh, M.

1999-01-3530   *Measurement of the Number and Size Distribution of Particles Emitted from a Gasoline Direct-Injection Vehicle*; Hall, D.E.; Dickens, C.J.

1999-01-3585   *The Effect of Fuel Sulfur Content on the Exhaust Emissions from a Lean-Burn Gasoline Direct-Injection Vehicle Marketed in Europe*; Kwon, Y.; Stradling, R.; Heinze, P.; Broeckx, W.; Esmilaire, O.; Martini, G.; Bennett, P.J.; Rogerson, J.; Kvinge, F; Lien, M.

Gasoline Direct-Injection Engine – Fuels and Deposits

982701   *Influence of Fuel Volatility on Emissions and Combustion on a Direct-Injection, Spark-Ignition Engine*; Sandquist, H.; Denbratt, I.; Ingemarsson, S.; Olsson, J.

1999-01-1496   *The Evaluation of Performance-Enhancing Fluids and the Development of Measurement and Evaluation Techniques in the Mitsubishi G-DI Engine*; Morris, S.W.

1999-01-1498   *A Comparison of Gasoline Direct-Injection and Port-Fuel-Injection Vehicles: Part 1—Fuel System Deposits and Vehicle Performance*; Arters, D.C.; Bardasz, E.A.; Schiferl, E.A.; Fisher, D.W.

1999-01-1499   *A Comparison of Gasoline Direct-Injection and Port-Fuel-Injection Vehicles: Part II—Lubricant Oil Performance and Engine Wear*; Bardasz, E.A.; Arters, D.C.; Schiferl, E.A.; Righi, D.W.

1999-01-3656   *A Method for Suppressing Formation of Deposits on Fuel Injector for Direct-Injection Gasoline Engine*; Kinoshita, M.; Saito, A.; Matsushita, S.; Shibata, H.; Niwa, Y.

1999-01-3663   *Emissions Response of a European Specification Direct-Injection Gasoline Vehicle to a Fuels Matrix Incorporating Independent Variations in Both Compositional and Distillation Parameters*; Kwon, Y.K.; Bazzani, R.; Bennett, P.J.; Esmilaire, O.; Scorletti, P.; David, T.; Morgan, B.; Goodfellow, C.L.; Lien, M.; Broeckx, W.; Liiva, P.